# Cascade-Olympic Natural History

D0964673

# Cascade-Olympic
# Natural History

**Daniel Mathews**

## Published by Raven Editions
### in conjunction with the Portland Audubon Society

**Library of Congress Cataloging-in-Publication Data**

Mathews, Daniel, 1948–
    Cascade-Olympic natural history.

    Bibliography: p.
    Includes index.
    1. Natural history—Cascade Range. 2. Natural history—
Washington (State)—Olympic Mountains. 3. Natural
history—Northwest, Pacific. I. Audubon Society of Portland.
II. Title.
QH104.5.C3M38    1988      508.795        88-61494
ISBN 0-9620782-0-4 (pbk.)

Printed and bound by InterPacific Printing Corp., USA/Hong
Kong. Typeset by Irish Setter, Portland, Oregon, in Meridien
and Frutiger, two typefaces designed by Adrian Frutiger. Book
designed by John Laursen. Cover designed by Martha Gannett. Photo by Daniel Mathews: monkeyflower/willow-herb
community in the bed of a seasonal rill, Glacier Peak. Back
cover photo by Tom and Pat Leeson: black bear fishing.

Eating wild plants and fungi is inherently risky. Individuals vary in
their physiological reactions, and may make mistaken identifications
regardless of their level of expertise or the accuracy of printed information they read. The publishers and the author cannot accept responsibility for their health. Readers eat wild foods at their own risk.

This book is for
the fourlegged people
    the  standing  people
        the  crawling  people
           the swimming people
               the  sitting  people
                  and the flying people

                    ◆   ◆   ◆

                    that  people  walking
                       with them may know
                         and honor them.

# Acknowledgements

First I thank Jim Jackson, who first thought of writing this book, but then had the good sense to abandon it to me. Esther Lev took a hand in the writing, including a first draft of the bird chapter.

Portions of the manuscript were reviewed, and valuable suggestions made, by Craig Collins, Carolyn Driedger, Eric Eaton, Steve Engel, Wally Englert, Richard Forbes, Christie Galen, Marshall Gannett, Rick Hafele, Paul Hammond, Mike Houck, Jim Jackson, Jimmy Kagan, Julie Kierstead, Esther Lev, Ann Littlewood, Ernest A. Mayer, Owen Schmidt, Mark Smith, Jeff Smith, Daphne Stone, Glenn Walthall, Hank Warren, and Steve Yates. Personnel of the U.S. Forest and Park Services, the Oregon Department of Fish and Wildlife, and the Multnomah County Library were also helpful countless times. Despite all their efforts, errors persist. (I have spotted far too many of them in other fine nature books to harbor illusions that others will not spot them in mine.) *Mea culpa*: errors are my responsibility alone.

The copy preparation stage demanded (and received) diligence on the part of many workers and advisors. Catherine Paglin and Ann Littlewood edited the copy; Christie Galen keyboarded it; Barry Smith helped with Macintosh matters; John Laursen gave the book its design as a whole; Irish Setter set it; Kris Elkin pasted it up; at Portland Audubon Society, supervisory teamwork was contributed by James Davis, Mike Houck, Philip Jones, Claire Puchy, Bob Wilson, and especially by Martha Gannett, who also designed the cover.

Special thanks for personal support through the years of writing go, in alphabetical order, to Gertrude Marshall, Louise Steinman, and Sabrina Ullmann.

Profuse thanks go to the contributors of illustrations and photographs.

Finally I thank, in memory, my first two botany teachers, Ruth E. Mathews and C. Leo Hitchcock.

## Photography Credits

Back cover by Tom and Pat Leeson. Front cover by the author. The author by Peter Dammann.

Plants (pages 115–46) by the author, except squawcarpet, ladies-slipper, buckbean, dogbane, woolly-sunflower, sweetpea, and yerba de selva by Julie Kierstead; and yellow pond-lily by Gordon Whitehead.

Fungi and lichens (pages 483–89) by the author, except destroying angel, honey mushroom, autumn galerina, and hedgehog mushroom by Kit Scates; fly amanita and king boletus by Preston Alexander; and admirable boletus by Kent Powlowski.

Mammals by Tom and Pat Leeson, except Douglas squirrel, flying squirrel, jumping mouse, and striped skunk by Richard B. Forbes; badger and mule deer by Geoff Pampush; ground squirrel and chipmunk by Nancy A. MacDonald; red tree vole by Murray L. Johnson; marmot by Roger Baker; and vole nest, gopher cores, porcupine, mountain goat, and bear and bobcat slashings by the author.

Birds by William E. Hoffman, except osprey, red-tailed hawk, kestrel, prairie falcon, ptarmigan, and dipper by Tom and Pat Leeson; Mallard, harlequin duck, Vaux's swift, and downy woodpecker by Richard B. Forbes; grebe, heron, vireo, and warbler by Harry Nehls; goshawk, tree swallow, and Cassin's finch by Tom Crabtree; northern flicker and western flycatcher by Nancy A. MacDonald; golden eagle and blue grouse by Geoff Pampush; Townsend's solitaire by Jeff Gilligan; song sparrow by Roy Gerig; and bald eagle by Ethel Paschal.

Amphibians and slug by Richard B. Forbes, except Larch Mountain salamander, Oregon slender salamander and roughskin newt by Alan D. St. John.

Butterflies by John Hinchliff with the author.

Rocks by Gene Pierson with the author.

# Drawing Credits

Raven logo and mosses (pages 249–60) by Barbara Stafford.

Maps and graphs; flower parts (pages 147, 182); leaves (pages 93, 188–89); fish parts (page 431), mammal tracks, antlers, Giardia, and volcanoes by Kris Elkin.

Insects (pages 448–59) by Eric Eaton with Kris Elkin; bark beetle galleries (page 461) by the U.S.D.A. Forest Service.

Conifers and Flowering Trees and Shrubs (pages 11–114) by the U.S.D.A. Forest Service (artists not known); except yew, ponderosa pine, whitebark pine, cottonwood, aspen, ninebark, bitterbrush, elderberries, twinberry, snowberry fruit, salal and pyrola by Willard Jepson, from *A Manual of the Flowering Plants of California*, courtesy of the Jepson Herbarium; and Douglas-fir cone, cedar twigs, Scouler willow, Oregon-boxwood, and crowberry by Jeanne R. Janish, from *Vascular Plants of the Pacific Northwest* (Hitchcock *et al.* 1955–69), used by permission of University of Washington Press.

Flowering Herbs and Ferns (pages 148–246) by Jeanne R. Janish, used by permission of University of Washington Press; except rushes, bottlebrush squirreltail, groundsel, cow-parsnip, poison-hemlock, and water-hemlock by Willard Jepson, courtesy of the Jepson Herbarium; and meadow-rue by Barbara Stafford.

Mammals, Birds, Amphibians, and fishing flies and hatches (pages 440–41) by Sharon Torvik, except tailed frog by Pat Hansen; all used by permission of the Oregon Department of Fish and Wildlife.

Reptiles by Alan D. St. John; used by permission.

Fishes by Ron Pittard, used by permission of Ed Lusch/Windsor Publications (courtesy also of Frank Amato); except shorthead sculpin by Reeve M. Bailey (from Bailey and Bond 1963), used by permission.

# Abbreviations and Symbols

| " | inches |
|---|---|
| ' | feet |
| ° | degrees Fahrenheit |
| ± | more or less |
| × | by (as in length by width) |
| **C** | central |
| **Cas Cr** | Cascade Crest |
| **diam** | diameter (diameter at breast height in the case of trees, defined as 4'6" above the ground) |
| **E** | east(ern) |
| **elev(s)** | elevation(s) above sea level, in feet |
| **E-side** | the area east of the Cascade Crest |
| **esp** | especially |
| **exc** | except |
| **incl** | including |
| **mm** | millimeters |
| **mtn(s)** | mountain(s) |
| **N** | north(ern) |
| **Oly(s)** | Olympic(s) |
| **p(p)** | page(s) |
| **PNW** | the Northwest: Ore, Wash, W British Columbia, and SE Alaska |
| **S** | south(ern) |
| **spp.** | species plural: any and all of the species of a genus, not distinguished at the species level |
| **W** | west(ern) |
| **ws** | wingspread: the measurement across outspread wings |
| **W-side** | the area west of the Cascade Crest |

# Contents

**1**   **The Cascades and Olympics**   *3*

**2**   **Conifers**   *11*

**3**   **Flowering Trees and Shrubs**   *59*

      COLOR PAGES   *115*

**4**   **Flowering Herbs**   *147*

**5**   **Ferns, Clubmosses, and Horsetails**   *237*

**6**   **Mosses and Liverworts**   *247*

**7**   **Fungi**   *261*

**8**   **Mammals**   *295*

**9**   **Birds**   *365*

**10**   **Reptiles**   *417*

**11**   **Amphibians**   *422*

**12**   **Fishes**   *429*

**13**   **Insects**   *445*

**14**   **Other Creatures**   *477*

      COLOR PAGES   *483*

**15**   **Geology**   *515*

**16**   **Climate**   *557*

      **Appendixes**   *569*
      *Pronouncing Scientific Latin*
      *Chronology of Early Naturalists*
      *Five-Kingdom Taxonomy*
      *Abbreviations and Symbols*

      **Glossary**   *581*

      **Selected References**   *597*

      **Index**   *611*

## Organization of Chapters 1–6 (the Plants)

Seed plants
  Conifers, *11*
    single needles, *16*
    bunched needles, *42*
    tiny scalelike leaves, *52*
  Flowering plants
  Trees and Shrubs, *59*
    deciduous
      over 6 feet tall
        various
        many stamens (rose family), *75*
      under 6 feet tall
        many stamens (rose family), *80*
        5 stamens, *86*
        8 or 10 stamens, *93*
    broadleaf evergreen
      over 10 inches tall, *98*
      under 10 inches tall (includes subshrubs), *106*
  Herbs (and deciduous subshrubs), *147*
    monocots
      grasslike plants
        triangular stems, *149*
        round pith-filled stems, *152*
        round hollow stems, *154*
      showy monocots
        6 stamens (lilies), *162*
        3 stamens (irises), *173*
        irregular flowers (orchids), *174*

Seed plants (continued)
   Flowering plants (continued)
      Herbs (continued)
         dicots
            without green parts, *179*
            with green parts
               composite flower heads, *182*
                  with disk and ray flowers (daisylike), *183*
                  without rays (thistlelike), *186*
                  without disks (dandelionlike), *191*
               flowers not in composite heads
               without visible petals or sepals, *192*
               with petals and/or sepals
                  strongly irregular flowers, *194*
                  radially symmetrical flowers
                     5 petals
                       5 or more sepals
                         5 stamens
                             fused petals, *205*
                             separate petals, *211*
                         10 stamens, *212*
                         15 or more stamens, *216*
                      2 sepals, *220*
                      petals and sepals total 5, *222*
                  4 petals, *228*
                  3, 6, or several petals or sepals, *232*
Spore plants
   Vascular
      Ferns, *237*
      Clubmosses and spikemosses, *244*
      Horsetails, *245*
   Nonvascular, *247*
      Mosses
         upright, fruiting at the tip, *249*
         sprawling or hanging, fruiting at midstem, *255*
      Liverworts, *259*

# Cascade-Olympic Natural History

*The high mountains in the neighbourhood, which are for the most part covered with pines of several species, some of which grow to an enormous size, are all loaded with snow; the rainbow from the vapour of the agitated water, which rushes with furious rapidity over shattered rocks and through deep caverns produc[es] an agreeable although at the same time a somewhat melancholy echo through the thick wooded valley; the reflections from the snow on the mountains, together with the vivid green of the gigantic pines, form a contrast of rural grandeur that can scarcely be surpassed.*

—David Douglas
March 20, 1826

# 1

# The Cascades and Olympics

This book is designed for the rucksack that can't hold a library. It treats most field guide subjects—plants, mammals, birds, lower animals, and the land itself—in a single volume. Included here are not only plants and animals you can't miss, and unseen creatures that see you, but also many equally vital, but often overlooked, smaller organisms.

*Cascade-Olympic Natural History* is much more than names, identifications, and pictures. It describes the behavior of living things, and their relations to each other, to habitat, and to people.

The Cascades and Olympics comprise five visibly different "physiographic provinces" described in this chapter. The varied landforms are unified by their inhabitants; along similar east-to-west cross-sections they support similar communities of plants and animals. Though the wet Westsides contrast dramatically with the dry Eastsides, north-to-south changes are slight and gradual between southeastern Alaska and the far tip of the Sierra Nevada. The one exception is an abrupt vegetational shift from Central to Southern Oregon, largely caused by lengthier summer drought.

This book, therefore, keeps a unified range by drawing a boundary at the Willamette/Umpqua divide—a standard practice in Northwest plant books. That means species are included based on their prominence in "our range" rather than in the Cascade Range as a whole. And it means generalizations

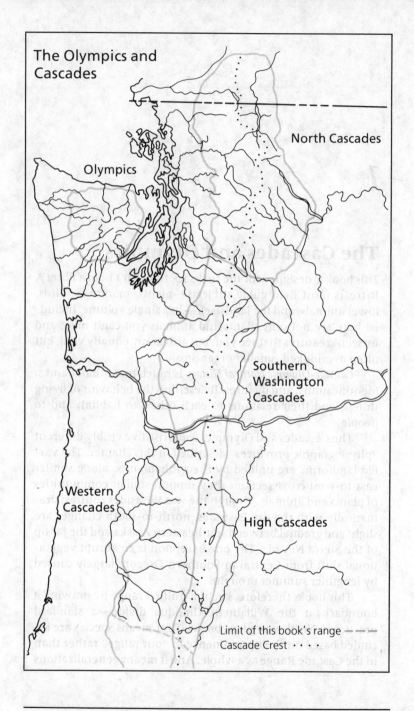

The Olympics and
Cascades

North Cascades

Olympics

Southern
Washington
Cascades

Western
Cascades

High Cascades

Limit of this book's range  — — —
Cascade Crest  · · · ·

about "here" may not apply to the Southern Oregon and Northern California Cascades. Nevertheless, the book can be used outside of its range, especially in the Coast and Insular Ranges of British Columbia and Southeast Alaska, and in the Oregon and Washington Coast Ranges. It should be useful also in the mountains of Northeast Washington, Southeast British Columbia, and adjacent Idaho and Montana.

## The Olympics

The Olympics are unusual in being a nearly round mountain range. No strong grain is discernible in the layout of their peaks and ridges, and their drainage patterns are often described as "radial"—the main streams radiating from the center—suggesting that they result from the uplift of the round range, rather than predating it. This radial pattern is less clear at the center of the range than at its perimeter.

With help from a geologic map (page 541) and some imagination, you can see a deeper pattern—NW-to-SE arcs, bowed out to the northeast. The greatest of these is the Olympics Basaltic Horseshoe, a belt making up the south, east, and north flanks of the range and running WNW out to Cape Flattery. Just inside this arc lie a number of concentric arcs—ridges, valleys, faults, additional slices of basalt. Each basaltic arc includes high peaks, evidence of this rock's relative resistance to erosion. The rest of the rocks in the Olympics derive from marine sediments—shale, sandstone, graywacke, and their metamorphic products slate and schist.

---

## We, Here

"I" am the writer, "you" are the readers, and "we" are late-twentieth-century people afield in the Olympics and Cascades (as limited above), also known as "our mountains," "our region," "our range," and "here." "Our" lilies, "our" volcanoes, etc., are those occurring in natural habitats here.

"I" don't appear often, since this is not primarily a work of personal reportage. Major sources are listed in the Bibliography.

---

During the last Ice Age, major ice-sheet tongues grinding west through the present-day Strait of Juan de Fuca and south through Hood Canal shaved the abrupt north and east flanks of the Olympics. These megaglaciers carried huge volumes of rock from Canada, leaving individual boulders as "erratics" at elevations up to 4,500' on Olympic slopes—impressive visual aids for imagining the enormity of the ice tongues. The valleys have since been thoroughly recarved by smaller alpine glaciers, of which today's glaciers are the uppermost remnants. Lower portions of valleys were left with broad bottoms lined with outwash gravels and cobbles, and later terraced by minor glacial advances and retreats of the last 1,000 years.

The same processes worked on many Cascade valleys, but here in the western Olympics some poorly understood combination, including ocean fog and heavy selective browsing by elk, has produced a unique style of forest (see page 34). The so-called Olympic Rain Forests are famous for huge conifers, both standing and down, and an abundance of tree-draping mosses, lichens, and ferns unequaled outside the subtropics. But in comparison to other Northwest old-growth these forests are parklike and open—to sunlight, and to people on foot.

Timberline—the transition from closed forest to meadow vegetation—begins below 4,000' in parts of the Olympics, yet a few trees grow on 6,000' crags nearby. This situation doesn't fit concepts of timberline developed in other mountain regions. Elsewhere, timberlines are a function of temperature; up to a certain elevation there are trees, and above that grow smaller, more cold-tolerant plants. Such a timberline would lie well above 6,000' in the Northwest. Our timberlines are low thanks to the short growing season; the sheer quantity of snow takes a long time to melt. (Annual precipitation, most of it snow, probably exceeds 200 inches—the greatest total in the 48 states—somewhere high on Mt. Olympus, but the difficulty of servicing rain gauges up there leaves both the location and the amount imprecise. Precipitation decreases sharply northeastward; see p 560.) Additional effects on timberlines in our mountains, especially the western Olympics, are that they are broad elevational belts; they are unstable over time; and since they enjoy a warmer climate than subalpine areas elsewhere, their meadows are extraordinarily luxuriant.

# The North Cascades

Extending from the Fraser to the Snoqualmie and Yakima Rivers, the North Cascades are the topographic *ne plus ultra* of our range, if not of the entire lower 48 states. Local relief, measured as the total ups and downs on a fifty-mile line crossing the range, exceeds that of any comparable breadth of U.S. Rockies or Sierras, and approaches that of the Alps. A majority of the active glaciers in the lower 48 states are here; most of the remainder are on Cascade volcanoes or Olympic peaks.

These glaciers, and their much longer, deeper incarnations that waxed and waned over the last million years or so, eroded the mountain walls to their present precipitous state. Filling the stream valleys nearly brimful with ice, glaciers deepened them and reshaped them from V to U cross-sections. Heads of major valleys in the heart of the range were eroded to within 3,000' of sea level, leaving such a low gradient (slope of descent) out to sea that current stream cutting is relatively slow. With the ridgetops eroding as fast or faster than the valley bottoms, relief is about as high as it can get—until uplift speeds up or the glaciers grow down into the valleys again.

From a high central viewpoint on a clear day, you can't miss the resemblance of North Cascade topography to waves on a stormy sea—wild array, whitecaps, vastness, and generally equal height. The equality of peak heights is attributable primarily to frost shattering (expansion of ice in cracks), which breaks down rocks most efficiently at elevations where the temperature goes from well above to well below freezing on many, many nights each year. At higher and lower elevations the freezing point is crossed less frequently.

Most peaks near the Cascade Crest, from Canada to just southeast of Glacier Peak, are 8,000–9,000' high. To the north, south, and west of the Crest, elevations of peaks and valleys decrease together, so that local relief is nearly as great; 6,000' peaks on the Western range-front and near Snoqualmie Pass are as impressive as 8,500' peaks on the Crest.

Eastward there is a marked decrease in ruggedness (with one dramatic exception, the 9,415' Stuart Range) even though many peaks are over 8,000'. Frost shattering and stream erosion, rather than alpine glaciers, were the chief sculptors of the eastern North Cascades, and they produce gentler topography.

The relative scarcity of alpine glaciers on peaks east of the Crest is caused by the rain shadow effect (page 560). Where Ice Age glaciation had the greatest effect on eastern topography—in the Pasayten—it was more gentling than sharpening. The Okanogan Lobe of the Cordilleran Ice Sheet ground across the entire area east of Ross Lake for a few centuries just prior to its abrupt retreat 12,000 years ago.

The rain shadow effect also populates the East Slope with strikingly different plant and animal species. The drainage crest marked by county lines and the Pacific Crest Trail deviates from the true Cascade Crest evidenced by sharp shifts in precipitation, glaciation, and species composition. The true Cascade Crest follows the Picket Range and the center of North Cascades National Park, before converging with the drainage crest from Boston Peak south. In this book, "Cascade Crest" and "Eastside" refer to this true, albeit unconventional crest.

The geology of the North Cascades is a bewildering variety of rock formations of widely differing ages. Some geologists suspect this is an exotic terrane or "microcontinent" that has been joined to North America for only 40 or 50 million years (see page 520). Most of its rocks are much older than that, unlike the rocks of the rest of the Cascades. Major North Cascade peaks are made of gneiss, schist, or granitic rocks, all rare in other parts of our range. Conversely, volcanic rocks, which comprise the other Cascades almost exclusively, constitute a smaller portion of North Cascade rocks—primarily, two High Cascade-age volcanoes (Mt. Baker and Glacier Peak), a few much older erosional peaks scattered between Snoqualmie Pass and Darrington, and layers of sediments eroded from long-gone volcanoes somewhere above the present peaks. The deep roots of those old volcanoes remain at today's surface in the form of granitic intrusions.

## The High Cascades

The High Cascades Province encompasses Oregon's famous snowcapped volcanoes, with elevations of 7,500 to 11,235 feet. Building upon a long, 5,500–6,500' plateau of overlapping volcanoes from the last eight million years, the conspicuous cones grew within relatively recent times—the last three-quarter-million years. Some will erupt again.

All but the youngest (South Sister, Bachelor) are deeply eroded. Glaciers have breached their craters (Hood, Jefferson, North Sister) or stripped them down to cores of their former selves (Three-Fingered Jack, Washington, Thielsen), with "necks" or central magma columns remaining after the more fragmental flanks have eroded away. The base plateau, in contrast, is sharply incised by streams or glaciers in very few places. It includes most of the mildest topography in our range, and many large lakes.

A different High Cascades eruptive style produces basaltic cinder cones and lava flows much smaller and more numerous than the high cones. Many low-profile volcanoes, such as those of the McKenzie Pass area, appeared within the last 20,000 years. The volcanoes of Washington, Southeast British Columbia, and Northern California are also of High Cascade age and geologic style. They differ from the High Cascades physiographic province in being geographically isolated within areas of contrasting older material, rather than interconnected by a young (High Cascade-age) lava plateau.

## The Western Cascades

In Oregon, the Cascades comprise two parallel mountain provinces—the High Cascades and the Western Cascades. Few Oregonians and fewer maps identify the two separately, but you can easily see them on a good relief map, or from several viewpoints. For much of their length they are separated by north-south stretches of the major westward-draining river valleys, the Clackamas, North Santiam, and McKenzie.

The Western Cascades are a heavily wooded jumble of 4,000–5,600' ridges left standing between valleys cut by streams and glaciers into a great mass of volcanic material. Few volcanoes remain as salient features; those that do, like Battle Ax, are outliers of the later High Cascade volcanism. What we have here is the eroded base of an old volcanic range. The locale of activity shifted and narrowed eastward about eight million years ago to commence the High Cascades volcanic episode.

The Western Cascades bear no glaciers, and lie entirely below timberline—though "grass balds" and "hanging meadows" persist on exposed, thin-soiled sites. Less spectacular than the High Cascades and with far greater timber value, they have received scant and belated protection as Wilderness Areas. On their lovely, riffled rivers, fishing and recreation have suffered in competition with hydroelectric potential.

## The Southern Washington Cascades

Here we find the same elements as in Oregon's High and Western Cascades Provinces, but not separated into two lines. The youngest and most active of our stratovolcanoes (St. Helens, less than 3,000 years old) is near the western edge. Far to the east lie the Cascade Crest and our two highest peaks (Rainier and Adams, built over the last half-million years) as well as the Goat Rocks, remnants of a huge, somewhat older volcano. Still more voluminous, though less towering, are Indian Heaven and the Simcoes, two areas of basaltic shield volcanoes southwest and southeast of Mt. Adams. All the above volcanoes are of High Cascade age—younger than eight million years.

The remainder of the province, like Oregon's Western Cascades, consists of erosional forms carved out of the base of a 40- to 8-million-year-old volcanic mountain range. North Cascades–style elements also emerge toward the north end of the province, where the slightly higher latitude and altitude (5,000–7,750') produced more and bigger Ice Age glaciers, leaving as their legacy a few remnant glaciers and a great deal of glaciated topography. The old volcanic rocks around Mt. Rainier have been uplifted thousands of feet since they were formed. Peak elevations are only slightly higher than in Oregon's Western Cascades, since erosion accelerated as well, nearly keeping pace; but much deeper parts of the old volcanoes are now exposed here, including several granitic intrusions. If uplift were to continue and Ice Ages to recur, they would probably expose more granite and turn the area into an extension of the North Cascades. Someday that may well come to pass.

# 2

## Conifers

If the thought of hiking here in the Pacific Northwest brings to mind cool, dark, mysterious forests of huge conifers, you've got the right picture. The area made rainy by the Cascades, Olympics, and other Pacific coastal ranges is the Conifer Capitol of the World. This is the only large temperate-zone area where conifers utterly overwhelm their broadleaf competitors. Conifers grow bigger here than anywhere else, and the resulting tonnage of biomass and square-footage of leaf area per acre are among the world's highest, rivaling tropical rain forests and easily outstripping temperate deciduous and boreal coniferous forests.

While "area made rainy" by these mountains may demark the region of peak coniferous growth, it's *when* the rain falls that makes the region unique. Generally, a wet temperate climate on this planet either supplies rainfall throughout the

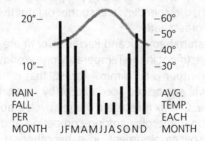

Quinault R.S., Washington

year, or concentrates it in the warmer months. But here, the warm growing season becomes a season of low humidity and frequent drought, so severe for weeks at a time, most summers, that conifers and broadleaf trees alike close their leaf pores, shutting down photosynthesis, rather than risk serious drying through open pores. For a deciduous broadleaf tree, whose main life functions are confined to the half-year when it has leaves, this is a great handicap. Our evergreen conifers, in contrast, get more than half their photosynthesis done during spring, fall, and even winter, when sunlight and temperature levels limit growth but moisture, to say the least, does not.

## Succession and Tolerance

In the forest you see slow changes continually taking place—different species of conifers increasing or decreasing in number, stature, and health or vigor. They often look like this: deep forest; canopy foliage away up out of clear view; overstory mostly groove-barked Douglas-fir; a few fibrous-stripped trunks of redcedar and the checkery bark of western hemlock; ground profusely littered with cones, Douglas-fir cones with their three-pointed bracts conspicuous among them; but the saplings are hemlock—not one Douglas-fir! If all these trees were to age a few centuries without other forces coming into play, the Doug-firs would die and hemlocks, with some cedars, would replace them "in succession" as forest canopy dominants.

This would happen because hemlocks are "tolerant" of the constraints of understory status (primarily shade) while Douglas-firs are relatively "intolerant," i.e., young Doug-firs lose their vigor, their health, their needles, and sooner or later their lives when kept in deep forest shade.

Since western hemlock and Pacific silver fir are the most tolerant trees here (if not in the Temperate Zone worldwide) a forest dominated by either is a "climax" forest; that is, the dominant tree species will not be replaced except by forces less constant than tolerance. Climax forests are the relatively stable product of slow successional interplay among competing and cooperating species, and soil development, in a hypothetical unchanging,

Nutrient uptake during the cold season is likewise crucial, since summer drought shuts down the bacterial and fungal decay that liberates nutrients. Evergreen conifers, though slower than flowering trees in acquiring nutrients, are able to acquire them all year long. They are also more efficient in using them—for instance, in keeping their needles rather than jettisoning them each growing season.

Our conifers' sheer size is an advantage here, providing storage space for water to supply them through summer. In many other temperate regions, typhoons and hurricanes blow through often enough that any genetic potential for great

---

"undisturbed" environment.

In some hard-to-burn environments like our soggy Westsides, essentially climax forests exist. Normally, however, destructive elements like fires, floods, winds, landslides, avalanches, people, and animals intervene, turning the successional sequence into a tangle of feedback loops and whorls instead of a stately linear advance toward climax.

Though called "disturbances," these forces as a group are the rule, not the exception. The fall of a forest giant and the lesser trees it brings down with it, followed over the years by nearby trees falling victim to wind or frost, creates a sunny gap where intolerant species can grow, sustaining their presence indefinitely. Fires, large and small, are such a persistent element of most ecosystems that it might be more accurate to think of fire *suppression* (a major force only in this century) as the disturbance of the natural order. Avalanches, too, are a "disturbing" but reliable element of many sites. Plants themselves may hold wild cards, such as the bog sequence in which sphagnum mosses kill and replace trees. Climate—too gradual to call a disturbance—fluctuates in cycles briefer than many successional sequences.

All of these relationships were underestimated by early theorists of plant succession; as they become better understood, the climax concept becomes less clear. Either climax forests are rare, or the climax concept needs to include the fires, gaps, climate shifts and other changes formerly considered to define climax by their exclusion. Succession—gradual, species-interactive change

height and longevity would go unrealized.

"Conifer" is a common name for members of the Coniferophyta, a division of trees and shrubs many (but not all) of which bear needlelike leaves and woody "cones." The largest family of conifers, the Pines, have both needles and cones. The Yew family (page 40) has needles but carries its seeds singly, in juicy berrylike orbs. The Cypress family (page 52) has cones, but has sprays of short, crowded scalelike leaves—except for junipers, which may have either scalelike or needlelike leaves, and bear dryish, several-seeded "berries" instead of woody cones. The four other conifer families are not native here.

---

in the plant community—is nonetheless real, and tolerance differences are a real force driving it.

Succession following common disturbances is "secondary;" "primary succession" proceeds from utterly barren ground— fresh volcanic rock, or terrain from which a glacier, sea, lake, or river has just retreated.

The "pioneers" after a devastating fire either sprout from charred stumps or root crowns, or grow from seeds well adapted to withstand intense heat or to move in abundantly from nearby. Their seedlings are quick to tap available water, nutrients, and light. The shade and transpiration of the pioneers make new microclimates. Their roots, in symbiotic association with fungi and bacteria, work over the soil physically and chemically, depleting some nutrients and accumulating others. Many pioneers are fast-growing annuals that donate their entire corpses to the humus fund in the fall; perennials and shrubs contribute leaves. The seeds of more diverse and subtle competitors, trickling in on wind and fur and feces, soon find the environment more congenial than it was at first. The quick-and-grabby strategy no longer ensures survival.

Plants succeed each other rapidly at first but, barring more disturbances, later succession gets slower and slower. We see hemlock replacing Douglas-fir, but we never see Douglas-fir 100% replaced, thanks to its size, longevity, and thick fire-resistant bark, and an environment that always includes disturbances. Intolerant species have ways of keeping seed sources around for

---

(Higher taxonomy, including this division, is endlessly disputed; the version this book follows is on page 000.)

All conifers are woody, i.e., trees or shrubs. They produce true seeds by sexual fertilization, but they lack true flowers. The young cones are female flower counterparts, receiving airborne pollen from small male "staminate cones." Seed plants other than conifers are "flowering plants" or Angiosperms.

Confusion abounds over the many terms for conifers and flowering plants. Flowering trees and shrubs are "broadleaf" even though a few, like heather, have needle-thin leaves, and

---

when disturbances open new seedbeds; long-lived individuals are Douglas-fir's way.

Our two winners in the shade tolerance sweepstakes, western hemlock and Pacific silver fir, are so well-matched that their relationship has been a puzzle. Though silver fir gets its fastest and fattest growth near its lower elevational limit here, it tends to be replaced below 2,500' elevation by western hemlock, which grows faster. But as we ascend above 3,000' we see more and more stands where silver fir saplings prevail under a hemlock/Doug-fir canopy, showing that silver firs will replace hemlocks and be the climax dominants in higher Westside forests.

One study of hemlock-silver fir succession produced a surprising explanation. The massive snowpack in the silver fir zone lasts six to eight months. During spring snowmelt, most of a year's accumulated tree litter—dead needles, twigs, bark, moss, and lichens—is deposited all at once, in an almost feltlike mat. Under a mature forest (which generates more litter) this is just too much for the supple hemlock seedlings, bending them to earth in their first or second spring, never to rise again. Silver fir seedlings are stiffer, and do not succumb; they also produce a longer taproot their first season. By this theory, the balance between hemlock and silver fir could be tipped by snowpack duration (time for litter to accumulate) which in turn is a function of elevation, latitude, climate cycles, canopy coverage, airflow patterns, and steepness and aspect (compass direction) of slope. Successional mechanisms are subtle; many remain to be discovered.

some conifers, like Australia's bunya-bunya, have rather broad ones. To a forester or lumberman, conifers are "softwoods"—even those few that are very hard, like yew. The word "evergreen" and its opposite, "deciduous," refer to whether the foliage remains alive through more than one growing season; people think of these words as synonymous with conifer and broadleaf, but in fact there are several deciduous conifers, like larch, and many broadleaf evergreens.

---

Conifers: single needles

---

# Douglas-fir

*Pseudotsuga menziesii** (soo-doe-tsoo-ga: false hemlock; men-**zee**-zee-eye: after A. Menzies, p 94). 72" diam × 250';† needles ½–1½", varying from nearly flat-lying to almost uniformly radiating around the twig, generally with white stomatal stripes on the underside only, blunt-pointed (neither sharp to the touch nor notch-tipped nor broadly rounded); cones 2½–4" × 1½", with a paper-thin 3-pointed bract sticking out beneath each woody scale; soft young cones sometimes crimson or yellow briefly in spring; young bark gray, thin, smooth with resin blisters; mature bark dark brown, deeply grooved (up to 12" thick with grooves 8" deep), made up of alternating tan and reddish brown layers visible in cross-section slice; winter buds ¼" long, pointed, not sticky; trunk usually very straight, tapering little. Ubiquitous below 4,000'. Pinaceae (Pine family). Color p 115.

As a first rule for recognizing Northwest trees, I can propose with only slight exaggeration that if you see a big evergreen tree, it's probably a Douglas-fir. This is far and away our most abundant and widespread tree species, and one of our biggest. It's the Northwest's major crop and export and currently the top commercial lumber species in North America, if not the world. If you see a board, or a sheet of plywood, in the Northwest, it's probably Doug-fir. It would be a strong candidate for World's strongest, straightest, fastest-growing tree. Rarely can so many superlatives be heaped on one species.

*This name replaced *Pseudotsuga taxifolia* some years ago when a researcher found some flaw in the original publication of that name.
†Descriptions of tree species start with the dimensions typical of mature specimens on their most favorable sites. "dbh" is diameter at breast height, standardized as 4'6" above the ground.

Before white men brought steel tools to the Northwest, Douglas-fir was economically unimportant. Red-cedar was much preferred both for aesthetics and for ease of working with stone tools. As for Douglas-fir, the thick bark was gathered for fuel, while the trident-bristling cones, either tossed into the fire or gently warmed next to it, fortified people's hopes for a break in the weather. The seedlings are a winter staple for deer and hares, and the seeds are eaten by small birds and rodents. Bears sometimes strip the bark to eat its succulent inner layer; this wounds the tree, often fatally, by making an opening for invasions of insect larvae or rotting fungi.

The tree is named for David Douglas, popularly but erroneously known as "the Discoverer of Douglas-fir." It is chauvinistic and silly to call any paleface the discoverer of America or of any conspicuous feature on it; and even to Western science, this species was described before Douglas was born, by

another Scot, Archibald Menzies, who was surgeon and botanist on Captain Vancouver's ships in 1891. The tree didn't escape Lewis & Clark's notice either, in 1806. All that was left for Douglas to do in 1825 was to sing its praises, and ship its seeds, to a waiting England. Douglas called it a pine; later taxonomists tried out "yew-leafed-fir," "spruce" and finally "false-hemlock," while sticking with fir for the common name. In truth it is none of the above.

Like our hemlocks and cedars (also misapplied European names) it is in a Pacific Rim genus, having three congeneric species in Japan and China and one in a tiny mountainous area of Southern California. Since 1826 it has been planted abroad with great success, especially in New Zealand and Scotland; Douglas and Menzies would be pleased.

Here in the Northwest, Doug-fir's commercial reputation and much of its use pattern rest on our inheritance of old-

---

growth fir—trees from one to five centuries old—now on the verge of total depletion. Top grade fir plywood and lumber is the close-grained, almost knotfree stuff the tree makes after its first century, after the scars of self-pruned limbs heal over on the lower part of the trunk. This part of the tree's volume is added much more slowly than the wide-grained, knotty wood of its youth, so rotations of longer than 50 to 90 years aren't considered cost-effective. Today, it's hard to imagine 300-year-old fir commanding a high enough price to pay for the extra centuries of capital investment. Nevertheless, some top forest scientists have begun to promote the concept of periodic 300-year rotations as a necessity for maintaining forest health and productivity. Before that vision can become reality, clear

---

**David Douglas** should be canonized as the patron saint of Northwest backpackers. Time and again he set off into the wilderness, usually with Indian guides or Hudson's Bay Company trappers, but also often alone. He packed a cast-iron kettle, a wool blanket, lots of tea and sugar, trade items such as tobacco and vermillion dye, his rifle and ammunition, and pen, ink, and reams of paper for wrapping plants, seeds, and skins—no shelter usually, no dry change of clothes, no waterproofing but oilcloth for the papers and tins for the tea and gunpowder. Often without food in his pack, he might eat duck, venison, woodrat, salmon, or wapato roots; other days he consoled himself with tea, and berries if he was lucky. Once while boiling "partridge" for dinner, he fell asleep exhausted and, waking at dawn to a burnt-through kettle, counted himself clever to boil up a cup of tea in his tinderbox lid.

He seems to have approached each Indian as a potential friend, accepting his dependence on Indians for food, information, or portage while also knowing some of them would rather kill him or steal than barter for his goods. "They think there are good and bad spirits, and that I belong to the latter class, in consequence of drinking boiling water and lighting my tobacco-pipe with my lens and the sun." No doubt they didn't intend to kill Man of Grass quite as many times as he thought.

Douglas figured he walked and canoed 6,037 miles of Washington and Oregon in 1825 and 1826; the next year he

vertical-grain Doug-fir lumber may be very scarce for a few (human) lifetimes.

Douglas-fir predominates in natural forests here almost as much as in managed timber. In the major population study of Cascade forests to date, it was the dominant tree species in two-thirds of the sample plots; in the lower-elevation half of the study, it was the only plant species common to all 202 plots. It's our only tree equally abundant west and east of the Cascades; it reaches the east slope of the Rockies; and it is abundant over a wider elevational range than our other trees, being uncommon only near upper and lower timberlines.

On the lowest, driest parts of its Eastside range, Douglas-

crossed the Canadian Rockies to Hudson's Bay to catch a ship back to England. For all that, the Royal Horticultural Society paid him their standard collector's salary of £100 a year, plus £66 for expenses. His mission was to ship them seeds or cuttings to grow exotics in English gentlemen's gardens—a lucrative market. He enjoyed some celebrity in London at first, but soon ran into the proverbial difficulty keeping his head above water in high society.

He proposed a second voyage to extend his explorations to Northwest Mexico, Southeast Alaska, and the length of Siberia. This time there was no salary, as he never reached England again to collect it. He explored central California, the Columbia region again, and the Fraser River (losing all his notes, journals, and instruments in the rapids) before sailing to Hawaii where, at the age of 34, he came to a gruesome end. While out walking alone with Billy, his faithful Scotty dog, he fell into a pit trap for feral bulls, and was gored and trampled.

The son of a Scottish stonemason, Douglas had left school at 11 for an apprenticeship in gardening. As his interest in botany grew, he garnered such bits of education as he could, auditing lectures by William J. Hooker (page 169) while employed as an undergardener in Glasgow. Hooker was impressed, took young Douglas on field trips in the Highlands, and sent him off to London and eventual fame. Despite the spotty education, Douglas wrote well, sometimes even eloquently, in his journals—and their subject is fascinating.

fir competes mainly with pines and junipers whose shade tolerance is very low, so it is a likely climax dominant. Elsewhere it must compete with the more tolerant hemlocks and true firs, which will generally increase at Douglas-fir's expense if forest succession is undisturbed. Most forests of the Eastern Olympics are largely Doug-fir today, but they date from a rash of fires about 300 years ago, and should see increasing numbers of western hemlocks despite the regionally modest rainfall. Fire-fighting as practiced in this century favors fire-susceptible species like hemlocks and grand fir over Douglas-fir, and Douglas-fir in turn over the fire-adapted pines and larches. In the long run, Douglas-fir's longevity (commonly 750 years) in combination with small disturbances would give it at least a minority role in virtually any climax forest here.

## Western Hemlock

*Tsuga heterophylla* (tsoo-ga: the Japanese term; hetero-fill-a: varied leaves). Also **lowland hemlock.** 42" diam × 200'; needles of mixed lengths, ¼–¾", round-tipped, flat, slightly grooved on top, with white stomatal stripes underneath only, spreading in ± flat sprays; most cones just ¾–1" long, thin-scaled, pendent from branch tips; mature bark up to 1" thick, platy, checked (almost as much horizontal as vertical texture); inner bark streaked dark red/purple; branch tips and treetop leader conspicuously drooping. Pinaceae.

My image of western hemlock is of a sapling's limbs, their lissome curves stippled a soft green made incandescent, in the understory dimness, by a stray swath of sunlight. While less abundant than Douglas-fir as a canopy tree, western hemlock is far and away our commonest understory sapling, and owes that to its efficiency at utilizing those scant filtered rays. These hordes of saplings are the wave of the future; western hemlock is a chief climax tree of nearly all sites below 3,500', in and among mountains west of the Cascade Crest, and also on some Eastside slopes at around 2,600–4,000'.

This leads to a paradox reminiscent of the tortoise outrunning the hare. Western hemlock is not only slower-growing, but also shorter-lived and hence inevitably smaller than the behemoths it replaces—Douglas-fir, Sitka spruce, and even the coast redwood in southwest Oregon. A Westside forest succeeding to hemlock is likely shifting toward smaller stature and younger average age. Size and longevity may impress humans, but as competitive strategies they are most useful to tree species that tend to die out in succession, and hence need old seed trees to hold out until the next postdisturbance opportunity. In contrast, hemlock's terrific shade tolerance is a slow but sure tactic assuring a major climax role.

Thanks to shade tolerance, hemlocks can share an acre of sunshine among an exceptional number of individuals, leading them to a kind of speed record. Where there's plenty of moisture, as on our coastal fog-belt lowlands, a very young pure hemlock stand can produce organic tissue ("biomass") at a phenomenal rate—the fastest yet measured in the world. A densely stocked acre of mature hemlocks often yields more board feet than a like-aged stand of larger but necessarily sparser Douglas-firs. Hemlocks achieve their efficiency partly by sheer leafiness—a six-inch trunk typically supports over 10,000 square feet of leaf surface area, almost twice as much as Douglas-fir. While the greater leaf area catches more light, it also loses more moisture; shade tolerance seems to be a trade-off against drought tolerance.

Western hemlocks also excel in seed production. Notice the profusion of little hemlock cones on the forest floor, or on the tree, lending it a purplish cast in the distance. Cones are produced copiously every year—unlike most other conifers that drastically vary their seed production in order to limit the numbers of seed-eating creatures. Each year, a mature hemlock drops more than one viable seed per square inch of ground under it. Only an infinitesimal fraction, of course, will make it to tree size. In many areas, successful reproduction is confined to "nurse log" substrates—rotting logs, snags, rootwads, etc. (See page 34.)

Mature hemlocks, with thin bark and shallow roots, are frequent victims of fire or wind. They are also prone to heartrot, the usual cause of death among aging hemlocks and the first strike against them as lumber. When sound, however, the wood deserves a better reputation than it enjoys. Though moderately soft, weak and prone to splitting, it is resilient, flexible and easily worked. It's excellent for gym floors, for example, displaying good sportsmanship under softsoled pummeling.

Northwest tribes used tannin-rich hemlock bark to tan skins; to dye and preserve wood (sometimes mashed with salmon eggs for a yellower dye); to shrink spruce-root baskets for watertightness; to make nets invisible and even alluring to fish; and on their own skins to stop bleeding. They smeared the pitch on their faces to prevent chapping or to provide a dark sticky base for face paint.

Under the bark lies a soft layer which some tribes ate to tide them over the lean times of late winter, after the dried salmon was all eaten or putrid. Countless hemlocks (and some Sitka spruces and other trees) died, their bark stripped to keep the Indians from starving. Though edible fresh, the "slimy cambium" was preferred steamed in pits over heated rocks laden with skunk-cabbage leaves, then pressed with berries and dried in cakes for later consumption with the universal condiment, eulachon fish oil.

The word "hemlock" traces back to A.D. 700 as the English word for the deadly parsleys notorious as the agent of Socrates' execution. The English somehow saw parsley in the lacy foliage of certain New England conifers new to them, and called them hemlock spruce—later shortened to hemlock.

## Why Deep-forest Leaves Lie Flat

Flat leaf arrays—an adaptation to deep forest habitats—are conspicuous on both herbs (vanillaleaf, bunchberry, twisted-stalk) and conifers (western hemlock, silver and grand firs). This arrangement makes the most of sunlight that filters weakly from above, never beaming in from east or west.

Knowing this helps us connect the descriptive traits of trees with their habitats. In the long run, though, it doesn't make distinguishing among the true firs or the hemlocks easy so much as it explains why it can be hard. When a western hemlock grows in the open (along Eastside streams, for example) it adopts a round bottlebrush leaf array just like that of mountain hemlock. Luckily, we can check for bloom on the top *and* bottom of mountain

**Western hemlock**

**Mountain hemlock**

## Mountain Hemlock

*Tsuga mertensiana* (mer-ten-see-ay-na: after K. H. Mertens, p 210). Also **black hemlock.** 36"diam × 110' (average much smaller); needles ½¾", bluish green with white stomatal stripes on both top and bottom sides, ± ridged, thus somewhat 3- or 4-sided, radiating from all sides of twig, or ± upward- and forward-crowding on exposed timberline sites; cones 1–2½", light (but coarser than spruce cones),

hemlock's needles, or for its much larger cones on the ground.

To identify true (*Abies*) firs we sometimes have to see cones, since both leaf array and stomatal bloom (our handiest identifying characters) vary not only from site to site, but from branch to branch on the same tree. Topmost branches in the forest canopy are always in full sun, and fail to display the flat leaf arrays and bottom-only stomata that typify the shade-tolerant silver and grand firs. These same nondescript top branchlets are the very ones that turn up on the forest floor, since they're the ones in the wind and, more importantly, the ones that bear cones, which squirrels harvest. But most lower branchlets in a mature forest are so high, not even a telescope could bring them close enough for us to see stomatal bloom. We can only hope to identify a harvested cone before the squirrel collects it.

often purplish, borne on upper branch tips; bark much furrowed and cracked; mature crown rather broad; also grows as prostrate shrub at highest elevs. Subalpine; abundant near and W of Cas and Oly Crests; lower limit 4,500–5,500'. Pinaceae.

The compact, gnarled shoulders of mountain hemlocks shrug off the heaviest snow loads in the world.* At every age, this species' form is brutally determined by snow. The seedlings and saplings are gently buried by the fall snows, then flattened when the snowpack, accumulating weight, begins to creep downslope. When tramping across the subalpine snowpack on a hot June afternoon, you can almost hear the tension underfoot of all those young trees straining to free themselves and begin their brief growing season. The stress of your foot on the surface may trip some unseen equilibrium, snapping a hemlock top a few feet into the air. After the trees grow big enough to take a vertical stance year round, they may keep a sharp bend at the base ("pistol-butt") as a mark of their seasons of prostration. Their crowns grow ragged from limbs breaking; some, after being encased in snow the better part of the year, spend the remainder matted with a weird black fungus called snow mold, *Herpotrichia nigra* (meaning "black creeping hair," but much less hairlike than old-man's-beard, page 291). Luckily, snow mold isn't as deadly as it looks.

Mountain hemlock abounds within subalpine tree clumps over most of our range. It is a climax dominant in the highest closed (i.e., below timberline) forests along the Crest in Central and Southern Oregon, out of range of Pacific silver fir. North of Mt. Jefferson, silver fir may gradually replace mountain hemlock on sites favorable enough for closed forests; otherwise, the hemlock is more shade-tolerant than its competitors. Mountain and western hemlocks seldom grow in the same place; where they do, they may hybridize naturally.

*Mt. Baker Lodge has the record annual snowfall average, at 550', while Paradise on Mt. Rainier chalked up the single-season record of over 1,000" in 1974–75. But these are merely the best-buried year-round weather stations. The SW flanks of Mts. Baker and Olympus, to say nothing of Alaska's Wrangell and St. Elias Ranges, probably get more snow. All of these locations grow mountain hemlocks at timberline. Most of the world's other snowiest places lack an annual snowmelt to equal their snowfall—a situation that grows glaciers, not trees.

# Subalpine Fir

*Abies lasiocarpa* (ay-bih-eez: the Roman term; lazy-o-**car**-pa: shaggy fruit). 24″ diam × 100′; needles ¾–1½″, bluish green with one broad white stomatal stripe above and two fine stripes beneath, usually curving up to crowd the upper side of the twig, tips variable; cones purplish gray to black, barrel- to cigar-shaped, 2½–4 × ¾–1¼″, borne erect on upper branches, dropping their seeds and scales singly while the core remains on the branch; bark thin, gray, smooth exc on very old bases, without superficial resin blisters, resin in pockets throughout inner bark; upper branches very short, horizontal, lower branches at ground level, long; or shrubby, prostrate. Abundant at timberline, esp eastward; rarely down to low elevs. Pinaceae.

The peculiar narrow spires of subalpine firs are ubiquitous at timberline here and in the north and central Rockies; they are the archetypal form of a subalpine tree. The upper limbs are short and stubby because, being true fir limbs, they're stiffly horizontal and brittle; if they were long, they wouldn't hold up to the snow and wind in the subalpine zone. The long lower limbs escape those stresses by spending the winter buried in snow. Hugging the ground also enables them to "layer," or reproduce by sprouting new roots and stems from branches in contact with soil. Subalpine fir is our best layerer, and hence our most aggressive tree species at tree line and scrub line. At scrub line it grows in krummholz (prostrate) form, and spreads almost exclusively by layering.

Occasionally it produces a combined tree/shrub—a small, asymmetrical, half-dead-looking upright tree with voluminous krummholz "skirts." Such a tree was confined for years to the shape of the snowpack; any foliage above the

**Subalpine fir**

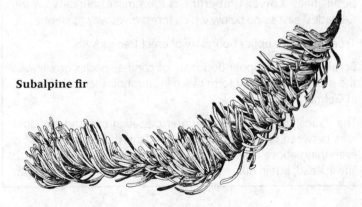

---

snowpack was killed during the winter by a combination of wind desiccation, frost rupturing, and abrasion by driven snow crystals; this is the krummholz way of life. Then for a couple of winters there was deeper snow, providing several inches of growing room for half a dozen little vertical shoots. The next time a normal-snowfall winter came, one of the shoots managed to survive with some needles on its downwind side—the side more protected from drying and abrasion. Years later, the little tree is likely "flagged," its surviving limbs positioned downwind and above the snow abrasion level (the first 8–12" above the snow/krummholz level).

Though a little bit of lasting snow may be a conifer's best friend on windswept alpine ridges, down in the subalpine parkland the thick, long-lasting snowpack is the main force blocking tree establishment. Other forces include soil too sodden, arid, or shallow, and sedge turf too dense. On sloping meadows, seedlings may be wiped out by snow creep.

---

## Timberlines

"Timberline" (as used in this book) is not a line, but a belt encompassing three successive "lines," all of them irregular:

Forest line is the uppermost boundary of continuous closed forest growth. Forest-enclosed meadows also occur sporadically below forest line at any elevation, usually owing to patch fires or soil peculiarities. Lower timberline is the similar boundary on the Cascades' east slope below which forest gives way to steppe.

Tree line is the upper boundary of erect tree growth.

Scrub line is the upper boundary of conifer species growing in the prostrate, shrubby form called krummholz ("crookedwood" in German).

The "subalpine zone" is the interspersed grove-and-meadow area between forest line and tree line, and the "alpine zone" is everything above tree line. Some writers divide the two at scrub line instead, but in our region krummholz usually associates with

---

Conifers aren't as well adapted to a short growing season as subalpine herbs and shrubs. Once conifer seedlings are several feet tall they can start photosynthesizing long before the snow is gone, but to make it past the seedling stage in the open they require a run of longer than average snowfree seasons. Such a run of long summers between 1920 and 1945 started a generation of young trees, mostly subalpine firs, scattered in subalpine meadows. This invasive growth contrasts with the normal style of slow tree-clump expansion.

Since 1957, when the European balsam woolly aphid reached our area, subalpine fir has proven badly susceptible, with mortality as high as 80% in a few mid-elevation stands. You can recognize the aphid's victims (among any of our true fir species) by their extremely swollen branch tips. Fortunately, the insect has shown no vitality at subalpine elevations, where this fir is an irreplaceable element of the landscape.

Throughout the northern Rockies, subalpine fir is the ma-

characteristic alpine species, so we will consider it alpine.

Timberline visibly expresses large, often invisible forces. It responds sweepingly to slight changes in climate—but so slowly that the climate is often swinging the other way by the time tree succession gets into gear; so slowly, in fact, that our mountain soils and communities today are still recovering from the Ice Age. With our glaciers in retreat this century, you can find at the foot of each of them a primary successional sequence, starting from bare rock as it emerges from a "Little Ice Age."

Nevertheless, charcoal in meadow soil profiles reveals that most meadows below tree line have reached a forest stage at least once since the Ice Age. The present tree line also shows how high forest growth is possible. The climax vegetation for our entire subalpine zone is thus, arguably, forest.

Timberline successional patterns move horizontally in space, as well as in time. Once a tree is established in the open, it's easier for others to get a start next to it, for two reasons. First, most subalpine trees are adept at "layering," or propagating a new

jor climax dominant along with Engelmann spruce, but in our range it can be replaced in succession by the more shade-tolerant hemlocks, silver and grand firs, or Douglas-fir. It reaches its lowest elevations here—3,000' or even 2,000'—in cold air pockets and scant-soil sites like lava flows and talus.

## Pacific Silver Fir

*Abies amabilis* (a-ma-bil-iss: lovely). Also **lovely fir.** 40" diam × 165'; needles of two sizes: some ¾–1¼", flat-spreading, others ¼–¾", pointing forward and upward along the twig; deep glossy green on top, with 2 strong white stripes beneath; notch-tipped exc on cone-bearing branches; cones dense, heavy, barrel-shaped, 3–5" × 1½–2", green maturing to brown, borne erect on upper branches, dropping their seeds and scales singly while the cores remain on the branch a year or more; bark gray to silvery white, resin-blistered, smooth exc on some old trees; branches horizontal. Dense mature forests near and W of Crests, mainly 3,000–5,000'. Pinaceae.

The handsome dark needles of silver fir lie mostly in a flat plane; an additional series of shorter ones presses forward in a

plant by extending roots from branches in contact with earth; the parent limb feeds the new shoot intravenously for a while, a big advantage over growing from seed. Second (the "black body effect"), during spring thaw, dark tree foliage absorbs more of the sun's heat than white snow does, and melts a little well in the last few feet of snow. The well is a microsite with a growing season several weeks longer than the surrounding meadow— just what seedlings need. Hence trees in subalpine parkland typically grow in tight, slowly expanding clumps, often elongated downslope into a teardrop shape.

The founding mother of a clump is often a subalpine fir or whitebark pine. As the clump gets big enough to have a shaded, clear floor in the center, a few mountain hemlock seedlings appear there, eventually growing to replace the oldest firs or pines. As the broader, denser hemlock crowns mature, the deepening shade favors silver fir, provided there is a nearby seed source and a chipmunk or other animal to bring the seeds. Subalpine fir continues to expand the clump by layering at the perimeter, often

herringbonelike pattern that neatly hides the twig from directly above. This unique arrangement is Clue Number One both to which species this is and what it's up to—shade tolerance. Hiding the twig from above means not letting any overhead sunlight go to waste on a nonphotosynthesizing surface. The dark surface also maximizes light absorption.

Because it's our most shade-tolerant tree (along with western hemlock) you can expect silver fir to be a major climax tree everywhere you see its saplings. The climax silver fir zone is a Westside midslope belt—3,000–4,000' in southern Washington, a little higher in Oregon, and lower northward all the way to sea level at the southeast corner of Alaska. Silver fir is rare in our lowlands, though the largest known specimen (7'10" diameter by 203' tall) grows below 800' near the Olympic seashore. It reaches tree line in the protection of other trees, but rarely thrives in the open, being prone to windthrow and drought. It grows only in our range and northward.

Being alert to the first appearance of these forest-green saplings can bolster your sense of progress during slow hours

behind a vanguard of low-to-tall heath family shrubs. At an advanced stage, our typical tree clump might display in slowly expanding, roughly concentric rings, this successional sequence: red heather → black huckleberry → white rhododendron → subalpine fir → mountain hemlock → silver fir. Or sometimes only shrubs fill the center after the pioneer trees die, leaving a hollow tree clump or "atoll."

The timberline species makeup varies along our wet-to-dry west/east gradient. Whitebark pine, Engelmann spruce, and subalpine fir increase in importance from west to east, while silver fir, Alaska-cedar, and mountain hemlock do the opposite. In the eastern North Cascades, subalpine larch as a "mother tree" enables other species to grow upright at higher elevations than they could without its protection.

In comparison with other mountain ranges, ours have a wide subalpine parkland belt and only a narrow fringe of alpine vegetation. This is largely attributable to our paradoxical combination of mild temperature with a very long snowbound season.

**Pacific silver fir**

lower branch

upper branch

of switchbacking up from valley floor toward high basins; vistas unfold even while you can't see out of the forest. The shrub layer is thinning, perhaps showing off charming montane herbs—bunchberry, bead lily, false-Solomon's-seal, coralroot, and wintergreens; other silver fir communities have neither shrubs nor herbs to speak of, except beargrass or (in Oregon) rhododendron—each rather dull most of the time but spectacular in a good flowering season.

Midslope forests can be incredibly quiet, with only faint fricative sounds sifting up from torrents far below. Animals are relatively scarce and mainly diurnal here. Grouse flushing under your nose may raise more adrenalin than "BOO!" in an empty house. Don't be alarmed if an unseen assailant high in the trees bombards you in September; it's only a Douglas squirrel harvesting big, thudding silver fir cones for his winter stores.

After deep shade, late-lying deep snow is the most critical challenge here. The dry season is often underway before the snowpack disappears; after the meltwater is gone, the needle-duff seedbed may be bone dry. Silver fir seeds cope with this by germinating the moment they see the light of day after the winter chill—even though this means all too many of them germinate *on* snow and die. Luckier ones put most of their energy into their taproot for a few seasons, getting a tap on summer-long moisture, while their shoots raise themselves a mere inch or two, sprout a few needles, and form fat buds well protected for winter.

*Amabilis* is one of many names from the pen of David Douglas. Travelers of his day found that, of all boughs, silver fir made the loveliest bedding. Foam pads, invented for the age of low-impact camping, are much less trouble.

## Noble Fir

*Abies procera* (pross-er-a: noble or tall). 50″ diam × 210′; needles ¾–1¼″, bluish to silvery green with white stomata both above and below, typically* in 4 distinct stripes, blunt to pointed (not notch-tipped), thickish to ± 4-sided, crowding and curving upward from the twig, many with a sharp "hockey-stick" curve right at the base; cones dense, heavy, nearly cylindrical, 4–7″ × 1¼–2½″, green maturing dark red-brown, scales ± covered by papery green to straw-colored bracts with slender upcurved points; cones erect on upper branches, dropping their seeds and scales singly while the core remains; young bark gray, smooth, resin-blistered; mature bark red-brown, thin, flaking, cracked rectangularly; branches horizontal. Mostly 3,100–4,800′ on W-side, Ore and S Wash Cas. Pinaceae.

Nineteenth-century botanists gave the Pacific Coast's true firs Latin names meaning Grand, Lovely, Magnificent, and Noble —saving Noble for the largest, longest-living member of the

*One versus two topside stripes of stomata is the easy way to tell subalpine noble firs apart, but it isn't reliable. Once I picked twigs of the two species growing side by side and couldn't spot any difference in their needles, whether of stomata, color, size, curvature, or tip. Hitchcock and Cronquist supply a distinction visible in a clean cross-section of a needle under 10× magnification. You can go by the resin-blistered bark of noble fir, or by its lower-elevation, more southerly and westerly range. To be more positive, you have to find cones. Noble fir cones are green to brown and they bristle with hooked bracts; subalpine fir cones are smaller, gray to black, with straight bracts that stick out only from immature cones.

genus. Evenly spaced annual tiers of stiffly horizontal limbs are seen on all true firs, but especially this one; they are visible even from a quarter-mile away as a fine horizontal lined texture. Together with an evenly conical shape, they make young noble firs elegant Christmas trees—good to look at, but hard to bundle up for shipping, and slow to grow. Growth rate picks up impressively after the second decade, though. Hundred-year and older noble firs are often larger than the like-aged Douglas-firs they typically grow with. Noble fir also makes good lumber and is, after Douglas-fir, the tree most often planted in timberland west of the Cascades. On British tree farms it has been planted for over a century.

All true fir and western hemlock lumber is currently sold as "Hem-Fir" on the grounds that the milled woods are indistinguishable to the naked eye. This is unfortunate, since noble fir lumber is stronger. Earlier lumber merchants tried to raise noble fir and Doug-fir above the true-fir pack by labeling them "Oregon Larch" and "Oregon Pine." The misnomer filtered back via loggers to cartographers, and we ended up with two Larch Mountains—so named for their noble firs—near the west end of the Columbia Gorge. No larches grow there.

Noble firs are rather intolerant trees; they pioneer after fire, usually mixed with Douglas-firs and a few other conifers, and persist as canopy trees for many centuries. Lovely stands result, with massive straight trunks supporting a dense canopy somewhere above 100' up. As much as three-quarters of the total height may be limbless, a sign that intolerant foliage doesn't photosynthesize efficiently enough to earn its keep after it loses its canopy position in direct sun. These stands have few shrubs and fewer saplings—mostly silver fir, the climax species beginning to take over. The herb layer may also be impoverished, but more often it is lush, featuring vanillaleaf, dwarf raspberry, and small false-Solomon's-seal.

Noble firs are unfamiliar to most people, partly because their range is small. Their northern limit, near Stevens Pass, is abrupt, but they are vigorous enough there to suggest a potential range well into British Columbia. Fossil pollen shows that noble firs grew far to the north before the Ice Age, but they have yet to regain the border in following glacial retreat. The slow migration is blamed on heavy seeds (poor wind carrying distance) combined with shade intolerance; each time the northernmost noble firs are replaced by climax silver firs, the migration is pushed back until fire clears the path again. Noble firs have also failed to cross the lowlands to the Olympics, but they have reached Oregon's Coast Range.

The southern limit of noble fir range is not abrupt at all; the species and its close kin, California red fir, *Abies magnifica*, hybridize and intergrade where their ranges meet, producing a continuum of characteristics between typical noble fir near the Three Sisters and typical red fir in the Trinity Mountains. A section of the continuum (south of the Sisters) is sometimes distinguished as Shasta red fir, *A. magnifica shastensis*.

## Grand Fir

*Abies grandis* (gran-dis: big). Also **lowland white fir.** 44" diam × 200'; needles ¾–2", quite broad and thin, spreading in a flat plane from the twig, notch-tipped to rounded, dark green above, 2 white stomatal stripes beneath; (needles of topmost branches often neither flat-spreading nor esp dark); cones dense, heavy, ± cylindrical, 2½–4 × 1½", greenish, borne erect on upper branches, dropping their seeds and scales singly while the core remains; bark gray to light brown, resin-blistered, becoming ± ridged and flaky with age; branches horizontal. Common E of Cas Cr, 3,200–5,000'; scattered on W-side. Pinaceae.

The foliage on a grand fir sapling catches your eye, the tidy flat array of long, broad needles showing off their glossy green. Flat leaf arrays (page 22) imply shade tolerance. Grand firs are only slightly less tolerant than western hemlocks and silver firs, and prefer less rainfall—30–45" per year, especially where the summer drought is ameliorated by streamside groundwater or by mountain coolness. That "either/or" preference makes the species ecologically two-faced, with two geographically and genetically distinct ecotypes.

"Typical" grand fir grows below 2,000' west of the Cascades, mostly in partially rain-shadowed areas like the Willamette Valley (where it is apparently a climax dominant; see page 70) and the northeastern Olympics. It is commonest near streams, often with black cottonwoods and red-cedars.

"Montane" grand fir, on the other hand, grows at mid elevations in the Cascades, mainly Eastside. It is a climax dominant in a 3,200–5,000' belt (the grand fir zone) on the East Slope in Oregon and most of Washington. Douglas-fir, sub-alpine fir, and western hemlock are also climax dominants in many environments within this very mixed zone, and pines and larches also persist, given a natural regime of recurring fires. In fact, mature grand fir stands are not extensive here, though they are on the increase; the fire-prone grand fir often benefits (at the expense of ponderosa pine) from Smokey's fire-fighting efforts.

South of our range, in the Klamaths and Sierra Nevada, grand fir's close relative white fir, *A. concolor*, plays a similar role. Where the two species meet they "intergrade," or hybridize in proportions that vary along a continuum. The slight differences of form between typical and montane grand firs may represent a genetic trace of white fir in the montane type—longer needles, more stomatal bloom in the upper

## Nurse Logs

By far the most favorable seedbed for a conifer in our wet forests is the rotting trunk, stump, or upended rootwad of a fallen tree. In some stands, especially on the Olympics' Western rivers, these "nurse logs" support hundreds of seedlings and saplings while the ground in between has none.

To the casual passer by, this odd distribution of seedlings is less striking than its effects on the shape of grown trees. By age 10 or sooner, the seedling extends fine rootlets down the nurse log's sides to mineral soil, even from perches 20' up on snag tops. Over the decades, if the seedling survives, the rootlets grow into sturdy "prop roots" while the nurse log slowly rots out from underneath. You can see this process at all stages; at some point the support relationship may reverse, a few chunks of rotting nurse now dangling from the prop roots. Over the course of two to six centuries, the nurse log disappears and the tree's roots fill in, remaining as "buttresses" on the lower trunk of a huge

groove, more upward curve from the twig, and less notch at the tip. Variable needles and lots of intergrading make our true firs notoriously hard to tell apart.

## Sitka Spruce

*Picea sitchensis* (pis-ia: Roman term, derived from "pitch," for some conifer; sit-**ken**-sis: of Sitka, SE Alaska). Also **tideland spruce.** 90″ diam × 235′; needles stiff and very sharp, ± 3-sided (flat on top), ½–1″ long, projecting equally on all sides of the twig, light green with 2 stomatal stripes on top, bloom scant or lacking beneath; young twigs smooth, old defoliated twigs rough and scratchy with the peglike bases of the fallen needles; cones 2–3½″ long, light, scales thin and finely, irregularly toothed; bark scaly, thin (less than ¾″ even on huge trees); mature trunks very straight, round, and untapering, though often buttressed. W-side lowlands; abundant in Olys, uncommon in Wash Cas and absent from Ore Cas. Pinaceae.

Sitka spruce, the world's fourth-tallest species, occupies a 2,100-mile coastal strip bounded by the reach of ocean fog. The next 500 fog-shrouded miles to the south support the tall-

---

spruce, hemlock or Douglas-fir. (Red-cedars develop flaring buttresses without prop root origins.) Where adjacent trees' buttresses head straight toward each other, more or less meshing, they were once prop roots on the same nurse log. A bygone nurse log is also implied by a row or "colonnade" of similar-aged trees with buttresses accentuating their alignment.

A nurse log plainly offers great advantages to a seedling, but what these are is less clear. No doubt there is truth in several of those proposed:

**Water:** Well-rotted wood stays sodden all through the summer, not only retaining rainfall but actually manufacturing water as a decomposition product. Ironically, this could explain a "rain forest" quality as an effect of summer rainlessness. In parts of the Hoh Valley, for example, conifer and other seedlings proliferate on the ground on lowest and uppermost terraces, but are scarce, except on nurse logs, on middle terraces. Possibly, first-year seedlings root into the water table on lowest terraces, and into well-developed soil on the highest (600-year-old) terraces, while those

---

est tree, the coast redwood, while Douglas-fir, the third tallest, joins them along the entire stretch.

The successional status of Sitka spruce is controversial. Though measurably less shade-tolerant than western hemlock and red-cedar, and considered successionally transitory to them by some ecologists, it holds its own in competition with them even in a sodden range that rarely burns. If these spruce/hemlock "rain forests" are not climax, then what has been disturbing them? Elk, for one thing; Olympic elk congregate in these valleys every winter, selectively browsing hemlock seedlings, avoiding the prickly spruce. Elk browsing also helps to keep the forest open, and so does each huge tree that falls. This makes shade tolerance less crucial.

The river bottoms are young soils dating from minor, post–Ice Age glacial advances, and rarely as much as a century older than the oldest spruces upon them. The glaciers themselves did not reach the lower river bottoms that recently, but each time they retreated, meltwater flooded valleys downstream from the glacier snouts, dumping deep layers of cob-

---

germinating on the heavily drained gravel soil of middle terraces are killed by summer drought Big spruces on middle terraces originally grew when these were lowest terraces close to the water table, which later dropped as the river cut. A wet nurse log would be at least as valuable on midmontane slopes, where excessively drained soils are typical.

**Nutrients:** Nutrients for a young conifer are present in a dead conifer, though actually not in exceptional quantities. Available nitrogen increases as decomposition—by nitrogen-fixing bacteria—advances.

**Mycorrhizae:** Quickly forming a mycorrhizal partnership (page 262) with a fungus is crucial to conifer seedlings. The fungus draws water and processes nutrients using enzymes that only fungi produce. Rotten wood is the preferred substrate of some mycorrhizal fungi.

**Disease:** Some evidence suggests that seedlings on nurse logs are less subject to some diseases.

---

**Cascade-Olympic Natural History**

bles and gravel. As the glaciers advanced again, the rivers cut through the "outwash fill" until the next glacial retreat, then partially refilled that cut, then began cutting again. Each new cut narrower than the previous one left a terrace. The spruce-dominated stands along the Hoh are on two terrace levels, 620 and 340 years old. Soil development may yet allow the stands to succeed to hemlock, but there is no clear evidence of that happening now. It's hard to say whether the abundance of spruce along Olympic rivers (relative to similar Cascade ones) owes most to elk, fog, or some difference in the terrace soils.

Sitka spruce's high strength-to-weight ratio was pounced upon for early aviation; spruce was the ideal aircraft material. It provided nearly all the airplane frames for America's side in World War I. Sitka spruce logging peaked then at more than twice the volume of any time since—except World War II, when airplane frames still used spruce.

Today the arts claim the best spruce on the market; spruces have the best resonance for piano sounding-boards, guitar tops, etc. Old-growth Sitka spruce provides far and

---

**Browsing:** Though many nurse logs are at a perfect height for deer and elk to browse from, the taller stumps, rootwads, and logs jackstrawed up on top of other logs are out of reach. This would be significant in valleys where elk herds winter.

**Litter:** One researcher found hemlock seedlings fatally flattened by a mat of tree litter coalescing in the melting snowpack in some montane forests. (See page 15.) Hemlock seedlings were restricted to nurse logs there, and he concluded that litter sloughs off the sides of nurse logs.

**Sunlight:** A superior position for catching light, up above fern, moss, and herb competition, is an explanation often proposed. A raised position may help in some cases, but in many groves almost all the competition is located on the nurse logs with the conifer seedlings. Seedlings have been found to favor spruce over hemlock nurse logs, and moderately rotted logs over either fresher or totally decayed ones; these preferences point to advantages other than physical position.

---

away the biggest clear, straight-grained sections. Even second-growth should produce acoustical spruce in the foreseeable future. Sitka spruce also produces the best pulp of any West Coast tree. It hybridizes with white spruce, *P. glauca*, the chief lumber species of Canada and interior Alaska, where their ranges meet in Alaska. This natural hybrid has been proposed for commercial planting in coastal Washington and Oregon because it seems resistant to the spruce weevil, which stunts and deforms young Sitka spruces.

The long, tough, sinewy (after pounding) small roots of Sitka spruce were vital to native Northwest culture. They supplied most of the exquisite and highly functional basketry, and also most of the twine and rope, including whaling lines. A spruce twig stuck in the hair was a charm for whaling, while the harpoon tips were protected, and the canoes caulked, with spruce pitch. The Makah and Quinault were fond of chewing spruce pitch; it's fragrant and spicy-sweet, turning bitterish as it is chewed. Try some.

# Engelmann Spruce

*Picea engelmannii* (eng-gell-mah-nee-eye: after George Engelmann). 40″ diam × 160′; needles ¾–1¼″, sharp, 4-sided, bad-smelling when crushed, crowding upward and forward from the twig or radiating ± evenly from it, deep blue-green with stomatal stripes ± equally on all sides; young twigs usually fuzzy; cones 1½–2½″, light, much like mtn hemlock cones but scales are thinner, closer, and irregularly toothed-to-wavy along outer edge; often with conelike galls from branch tips; bark thin, scaly; crown dense, narrow, with fringelike pendent branchlets; or prostrate, shrubby. 3,000–8,000′ E of Cas Cr, esp in N- to E-draining ravines; rare in NE Olys. Pinaceae. Color p 115.

Spruces are the second most northerly conifer genus (after larches). In the Rockies, Engelmann spruce and subalpine fir are the major climax trees in all higher forests. Here, this spruce specializes in cold and/or swampy Eastside sites, ranking about average in tolerance. Though it is large and distinctive among subalpine trees, its three stands in the Olympics went undiscovered until 1968. Take that as a challenge.

In addition to their cones, many Engelmann spruces bear curious conelike appendages—galls, or "houses" for aphid lar-

**Sitka spruce**

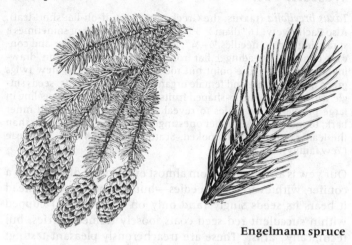

**Engelmann spruce**

vae (page 454). Gall tissue is secreted by a plant in response apparently to chemical stimulation, usually by a female insect laying eggs. The familiar spheres on oak twigs are galls of various creatures, especially gall wasps. The spruce gall aphid's gall terminates and envelopes new growth at the tip of a branch. The dead needles turn a tan color along with the gall; together they look much like a 1–2" cone with needle-tipped, melted-together "scales," each hooding an opening into a larval chamber. The gall may hang from the branch for years, long after the larvae mature and move on; other insects may colonize it. Though the spruce gall frustrates the growth of its branchlet, it scarcely harms the tree.

Like other spruces, the Engelmann spruce yields high quality lumber—light, acoustically resonant, and free of pitch, color, odor, and taste. Yet it isn't found in many barrels or guitar tops, nor logged much at all in our region, due to its relative scarcity and inaccessibility. Indian use of it (and of all our montane trees) was negligible here, because Northwest tribes didn't live or hunt in the high mountains. They regularly crossed the Cascades for trade between Coast and Interior tribes, but on such missions the rule was to pack light, keep party size small, and not waste time beating around the bush.

# Western Yew

*Taxus brevifolia* (tax-us: the Greek term; brev-if-**oh**-lia: short leaf). Also **Pacific yew.** 16″ diam × 35′ or (rarely) up to 80′; or sometimes a sprawling shrub; needles ½–¾″, grass-green on top, paler and concave beneath, spreading ± flat from the twig, broad and thin, drawing abruptly to a fine point but too soft to feel prickly; new twigs green all year; male and female organs on separate plants; seeds single within juicy red cup-shaped fruits ¼″ diam; bark thin, peeling in large purple-brown scales to reveal red to purplish, smooth inner bark; branches sparse, upper ones angled up, often much longer than the leader; trunk often crooked. Scattered below 4,500′. Taxaceae (Yew family). Color p 115.

Our yew is an anomaly from almost every point of view. It is a conifer, with evergreen needles—but without cones. Instead it bears its seeds singly (and only on female trees) cupped within succulent red seed coats loosely termed berries, but technically "arils." These are treacherously pleasant-tasting;

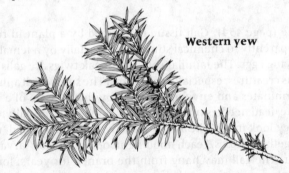

**Western yew**

the seeds of many yew species contain alkaloids capable of inducing cardiac arrest in humans. Attractive but poisonous fruits are few in our area; smooth, bright red berries are good ones to keep your kids away from (see Baneberry, page 234). Birds love yew berries, passing the toxic seeds undigested.

Woodworkers class conifers as "softwoods," by definition, but they've all heard of yew as the exception that proves the rule—it is among the hardest of woods. It can be worked quite well with power tools, or even carved to make extraordinarily durable and beautiful utensils; the sapwood is cream, the heartwood orange to rose. Yet few have worked it. Loggers burn it with the slash, finding it hard to market such small, allegedly scarce trees. But yew is not scarce here, and often reaches marketable size by hardwood standards.

The Indians knew better. They made it into spoons, bowls, hair combs, drum frames, fishnet frames, canoe paddles, clam shovels, digging sticks, splitting wedges, war clubs, sea lion clubs, deer trap springs, arrows, and bows. (Bows of yew species are almost universal. The Greek name for yew, *taxos*, is related to both "toxin" and *toxon*, meaning "bow.") Prizing yew for strength, elasticity, and hardness, young Swinomish men rubbed a yew's limbs on their own in the belief those qualities would rub off on them. They also sometimes added yew needles to their smoking mixtures, perhaps more for "toxins" than flavor.

Anomalous among conifer barks here, yew bark resembles the self-peeling bark of madroño. The smooth red underbark is tawny on madroño, while it can be almost cherry red on yew. In 1987, Western science suddenly wanted yew bark. An order was filled for 60,000 pounds of bark, from which to extract a tiny amount of something called "taxol"—a compound that shrinks tumors in preliminary research. If taxol proves valuable against cancer, the demand will be far too great for western yew bark to supply. It may be a great tragedy for the yew if trials of taxol in humans show promise before efforts at commercial synthesis of the drug succeed.

Western yew, the largest species of yew, is our only forest conifer that never achieves canopy stature; it belongs to the tall shrub layer, though it is normally a tree and outgrows the tall shrubs. Its distribution also inspires confusion; often described as a moist-site tree, yew is actually almost indifferent to climatic variation within low-to-midslope Westside forests, much like the ubiquitous vine maple, Oregon-grape, and twinflower. On the Eastside it specializes in streamsides below 4,000. Extremely shade-tolerant, it spends its entire life under the forest canopy. It may suffer when the shade is removed, turning orange all over but not necessarily dying. I saw one scrubby orange yew alone on a steep, burningly exposed southwest-facing site, with no evidence whatsoever of enjoying shade during its lifetime.

The needles are bunched differently in these two genera:

**Pines** of every species bear long evergreen needles in fascicles (bundles) bound together at the base by tiny membranous bracts. The number of needles per bundle (five, three, or two) is the easiest step in identifying pines; check several bundles, because individual trees may be inconsistent. Five-needle pines (pp 46–48) are loosely termed "white pines" and three-needle pines (ponderosa, below) are "yellow pines." The East has various two-needled "red pines," but no one calls our lodgepole a red pine. The Southwest has piñon pines with bracted fascicles of just one needle.

**Larches** bear soft deciduous needles, mostly in fat false whorls of 15 to 40 needles at the tips of peglike spur twigs about ¼" long and wide. However, on this-year's twigs the needles are single, and spirally arranged. Technically, the pegs and their whorls are also twigs—exceedingly short ones, with compressed spirals of single needles—hence "false whorls." But to the naked eye they are bunches.

## Ponderosa Pine

*Pinus ponderosa* (pie-nus: the Roman term; ponder-oh-sa: massive). Also **western yellow pine.** 44" diam × 175', needles 4–10", in bunches of 3, yellowish green, clustered near branch tips; cones 3–5" × 2–3", closed and reddish until late in their second year, scales tipped with stout recurved barbs; young bark very dark brown, soon furrowing, maturing yellowish to light reddish brown and very thick, breaking up into plates and scales shaped like jigsaw puzzle pieces. Dry habitats, mostly at low elevs E-side; very few in E Olys and Willamette Valley. Pinaceae. Color p 115.

Much of ponderosa pines' charm is in the parklike grassy spacing they maintain over the centuries, evoking the spirit of the great cowboy West, or at least the great cowboy Western. It makes you wish you were on a horse. That spacing pattern—and the prevalence of the pines and the grass underneath—result from frequent ground fires. A 316-year-old ponderosa east of Mt. Jefferson, for example, bears scars from 18 fires. Mature ponderosas are the most fire-resistant trees in their range, thanks to thick bark and high crowns, but young ones are vulnerable. The sapling that stands a good chance of surviving fires to reach immune size is the one growing away from other trees, because the two fuels likely to bring ground fire to it are other saplings and the needle litter under big trees; hence the parklike spacing.

As long as the stand is widely spaced, shade intolerance is no problem, but where fire-fighting and adequate moisture have allowed denser stands to grow, we find a strong successional trend away from the pine, in favor of Douglas-fir, grand fir, and incense-cedar, which are less fire-resistant but more shade-tolerant. Since ponderosa pine is the preferred lumber species in most of its range, this trend was a compelling reason for foresters to reconsider the virtues of fire. Another was the correlation of dense growth with bark beetle outbreaks (page 459). A third was that without fire, brush and dead branches may build up to a point where a fire during summer drought won't stay on the ground, but will ignite even the tallest pine crowns and kill them; this makes "prescribed burning" much riskier than it would have been fifty years ago.

Ponderosa pine ranks second nationally (after Douglas-fir) in lumber production. Most "knotty pine" paneling comes from young ponderosas. Clear mature wood is versatile, similar to white pine but nicely two-toned, with pale yellow to orange-brown heartwood and broad, nearly white sapwood.

The "pine nuts" of commerce come from American piñon and European stone pines, but the seeds of all pines are delicious, and prized by birds and rodents, who bury innumerable seeds in small caches. They intend to come back for them someday but inevitably overlook some caches, which then germinate. This manner of seeding benefits pines; it plants seeds where wind might never carry them, and plants them deeper, in mineral soil often in litter-free spots, sparing the seedlings from drought and the eventual saplings from ground fire. When you see a clump of several pine seedlings within a square half-inch or so, it's undoubtedly a forgotten cache.

The lowest belt of Eastside Cascade forest is the ponderosa pine zone, though fire-suppressive management has shown that other conifers could crowd ponderosas out of climax forest in most of that zone. Still, there is a narrow fringe of dry sites where ponderosa pine is the only large tree drought-hardy enough to grow. Irregular "lower timberlines" at the limit of drought tolerance often reflect soil texture more directly than climate; coarser soil permits easier, deeper rooting, and absorbs more of the rain or snowmelt. Central Oregon's Lost Forest is an isolated stand of ponderosas and sizable junipers neatly filling a patch of very sandy soil, surrounded by a 40-mile radius of sagebrush steppe on relatively clayey soil. It lives on 8.7" of annual precipitation—the driest regime that supports a forest anywhere in the American West.

**Ponderosa pine**

**Lodgepole pine**

## Lodgepole Pine

*Pinus contorta.* 20" diam × 100'; needles in 2s, 1½–2½" long, yellow-green; cones 1½–2" long, egg-shaped, point of attachment usually quite off-center, scales sharp-tipped; cones abundant, borne even on very young trees (5–20 years), persistent on the branch for many years either closed or open (empty of seeds); bark thin (less than 1"), reddish brown to gray, scaly. Common above 3,500 in E Olys and E Cas; scattered elsewhere. Pinaceae. Color p 115.

Lodgepole pines are tricksters on the ecological playing field. They don't seem to compete on the same grounds as our other conifers—size, longevity, shade tolerance, and fire resistance. They excel instead at profligate and gimmicky seeding habits, short-distance speed, and tolerance of poor soil.

Lodgepoles are prolific to a fault. They produce viable seeds in huge numbers year after year (a rarity among conifers); they release some of them at all times of year; they begin fruiting younger than other conifers; and their seedlings and saplings grow fastest. They unleash their most remarkable punch after a fire; some of the cones on many lodgepole pines are sealed shut by a resin with a melting point of 113°. The seeds inside, viable for decades, are protected through a fire by the closed cone. The fire kills the pines but melts their cone-sealing resin; afterward, the cone scales open slowly to release seeds upon a wide-open field. (These "serotinous" cones, typical of lodgepoles in the Rockies, are rare in our range.)

As in rabbits, prolificacy can lead to overpopulation—a stagnant stand of pines stunted by intraspecies competition. This looks dismal to both foresters and hikers, but isn't so bad in terms of species survival. In nature, the dwarfed stand often persists until fire comes to give the species another try.

Lodgepole pines are the commonest tree of the Rockies from Colorado north; here they are fewer, albeit widespread, because at maturity they fall short—short-statured, short-lived, and shade-intolerant. Rarely tall enough, once other conifers mature around them, to keep the bright light they need, they persist by stifling competition. They also dominate sites with microclimates and substrates that discourage other trees. One such site is a frost pocket with severe temperature fluctuations, on the lower east slopes. A second is a seashore with salt spray; the contorted "shore pines" that grow there are so unlodgepolelike that they were considered a distinct variety until it was found that a shore pine seedling transplanted among montane lodgepoles would grow just like them, and vice versa. A third lodgepole habitat is on chemically bizarre, sparsely vegetated soils made of serpentine minerals (page 545). Whatcom County's Twin Sisters, a rare large block of the rock dunite, which weathers into serpentine, support an unusual krummholz form of lodgepole pine at timberline—making lodgepole the only tree found at both sea level and timberline in Oregon or Washington. Fourth, the biggest concentrations of lodgepole in our mountains are on relatively recent lava flows, mudflows, and pumice deposits from Cascades volcanoes—in particular, the 100-mile stretch of Oregon's Cascade Crest and Eastside mantled with pumice from the huge Mt. Mazama (Crater Lake) eruption of 6,700 years ago. Mt. St. Helens' timberline was similar before 1980, and lodgepole pines can be expected to pioneer there again.

# Whitebark Pine

*Pinus albicaulis* (al-bic-aw-lis: white stem). 20" diam × 65'; needles in 5s, 1½–3" long, yellow-green, in tufts at branch tips; cones 1½–3" long, egg-shaped, purplish, dense, long persistent on the tree, usually falling scale by scale long after dropping the seeds; bark thin, scaly, superficially whitish or grayish. Alp/subalpine. Pinaceae.

With their broad crowns and tufted, paler foliage, whitebark pines are easy to tell from other high-country conifers—subalpine fir, Engelmann spruce, and mountain hemlock. They are shade-intolerant, dry-terrain pioneers, only occasionally found west of the Cascade Crest, and rarely in the Olympics.

Form varies with elevation. As krummholz (dense prostrate shrubs) whitebark pines reach the highest elevations of all our conifers—8,200' in the Stuart Range. At their lowest (5,000') they grow straight, resembling lodgepole pines. Their main range is the subalpine parkland, where they are usually contorted and multistemmed, but nevertheless erect up to 7,000'. Blue grouse find their dense crowns cozy in winter.

Whitebark pine "nuts" travel on surrogate wings. Their own wings, undersized to begin with, remain stuck to the cone scales when the seeds fall out; fat, heavy, and wingless, the seeds nevertheless fly far in the beaks of Clark's nutcrackers. Most whitebark pine seedlings grow from caches buried and then forgotten by these birds; you can tell when you see pine seedlings in tiny clumps.

**Whitebark pine**

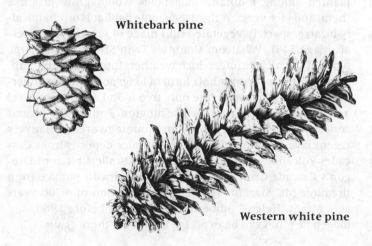

**Western white pine**

# Western White Pine

*Pinus monticola* (mon-tic-a-la: mtn dweller). 40" diam × 120' (larger elsewhere); needles in 5s, 2–4" long, blue-green with white bloom on inner surfaces only, blunt-tipped; cones 6–10" × 2–4", thin-scaled and flimsy for their size, often curved, borne by a short stalk from upper branch tips; young bark greenish gray, maturing to gray with a cinnamon interior, cracking in squares. Widely scattered at mid elevs. Pinaceae. Color p 115.

The Latin name "mountain-dweller" notwithstanding, western white pine is scattered throughout the Northwest from coastal bogs to low-subalpine forests. In the mountains of Northern Idaho it grows bigger, dominates climax forests, and is the pick of the lumber trade, and the State Tree. Here, it is a bit out of place—never more than a small minority and, sadly, dying out from that minor role. You would likely walk by without noticing it, if it weren't for the startlingly outsized cones on the ground among smaller cones from larger trees.

White pines are dwindling in our mountains, due partly to fire suppression but mostly to white pine blister rust. Commercial success brought this fate—an irony that's hardly unique. America's logging industry, after growing up on eastern white pine, *P. strobus,* until that species was depleted, was delighted to find a bigger white pine in the Northwest. Before long, so much western white pine replanting stock was needed that foreign nurseries entered the market. A shipment of French seedlings in 1906 brought blister rust, a European disease to which American pines have no resistance. Since the rust fungus requires a currant or gooseberry plant as an alternate host, currant extermination was attempted in parts of Idaho, but proved too laborious. Western white pines are dying off almost as inexorably as American elms and chestnuts. Naturally rust-resistant genetic strains are being bred, and should help restock the species some day. Though not as severely threatened, whitebark and sugar pines are also susceptible, as are virtually all five-needle or "white" pines.

Soft, easily worked, and fine-textured, western white pine is a good wood for moldings, windows, doorframes, and pattern stock in metal foundries. The hottest demand came from wooden match producers, who raised western white pine production to its highest levels back in the 1920s and '30s, before book matches took over.

---

## Sugar Pine

*Pinus lambertiana* (lam-ber-tee-ay-na: after Aylmer Lambert). 50″ diam × 180′; needles in 5s, 2–4″ long, blue-green with white bloom on all surfaces, sharp; cones 10–18″ × 3–6″ (the largest cones anywhere) hanging from upper branch tips by stout 1½–3″ stalks; seeds with wings 1–1½″; dark gray bark, soon maturing to reddish brown, deep-furrowed and scaly. Low to mid elevs in C Ore. Pinaceae.

**Sugar pine**

The biggest pine of all is this California tree, fairly common also in Southern Oregon but not so in our range; its northern limit is on the upper Clackamas. Western white pine, our common five-needled, big-coned pine, overlaps sugar pine's range. You aren't looking at a really big cone unless it's a foot long; an 8–10″ cone is probably western white. Sugar pine grows on hot, dry Westside sites below 3,000′, and less commonly on well-watered Eastside sites. It is heavily attacked by white pine blister rust (page 47).

The tasty, nutritious seeds from these giant cones are good-sized, but no larger than the pine nuts traditionally collected from little southwestern piñon and digger pines and European stone pines.

David Douglas counted the sugar pine the greatest of his discoveries in the Northwest. He trekked south from Fort Vancouver specifically to find it after being shown some seeds and told of the cones by Indians. When he found it he was alone in country almost totally unexplored by whites, somewhere near the Umpqua Divide. He retrieved cones by shooting them down with his rifle, leading forthwith to what he felt was his closest brush with death at the hands of Indians. He managed to divert them and run off with his precious specimens. No wonder he cherished them!

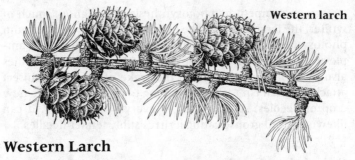

## Western Larch

*Larix occidentalis* (lair-ix: the Roman term; ox-i-den-tay-lis: western). Also **tamarack**. 52″ diam × 170′; needles deciduous, soft, pale green, 1–1¾″, 15–30 in apparent whorls on short peglike spurs, exc that needles are single and spirally arranged on this-year's twigs, and often winter-persistent on seedlings; cones 1–1½″, often persistent, reddish until dry, bristling with pointy bracts longer than the scales; young bark thin and gray, maturing yellowish to cinnamon brown, 3–6″ thick, furrowed and flaking in curvy shapes ± like ponderosa pine bark. Scattered, on ± moist sites 2,500–5,000′ E or (S of Mt Adams) slightly W of the Cas Cr. Pinaceae. Color p 115.

A larch is something many people mistakenly think of as a contradiction—a deciduous conifer.* The deciduous needles set it off visually, even from a distance—intensely chartreuse in spring, then a subtle but distinctive grassy-green through summer, smashingly yellow in October, and conspicuous by their absence for five or six months of winter. You can tell a larch in winter from a maple or cottonwood by its coniferous form (single, straight trunk; symmetrical branching) and from a dead evergreen by its warty texture (pegs on its twigs).

Ecologically too, western larch's leaf-dropping habit sets it apart. Among big Eastslope conifers, it is the fastest-growing, longest-living, most fire-resistant, shade-intolerant, and water-demanding—an unusual combination. Since its

---

*Other deciduous conifers are the bald-cypresses, genus *Taxodium*, of Southern swamps, and the dawn redwood, *Metasequoia glyptrostroboides*. The latter once dominated Northwest forests, judging by its fossils all over eastern Oregon and Washington. It was thought to have been extinct for millions of years, with the coast redwood its only descendent. Astonishingly, a few remnant stands turned up in 1946 in remote mountains of central China. Since then it has done well where planted in Northwest cities.

---

evergreen competitors photosynthesize throughout much of winter, the larch has to make up for lost time with maximum photosynthetic efficiency. This requires full sunlight and ample groundwater through the dry months. Deciduous needles also make larches resilient in the wake of defoliating insect attacks; a larch isn't fazed by having to produce a whole new crop of needles, since it does so every year anyway. It can likewise afford grouse eating its irresistibly tender needles.

## Subalpine Larch

*Larix lyallii* (lye-ah-lee-eye: after David Lyall, p 51). Also **woolly larch.** 32" diam × 70'; needles deciduous, soft, pale green, 1–1½", 30–40 in apparent whorls on short peglike spurs, exc that needles are single and spirally arranged on this-year's twigs, and are usually evergreen on lowest branches of saplings; cones 1½–1¾", bristling with pointy bracts much longer than the scales; this-year's twigs densely, minutely woolly; bark gray, rarely more than 1" thick; tree broad-crowned, heavily branched and/or multistemmed; or occasionally krummholz, though not layering. 5,800–7,500' in the N Cas, from Mt Stuart N and from Cas Cr E. Pinaceae.

A slight paradox: though evergreen conifers generally inhabit colder climates than broadleaf trees, the most cold-loving of all trees are deciduous conifers, the larches. They are the most northerly and the most alpine genus of trees all around the Northern Hemisphere. Where it is so cold that plants go for months without liquid water for their roots, the winter wind sucks all the moisture (even frozen) out of needles, killing them. Any foliage caught showing above the snow in midwinter gets nipped, so the outlines of evergreen krummholz show summer hikers the depth and shape of the winter snowpack. But in places, we find subalpine larches standing tall *above* the krummholz; their bare branches are relatively safe in winter from both cold desiccation and storm breakage.

Sometimes an early frost "freezes" the needles in place through winter; they drop when they thaw in spring, and are soon replaced. While the tree's base is still deep in snow, its upper branches leaf out, providing spring's first greens—a treat for grouse that survived winter on a diet of tough old fir needles. Larch needles taste like tender young grass, with an initial spicy resinous burst. They're a visual treat, too, contrasting with other needles twice a year—bright grass-green in June, yellow in late September (or August in dry years).

Subalpine larches keep no low limbs, a habit with one big disadvantage—they cannot layer—and one big advantage—they are relatively invulnerable to ground fires. They outlive all their associates, and possibly the cedars of the Westside as well. It is no coincidence that some of them resemble California's bristlecone pines, another dry-country timberline species thought to include the world's oldest trees. Year after year, century after century, they put out a tiny bit of growth here while they die a little over there. On more congenial sites, though, they reach impressive size. The current champion subalpine larch, found by the author at 6,500' in the Entiat Mountains, is 95' tall and 6'7" in diameter. Typical sites are bedrock or talus outcrops, since they need to be snowfree by July; that's on the early side for the Cascades, especially near the Crest, which is the tree's westerly limit. Subalpine larches on Luna Peak are among the many botanical signs that the Picket Range, not the Skagit/Pasayten divide, is on the true Cascade Crest (see pages 8, 561).

David Lyall was one of the last of the rugged Scots prominent in early exploration of the Northwest. After a pioneering botanical trip to the San Juan Islands in 1853, he was appointed surgeon-naturalist to the British contingent (Canada being a British possession) of the Northwest Boundary Survey of 1857–62. The American survey party included George Gibbs as naturalist, geologist, and ethnographer, and Henry Custer, a topographer who was the first explorer to write passionately on the beauty of the North Cascades. The two parties did the same job separately, often coming up with different boundary locations. (The boundary was more conclusively surveyed in 1901–08.) The task of following an arbitrary beeline across the unknown, precipitous North Cascades was heroic, and they did well to find energy and enthusiasm for scientific studies along the way. Lyall published a *Botany of Northwest America* in London in 1863.

This group consists of the Cypress family (Cupressaceae). The foliage is typically compressed and scalelike at maturity; most family members also have a juvenile phase of sharp, spinelike needles up to ¾" long, closely packed along the stem. These differ from other conifer needles in being arranged (like the mature scales) in opposite pairs or in whorls of three. In the three genera we call cedars, juvenile foliage grows only for the seedlings' first year or two, but in many junipers it continues for years—through sapling size and even into maturity on lower branches. Common juniper grows only the "juvenile," spinelike needles; it is placed in these pages with the rest of its family by default, for none of its leaves are truly scalelike.

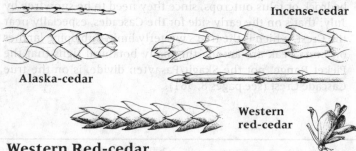

Incense-cedar

Alaska-cedar

Western
red-cedar

# Western Red-cedar

*Thuja plicata* (**thoo**-ya: a Greek term for some aromatic tree; plic-**ay**-ta: pleated). Also **giant arborvitae.** 84" diam × 200'; leaves tiny, yellowish green, in opposite pairs, tightly encasing the twig, strongly flattened, the twig (incl leaves) being 4–8 times wider than thick; foliage dies after 3–4 years, turning orange-brown but persisting several months before falling; cones about ½" long, consisting of 3 opposite pairs of seed-bearing scales, plus a narrow sterile pair at the tip and 0–2 tiny sterile pairs at the base; bark reddish, thin (up to 1"), peeling in fibrous vertical strips; leader drooping; trunk very tapered, its base fluting and buttressing with age. Moist to wet sites below 4,200'. Cupressaceae (Cypress family). Color p 115.

Our cedars are manifestly a breed apart, easily recognized (as a family) by their droopy sprays of foliage and vertical-fibrous bark. The bark is usually clean, being too acidic to encourage lichens, fungi, or moss. Though slow growing, our cedars resist windthrow, rot, and insect attack, and commonly live over

1,000 years. Many western red-cedars, the family's largest trees, develop buttressed waistlines 30 to 60 feet around.

Pure stands of red-cedar occur only on ground too wet for other big trees. Scattered individuals, though, are widespread; once established, they are too tolerant to shade out. East of the Crests, their soil moisture requirement—about 12%, even through August—limits them to year-round seeps, typically in ravines; but they can venture into drier climates than hemlocks or silver firs as long as their roots are wet.

Red-cedar was the Northwest Indians' most important plant. The inner bark, woven after laborious shredding with deer bone, made warm clothing. Unwoven, it was soft enough for cradle lining or sanitary pads; torn in strips and plaited, it became roofing, floor mats, hats, blankets, dishes, or ropes.

Buds, twigs, seeds, leaves, and bark each had medicinal uses. Cedar charms sanctified or warded off spirits of the recent deceased. Cedar bough switches were skin scrubbers for both routine and ceremonial bathing. A Lummi boy would rub himself with cedar switches, then tie them to the top of a cedar tree, in preparing for his guardian-spirit quest.

The wood was used for every purpose that didn't require hardness or great strength. Easy to work with stone tools and fire, it made up in durability and aesthetics what it lacked in strength. Structural timbers, totem poles, and dugout canoes (up to 60' long by 8' wide) were almost always cedar. (Red-cedar and Alaska-cedar were interchangeable, but Alaska-cedar was available to coastal people only in Alaska).

Cedar heartwood is warm red, weathering to silver-gray; it smells wonderful, resists rot, and splits very straight, especially if it comes from an old, slow-grown tree. Split cedar shakes shed water well, having scarcely any cut-into cells or vessel ends to absorb it; in dry weather the shakes shrink, enlarging the spaces between them to let the roof breathe. In pioneer days, most Northwest buildings had shake roofs, and many had shake or shingle siding. High quality shake wood is a rarity today; crookeder shakes fetch higher prices because they are more conspicuous as a status symbol. Salvaging shake bolts from old snags and down cedar is a way to make a (usually) honest living independently in the Northwest woods, requiring little investment but a pickup truck and a saw.

# Alaska-cedar

*Chamaecyparis nootkatensis* (cam-ee-**sip**-a-ris: dwarf cypress; noot-ka-**ten**-sis: of Nootka, B.C.). Also **yellow-cedar.** 50″ diam × 130′; leaves tiny, bluish green, encasing the twig but somewhat divergent from it at the sharp tips, hence prickly-feeling, spray twig (incl leaves) from slightly wider than thick up to twice as wide; foliage dies after 2 years, turning brown but persisting usually a year before falling; cones round, less than ½″ diam, like hard green bumpy berries their first year, becoming brown and woody the second and opening into 4–8 scales like tiny shields with a prickle in the center; bark thin, silver-gray, red-brown inside and on saplings, flaking off in thin strips, but not especially easy to tear vertically; leader and branch tips drooping extremely; mature bases ± fluted. Mainly above 3,000′, esp in avalanche tracks and wet N aspects; locally abundant as alpine krummholz. Cupressaceae (Cypress family). Color p 115.

The name "weeping cedar" may not have seen print until now, but it comes quickly to mind. The willowy branches slough off the snow whenever it gets too heavy; their great flexibility minimizes snow breakage, whether from accumulation on the limbs, snow creep on a steep meadow, or the sudden cataclysm of an avalanche. Alaska-cedar is the main avalanche-track community dominant at highest elevations;

---

## Cedar aromatics

Each of our "cedar" woods has its own pungent smell, emitted by rot- and insect-repellent chemicals they have evolved. The traditional moth-repellent Cedar Chest is Eastern red-cedar—not a *Thuja*, but a *Juniperus*—but Alaska-cedar and incense-cedar were used in similar ways in nineteenth-century China.

Human response to these smells is highly individual. Speaking for myself, I find Port-Orford-cedar so exquisite that I would wear it for perfume if I could. But Alaska-cedar (also a *Chamaecyparis*), which charms my eye, makes my nose uneasy. Millworkers subject to eight-hour doses of *Chamaecyparis* volatiles suffer everything from headaches and laxative effects to kidney complications. Western red-cedar's smell is among the virtues of cedar cabins and saunas; I poke my fire with split cedar for its smouldering fragrance.

---

lower (often within the same track), it is far outnumbered by faster-growing Sitka alders.

A spell of sunny weather while the soil is frozen may kill upper parts of the tree, resulting in dramatic bleached white "spike-tops." The tree survives this and other hardships—thanks largely to the rot resistance typical of its family—often for over 1,000 years. The maximum age may be less than the 3,500 years some foresters have asserted, but is probably greater than that of any other Northwest plant.

Alaska-cedar is commonest on steep north slopes with wet-ground associates like devil's club and sword ferns, but it is also found on dry, rocky ridgetops; this implies it selects not wet sites but poor ones, where competitors are handicapped. It grows too slowly to survive vigorous competition.

Little Alaska-cedar is logged here, but much is logged in coastal British Columbia and Alaska. Almost all of it, and of Port Orford-cedar (*C. lawsoniana*), is exported to Japan because of their close resemblance to *hinoki* (*C. obtusa*). The Japanese revere their *hinoki*, and pay high prices to get their *Chamaecyparis* wood from us and save their own from the axe.

---

The Europeans who first named America's scaly-leaved trees "cedars" were either very confused or at a severe loss for words. These trees resemble their relatives the cypresses, genus *Cupressus,* and not the true cedars, which have long needles in whorls (like larches, only evergreen) and fat, solid, upright cones (like true firs). The true cedars (*Cedrus,* a Pine family genus) are much planted in Northwest cities; they are three species native to Lebanon, Israel, and the Atlas and Himalaya Mountains. Each of our "false" cedars represents a Pacific Rim genus, with other species in Northeast Asia. *Thuja* and *Chamaecyparis* each also have an Eastern North American species called a "white-cedar," while "red-cedar," back East, means a juniper tree.

Foresters and entomologists (see page 449) follow a rule of hyphenating or compounding all common names that are taxonomically misleading; though botanists do not follow it as consistently, the rule is responsible for most of the hyphens in common names in this book.

---

Alaska-cedar wood is clear pale yellow, straight-grained, with very close, faint, annual rings, and a heavy smell. Its great durability even when soaked led Alaskans to use it for fishing boats, Japanese for temples, Oregonians for hot tubs, and the Haida, Tlingit, and Tsimshian for canoes, paddles, and totem poles. Indians found the inner bark even softer and finer than western red-cedar's; they stripped, soaked, and pounded it for weaving and plaiting into clothing, bedding, or rope.

## Incense-cedar

*Calocedrus decurrens* (cal-o-**see**-drus: beautiful cedar; de-**cur**-enz: running down). 40" diam × 140'; leaves small, average ¼" but up to ¾" on larger twigs, yellowish green, tightly encasing the twig, flattened, the twig/leaf 3–6 times wider than thick, the opposite pairs of leaves so nearly neck-and-neck as to make apparent whorls of 4; cones about 1", of apparently only 3 scales but actually of 6 (a sterile fused pair in the center flanked by an equally long fertile pair, and then by a tiny, recurved sterile pair); seeds with 2 unequal wings; bark red-brown weathering grayish, fibrous but smooth, furrowed, up to 4" thick; leader erect; crown very regular, dense, narrowly conical. Sunny E-side slopes below 4,000' S of Mt. Hood; rarely, on W-side below 3,000'. Cupressaceae (Cypress family). Color p 115.

Incense-cedar's aroma has been familiar ever since we first sharpened pencils in school. Most wooden pencils are still made from incense-cedar. The lumber was formerly downgraded by the fine parallel, linear holes often left in it by a "dry-rot" shelf fungus, *Tyromyces amarus*, but recent fashion in paneling regards this defect as decorative. Most other fungi and insects find the tree as repellent as our other cedars.

Outliers of a Californian range, our incense-cedars grow on the hottest, driest sites in the Oregon Cascades. They are slightly more tolerant (but less fire-resistant) than Douglas-fir and Ponderosa Pine, and can be expected to persevere and perhaps increase, especially with the help of fire-fighting.

A year-old incense-cedar seedling exemplifies the gamut of juvenile leaf styles typical of the cypress family. First come the two cotyledons, or "seed leaves," about 1" by ⅛". Above

*Some authorities retain *decurrens* in a larger genus *Libocedrus*. Attempts to split that diverse genus date back at least to 1926, but have yet to produce unanimity.

these grow needles about half as large, in whorls of four. Then, still within the first year, an incense-cedar branches and grows its first scalelike, close-packed foliage—still slightly more spreading and prickly than the mature foliage, which usually comes in the second growing season.

**Western**       **Rocky-mtn.**

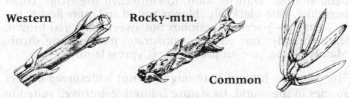

**Common**

## Western and Rocky-Mountain Junipers

*Juniperus* spp. (jew-nip-er-us: the Roman term). 18″ diam × 30′; mature leaves tiny, scalelike, yellowish green, tightly encasing the twig, not flattened; juvenile leaves (seedlings, saplings, lowest limbs of young trees) needlelike, average ¼″, prickly; cones berrylike, blue-black, dryish, resinous, 1–3-seeded, ¼″ diam; a few plants have only male (inconspicuous) or only female fruits; bark red-brown, fibrous, shreddy; dense small pyramidal trees with limbs nearly to the ground, or sprawling shrubs. Cupressaceae (Cypress family).

**Western juniper,** *J. occidentalis* (ox-i-den-tay-lis: western). Leaves in whorls of 3, each whorl rotated 60° from the next. Dry low E-side from Mt Adams S.

**Rocky-Mountain juniper,** *J. scopulorum* (scop-you-lor-um: of crags). Leaves opposite, each pair rotated 90°. Dry low elevs E of Cas Cr from Mt Adams N, incl upper Skagit Valley.

These are the juniper trees that have taken over much of the steppes of Eastern Oregon and Washington, respectively, thanks to fire suppression on the rangelands. A few slip into Eastside Cascade forests, mainly in drought-induced clearings on rocky ground. They achieve their best growth far to the south. In the Sierra Nevada there stands a western juniper 14′ in diameter and 87′ tall, thought to be well over 3,000 years old; that would place the species second or third in world Aging championships. A nearly equal Rocky-Mountain juniper grows in Utah's Wasatch Range.

The Swinomish boiled juniper leaves to steam sickness out of a house, or to bathe a sick person in juniper-leaf tea. Back east, juniper trees called "red-cedars" (mainly *J. virginianus*) are milled for moth-repellent cedar chests and closets.

# Common Juniper

*Juniperus communis montana* (com-you-nis: common; mon-tay-na: montane). Leaves all ¼–¾", sharp, curved upward, closely packed along the twig in whorls of 3, from a ± distinct joint at each leaf base (unlike juvenile *J. occidentalis,* whose 3-whorled needles bend sharply, with no joint, to run down the twig); cones berrylike, blue-black with bloom, round and quite fleshy, ¼–⅜" diam, 1–3-seeded, resinous but sweet; bark red-brown, thin, shreddy; our variety a prostrate, mat-forming shrub. Mainly alpine here. Cupressaceae (Cypress family).

The common juniper is among the most widespread conifer species in the world. Its stature is humble, but well suited to cold, windswept ridges and slopes, where even tall conifer species grow as low creeping krummholz. Dry alpine slopes— south-facing, rain-shadowed, rocky, almost devoid of real soil —are where we typically find common juniper alongside evergreen shrubs like kinnickinnick and Oregon-boxwood. All look beaten down, confined within the shape of whatever little snowbank they can keep through winter; exposure to wind while frozen would dry and kill their leaves.

Junipers (including the erect species, above) are anomalous among conifers in enclosing their seeds in fleshy, edible "fruits." (Yews cup—but do not enclose—their poisonous seeds in fleshy "berries.") Properly speaking, a berry is a fruit and a fruit is a thickened ovary wall, so neither junipers nor yews have berries; a juniper "berry" is technically a cone of very few, fleshy, fused scales. It has a sweetish resiny flavor of suspiciously medicinal intensity. Used with restraint, juniper berries are delicious in teas, stuffings, gin (a word derived from the French *ginevre* for juniper) or your water-bottle, if your water has been tasting plasticky in hot weather. Those with inquisitive palates will try them straight off the bush. Birds love them, and disseminate the indigestible seeds.

**Common juniper**

# 3

# Flowering Trees and Shrubs

The distinction between trees and shrubs is neither absolute nor taxonomically meaningful, but simply descriptive. Many species in this chapter grow as either trees or shrubs depending on their environment. You would be safe to call anything a tree if it has a single, woody, upright, main stem at least 4" thick or 26' tall at maturity. All shrubby plants, regardless of height, with stems more canelike or vinelike than truly woody and self-supporting, are in the "under 6 feet" category, which begins on page 80.

---

**Trees and Shrubs:** deciduous, over 6'

---

## Bigleaf Maple

*Acer macrophyllum* (ay-sir: the Roman term; macro-fill-um: big leaf). 30" diam × 65'; leaves opposite, deeply 5-lobed, 5–12" wide and long, on equally long leafstalks, turning rich yellow in fall; flowers small, yellow-green, 10–50 in long pendent clusters conspicuous in Apr-May when leaves are just emerging; 1½–2" winged seeds in acute-angled Siamese-twin pairs; bark gray-brown and smooth in youth, later furrowed and often mossy; trunk and limbs ± knobby and crooked. Widespread W-side below 2,000'; less common 2,000–4,600', and in E-side Canyons. Aceraceae (Maple family).

This, the largest of all maples, is the broadleaf tree taking the largest role in our mostly coniferous landscape. Not confined, like cottonwood and red alder, to streamsides and disturbed sites, bigleaf maple is nevertheless favored by sites where conifers have failed, for one reason or another, to form a deep shading canopy. Prime examples of these are among the so-called rain forests of Westside Olympic valleys. Most bigleaf maples there grow on old rockslides fanned out across river terraces; rocky soil may help answer the mystery of why these valley bottoms, which obviously grow conifers to perfection, don't grow them to maximum density.

"Halls of Mosses" in these valleys are named for their extravagant raiment of epiphytic plants—lichens, clubmosses, some ferns, and higher plants, as well as mosses proper—all of which seem to find bigleaf maple bark ideal in its alkalinity and porosity. Though they draw no nutrients out of the trees supporting them, epiphytes do compete with the trees for nutrients that percolate down the tree trunk. The dripwater, which originates as pure rain and fog intercepted by the trees, carries nutrients contributed by nitrogen-fixing lungwort lichens and by insects and mites that graze in the tree canopy. Maples, alders, and cottonwoods extend small roots among the epiphytes on their own bark, to catch a share of the nutrients before they reach the ground.

Deer and elk relish maple leaves when they can reach them. People relish maple syrup, which is maple sap with 98% of its water and volatiles boiled away. Oregon maple syrup is a commercial possibility; bigleaf sugar content and flavor are adequate by most opinions, and plenty of sap flows, though without the strong spring burst (following an icy winter) exploited in New England's sugar maples. Even in New England, syrup making is nowhere near as profitable as it is picturesque.

Maple wood is very hard, and makes good flooring. Some trunks near the ground develop "curly" grain patterns prized in furniture and stringed instruments. Though difficult to cut and carve without metal tools, maple wood was important in Northwest Indian culture for utensils, tools, and ornaments. The inner bark was plaited into straps or woven into baskets, and the leaves were used to line berry baskets and baking pits, or as rags for cleaning fish.

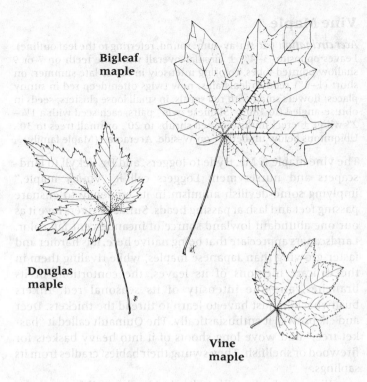

**Bigleaf maple**

**Douglas maple**

**Vine maple**

## Douglas Maple

*Acer glabrum douglasii* (glab-rum: smooth; da-**glass**-ee-eye: after David Douglas, p 18). Also **Rocky Mountain maple.** Leaves opposite, 3- (or ± 5-) lobed, toothed, red-orange in fall, 2¼–5", on equally long, often reddish leafstalks; twigs reddish; flowers small, green, in small clusters; winged seeds rarely over 1" long, straight-backed, in pairs at ± right angles; bark gray to purplish; tall shrubs or trees up to 40'. Shrub thickets and open forests E of Cas Cr; uncommon in W-side forest openings. Aceraceae (Maple family).

The common maple of eastern Washington and Oregon, Douglas maple is vital winter browse there for deer, elk, and sheep, who eat the twigs.

Maples have a lovely helicopterlike way of extending their seeds' range in a breeze. Their fruits (called samaras, or winged achenes) provide each seed with a long wing, sending it into a fast spiral for a slow fall.

---

# Vine Maple

*Acer circinatum* (sir-sin-ay-tum: round, referring to the leaf outline). Leaves opposite, 3–4", ± circular overall with fine teeth on 7 or 9 shallow pointed lobes, coloring intensely in fall or late summer, on short (1–3") reddish leafstalks; new twigs often deep red in sunny places; flowers small with red sepals, in small loose clusters; seeds in obtuse-angled (to 180°) "Siamese twin" pairs, each seed with a 1¼–2" wing; ± erect multistemmed shrubs to 20', or small trees to 30'. Ubiquitous below timberline on W-side. Aceraceae (Maple family).

The vine maple is Mr. Hyde to loggers, and Dr. Jekyll to landscapers and nurserymen. Loggers call it "vin*ing* maple," implying some devilish animism in its tendency to ensnare passing feet and lash at passing heads. Sunday drivers love it as our one abundant lowland source of incandescent fall color. Landscapers appreciate that being native here, it is hardier and faster-growing than Japanese maples, while rivaling them in the delicate incisions of its leaves, the contortions of its branches, and the intensity of its seasonal red. Hikers bushwhacking just have to learn to thread the thickets. Deer and elk browse it enthusiastically. The Quinault called it "basket tree," and wove long shoots of it into heavy baskets for firewood or shellfish. They swung their babies' cradles from its saplings.

Vine maples flourish indiscriminately at all forested elevations west of the Crest; they are far and away our most ubiquitous and abundant tall shrubs. At higher elevations they shy away from dry ridges that favor beargrass; in Western Olympic valleys elk seem to hold them back. For unknown reasons they have yet to invade Vancouver Island.

# Red Alder

*Alnus rubra* (al-nus: the Roman term; **roo**-bra: red). 24" diam × 110'; leaves 2–6", oval, pointed, flat green above, pale gray beneath, margins coarsely toothed, ± wavy and/or rolled under; flowers or "catkins," male and female on the same tree, appear in summer and mature the next spring before leaves appear; female catkins ½–1" long, woody, like miniature spruce cones; bark ± smooth, pale lichen-coated. Abundant W-side on logged land, burns, and streamsides. Betulaceae (Birch family).

The splotchy whiteness of red alder bark, ranging from paper birch white through various pearl grays, is really a crust of lichens that the bark reliably hosts. The redness of red alder might be the gorgeous antique rose cast lent to hillsides in early spring by millions of alder catkins and buds. More likely the name refers to the deep stain that appears on freshly cut alder bark or green wood. Indians boiled, chewed, or urinated on cut bark and wood in various techniques to make dye; one use for the dye was to make nets invisible to fish. Alder's easily carved, flavor-free wood was good for spoons and dishes, and its rich, oily smoke for smoking salmon. Try alder-smoked salmon when you get the chance.

Northwest alders serve an invaluable role in nature; they pioneer on wet gravel exposed by retreating glaciers and by shifting, downcutting rivers. Fresh or actively migrant gravel bars, with little or no sand between their cobbles, support mainly willows and annuals; older, sandier bars support alder flats. Recently deglaciated terrain here is above red alder's limit (2,500') but often grows Sitka alder. Alder seedlings prefer poor mineral soil, since they don't need soil nitrogen, but do need so much water that they suffer on duff, which gets very hot and dry. They grow fast, and rapidly improve soil structure and fertility. Like legumes, they host bacteria in nodules on their roots that convert atmospheric nitrogen for plant use, fixing some in the soil and some directly into the plant. Alder leaf litter is plentiful, fast-decomposing, and nitrogen-rich.

In nature, red alders in our region grow mainly along lowland streams and swamps; but after white settlement and logging, red alder was seen mainly as an invader of logged land. The better the land for timber farming, the more likely it was to be taken over by alder. "Scarifying" or scraping the earth down to mineral soil, which in theory enhances Douglas-fir seedling survival, often did the opposite by aiding the alder competition more. Nitrogen fixation—long appreciated in legumes—went undiscovered in alders. As long as there was plenty of other wood, alder was used for little but firewood, being soft, brittle, and relatively short-fibered. For decades, a primary task of forestry in the Northwest was to find an easy way to massacre alders without harming Doug-firs. Broadleaf herbicides provided an apparent solution, but lost several court battles on grounds of toxicity to humans.

Fortunately, foresters are turning the ugly duckling back into a swan. A planting of alders sparsely intermixed with Douglas-firs may be the most economical way of feeding firs nitrogen; the cost edge can only improve as urea fertilizer, made from natural gas, becomes more expensive. Red alder is also prescribed for sites infested with laminated root rot, an incurable disease; alder roots secrete a toxin for that fungus.

Processing and marketing of red alder are also improving. Pulp mills bleach out the reddening, which used to be a problem; three-fourths of alder production gets pulped. Alder chipboard faced with Douglas-fir veneer is replacing plywood in many uses, as veneer-grade (old-growth) fir gets scarce. Alder lumber, now used in midpriced furniture, may pose as a fine hardwood, perhaps under a trendier name. A futuristic vision for alder is of "energy plantations"—crops of red alder saplings cut, stump-sprouted, and recut on three- to five-year rotations, then burnt to generate electricity or converted to methane. The saplings can reach 20' in five years.

Every 10 years or so an outbreak of tent caterpillars (page 463) makes the region's alder stands look wretched, consuming billions of leaves down to the midvein and "tenting" the remains under cobwebby shelters. This stunts the alders' growth for the year, but does no lasting damage.

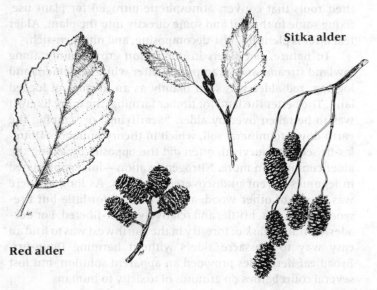

**Sitka alder**

**Red alder**

# Sitka Alder

*Alnus sinuata** (sin-you-ay-ta: bent). Also **slide alder.** Leaves 1½–4", oval, pointed, shiny, margins sharply doubly toothed, ± wavy but not rolled under; flowers or "catkins," male and female on same tree, appear in spring simultaneously with the leaves; conelike woody female catkins ½–¾" long; bark gray-green, warty-textured, often lichen-coated; erect to sprawling shrubs 5–12", or rarely trees up to 25'. Abundant on mtn streamsides, seepy slopes, avalanche basins and tracks; rare below 2,600'. Betulaceae (Birch family).

The Sitka alder we know best is at its worst—the downhill-sprawling "slide alder" of avalanche tracks. What the roaring avalanche finds accommodating, the sweating bushwhacker finds maddeningly intransigent, a tangle of springy, unstable stems always slipping us downslope like flies in the hairy throat of a pitcher-plant. Too often, what we slip onto is a neighboring devil's club. If, like the avalanche, we could stick to downhill travel, we might have no problem with slide alders. Or, if we can't beat'em, we can join'em Indian-style—by cutting a stem to wear for perfume.

A Sitka alder has the genetic information for growing upright, as a tree, but it also knows how to flex and bow and sprawl where its environment demands. To this yielding nature it owes its dominant position in three-fourths of the avalanche tracks and basins in the Cascades and Olympics. Few other ranges in the world have so much snow combined with such steep valley walls; as a result, few have so many avalanche tracks. A typical North Cascades track has a superficial avalanche (one just trimming the top of the plant community) every few years, or every year in some places. A full-depth avalanche comes at 6–20-year intervals, snapping some of the larger Alaska-cedars, mussing up the Sitka alders a bit, and demolishing nearly every other woody plant in its path. Alaska-cedar is the only robust tree adapted for life in the avalanche track, and it is a slow grower, competitive with Sitka alder only on severe, cold, high sites.

*Some authorities consider *sinuata* a subspecies of *Alnus viridis.*

# Hazel

*Corylus cornuta* (cor-il-us: the Roman term; cor-**new**-ta: horned). Also **filbert**. Leaves 2–4″, broadly oval, doubly toothed, ± minutely fuzzy all over when new; female flowers tiny buds with red stigmas, appearing very early in spring along with the male catkins; nuts much smaller than cultivated filberts, in heavy shells about ½″ diam, in hairy long-necked husks; tall shrubs 4–18′, rarely treelike in our mtns. Hot, dry sites (often S- to W-facing slopes) at low to mid elevs, mainly W-side. Betulaceae (Birch family).

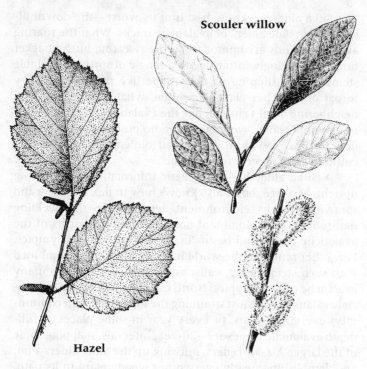

**Scouler willow**

**Hazel**

Though our native subspecies is called *californica*, hazels truly belong to Oregon, which has the most—both of this native shrub and of the orchard tree derived mainly from European hazels. Accustomed to commercial nuts, we find our native nutmeats puny, albeit tasty. Lewis and Clark were more than grateful for the ones they could buy from the Indians. The nuts grow in back-to-back pairs, each nutshell encased in a fuzzy leaf shaped like a fringe-topped, long-necked vase.

# Scouler Willow

*Salix scouleriana* (say-lix: the Roman term; scoo-ler-ee-ay-na: after John Scouler, below). Also **fire willow**. Leaves 2–4", narrowly oval, widest past midlength, typically reddish-velvety underneath, flanked at first by pairs of ¼" ear-shaped bracts; female ("pussy willows") and male catkins on separate plants, flowering very early, occasionally before snowmelt; bark bitter, bad-smelling when crushed; shrubs or small trees in clumps, 3–40'. Widespread. Salicaceae (Willow family).

Scouler willow is a pioneer on two radically different habitats. On gravel bars and banks of braided mountain rivers where little soil has accumulated, it is often the only shrub able to grow, and rarely exceeds 4' high. In clearings (typically, recent burns) on the dry east slope of the Cascades, it invades along with snowbrush. Thriving independently of watercourses is unusual among willows. Umpteen species and hybrids of *Salix* line streams throughout our range; they are notoriously hard to identify. Most look much like Scouler willow but have narrower, pointier leaves. Two other *Salix* species (page 192) are mat-forming alpine shrubs about 4" high.

Northwest tribes all twisted willow bark into twine for many vital uses—fishnets, baskets, tumplines, even harpoon lines for sea lions. Poles cut from willows were chosen for supporting fishing platforms or weirs. They would often take root where implanted in the riverbed.

**John Scouler,** as a Scottish lad of 19, signed on as ship's surgeon to H.M.S. *William and Ann* for the same 1824 sailing that brought David Douglas (page 18). Douglas was delighted, since the two had been school pals in Glasgow. Scouler spent only seven months in the Northwest, taking a few walks with Douglas, and botanizing on his own when the ship visited the San Juan, Queen Charlotte, and other Islands. The same ship and all aboard were lost entering the Columbia River a few years later, but Scouler was safely back in England, where he completed his M.D. and lived to a ripe old age.

# Black Cottonwood

*Populus trichocarpa** (pop-you-lus: the Roman term; try-ko-**car**-pa: hairy fruit). 40" diam × 150'; leaves 3–6", long-pointed, broadly round-based or somewhat heart-shaped; glossy dark green above, dull light gray beneath, coloring bright yellow in fall; leaf buds (and their fallen scales in spring) sticky and richly sweet-scented; female and male catkins on separate trees; round seed pods, in long strings, split 3 ways to release many tiny seeds with cottony fluff; bark gray, grooved (see below); tall, fairly straight trees with V-shaped crowns. Low to mid-elev streamsides. Salicaceae (Willow family).

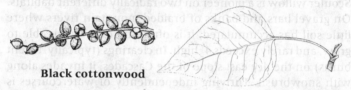

**Black cottonwood**

The elegant trunks of black cottonwood are straight and limb-free to such a height that they often go unnoticed among red cedars and hemlocks. You spot them instantly, though, once you've learned their distinctive bark—clean pale gray with little flakiness, moss or horizontal texture, but vertically fissured with heavy, dark, rough grooves. In spring the luscious honey fragrance alone is enough to alert you.

Black cottonwoods are easily our largest broadleaf trees, and our fastest-growing trees of any kind. They can reach 100 feet in as little as 20 years—twice as fast as red alder and Douglas-fir. The first tree farms planted by white settlers in Oregon were of black cottonwood, but the "instant wood" made lousy lumber and worse firewood. Today it is used for crates, pulp, and cheap plywood cores. Indians used younger trees as posts, and thin saplings for sweatlodge frames.

Cottonwoods rarely grow far from water, though they can in our rainy climate. On Westside river flats they join Sitka spruces and maples in replacing red alders; they will be outlived by the spruces and eventually replaced by hemlocks. Their location, size, and branching habit (providing stout crotches 120+' above ground) make them choice nesting trees for river-fishing bald eagles and great blue herons.

*Some texts consider this tree a variety, *P. balsamifera trichocarpa,* of the balsam poplar of E North America.

# Quaking Aspen

*Populus tremuloides* (trem-you-lo-eye-deez: trembling—). 10″ diam × 40′; leaves 1–2½″, broadly heart-shaped to round, point-tipped, bumpy-edged to fine-toothed, on leafstalks 1–2½″ long and flattened sideways; female and male catkins on separate trees; ¼″ conical seedpods, in long strings, splitting in 2 to release minute seeds; bark greenish white, smooth, dark and rough on old trees only. E-side streamsides and avalanche tracks. Salicaceae (Willow family).

 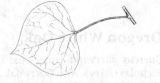

**Quaking aspen**

Quaking aspen is the widest-ranging American tree, covering much of Alaska, Canada, the Northeast and North Central states, the Rocky Mountains, and the Sierra Madre south to Guanajuato. Around the Great Lakes it invades after conifers are logged, as red alder does here. In the Rockies it provides most of the fall color—yellow. In Washington's Okanogan Highlands it fills avalanche tracks, like Sitka alder. But our range, atypical as it is of North American forests, offers quaking aspen only a small role, along streams east of the Crest. Wildlife, from elk and beaver to grouse and pika, make the most of it; aspen leaves are choice browse.

Most aspen trunks grow from root suckers, a type of vegetative reproduction. Whole groves called "clones" are genetic individuals, each with a single huge root system. An aspen clone is all of one sex, and may spread over acres and acres, further stacking the odds against sexual (seed) reproduction. The extent of each clone is dramatically visible in autumn, since the timing and hue of fall color are identical throughout a clone, but vary from one clone to the next.

Aspen leaves quake or flutter in the lightest of breezes because their flat leafstalks are flexible in the lateral direction only. (Try rolling one between your fingers.) This "tremulousness" suggested the name *tremula* for the European aspen, and thence *tremuloides* for its American cousin. The genus also includes poplars and cottonwoods.

---

**Trees and Shrubs:** deciduous, over 6′

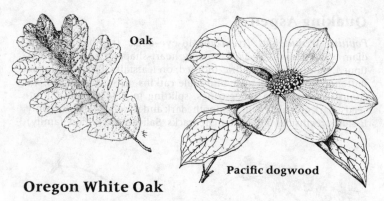

**Oak**

**Pacific dogwood**

# Oregon White Oak

*Quercus garryana* (**quer**-cus: the Roman term; gary-ay-na: after Nicholas Garry). Also **Garry oak**. 36″ diam × 80′; leaves 3–6″, deeply pinnately blunt-lobed; male catkins on same tree with females, which are single or paired in leaf axils, each consisting of a 3-styled pistil in a cup-shaped involucre which later hardens into the cap of the acorn; acorns ¾–1½″ long; bark gray, furrowed; limbs gnarly. Dry rocky spots near the Columbia River. Fagaceae (Beech family).

When whites first saw the Willamette Valley, it was a savannah of grasses and scattered, huge white oaks, thanks to centuries of prairie-burning by the natives. White settlers cleared the flats for farming, leaving the hills that dot and line the Valley to be taken over, in the absence of fire, by closed forests of relatively puny white oaks; a few scattered old giants remain. The oaks are long-lived (300 years or so) but intolerant of shade, so they are being replaced by grand fir and Douglas-fir saplings. The Puget-Willamette Trough has a much drier climate than the mountains on either side of it; residents may have a hard time believing that, but the oaks prove it. In our mountains, oaks grow only on the dry margins of pine forests near the Columbia, and on rocky slopes of the Gorge. Oak galls—hard, hollow round structures on oak twigs—show that oaks are favored hosts of gall wasp larvae.

California Digger Indians who subsisted heavily on acorn meal from various oak species developed lengthy processes for leaching out the tannic acid. Northwest tribes ate white oak acorns raw or cooked, usually without prior processing except sometimes burial in maple bark wrapping until rotten; presumably they ate few enough acorns for their bodies to tolerate the tannin. White people's tastes generally can't.

# Pacific Dogwood

*Cornus nuttallii* (cor-nus: the Roman term; nut-all-ee-eye: after Thomas Nuttall, below). 8" diam × 30'; leaves opposite, 3–5", bright green, elliptical, pointed, wavy-edged, with veins curving around to merge along the leaf margin; flowers tiny, greenish white, 4-merous, in a tight head surrounded by 4–7 large (1–2½") white ± parallel-veined bracts; berries bright orange-red, mealy, dry, bitter, tightly crammed hence irregular in shape; trees or tall shrubs. Scattered in lower W-side forests. Cornaceae (Dogwood family).

The function of showy petals is to provide a visible target for nearsighted insects as they buzz around gathering nectar and inadvertently strewing pollen. There is no reason this function couldn't just as well be performed by, say, leaves close to the flower; and sure enough, specialized leaves called "bracts" are the showy parts of flowers like paintbrush and dogwood. (On many tiny-flowered plants, a chemical attractant does the job, or wind pollination obviates nectar and insects both.) The only trouble with showy bracts is that they confuse people's flower descriptions; dogwoods have flower parts in fours, but the showy bracts (which are not flower parts, and surround a head of flowers, not a flower) are often five, six, or seven.

Dogwood leaves put on a second show in the fall, a fine painterly smear of plum, bronze, russet, and magenta. A modest batch of white-bracted blooms may appear then as well,

**Thomas Nuttall** probably collected and named more new species from west of the Mississippi than anyone else. He came along at the right time, after travel had become easier and safer than it had been for Lewis and Clark, or Douglas, but while there were still plenty of conspicuous species left to describe. He made his reputation botanizing on the Great Plains and writing a major flora, *The Genera of North American Plants* (1818). In 1834, the visionary settler-entrepreneur Nathaniel Wyeth persuaded Nuttall to quit his prestigious position at Harvard and join an expedition to the Oregon Territory. Nuttall collected along the Columbia, then sailed to Hawaii, California, and home via Cape Horn. Having exhausted his savings on the expedition, Nuttall retired to an inherited estate in his native England.

alongside red-orange berries from the spring flowering.

Pacific dogwood can grow up to 60' tall, but is more often suppressed by the shade of conifers, joining vine maple, yew, chinquapin, etc., in a tall shrub understory layer.

The the tough wood was used here for bows and salmon spears. Europeans used their species for mallet heads, tool handles, and weaver's shuttles.

**Red-osier dogwood**

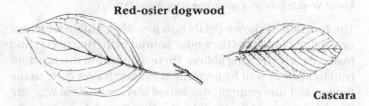

**Cascara**

# Red-osier Dogwood

*Cornus stolonifera* (sto-lon-if-er-a: with many shoots). Also creek dogwood. Leaves opposite, 2–5", elliptical, pointed, wavy-edged, with veins curving around to merge along the leaf margin, coloring richly but inconsistently in fall; petals 4, white, ⅛", stamens 4, as long as petals, sepals 4, minute, flowers in flat-topped clusters; berries dull pale bluish or greenish, ¼" diam, single-seeded, unpalatable; new twigs deep red or purplish; shrubs 6–16'. Widespread in wet places. Cornaceae (Dogwood family). Color p 116.

It is hard to think of this shrub as a dogwood because its flowers lack the large white bracts we think of as dogwood flowers. The tiny true flowers of all dogwoods are alike in structure, though, and red-osier and Pacific dogwoods are also alike in the outline, venation, and fall coloring of their leaves. After the leaves have fallen from this species, rich-red young stems (osiers) remain. They make the plant easy to spot in every season except summer, since it is usually found (or lost?) in swampy or streamside thickets with willows and salmonberry. "Osier" is an old word from the French, meaning a long slender new shoot, originally of willow, suitable for wicker. "Dogwood" derives from the Scandinavian *dag*, for "skewer." The Okanagan and Shuswap also used this species for skewers, roasting racks, and drying stretchers for salmon, crediting it with a nice salty flavor. Other Indian uses revolved around food, fire, or fibre—fishing weirs, sweatlodge frames, fuel to smoke fish or dry berries, and pipestems.

# Cascara

*Rhamnus purshiana* (ram-nus: the Greek term; pur-she-ay-na: after Friedrich Pursh, p 81). Also **cascara buckthorn**. 12″ diam × 35′; leaves 2–6″, oval, with recessed, strikingly regular pinnate veins, dark glossy green above and sometimes rather leathery but deciduous (exc ± persistent on saplings); flowers tiny, greenish, with 4 or 5 calyx lobes and minute petals, clustered in leaf axils; berries up to ½″, 1-seeded, yellow or red ripening to black; bark thin and smooth, or scaly on mature trees only, numbingly bitter; trees or shrubs. Scattered, W-side below 2,500′. Rhamnaceae (Buckthorn family).

Settlers in the Northwest learned from the natives that the bark of this buckthorn, after curing for many months, is a most potent laxative. The medicine leapt to commercial success under the high-sounding name Cascara Sagrada ("Holy Bark" in Spanish) and before long the species was well-nigh endangered. Though a felled cascara regenerates luxuriantly from stump sprouts, a bark-peeled tree dies. Today the tree is far from abundant, and mature specimens are rare. A few decades ago the bottom fell out (fittingly enough) of the "chittam-bark" market, and populations are recovering, while some bark is still gathered. Cascara prefers south aspects with conifers, or swampy lowland clearings with alder and vine maple; it tolerates shade, unlike most broadleaf trees here. The berries are edible, nutritive, and nonlaxative, yet unliked by humans.

# Elderberries

*Sambucus* spp. (sam-bew-cus: the Greek term). Also **elders.** Leaves opposite, pinnately compound or (rarely) twice-compound, 5–12″ long; leaflets narrowly elliptical, pointed, fine-toothed, ± asymmetrical at base; flowers cream-white, 5-merous, tiny, in dense clusters;

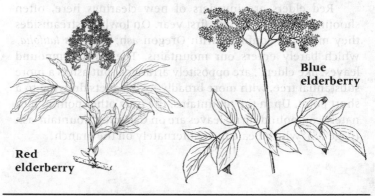

**Blue elderberry**

**Red elderberry**

berries ¼" round, 3–5-seeded; stems pith-filled; shrubs or shrubby trees 3–20' tall. Roadsides, streamsides, clearings, thickets. Caprifoliaceae (Honeysuckle family).

**Blue elderberry,** *S. cerulea* (see-**rue**-lia: skyblue). Leaflets usually 7, 9, or 11; leaf undersides and twigs whitened; flowers and (later) berries in ± flat-topped clusters without a single central stem; berries blue (blue-black with a waxy bloom). Common E of Cas Cr, less so W.

**Red elderberry,** *S. racemosa arborescens* (ras-em-oh-sa: bearing racemes; arbor-**es**-enz: treelike). Leaflets usually 5 or 7, not whitened; flowers/berries in ± conical clusters (racemes) with a main central stem; berries red, bitter. Common W-side, less so E. Color p 116.

**Black elderberry,** *S. racemosa melanocarpa* (melon-o-**car**-pa: black fruit). Like red elderberry, but berries black to very dark red; shrubs mostly 3–6'. E edge of Cascades; commoner farther E.

Where they grow together, red elderberry blooms, sets fruit, and ripens a good month ahead of blue elderberry. The blue berries are more popular for jelly, wine, or eating fresh than our red or black ones; legendary elderberry wine comes from an Old-World black elderberry, *S. nigra*. Ugly rumors crop up to the effect that our red elderberries are poisonous, but this is normally untrue of the fully ripe fruit. Northwest tribes ate berries of both species, fresh or more often steamed and stored until lean times. Elder bark, leaves, twigs, and roots, however, have been regarded as toxic and/or medicinal. Infusions of them were used to induce lactation, perspiration, or vomiting, or alternatively to reduce swelling, infection, or diarrhea—an odd mix of prescriptions. The soft pith of elder stems was hollowed out to make "pea shooters" for little boys, drinking straws for girls during the ritual restrictions of puberty, pipestems for men, and whistles to lure elk.

Red elders are invaders of new clearings here, often shooting up to 12' in their first year. On lowland streamsides they might be confused with Oregon ash, *Fraxinus latifolia*, which barely enters our mountains. The ash's compound leaves, like elders', are oppositely arranged; but ash is a more substantial tree, with more broadly oval leaflets drawing to a short point. Up in the mountains, the only other pointy, pinnately compound shrub leaves are on Cascades mountain-ash (page 78), which bears them alternately on the branch.

## Chokecherry

*Prunus virginiana* (**prune**-us: the Roman term; vir-gin-ee-**ay**-na: of Virginia). Leaves 2–4″, oval, pointed, fine-toothed; flowers ⅜″, white, with 20–30 bright yellow stamens, very showy, in long dense racemes; cherries oval, ¼″, sweetish but astringent, crimson to black; shrubs or (rarely) small trees up to 20′. Thickets and edges of clearings; common on E edge of Cas; also (subspecies *demissa*) in W-side lowlands. Rosaceae (Rose family). Illustrated p 76.

The genus *Prunus* includes all cherries, prunes, and plums. Chokecherries are named for powerfully puckering the mouth and throat. Fortunately, their sweetness preserves better than their astringency; plains tribes pounded them into pemmican, settlers boiled them into jelly. Though pemmican contained quantities of pounded chokecherry pits, some modern texts advise scrupulous avoidance of all cherry, peach, and apricot pits, since they contain amygdalin which can break down into cyanide. Chokecherry leaves, though choice browse among deer and elk, have in large quantities killed cattle.

## Bitter Cherry

*Prunus emarginata* (ee-margin-**ay**-ta: notchtipped). Up to 18″ diam × 80′; leaves 1–3″, elliptical, round-tipped or occasionally pointed, fine-toothed; flowers ½″, white with 20–30 yellow stamens, in loose clusters of 2–4, with several clusters grouped together; cherries ¼″ round, bitter, bright translucent red, drying dark; shrubs or trees. Uncommon; moist lowland woods. Rosaceae. Color p 117.

These cherries are both astringent and bitter, but somehow tasty to birds and rodents. The bark, of a lustrous bronze color, peels in horizontal strips almost like birch bark. It played a key decorative role in spruce-root basketry patterns, and was used medicinally. The wood is dark, lovely, and ravishingly aromatic when fresh-split, or while burning.

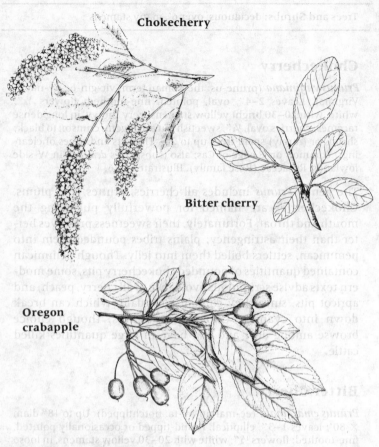

**Chokecherry**

**Bitter cherry**

Oregon
crabapple

## Oregon Crabapple

*Pyrus fusca*\* (pie-rus: Roman term for pears; **fus**-ca: dusky). Leaves 1½–3", ± oval, pointed, fine-toothed, often with a small lobe on one or both sides; flowers white to pink, ½–1" diam, broad-clustered; apples ½", egg-shaped, greenish yellow to red; twigs armed with stout thornlike spur twigs; tall shrubs, or trees to 40'. Uncommon; wet W-side lowlands. Rosaceae.

"Pear" as a generic name for the apple isn't as perverse as it sounds; it simply asserts that the definitive difference between

\*Some authorities retain apples as a genus apart from pears, making this tree *Malus fusca*.

apples and pears—chiefly, the grittiness in pear flesh—is too slight to warrant genus-level separation. Our crabapples aren't gritty, but they sure are hard and sour. Some tribes found them sufficiently softened and sweetened to eat after several months' storage; maybe that really works, as it does with Anjou pears, or maybe it's just that several months on winter rations could make anything taste sweeter. In any case, pioneer women found it more effective to cook them with lots of sugar. Crabapple bark tea was used for all kinds of stomach troubles among the Indians, and the hard, tough wood made seal-harpoon points and other tools.

## Indian-plum

*Oemleria cerasiformis*\* (ohm-lee-ria: after August Oemler; ser-ass-if-or-mis: cherry shaped). Also **osoberry**. Leaves 2–5", elliptical, smooth-edged, pointed; flowers ± white, loosely bell-shaped, borne very early in pendent racemes below the unfolding new leaves, stamens 15 or lacking, since male and female flowers are similar but on separate trees; berries ⅜" diam, 1-seeded, bitter, peach-colored ripening to blue-black; shrubs or small trees 5–15'. Low elevs W-side. Rosaceae. Color p 116.

Indian-plum's tassels of blooms below upraised sheaves of new leaves would be welcome any time of year, but are all the more so in February and early March. The cherrylike fruit is no plum; legend has it the Indians used to know where the sweet ones grow. Maybe it's just that we never taste ripe ones, since birds like them so much they eat them all first.

## Mountain-ash

*Sorbus* spp. (**sor**-bus: the Roman term). Leaves pinnately compound, leaflets 7–13, 1¼–2½", oblong to elliptical, fine-toothed at the tip but not at the base (note similarly placed teeth on serviceberry, p 78, and wildroses, p 85); flowers white, ½" diam, in dense, flat-topped clusters 1¼–4" across, fragrant; berries red to orange, ⅜" diam, several-seeded, mealy, bitter; leaves color brightly, mostly yellow, in fall, and berries may turn purplish; dwarfed to tall shrubs or occasionally small trees. Subalpine. Rosaceae.

\*Older texts list this as *Osmaronia cerasiformis*.

**Sitka mountain-ash,** *S. sitchensis* (sit-ken-sis: of Sitka, Alaska). Leaf tips toothed, round in outline. Common in Wash. Color p 116.

**Cascades mountain-ash,** *S. scopulina* (scop-you-lie-na: of crags). Leaf tips toothed, pointed. Common in Ore.

Mountain-ash berries offer little but fall color to humans, but are important forage for birds and subalpine small mammals, especially since they stay on through winter. Apparently they have enough sugar to ferment on the bush, judging by reports of birds flying "under the influence." A reputable wine is made from the sweeter berries of the European mountain-ash or rowan tree, *S. aucuparia*, a familiar ornamental on Northwest streets. Our two mountain-ash species, true to their name, are strictly montane here, ranging from timberline up to windswept ridgeline thickets.

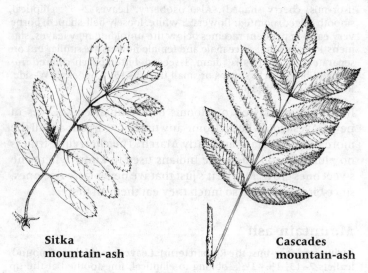

**Sitka mountain-ash**

**Cascades mountain-ash**

## Serviceberry

*Amelanchier alnifolia* (am-el-an-she-er: the archaic French term; al-nif-oh-lia: alder leaf). Also **sarvis**. Leaves 1–2″, broadly oval, toothed at the tip but not at the base; petals white, ½–1″ long, narrow, often so widely spaced and twisted that the inflorescence of 3–15 flowers is a jumble of petals; berries ½″, several-seeded, red ripening to purplish black, with bloom; low spreading shrubs to 30′ trees, though rarely over 10′. Exposed slopes, esp E-side. Rosaceae. Color p 116.

Meriwether Lewis said "sarvisberry," and many Easterners still do. The word derives not from "serving" but from *Sorbus*, the Latin name for mountain-ash. Serviceberry leaves look much like mountain-ash and wildrose leaf*lets*—oval, toothed only on the outer portion—but no confusion need result. Just remember that a compound leaf (ash, rose, etc.) terminates in a leaflet, while a stem of simple leaves (serviceberry) terminates in a bud, flower(s), or a growing shoot.

The Northwestern serviceberry is another edible fruit no longer much eaten in this fussy-tongued culture of ours. Birds and bears eat them, all the Northwest tribes ate them, Lewis and Clark ate them, and traditionalists on the Plains and in the Appalachians still turn local varieties into pies and jams.

**Serviceberry**

**Ocean-spray**

## Ocean-spray

*Holodiscus discolor* (ho-lo-dis-cus: entire disk; dis-color: variegated). Also **ironwood, arrowwood, creambush, rock-spiraea.** Leaves 1–2½", roughly oval, with coarse teeth and often fine teeth upon those; flowers tiny, whitish, profuse in 4–7" conical clusters; seeds single in tiny dry pods, the clusters of pods persisting in place through winter; shrubs 4–12'. Dry, exposed sites up to mid elevs. Rosaceae. Color p 117.

Blooming with masses of tiny cream-white flowers in parallel sprays, ocean-spray resembles an ocean wave breaking. "Ironwood" and "arrowwood" suggest uses Indians found for the straight branches after fire-hardening—arrows, fishing spears, digging sticks for clams and roots, roasting tongs, drum hoops, and baby's cradle hoops.

Common on both sides of the Cascades, ocean-spray on the Westside indicates hot, dry forest habitats which will shift toward hemlock dominance only at a snail's pace, if at all. On the Eastside it usually associates with ninebark.

If you don't find a plant you're looking for in this section, the next place to look would be in the preceding one (over 6') on the guess that you may have a young or environmentally suppressed specimen; categorization by size is unavoidably arbitrary. The present section begins in the middle of a continuum of larger-to-smaller Rose-family trees and shrubs.

Redstem Ceanothus, p 104, properly keys out to this section, but is mentioned under its close relative Snowbrush.

Robust herbs sometimes mistaken at first glance for shrubs include Corydalis, p 203; Spreading Dogbane, p 207; Goatsbeard, p 218; and Baneberry, p 234.

Pacific
ninebark

Bitterbrush

# Ninebark

*Physocarpus* spp. (fie-zo-car-pus: bladder fruit). Also sevenbark. Flowers small, white (exc stamens often pink), in dense hemispheric heads 1¼–2" diam; leaves 1–3", palmately veined and 3-lobed (occasionally 5-), coarsely toothed; bark flaking away, reddish or yellowish brown. Rosaceae.

Mallow ninebark, *P. malvaceus* (mal-vay-see-us: mallow—). Usually 2 pistils per flower, becoming 2 single-seeded seed pods per cluster; erect shrubs mostly 2–8' tall. E-side.

Pacific ninebark, *P. capitatus* (cap-it-ay-tus: headed). 3–5 pistils; 3–5 seed pods; erect shrubs mostly 6–13' tall. Moist sites, lower W-side. Color p 118.

Apart from stature, and which side of the mountains they grow on, these two ninebarks are much alike. On all ninebarks the bark of large stems shreds and peels in layers—but rarely, if ever, so many as nine, or even seven.

# Bitterbrush

*Purshia tridentata* (pur-shia: after Friedrich Pursh, below; try-den-tay-ta: 3-toothed). Also **antelopebrush**. Calyx funnel-shaped, fuzzy, 5-lobed, petals yellow, flat-spreading, ¾" diam, stamens protruding; capsules 1-seeded, long-pointed; leaves aromatic, ¾" long, very narrow, 3-pointed and -veined, grayish, white-woolly underneath, edges rolled under; stiff bushy shrubs 3–6' (rarely up to 10'); leaves persist through some winters in some areas. Open forest and steppe E of Cas. Rosaceae. Color p 118.

Bitterbrush is a major shrub of the Intermountain West, thriving on sites marginal for forest growth, such as lower timberlines. On the east Cascade slope, it dominates shrub layers under ponderosa pine and over bunchgrasses. Though bitter to us, the leaves are delicious and nutritious to deer, elk, and pronghorn antelope. Years of heavy browsing sometimes give the shrub a low, mounded form.

Bitterbrush shares with big sagebrush not only the specific name but similar-shaped (tridentate) grayish, fuzzy leaves, and many of the same sites. Their flower structure, however, couldn't be more different, and the plants are unrelated; the leaves evolved convergently in adapting to similar habitat. (Sagebrush leaves do not roll under at the edges, and they're equally woolly on top and bottom surfaces; see page 90.)

---

**Friedrich Pursh** (also spelled Frederick), a German botanist, spent several years in the eastern U.S., where he came into possession of the prize set of hitherto unpublished plant specimens of the era—those from the Lewis and Clark Expedition. He published them in his *Flora Americae Septentrionale* (1814), the first attempt at a coast-to-coast American flora. Detractors claimed he had no right to first publish these specimens, let alone to take many of them to Europe, as he did. In fairness, we may say that overacquisitiveness of this kind was rife among naturalists of that century, even some great ones, and Pursh at least did a creditable job of naming and describing. He spent 12 years on an intended second magnum opus, a flora of Canada, only to see it totally lost in a fire.

---

# Salmonberry and Thimbleberry

*Rubus* spp. (roo-bus: the Roman term). Rosaceae.

**Salmonberry,** *R. spectabilis* (spec-tab-il-is: showy). Leaves 3-compound; leaflets 1–3", pointed/oval, fine-toothed, often ± lobed; petals red or hot pink; fruits like large juicy raspberries in form, in mixed shades of yellow to scarlet irrespective of ripeness; stems erect, 4–10', ± woody but weak, sparsely thorny or occasionally thornless. Abundant on wet slopes and bottoms, W-side. Color p 117.

**Thimbleberry,** *R. parviflorus* (par-vif-lor-us: small- or few-flowered). Leaves 3–8", at least as broad as long, palmately 5-lobed (maplelike), fine-toothed, soft and fuzzy; petals white, ½–1", nearly round, crinkly; berries red, like thin fine-grained raspberries; stems erect, 4–7', ± woody but weak, thornless. Widespread.

A hitherto unreported controversy, apparently rooted in the collective Caucasian subconscious, concerns the palatability of salmonberries and thimbleberries. Most authors venture a bit of mealymouthed praise for one or the other—but never for both. I belong to the thimbleberry cult. I often try two or three salmonberries just to attune myself with the bears, or to see once again if the yellow ones are really any better than the red ones, or was it vice versa, or maybe because their peculiar vapidity challenges my powers of gastronomic recall. But thimbleberries! How anyone could call the thimbleberry "insipid" defies comprehension. Perhaps it's a kind of color-blindness of the tongue. I concede thimbleberries their grittiness, their sometimes dryness, their cankering acidity, and their exasperating sparseness on the bush, yet I find in a good thimbleberry the most exquisite berry flavor on earth.

Indians ate both salmonberries and thimbleberries, of course, as well as the young shoots whose astringent flavor cut the greasiness and fishiness of all those salmon dinners. That's probably how salmonberries got their name.

Both shrubs require moisture and sun. They colonize roadsides, clearcuts, and burns west of the Cascades. In the mountains, low lobes of subalpine meadow are full of thimbleberry, fireweed, and bracken. Salmonberry prefers soggier spots—streambanks or marshy flats too saturated for full tree cover. At lower elevations it grows with red alder, and may replace it in succession, especially in the coastal fog belt.

Fruits of the genus *Rubus* can all be divided up into raspberries and blackberries. A raspberry, when picked, is cup-shaped, pulling cleanly from its receptacle, or core. A blackberry pulls its receptacle with it. Both salmonberries and thimbleberries are raspberries by this definition, though neither tastes much like a garden raspberry. Some of our dwarf *Rubus* berries (page 217) taste like raspberries, but don't have enough little drupelets per aggregate berry to form a cup. The blackcap, *R. leucodermis,* is a bland black raspberry native to lowlands and hills marginal to our range.

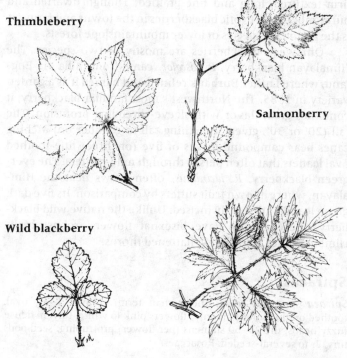

**Thimbleberry**

**Salmonberry**

**Wild blackberry**

**Evergreen blackberry**

## Wild Blackberry

*Rubus ursinus* (ur-**sigh**-nus: of bears, referring to the Bear constellations, i.e., northern). Also **dewberry.** Leaves 3- (or 5-) compound, sometimes persistent but not thick nor truly evergreen; leaflets elliptical, pointed, toothed; flowers white, usually with either stamens or pistils stunted and sterile, the functional males and females on separate plants (this species only); flowering stalks up to 16" tall, from long trailing stems with many slender thorns. Widespread in clearings and ± sunny forest. Rosaceae.

The Northwest is deservedly famous for growing mouthwatering blackberries like weeds. They *are* weeds. The most eaten ones are foreign invaders, backyard nuisances, archetypal briarpatch—but mouthwatering. They sometimes sweeten hot, dreary approaches over logging roads.

The Pacific "wild" blackberry, our only native blackberry, is not the sweetest but the most elegant in flavor, distinctively firm-textured, long and fine-grained. Though dwarfed and outnumbered by exotic blackberries in the towns and farms, it is the abundant species of lower mountainslope forests.

Our weed blackberries are mostly of two species. The Himalayan blackberry, *R. discolor*, came from India via England, where Luther Burbank refined and named it as a garden variety in 1885. The Northwest's most familiar blackberry, it combines great flavor with sleeve-shredding profusion. The tall (20' or 30', given something tall to clamber on) arching canes bear compound leaves of five (or three) fine-toothed oval leaflets that often persist through mild winters. The evergreen blackberry, *R. laciniatus*, often grows near the Himalayan, where its own fruit suffers by comparison. Its five dark green leaflets are deeply incised. Unlike the native wild blackberry, these exotics have bisexual flowers, robust high-climbing canes, and heavy, flattened thorns.

## Spiraeas

*Spiraea* spp. (spy-ree-a: the Roman term). Leaves 1–3", oval, toothed on the outer half only; flowers pink to white, tiny, in dense fuzzy heads, the 25–50 stamens (per flower) protruding; seed pods tiny, 2- to several-seeded. Rosaceae.

**Subalpine spiraea,** *S. densiflora* (den-sif-lor-a: dense flowers). Flowers pink, in slightly convex-topped heads about 2" across; prostrate to low shrubs, up to 3'. Moist thickets and meadows, commonest at 3,000–5,500' W of Cas Crest. Color p 118.

**Birchleaf spiraea,** *S. betulifolia* (bet-you-lif-oh-lia: birch leaf). Flowers white to very slightly pink; otherwise as above. Widespread on E-side only.

**Hardhack,** *S. douglasii* (da-**glass**-ee-eye: after David Douglas, p 18). Also **steeplebush.** Flowers pink, in conical to spikelike heads often over 3″ tall; leggy erect shrubs 2–7′ tall. Streamside and marshy thickets, mainly W-side. Color p 118.

**Nootka rose**

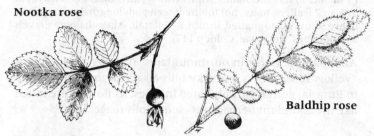

**Baldhip rose**

# Wildroses

*Rosa* spp. (ro-sa: the Roman term). Leaves pinnately compound; leaflets 5 to 11, oval to elliptical, toothed except at the base; flowers pink, ¾–3½″; fruit orange, turning red or purple, many-seeded, dry and sour; ± thorny shrubs 1½–8′ tall, rarely climbing. Rosaceae.

**Baldhip rose,** *R. gymnocarpa* (gym-no-**car**-pa: naked fruit). Flowers ¾–1″; fruits ⅜″, unique among our rose hips in not retaining the crown of 5 sepals—hence "bald"; plants 1½–4′, bristling all over with fine thorns; common W-side; also E-side mid elevs.

**Peahip rose,** *R. pisocarpa* (pis-o-**car**-pa: pea fruit). Flowers (1–2″) and fruits in small clusters; fruits small (¼″), pear-shaped, purplish; thorns usually few; plants up to 8′; strictly W-side, moist places.

**Nootka rose,** *R. nutkana* (noot-**kay**-na: of Nootka, Vancouver Island). Flowers (2¼–3½″) and fruits large, single; thorns variable, typically in stout pairs at leaf nodes; plants usually 2–6′. Both sides Cas, commoner E. Color p 116.

Looking at a wildrose, one wonders whether Europe's horticultural wizards had been at work multiplying rose petals and colors before her poets ever invested The Rose with all its mythic and symbolic weight. In fact, they had. "Hundred-petaled" roses were in cultivation at least by 400 B.C. In comparison, the five-petaled wild rose ranges from unassuming to downright ragged, its virtues humility, delicacy, and tender fragrance. Its fruit, the rose "hip" or "apple," has ruggeder qualities—perseverance on the bush, lots of vitamin C, and a halfway pleasant flavor after a couple of frosts have broken it down. Rose hips are trailside breath fresheners, as the Indians knew, and good survival forage.

---

## Shrubby Cinquefoil

*Potentilla fruticosa* (potent-ill-a: small but mighty: fru-tic-oh-sa: shrubby). Petals bright yellow, ½" long; sepals apparently 10 (5 smaller bracts alternating with 5 true sepals); compound leaves of 3, 5, or 7 leaflets, hairy, not toothed; seed pods long-haired; typically dense, rounded to matted shrubs 6–24" tall. Alp/subalpine gravels and outcrops. Rosaceae. Color p 117.

An alpine specialist in our mountains, shrubby cinquefoil or "yellow rose" grows on rocky hills all across the country and in Eurasia, and is also cultivated in cities. Its flowers are much like the abundant herbaceous cinquefoils (page 218).

---

**Shrubs:** deciduous, under 6', 5 stamens

---

## Currants and Gooseberries

*Ribes* spp. (rye-beez: Arabic term for rhubarb). Flowers ± tubular, the 5 sepals united for about half their length; 5 petals smaller and less colorful than sepals, attached just inside calyx mouth alternately with the 5 stamens; leaves palmately lobed and -veined; weak-stemmed shrubs mostly 2–6'. Grossulariaceae (Currant family).

**Red-flowering currant,** *R. sanguineum* (sang-gwin-ee-um: bloody). Flowers ¾" long, red to pink, 10–30 in dense pendent clusters; berries black with a heavy white bloom; leaves 1½–4"; shrubs 5–12'. Lowland woods. Color p 119.

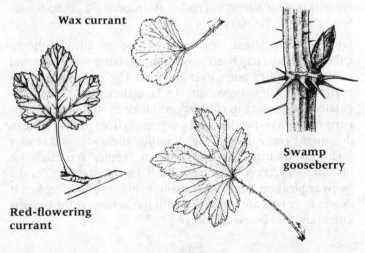

Wax currant

Swamp gooseberry

Red-flowering currant

**Swamp gooseberry, R.** *lacustre* (la-cus-tree: of lakes). Also **prickly currant.** Flowers saucer-shaped, dull pinkish, 10–15 in pendent clusters; berries hairy, dark purple; leaves ½–2½"; stems covered with tiny prickles, larger spines whorled around stem nodes; straggly or sprawling shrubs 3–5'. Forested wet spots, esp mid elevs W-side.

**Maple-leafed currant, R.** *howellii* (how-el-ee-eye: after Thomas Howell). Sepal lobes flat-spreading, dull red; berries black; leaves 2–3"; shrubs 2–4'. Mid elevs to subalpine, esp in streamside thickets.

**Wax currant, R.** *cereum* (see-ree-um: wax). Flowers yellowish white, tubular, ¾" long, ending in short spreading calyx lobes; berries red; leaves grayish, ± round, ½–1¼" diam, toothed, gummy-surfaced, muskily aromatic when crushed; bushy shrubs 2–6'. Sunny slopes E of Cas Cr. Color p 119.

The gooseberry has nothing to do with geese; the word (along with the family name) comes straight from the French *groseille*. Gooseberry plants are distinguished from currants by having prickles. All Northwest *Ribes* fruits are edible and nutritious, and most were widely eaten both fresh and dried by the Indians, but today they are recommended only to the starving or insatiably curious. Various authors suggest that certain species are worth eating, but don't agree as to which ones. Some will take your mouth to the movies, running sweet moments tightly sandwiched between sour and bitter episodes; others are only insipidly bitter. Many thickets offer side-by-side currant tastings.

All *Ribes* species prefer moist habitats, but there are as many Eastside as Westside species. All are prey to white pine blister rust, *Cronartium ribicola,* an introduced fungus that devastates western white pines (page 47). Since the fungus' life cycle alternates between *Pinus* and *Ribes* hosts, foresters have tried to eradicate currants from some prime white pine areas.

Our showiest currant is the red-flowering, which blooms deep pink profusely in early spring, when it may be the only burning color in the forest. It is an esteemed ornamental outside of the Northwest; David Douglas' original exportation of this species alone is said to have paid off his sponsors' investment in his two-year expedition. Occasional white-flowered specimens have been prized as breeding stock.

Large gooseberry thorns were used by some tribes as fishhooks and as needles for tattooing or for removing splinters.

---

# Poison-oak and Poison-ivy

*Rhus spp.* * (rooce: the Greek term for sumac). Leaves compound, leaflets 3 (rarely 5), 2–5", with highly variable lobes, the 2 lateral leaflets usually less lobed on their inner edge than on either their own outer edge or the 2 symmetrical edges of the terminal leaflet; flowers ¼" diam, greenish white; berries white, up to ¼", single-seeded, ± striped longitudinally. Anacardiaceae (Cashew family).

**Poison-oak,** *R. diversiloba* (diverse-il-oh-ba: varied lobes). Leaves coarsely lobed (oaklike); flowers/berries in loose pendent strings; straggly shrubs 2–6' or sometimes climbing as a vine up to 40' on trees. Low elevs in Columbia Gorge. Color p 118.

**Poison-ivy,** *Rhus radicans* (rad-ick-enz: rooting). Leaves narrowly pointed with few lobes; flowers/berries in dense, erect clusters. Low E-side canyons.

**Almost unknown** in our mountains except in the Columbia Gorge, where abundant, and a few hot, dry lowland sites peripheral to the range.

**Leaflets in 3's** almost always.

**Some leaflets crimson** by midsummer; others merely yellowing just before dropping in fall.

**New foliage reddish** and glossy in spring; later, foliage is not often strikingly glossy in our region.

**Translucent white berries** (like blisters, for mnemonic aid) in bunches, present by late summer and into winter, long after the leaves; many plants may fail to fruit.

**Often on or near white oak**, its leaflets seeming to mimic nearby oak leaves in shape, sheen, and shade.

Unlike nettle stingers—an elaborate and effective defense against browsing—poison-ivy/oak/sumac poison has little survival value to the plant. Call it an accident of biochemistry, or one of the commonest allergies in *Homo sapiens.* Other species don't seem to be susceptible; they gather the nectar, or browse the leaves with pronounced indifference. In humans, susceptibility can be acquired but rarely shed. Many people sure of their immunity have tried to show it off, only to get a

---

*Some texts put both these species in a separate genus *toxicodendron* ("poison tree") rather than lumping them with the harmless sumacs of genus *Rhus.*

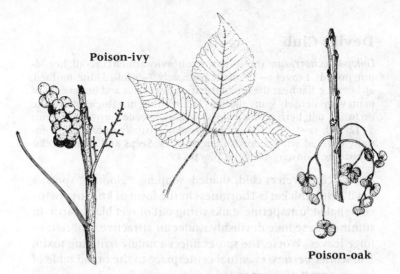

**Poison-ivy**

**Poison-oak**

rude shock a couple of days later. Apparently this never happened to Leo Hitchcock, senior author of *Flora of the Pacific Northwest,* who made a practice of picking specimens with his bare hands on class field trips. Lore has it that immunity can be cultivated by eating tiny leaves over the period of their development in spring. **Don't try it.**

Symptoms usually appear, if at all, between 12 and 72 hours after contact. We often start itching within minutes of laying eyes on poison-oak, but in most people this is due to other irritants, like anxiety. If you can wash your exposed parts within ten minutes, preferably with strong soap and hot water, and quarantine your exposed clothes (and dogs) until laundry, your chances of escaping are still excellent. The allergen, urushiol, is in the sap; a light brushup hardly brings any forth, and is harmless to most people. A few people, though, are alarmingly sensitive. There have been fatalities following poison-ivy smoke inhalation—drownings, technically, in a sea of blister fluid in the lungs. (The toxin is destroyed by complete burning, but smoke may carry unburned particles.)

For most of us, the only defense needed is to know when we are going into areas where poison-oak or ivy grows, and then to know the characteristics, spot the plants without fail, and circumvent them. All in all, the Olympics and Cascades are almost as good a place for staying away from poison-oak and ivy as from poisonous snakes and grizzly bears.

---

# Devil's Club

*Oplopanax horridum* (op-lo-**pan**-ux: heavily armed cure-all; **hor**-id-um: horrid). Leaves 6–15" diam, palmately 7–9-lobed, fine-toothed, all borne ± flat near the top of the stem, leafstalks and undersides of main veins densely spiny; flowers ¼", whitish, in a single erect spike up to 10" tall; berries bright glossy red, 2–3-seeded, up to ¼"; stems 3–12' tall, ½–1½" thick, punky, crooked, usually unbranched, entirely covered with yellowish tan prickles. Seeps and small creeks. Araliaceae (Ginseng family). Color p 119.

Devil's club prefers cold, shaded, sopping, "gloomy" spots; a devil's club thicket is Thorniness in the form of knobby, twisted, tangled, untapering stalks rising out of wet black earth. In summer these hide devilishly under an attractive umbrella of huge leaves. Worse, the spines inject a mildly irritating toxin. The scarlet berries, eventual centerpiece to the broad table of leaves, aren't recommended either.

 *Oplopanax* is an oxymoron; *oplo* implies weaponry, while *panax* is a cure, as in *panacea*. Devil's club may seem more weapon than cure; the "cure" half of its name refers to its relative ginseng (genus *Panax*), perhaps the most cross-culturally recognized of all herbal panaceas. Under devil's club thorns lies a thin bark which Puget Sound tribes used medicinally, magically, and cosmetically. It was thought to alleviate such diverse ailments as colds, rheumatism, excessive milk, amenorrhea, and bad smells. A twig on the wall was a household charm. The Lummi mixed the bark ash with bear grease to make black or sepia face paints and tattooing inks. Coastal tribes used the thorny stems as fishing snags and lures.

# Big Sagebrush

*Artemisia tridentata* (ar-tem-**ee**-zhia: the Greek term, honoring either a goddess or a queen; try-den-**tay**-ta: 3-toothed). Composite flower heads drab yellowish, tiny, in loose spikes; leaves spicy-aromatic, grayish-woolly, wedge-shaped, ± shallowly 3-lobed at the tip (average ½" wide), tapering gradually from there to base; bark shreddy; average 2–6' tall where growing abundantly at lower tim-

berline (and eastward *ad nauseam*); or dwarfed to 1–1½′ on subalpine ridges. Asteraceae (Aster family).

Most parts of Wild Oregon that aren't covered with Douglas-fir seem to be dotted with sagebrush. Those parts lie east of the forested Cascades, however; this legendary bush could have been left out of this book but for its importance in subalpine communities on ridges east of the Crest in Washington. The form growing here is a dwarfed, more or less isolated ecotype sometimes regarded as a distinct species. Sagebrush is closely related not to culinary sage (*Salvia,* in the mint family) but to culinary tarragon and wormwood (page 188).

**Black twinberry**

## Black Twinberry

*Lonicera involucrata* (lo-nis-er-a: after Adam Lonitzer; in-vo-lu-cray-ta: with involucres). Also **bush honeysuckle, inkberry.** Petals pale yellow, ½–¾″, fused (tubular) over ½ their length, stamens just appearing at tube mouth, calyx tiny and scarcely lobed, flowers paired, as are the glossy black ¼″ berries; leaves opposite, elliptical, pointed, 1½–5″, hairy especially at margins and under veins; shrubs 3–6′ (rarely to 10′). Widespread on streamsides. Caprifoliaceae (Honeysuckle family).

Northwest tribes called twinberries "crow food," Crow being the only spirit crazy and black enough to relish such bitter, black fruit. "Inkberry" juice was face paint for dolls, or dye for graying hair.

The twinberry plant carries twinning (the opposite-leaf style) to an extreme; opposite leaf axils bear long opposite stalks, each bearing a pair of flowers between two crossed pairs of hairy bracts, two of which are two-lobed. Two or all four of the bracts (collectively an "involucre") typically turn deep magenta and reflex downward over time to better offset the paired, purplish black berries that replace the pale flowers. This display usually hides in a wet thicket.

Purple-flowered bush honeysuckle, *L. conjugialis,* grows

---

**Shrubs:** deciduous, under 6′, 5 stamens

less abundantly, in the Oregon and Southern Washington Cascades. Its flower pairs are "conjugally" fused at the base, and produce just one, two-tipped berry between their fused ovaries. (Contrast with orange honeysuckle, below, with multiple flowers above fused leaf-pairs.)

## Orange Honeysuckle

*Lonicera ciliosa* (silly-oh-sa: fringed). Petals orange to almost red, 1–1½", fused (tubular) over ¾ their length, one petal lobe drooping ± apart from the upper 4, hairy inside tube; calyx insignificant; flowers and red ½" berries in terminal clusters ± nestled in a perfoliate leaf (see below); leaves opposite, oval, 1½–3", finely hairy-margined; perennial vines climbing 5–20'. Low-elev openings. Caprifoliaceae (Honeysuckle family). Color p 118.

This honeysuckle attracts its pollinators—hummingbirds—visually, whereas sweet-scented honeysuckles attract night-flying moths. Its flowers, long and narrow like a hummer's bill, are crammed together to make a bright orange bullseye in the center of a target comprised of the uppermost pair of opposite leaves modified into a single fused leaf—often shaped like a very full pair of lips—with the stem passing through the center. The next lower one to three pairs of leaves are decreasingly fused. Northwest tribes saw honeysuckle as women's medicine, using tea or steam from the leaves to encourage lactation, discourage conception, ease cramps, or add luster to little girls' hair. The vine occasionally clasps its host tightly enough to kill it by allowing no room for growth.

## Snowberries

*Symphoricarpos* spp. (sim-for-i-car-pus: gathered fruit). Also **waxberries.** Flowers pinkish to white, bell-shaped, less than ¼", petals fused over ½ their length; berries pure white, tightly clustered, pulpy, 2-seeded; leaves and twigs opposite, most leaves (see text) oval to elliptical, 1" long. Caprifoliaceae (Honeysuckle family).
*S. albus* (al-bus: white). Berries ½"; shrubs 3–7' tall. Mainly E-side.
*S. mollis* (mol-iss: soft). Berries ¼"; trailing shrubs with erect stems less than 20." Mainly W-side.

I've always thought of this as "popcorn plant," which seems to capture its likeness better than "snow" or "wax," though at the risk of falsely encouraging hungry hikers. The lightweight berries are utterly unlikable as people food; Indians appreciated them ironically as "good for the kids to throw at each other." Maybe that's the mode of dissemination they evolved for. (Actually, birds and rodents eat and disseminate them.)

Our two common species differ little, aside from growth form and range. *S. albus,* a sizable bush, dominates shrubby understories with wildrose and spiraea under sparse Eastside canopies; the northeast Olympics and Puget/Willamette Trough are also dry enough for it. *S. mollis,* a trailing vine, is scattered in Westslope forests. Fast-growing juveniles of both species have highly variable leaves, sporting all numbers and sizes of odd-shaped lobes, symmetrical or not; with maturity they calm down to nondescript inch-long ovals.

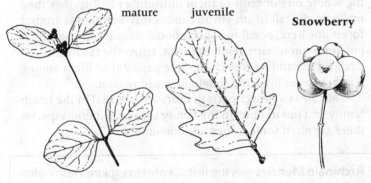

mature     juvenile     **Snowberry**

**Shrubs:** deciduous, under 6′, 8 or 10 stamens

# White Rhododendron

*Rhododendron albiflorum* (roe-doe-den-dron: rose tree, a misleading name in both respects; al-bif-lor-um: white flower). Also **Cascades azalea.** Flowers white, ¾" diam, broadly bell-shaped, petals fused no more than ½ their length, 1–4 in clusters just below the whorllike leaf clusters at branch tips; leaves elliptical, 2–3½", glossy, bumpy, slightly reddish-hairy; capsules 5-celled; bark shredding; leggy shrubs 3–6′. Locally abundant around timberline, especially on colder sites. Ericaceae (Heath family). Color p 119.

It's hard to see these blossoms—so modest by garden "rhodie" standards—as rhododendrons. But few who have reached timberline by foot would question their showiness.

# Fool's-huckleberry

*Menziesia ferruginea* (men-**zee**-zia: after A. Menzies, below; fair-u-**jin**-ia: rust-red). Also **false-azalea, rusty-leaf**. Flowers pale rusty-orange, jar-shaped, pendent on sticky-hairy pedicels, ¼", calyx and corolla shallowly 4-lobed, stamens 8; capsules 4-celled; leaves 1½–2½", elliptical, seemingly whorled near branch tips, often brown-hairy, coloring deeply in fall; bark shreddy; leggy shrubs 3–6'. Mid elevs to subalpine. Ericaceae (Heath family). Color p 119.

This plant has no berries to tempt anyone, no matter how foolish, so I'd call it a fool's fool's-huckleberry. True, on the basis of summer foliage alone it could be carelessly mistaken for its close relatives, black huckleberry and white rhododen-dron. More often it simply goes unnoticed among them, grow-ing where one or both of them outnumber it. Together they make up the tall heath-shrub community ubiquitous around forest line here, possibly a successional stage leading to subal-pine fir or mountain hemlock forest. Often the tall heaths en-circle an expanding tree clump (see page 28) or fill in among scattered trees invading subalpine meadowland.

*Menziesia*'s flower parts in fours are unusual in the heath family, but not unique. In the family's many 5-merous species there are often some 4-merous individuals.

Archibald Menzies was the first scientist to explore Washington or Oregon, so it's fitting that more species (six, plus this genus) in this book honor him than anyone else. After spending a month at Nootka on Vancouver Island in 1787, he was appointed surgeon-naturalist on H.M.S. *Discovery* under Captain **George Van-couver**. Menzies' 1792 journal records the first Washington landing at Discovery Bay and the naming of many other land and water features after officers on the ship (Puget, Whidbey, Baker, Vancouver) and Englishmen the captain admired (Rainier, St. Hel-ens, Hood). The ship's mission didn't allow Menzies time for much exploration ashore, and his live collections all died on board, but his plant descriptions and dried specimens that reached England aroused intense interest in the Northwest, eventually leading to voyages by David Douglas and others. Men-zies was an old man by 1824, when young Douglas visited him for a briefing on Northwest American plants.

# Huckleberries/Blueberries

*Vaccinium* spp. (vac-sin-ium: the Roman term). Leaves elliptical; flowers pinkish, small, globular, with 5 (rarely 4) very short, bent-back corolla lobes, and similar calyx lobes on the tip of the berry. Ericaceae (Heath family). Individual species described on following pages.

This diverse and well-distributed genus has always been of intense interest to bears, birds, Indians, and hikers. Toward summer's end, bear scats are often little more than barely cemented heaps of recognizable huckleberry leaves; the cement is first to decompose, so you may wonder, a week or two later, how these heaps of dry leaves came to be so neatly molded. The bears lack the patience or the dexterity to pluck the berries singly, and who can blame them, with only a month or two left to fatten up for hibernation?

Indians used the most sophisticated known huckleberry management technique—burning—to maintain areas of berries to pick, including the 8,000-acre Twin Buttes field southwest of Mt. Adams. That particular legacy has dwindled to a mere 2,500 acres under Smokey's antifire administration. The Forest Service is now rediscovering the beauties of a low blaze; one USFS report assigned twice as high a cash value per acre per year to the berries at Twin Buttes as to the trees that have been replacing them. Maybe they shouldn't have made Smokey out to be a bear all along—bears undoubtedly preferred Native American management at Twin Buttes to the huckleberry-squelching that goes on in Smokey's name.

"Huckleberry" vs. "blueberry" is a messy issue. Folks around here mostly use the terms for "wild" and "cultivated" *Vaccinium*, respectively, but back East a related genus (*Gaylussacia*) with seedy black berries has a prior claim on the name huckleberry. Some authors, to resolve this dilemma, try to reinstate the old English names "whortleberry" and "bilberry"; these may be a short etymological hop from "huckleberry" and "blueberry," but could you use them with a straight face? We could call them all blueberries, right down to the "red blueberries." I could get used to that. Still, there's nothing wrong with calling them huckleberries—it just risks confusing Easterners. Cranberries and lingonberries, though also in genus *Vaccinium*, are never spoken of as huckleberries.

---

The Northwest has twelve species of huckleberry. The following six, taken in order of descending elevation, are each important in one or more montane community types.

**Cascades blueberry,** *V. deliciosum*. Berries bright blue due to a heavy coating of waxy bloom; flowers spherical; leaves slightly toothed; plants 2–18″ tall. Alp/subalpine. Color p 120.

Cascades blueberries provide about 90% of the gorgeous bronzy russet fall color of subalpine slopes, and can also be credited for a lot of excitement in the animal kingdom at that time of year. *Deliciosum* indeed!

On moist alpine sites, they often grow on a 2–3″ scale that makes the berries look hugely rotund. Both there and in the subalpine zone, where they average a foot in stature, they often grow with red, yellow, or white mountain heathers. This association, the low heath-shrub community, is characteristic of our range, in the sense of being both ubiquitous here, at suitable elevations, and absent in the same form from mountains east or south of the Cascades. It is found on rocky convex topography or next to trees, in either case getting a longer snowfree season than nearby herbaceous meadows. It is prone to takeover by trees during warm/dry decades.

**Grouseberry,** *V. scoparium* (sco-pair-ium: broom—). Berries bright red, tiny (⅛″); leaves ¼–½″; plants 4–14″ tall; twigs green, with angled edges, numerous, all ± vertical. Near and E of Cas Cr in Ore.

Grouseberry is locally abundant on dry habitats, often on overly drained (gravelly) volcanic soil. The fruit is scrumptious but hard to gather in quantity.

**Black huckleberry,** *V. membranaceum* (mem-bra-nay-see-um: thin). Also **thinleaf huckleberry.** Berries black to very dark red; flowers much longer than broad; leaves 1–2½″, very thin, pointed, minutely toothed; shrubs 2–6′. Low subalpine.

Top huckleberry for combined availability, flavor, and texture, the black huckleberry is the main attraction of the huge wild huckleberry fields southwest of Mt. Adams. Though it grows and bears fruit most lavishly in burns and other clearings, it also dominates the shrub layer of many midmontane forests. It mingles there with the following two species, or with Cascades blueberry around timberline. Beargrass is another frequent companion.

**Oval-leaved huckleberry,** *V. ovalifolium* (o-val-if-oh-lium: oval leaf). Berries blue with bloom; flowers slightly longer than broad, on curved stalks; leaves ¾–2″, smooth-edged; shrubs 1½–5′. Mid elevs.

**Alaska huckleberry,** *V. alaskaense* (alaska-en-zee: of Alaska). Berries with or without bloom, i.e. blue or purplish black; flowers at least as broad as long, on straight stalks; leaves 1–2½″, ± smooth; shrubs 1½–4′. Mid elevs.

These two species are hard to distinguish by either habit or habitat. Alaska huckleberry is scarcer in Alaska than oval-leaved huckleberry. Both grow at forested elevations (2,000–5,000′) and are typically straggly understory shrubs, but are prepared to burst forth after fire or clearcut. They tolerate just slightly less temperature and moisture stress than black huckleberry, but for all that, they produce seedier, sourer berries.

**Red huckleberry,** *V. parvifolium* (par-vif-oh-lium: small leaf). Berries bright red; flowers at least as broad as long; leaves ¼–1″, ± smooth-edged; twigs squarish, angled; shrubs 3–12′. W-side.

In lower Westside forests, red huckleberry seems indiscriminately widespread but never abundant. Its berries are sparse, too; if only they were easier to gather in quantity, that legendary red huckleberry pie would be better known. For eating fresh, they are juicy but sour.

**Oval-leaved huckleberry**

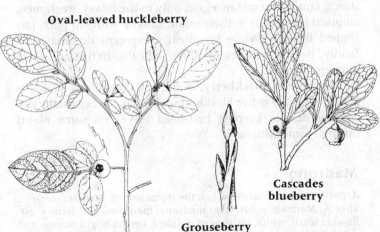

**Cascades blueberry**

**Grouseberry**

*\*V. alaskaense* is regarded by some taxonomists as a subspecific variety, *V. ovalifolium alaskaense*, or a naturally occurring hybrid, *V. ovalifolium* × *V. parvifolium*.

---

**Shrubs:** deciduous, under 6′, 8 or 10 stamens

Broadleaf evergreens are primarily a tropical and subtropical life form; where the growing season is year-round, the leaves naturally stay put. Though our region has winters, it offers advantages to evergreen plants that can continue their life functions, even at sharply reduced rates, through the cool seasons. This is largely because our summers are dry enough to curtail warm-season photosynthesis (see page 11). Our type of broadleaf evergreens, called sclerophylls ("hardened leaves"), have adapted with heavy, rigid leaves that dry out partially in summer, reducing transpiration and conserving water, without danger of wilt damage. Most of temperate North America's sclerophylls are found in Pacific coastal forests—many here in our range, but far more in southern Oregon and northern California. Down there, wet mild marine winters like ours alternate with summers longer, dryer, and hotter than ours.

This section begins with our only reliably tree-form sclerophyll, and moves on down through the tall and medium shrubs to the dwarf and prostrate shrubs and subshrubs. The Ericaceae (Heath family) dominate this section; they share a strong family resemblance, not only in the heavy, evergreen, elliptical leaves but in their small, usually white to pink, jar-shaped flowers. While broadleaf evergreens dominate the family, it also includes deciduous shrubs, herbs, and even non-green herbs.

**Evergreen Blackberry,** p 84, an introduced weed, is placed with the other blackberries, which are essentially deciduous despite keeping half-dead leaves on some plants through some winters.

## Madroño

*Arbutus menziesii* (ar-bew-tus: the Roman term; men-zee-zee-eye: after A. Menzies, p 94). Also **madrona, madrone.** 24" diam × 80'; flowers small, white, jar-shaped, 5-lobed, berries bright orange with pebbly skin, many-seeded, ¼–½" in large clusters; leaves heavy, glossy dark green above, silvery beneath, oblong-elliptical 2½–6"; bark flaking in thin sheets, very smooth and pale green when young, turning through bronze to deep red, finally becoming rough dark gray-brown, esp near base; irregularly branched trees. Sunny, rocky slopes, low NE Olys and W-side Cas. Ericaceae (Heath family).

I can't stop gazing at madroño bark, can't resist touching it, only barely resist stripping more and more of it—such elegant voluptuous limbs, so smooth and richly colored. Similar limbs on a smaller scale grow on manzanitas and kinnickinnick, "madroño's little sisters"; similar leaves on rhododendron. But of all the heath family, only *Arbutus* grows as a tree. It is by far the most northerly broadleaf evergreen tree on the continent, growing up and down the coast from mountains near San Diego to rocky shorelines up on the Georgia Strait. To cover that stretch it has to be indifferent to precipitation (18" to 166" a year) and temperature, requiring only that its own spot be well drained and sunny.

Around the Mediterranean grows a shrubby tree, *A. unedo*, called *arbutus* by the ancient Romans, *madro* by the Spanish, and "strawberry-tree" by the English. Spaniards in early California recognized its relative, calling it "big madro," *madroño*, later corrupted into madrone or madrona (the two usual spellings today) by Anglos. The sweetish berries of the strawberry-tree flavor a liqueur, *creme d'arbouse;* but madroño berries are bitter and slightly narcotic. The wood is heavy, hard and strong, but splits badly in drying, and isn't used much. Indians made spoons and ladles from it.

## Golden Chinquapin

*Castanopsis chrysophylla** (cas-ta-**nop**-sis: chestnut-like; cris-a-fill-a: gold leaf). Catkins clustered ± erect at branchtips, each bearing a series of male fuzzballs, with fewer female flowers near the base of the catkin; nuts ¼–½", in very spiny hulls ¾" diam; leaves 3–6", narrow, tapering, dark glossy green above, with a rough yellow-brown ("golden") coating beneath; new twigs also yellow-coated; bark white-splotched in youth, eventually thick and furrowed; shrubs or small trees (here) 10–30'. Drier forests and thickets, mainly W-side Ore. Fagaceae (Beech family).

*Some texts place *chrysophylla* in genus *Chrysolepis*.

In its best range—southwestern Oregon and northwestern California—the golden chinquapin is a big tree, up to 150' by 6' diam, which would top any of our broadleaf trees except cottonwood. But in our range we know it mainly as a tall shrub. It is easy to spot in much of Oregon's Cascades as the only big shrub with such long, narrow, evergreen leaves. It grows on dry lower-slope sites, joining rhododendron, salal and vine maple to make a very dense understory. In Washington it grows only in two places—one near the Columbia Gorge and one, very isolated, near the Hamma Hamma River.

Chinquapins are in a family with oak, beech, and chestnut trees. "Chinquapin," a vernacular name for chestnut trees (genus *Castanea*), derives from an Algonquinian name. Chinquapin nuts and chestnuts are both edible and both borne one to three in a bristly hull. Fortunately for our chinquapin, no chestnuts are native to the Northwest. If they had shared any range they would probably have shared a dismal fate—an exotic blight that wiped out the American chestnut. The wood is easily worked, and takes on a beautiful hue.

## Pacific Rhododendron

*Rhododendron macrophyllum* (roe-doe-den-dron: rose tree, two misleading words for this genus; macro-fill-um: big leaf). Flowers pink, 1–2", broadly bell-shaped, in large dense clusters, petals and sepals 5, fused at the base, 1 upper petal usually spotted; capsules woody, 5-celled, ½–¾"; leaves heavy, deep glossy green, 3–8" long, with smooth rolled-under edges; often straggly shrubs 3–15'. Abundant in some drier mid-elev forest, Ore and S Wash Cas, and E Olys. Ericaceae (Heath family). Color p 120.

The 600-odd species of genus *Rhododendron*—most of them Asian—have been recombined into thousands of garden "rhodies" and azaleas of all colors; they make overwhelming displays. Our native rhododendron is plain pink, spends most of its time in the shade, and runs to legginess. Still, when the woods are in bloom with them, like fat pink mirages on a gray day, or like randomly bursting fireworks in the shifting sunbeams of a sunny day, they have great power to enchant. The Washington legislature named them the State Flower, perhaps to make up for having fewer of them than Oregon.

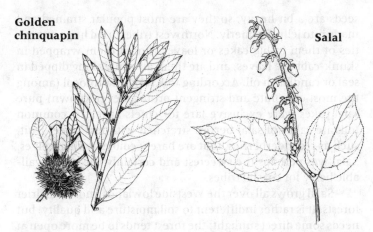

**Golden chinquapin**

**Salal**

## Salal

*Gaultheria shallon* (galth-ee-ria: after Jean-Francois Gaultier; shall-on: NW tribal term). Corolla pinkish to white, bell-shaped to nearly spherical, sepals red; flowers 6–15, pendent along a ± horizontal stalk; berries black to purple; entire inflorescence (incl flowers and berries) minutely hairy and sticky; leaves glossy, leathery, broadly oval, pointed, 2–3½"; branches zigzagging at leaf nodes; shrubs 1–8'. Abundant W-side. Ericaceae (Heath family). Color p 120.

April 1825. "Saturday the 9th in company with Mr. Scouler I went ashore on Cape Disappointment as the ship could not proceed up the river in consequence of heavy rains and thick fogs. On stepping on the shore *Gaultheria shallon* was the first plant I took in my hands. So pleased was I that I could scarcely see anything but it. Mr. Menzies correctly observes that it . . . would make a valuable addition to our gardens. . . ."

(1826) "Called by the natives 'Salal,' not 'Shallon' as stated by Pursh . . . Bears abundantly, fruit good, indeed by far the best in the country . . ."*

I wonder if David Douglas would have logged that appraisal if he had been in the mountains in huckleberry season; in any case, salal berries deserve a better reputation than they now enjoy. At their best, they rival huckleberries in flavor—less acid, with spicy resinous overtones—but their skins and

---

*See David Douglas, p 18, John Scouler, p 67, Archibald Menzies, p 94, and Friedrich Pursh, p 81.

seeds are a bit heavy, so they are most popular strained and made into jelly. Formerly, Northwest tribes dried huge quantities of them in big cakes or loaves, stored them wrapped in skunk-cabbage leaves, and ate them all winter long dipped in seal or candlefish oil. According to Kwakiutl protocol (among the most elaborate and stringent protocol ever known) pure salal cakes were exclusive fare for chiefs at feasts; common folk ate cakes of salal berries stretched with less sweet fruit, such as red elderberries, that are barely edible by themselves. Salal was perhaps the sweetest and certainly the most available berry for coastal tribes.

Salal grows all over the Westside lowlands, mostly in drier forests. It is rather indifferent to soil moisture and quality, but needs some direct sunlight; the forest tends to be more open at dry and wet extremes, so either extreme favors salal. Thickets of salal often indicate patches of poor soil with reduced competition from more demanding species. Several dwarf shrubs called wintergreen are recognizable as salal's "little brothers"; two (page 107) grow here, at higher elevations.

## Manzanitas

*Arctostaphylos* spp. (arc-tos-taf-il-os: bear grapes). Flowers pinkish to white, jar-shaped, 5-lobed, in small clusters; berries reddish brown, dry, mealy; leaves grayish, ± elliptical, 1–2"; twigs minutely hairy; larger branches smooth, red, with peeling flakes; dense shrubs 3–8'. Rocky exposed sites at all elevs. Ericaceae (Heath family).

**Green manzanita,** *A. patula* (patch-u-la: spreading). E-side from Mt Adams S, common on pumice. Color p 120.

**Hairy manzanita,** *A. columbiana* (co-lum-be-ay-na: of the Columbia River). W-side.

More than thirty species of manzanita conspire in the making of California's notorious chaparral, a dense brushfield community adapted for taking over mountain slopes after fires. Manzanita seeds lie dormant in the soil until awakened by groundfire heat; between fires, the plants spread by layering. When the chaparral burns, manzanita root crowns sprout again like crazy. The finishing touch in this monopolistic strategy, according to some researchers, is an "allelopathic" secretion which poisons the soil against competing species.

Our manzanitas are remote outliers of the chaparral, and fortunately they are less successful. Green manzanita often grows with snowbrush, which also has fire-germinated seeds. Hairy manzanita is the Westside lowland species, also pioneering on lava flows in Oregon's High Cascades. In the eastern Olympic foothills (and wherever their ranges overlap) it hybridizes with kinnickinnick (page 108), producing a red-fruited, low erect shrub popular in cultivation.

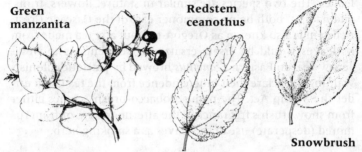

**Green manzanita**

**Redstem ceanothus**

**Snowbrush**

## Snowbrush

*Ceanothus velutinus* (see-an-oath-us: thistle, a misleading name; ve-lu-tin-us: velvety). Also **tobacco-brush, cinnamon-bush, sticky-laurel.** Flowers tiny, white, in dense fluffy ± conical 2–5" clusters; petals, sepals, and stamens each 5; seed pods of 3 separating, 1-seeded cells; leaves 1½–3½", broadly oval, fine-toothed, often tightly curling, heavily spicy-aromatic, shiny and sticky above, pale-fuzzy beneath (exc ± smooth on a variety found W-side below 2,600'), with 3 ± equally heavy main veins from base nearly to tip, otherwise pinnate-veined; robust shrubs 2–6'. Dry, sunny clearings, abundant E-side. Rhamnaceae (Buckthorn family). Color p 120.

Snowbrush invades burns, including slashburned clearcuts. Its seeds are activated by the heat of a fire and the increased soil warmth of a new clearing. The young shoots grow slowly, but easily crowd out annuals like fireweed in three or four seasons. The roots host nitrogen-fixing bacteria, a big advantage where the soil has lost most of its organic nitrogen. Snowbrush often grows dense enough to prevent conifer establishment, especially in southern Oregon where it rose high on foresters' hit lists. But like the similarly maligned red alder, it may aid eventual conifer growth by adding nitrogen. Snowbrush reproduction is largely vegetative, since most seeds are dormant for years or decades. The fire may come long after snowbrush has been shaded out of the forest, making the belated reappearance of a full-blown snowbrush community seem miraculous.

Some *Ceanothus* species are evergreen (e.g., squawcarpet, page 113) and others are not; redstem ceanothus, *C. sanguineus*, has distinctive smooth purple stems, but rather nondescript thin deciduous leaves—not fragrant, sticky, shiny, velvety nor so prominently three-veined as snowbrush leaves. The two species are similar in stature, flowers, fruits, and habitat, both being commoner east of the Cascade Crest. Redstem is also known as Oregon-tea, but any tea made from its leaves would be ersatz ersatz; the reference is to New-Jersey-tea, an Eastern *Ceanothus* brewed by patriotic colonists wanting to declare their independence from the tea taxed under the Stamp Act. The name "tobacco-brush" comes either from snowbrush's fragrance in the afternoon sun, or from rumored (desperate) use of the leaves as a smoking herb.

**Oregon-boxwood**

## Oregon-Boxwood

*Pachistima myrsinites* (pa-kis-tim-a: thick stigma; mir-sin-eye-teez: myrrh-like). Also **mountain- or myrtle-boxwood, mountain lover.** Flowers ⅛" diam, clustered in leaf axils, dark red petals and whitish stamens and sepals each 4; capsules splitting in 2; leaves opposite, ½–1¼", elliptical, shallowly toothed, glossy, dark above; twigs reddish, 4-angled; dense shrubs 10–40". Mostly dry sites, mid to high elevs; locally abundant E-side under true firs or western hemlock. Celastraceae (Staff-tree family). Color p 120.

Oregon-boxwood is our only sizable evergreen shrub whose leaves are opposite.* Otherwise it looks much like the heath family; florists use tons of it interchangeably with evergreen blueberry sprays from the Coast, and few customers notice the difference. Definitely a "foliage plant," it has pretty flowers if you squint down close enough to make them out. Deer browse it in winter.

*The evergreen dwarf shrubs, pp 106–14, are mostly either opposite-leaved or indistinctly alternate-leaved; also, don't mistake Oregon-grape's leaflets for opposite leaves.

# Oregon-Grape

*Berberis nervosa** (ber-ber-iss: the Arabic term; ner-vo-sa: veiny).
Also **holly-grape, mahonia.** Flowers yellow, in a terminal group of
3–7" spikes amid a cluster of sharp, persistent ½–2" bud scales; pet-
als/sepals in 5 concentric whorls of 3, outer whorl(s) ± green; berries
⅜", grapelike, purple with a heavy blue bloom; leaves compound,
crowded at top of stem, 10–16" long; leaflets 11–21, spiny-margined
(hollylike), pointed-oval, palmately veined; stems unbranched, 4–
18"; inner bark yellow. W-side forests, ± ubiquitous. Berberidaceae
(Barberry family). Color p 120.

Though Northwesterners today may take them for granted,
David Douglas rated Oregon-grape and salal each a smashing
find for the English garden. England proved him right, especially about Oregon-grape, which was all the rage there during
Victoria's reign. Though the leaves are evergreen, a few may
burst crimson at any time of year. The stamens snap inward at
the lightest touch to shake their pollen onto a bee.

The berries (not grapes by a long shot) have an exquisite
sourness not balanced by much sweetness. Both jelly and
wine from Oregon-grapes are traditional since pioneer days,
but decreasingly popular. Gourmands may gag, but those with
a penchant for wild plant foods still smack their lips. They're
juicy, and refreshing in their way. Indians mashed them with
sweeter berries for winter storage in dry cakes. They also gathered the roots for yellow dye, and for a tea to soothe a sore
throat or stomach.

This is the most ubiquitous understory plant of the western hemlock zone in Oregon; less so in Washington, it is absent
from Sitka spruce bottoms in the western Olympics.

Two close relatives in our mountains each have leaflets
with just one prominent midvein, unlike *nervosa.* Tall Oregon-
grape, *B. aquifolium,* the State Flower, grows on open lowlands on both sides of the Cascades. Its leaves, rarely of more
than nine leaflets, are borne on tall (2–10') woody stems.
Creeping Oregon-grape, *B. repens,* ranges up the Cascades'
east slope to middle elevations. Its leaves of five or seven leaflets stay close (6–30") to the ground.

*Oregon-grapes are known among horticulturists as genus *Mahonia.*
Taxonomists have lumped them with the barberries, genus *Berberis.*

---

A woody stem is hard to detect in very small plants; in fact, the difference between subshrubs and herbs is more a gradation than a gross visible character. In this chapter, evergreen leaves are taken as proof of aboveground perennial (i.e., shrubby) parts to support them; but there seem to be as many cases of borderline evergreenness as of borderline shrubbiness. Many small species whose closest relatives are clearly deciduous have adapted to long-lasting snowpacks by keeping leaves in place and green through one winter, and then letting them slowly wither as they are replaced by new foliage the next summer. This way there is green foliage ready to photosynthesize, however meagerly, from the first snowfree day to the last. Some alpine ecologists call foliage on this schedule "win-

## Plant Life Forms

Trees and shrubs are called phanerophytes ("conspicuous plants") in the life-form classes devised by C. Raunkiaer.

**Megaphanerophytes** are over 100' tall (30 m).

**Mesophanerophytes** are 27' to 100' (8–30 m).

**Microphanerophytes** are 6' to 27' (2–8 m).

**Nanophanerophytes** are 10" to 6' (.25–2 m).

**Chamaephytes** ("dwarf plants") are subshrubs less than 10" tall (25 cm).

**Hemicryptophytes** ("half-hidden plants") are perennial herbs with overwintering buds right at the ground surface.

**Cryptophytes** ("hidden plants") are perennial herbs with buds on subsurface tubers, bulbs, etc.

The remaining plant life form consists of annual herbs, which persevere from one growing season to the next only in the form of seeds.

tergreen," as opposed to "evergreen" leaves that function during more than two summers; the more usual, vaguer term is "persistent." In this book they are keyed with the herbs, according to flower structure: **Alpine Willows**, p 192; **Penstemons**, p 197; **Violets**, p 199; **Sibbaldia**, p 211; **Sandwort**, p 215; **Yerba de Selva**, p 216; **Partridgefoot**, p 216; **Dwarf Raspberries**, p 217; **Kittentails**, p 231; **Wild-ginger**, p 232; **Lewisias**, p 233.

When in doubt as to the evergreenness or the shrubbiness of a specimen you want to key out, try to think of other plants, either from memory or from the pictures, that it resembles and may be related to. This group (broadleaf evergreens under 10") has firm, dark, waxy-seeming leaves; all but the final three are in the Heath family. The only plant placed elsewhere despite having dark, firm evergreen leaves is **Rattlesnake-plantain**, p 177, which, as an orchid, is taxonomically remote; its flowers, if you look closely, are visibly orchids.

## Wintergreens

*Gaultheria* spp. (galth-ee-ria: after Jean-Francois Gaultier). Flowers white to pinkish, bell-shaped, about ⅛" long, from leaf axils; berries red, up to ¼", delicious; leaves oval, ± pointed; spreading shrubs 1–6" tall. Ericaceae (Heath family).

**Oregon wintergreen,** *G. ovatifolia* (o-vay-tif-oh-lia: oval leaf). Sepals densely hairy; leaves ¾–1¾"; mid-elev forest. Color p 121.

**Alpine wintergreen,** *G. humifusa* (hue-mif-you-sa: trailing). Sepals and berries smooth; leaves ½–¾"; around timberline.

That these wintergreens are dwarfed versions of salal is plain to see; you might guess them to be just stunted salal until you see and taste their fruit. The common name "wintergreen" has been used for both *Gaultheria* and *Pyrola*, but oil of wintergreen indisputably comes from leaves of *Gaultheria procumbens*, similar to these sweet-fruited shrublets. Though that eastern North American plant put "wintergreen" into common parlance, modern candy and gum makers substitute an extract of birch twigs or a synthetic flavor.

# Kinnickinnicks

*Arctostaphylos* spp. (arc-tos-taf-il-os: bear grapes\*). Flowers pinkish to white, jar-shaped, 5-lobed, clustered; berries ¼", dry and mealy, flat-tasting; leaves ½–1¼", widest past midlength; thin gray bark flaking revealing smooth red bark underneath. Rocky, exposed sites. Ericaceae (Heath family).

**Kinnickinnick,** *A. uva-ursi* (oo-va-ur-sigh: grape of bears). Also **bearberry.** Berries bright red; leaf tips rounded; shrubs prostrate/ trailing, rarely over 6" off ground. All elevs. Color p 121.

**Pinemat manzanita,** *A. nevadensis* (nev-a-den-sis: of the Sierra Nevada). Berries dull red; leaves pointed; shrubs dense, cushionlike, 4–10" tall. Above 4,000'. (Compare manzanitas, p 102.)

"Kinnickinnick" was an eastern intertribal trading word meaning "smoking-herbs;" Hudson's Bay Co. traders brought the word west and applied it to this plant the Northwest tribes taught them to smoke. Popular more for the high than the flavor, it was mixed with tobacco when people had the choice. The berries seem to please bears, but among Indians they were starvation fare or adulterants for sweeter berries like salal. These species are valuable pioneers on volcanic or glacial soils, and well-known ground cover ornamentals in the cities.

# Pipsissewas

*Chimaphila* spp. (kim-af-il-a: winter loving). Also **prince's-pine.** Stamens and pink-to-white petals flat-spreading, pistil fat, hublike; flowers ½" diam, nodding; leaves very dark, 1–3", narrowly ellipti- cal, saw-toothed, ± whorled on lower ½ of stem. Widespread in for- ests. Ericaceae (Heath family).

*C. umbellata* (um-bel-ay-ta: bearing flowers in umbels). Flowers 4– 12; leaves widest past midlength; plants 4–10" tall. Color p 122.

*C. menziesii* (men-zee-zee-eye: after Archibald Menzies, p 94). Flow- ers 1–3; leaves widest below or near midlength; plants 2–6".

\*Originally named *Uva-ursi uva-ursi;* the Latin genus name was trans- lated into Greek after an International Rules convention outlawed hyphens in names of genera, but not in species names.

These are among the most habitat-indifferent of our forest plants, growing in virtually every type of Westside forest and in closed forests on the Eastside as well. The chief difference between the two is that *menziesii*, as you might expect of a reduced version, is slightly commoner at higher elevations. (It is also confined to the Northwest, while big brother *umbellata* is "circumboreal," or common to all of the world's northern coniferous forests.) Though widespread, neither pipsissewa is reliably present anywhere in our range, nor ever an herb-layer dominant. Their success under heavy shade suggests strong mycorrhizal partnerships, which often accompany spotty distribution. Their leaves used to sit on apothecary shelves, labeled "Foliachimaphilae," as a remedy for bladder-stones; herbalists still harvest them, sometimes excessively, in the Northwest.

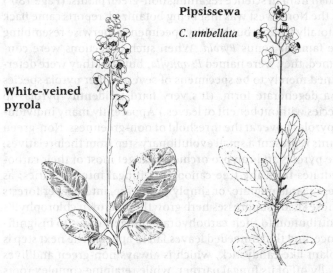

**Pipsissewa**

*C. umbellata*

**White-veined pyrola**

## Pyrolas

*Pyrola* spp. (pie-ro-la: small pear, referring to the leaf shape). Also **wintergreens, shinleafs.** In forest. Ericaceae (Heath family).

**White-veined pyrola,** *P. picta* (pic-ta: painted). Petals spreading, pale, greenish to purplish, style downturned; flowers 5–20 on a reddish, 4–12" stem; leaves all basal, egg-shaped, 1–3", dark green white-mottled along the ± pinnate veins (compare Rattlesnake-plantain, p 177). Color p 121.

**One-sided pyrola,** *P. secunda* (se-cun-da: with flowers all on one side). Flowers greenish white, bell-shaped, with long straight style, 5–15 all facing ± the same way; leaves 1–2½", variably egg-shaped, running up the lower half of the 3–7" stems from rhizomes.

**Heart-leaved pyrola,** *P. asarifolia* (ass-ar-if-oh-lia: wild-ginger leaf). Petals spreading, pink to red, style strongly downturned; flowers 8–25 on a 6–16" stalk; leaves basal, long-stalked, round to heart-shaped, 1–3".

**Single pyrola,** *P. uniflora*\* (you-nif-lor-a: one flower). Flowers single, ½–1", waxy-whitish, petals flat-spreading, pistil fat, straight, 5-tipped (like a chess rook); leaves oval, usually toothed, ½–1¼", from lower ¼ of 2–6" stem; often on rotting wood. Color p 121.

While the pyrolas are mysterious and interesting at first glance, they are all the more so when you know they embody a transitional moment in the evolution of non-green plants—and in human science regarding non-green plants. (Page 180.) As the Northwest was first being botanized, reports came back of totally leafless but healthy specimens otherwise resembling the familiar genus *Pyrola*. When such collections were confirmed, they were named *P. aphylla;* but later they were determined merely to be specimens of several other pyrola species in a degenerate form. (It's very hard to identify pyrolas to species without benefit of leaves.) Apparently many individual pyrolas hover at the threshold of non-greenness. Non-green plants represent a small evolutionary step from their relatives, like pyrolas and calypso orchids, that get most of their carbohydrates from the tree canopy via fungal intermediaries; as these plants migrate, or simply persevere, into deeper forests whose shade suppresses herb growth, their own chlorophyll's contribution to their carbohydrate budgets drops to insignificance, and the unneeded leaves fail to appear. The next step is a plant like candystick, which is always non-green and lives totally off of its fungal partner, while retaining complex roots it utilizes scarcely, if at all.

Six species of pyrola grow in our range. One-sided is the commonest on the Westside, white-veined commonest on the Eastside, heart-leaved the largest and most colorful, and single the smallest, but with the largest flower.

\*Some texts place *uniflora* in a separate genus, *Moneses.*

**Pink heather**

# Mountain-heathers

*Phyllodoce* and *Cassiope* spp. (fil-**od**-os-ee and ca-**sigh**-a-pee: characters in Greek myth). Ericaceae (Heath family). Color p 122.

**Pink heather,** *P. empetriformis* (em-pee-trif-**or**-mis: crowberry shaped). Corolla pink, bell-shaped; flowers 5–15 in apparent terminal clusters, erect in bud, pendent in bloom, then erect again as dry capsules; leaves needlelike, ¼–½"; dense matted shrubs 4–10" or up to 15". Alp/subalpine.

**Yellow heather,** *P. glanduliflora* (gland-you-lif-**lor**-a: glandular flower). As above, exc corolla cream-yellow to off-white, narrow-necked jar-shaped. Alpine.

**White heather,** *C. mertensiana* (mer-ten-see-**ay**-na: after Karl H. Mertens, p 210). Also **moss-heather**. Corolla bell-shaped, white; flowers pendent from axils near branchtips; capsules ± erect; leaves tiny (⅛"), densely packed along the stem in 4 ranks, thus square in cross-section; spreading, mat-forming shrubs 2–12". Alp/subalp.

Mountain-heather is an old friend, always there to welcome you back to the high country. Loosen your bootlaces, sit still, see how much you can take in. Innumerable scattered patches of pink and white heathers are a common denominator of the subalpine zone throughout the Northwest, British Columbia and Southeast Alaska; both reach the alpine zone as well. Pink reaches the lowest (forest fringe) elevations, Yellow the highest and driest (rare below tree line), while White has an in-between range; all three overlap between 6,000' and 7,000' in the North Cascades.

  In subalpine parkland, the mountain heather or low heath community usually grows on stony soils with fairly late snowmelt. This community is prone to tree invasion; trees encroach a few feet at a time, making the snow melt earlier with their "black body effect" (page 28). If fire in the subalpine were somehow prevented, all the heather might eventually be replaced by trees. (God forbid.) Unlike huckleberries, mountain-heathers rarely persist under a tree canopy, though they often encircle tree clumps.

*Cassiope* (white heather) foliage is almost like clubmoss; *Phyllodoce* (pink/yellow heather) foliage is more like common juniper or crowberry; still, heathers share an obvious resemblance in their flowers and habitat. Vast communities known as "heath" in Scotland are dominated by species of *Cassiope*, *Phyllodoce*, *Erica*, and especially *Calluna*—all called "heather" by the Scots. The "Scottish heather" of Northwest gardens is *Calluna vulgaris*.

**Alpine laurel**

## Alpine Laurel

*Kalmia microphylla** (kahl-mia: after Per Kalm; micro-fill-a: small leaf). Corolla pink, bowl-shaped, ½" diam; sepals tiny, green; flowers 3–8 in terminal clusters; capsules 5-celled, with long style; leaves opposite, ½–1" long, narrow, often with rolled-under edges; spreading subshrubs to 6". On saturated (at least in spring) subalpine soils. Ericaceae (Heath family). Color p 121.

These profuse pink blossoms do a lot to brighten up high seasonal bogs and soggy alpine slopes right after snowmelt. Look closely: ten little bumps on the odd-shaped buds hold the ten anthers (stamen tips). When the flower opens, the stamens are spring-loaded to throw their pollen on the first insect to alight.

This plant and its relatives are toxic to grazers. Laurel is a common name for *Kalmia* of all sizes, though they aren't closely related to the Laurel tree (genus *Laurus*) with which the ancient Greeks wreathed their champions. Pioneers across North America often called rhododendrons "mountain laurel"—at Laurel Hill on Mt. Hood's Barlow Trail, for example.

---

*Some texts consider *microphylla* a subspecies of *K. polifolia*.

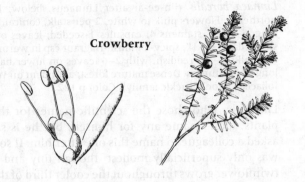

**Crowberry**

## Crowberry

*Empetrum nigrum* (em-pee-trum: on rock; **nye**-grum: black).
Flowers tiny, brownish purple, in leaf axils, ± 3-merous (maximum
of 3 stamens, 3 petals, 3 sepals, 3 bracts), stamens twice as long as
other parts, or sometimes stamens or pistil lacking; berries blue-
black, ⅛–¼", juicy, 6–9-seeded; leaves needlelike, crowded, ¼–½",
with rolled edges; mat-forming prostrate shrubs, erect stems to 6".
Rocky slopes, mostly 5,750–7,400'. Empetraceae (Crowberry family). Color p 122.

Crowberry grows in coastal bogs up north, and sometimes in
Oregon and Washington, but bypasses intermediate elevations in our mountains. The berries stay fairly sweet and juicy
all winter on the plant under the snow blanket, making them a
crucial resource for ptarmigan, grouse, bears, and Alaskan
Inuit. For my taste, they are minute, scarce, and insipid.

## Squawcarpet

*Ceanothus prostratus* (see-an-oath-us: thistle, a misleading name).
Also **mahala-mat**. Flowers whitish to blue, tiny, 5-merous, in round-
topped clusters from leaf axils; capsules of 3 1-seeded cells; leaves
opposite or in opposite-paired whorls, spiny-margined (hollylike),
½–1½"; prostrate, often mat-forming shrubs, freely rooting from
branch nodes. E-side Cas from Mt Adams S, uncommon in our range.
Rhamnaceae (Buckthorn family). Color p 121.

Squawcarpet has very different foliage from our other *Ceanothus* species (page 103). It makes an attractive ground cover in
cultivation.

---

# Twinflower

*Linnaea borealis* (lin-ee-a: after Linnaeus, below; bor-ee-ay-lis: northern). Flowers pink to white, 2 per stalk, conical, pendent, ½" long, 5-lobed, stamens 4; capsules 1-seeded; leaves opposite, very shiny, dark, ¼–1"; spicy- or anise-fragrant esp in warm sun; flowering stalks 3–5", reddish, with 2–6 leaves on lower half only; from long leafy runners. Dense mature forest; abundant on W-side. Caprifoliaceae (Honeysuckle family). Color p 122.

Linnaeus, who chose the scientific names for thousands of plants, didn't name any for himself, but he is said to have asked a colleague to name this one after him. If so, the choice was only superficially modest; though tiny and simple, the twinflower grows throughout the cooler third of the Northern hemisphere, and is universally admired.

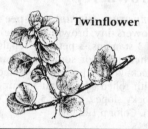

**Twinflower**

**Carolus Linnaeus** (born Carl) is considered the "Father of Systematic Biology" or taxonomy. In his day, naturalists often improvised Latin descriptions several words long for any plant or animal under discussion. Linnaeus successfully argued the proposal that this should be consistently reduced to a two-word ("binomial") name that all scientists would agree upon for each kind ("species") of organism, with the first name ("genus") being a broader category often embracing several species. During his lifetime (1707–78) he published hundreds of species, including most of our genera that occur in Europe. He usually took genus names from Theophrastus, Pliny, or other classical Greek or Roman literature. The name Linnaeus was coined by Carl's father by putting a Latin ending on the Swedish word for linden, a tree growing on the family farm. (Swedish peasants up until his day did not have surnames that passed from generation to generation.) Carl sometimes wrote his name in Swedish as "von Linne," but the Latin version is the name he was born with.

Douglas-fir, p 16.

Engelmann spruce, p 38.

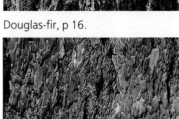

Western larch, p 49.

Western yew, p 40.

Ponderosa pine, p 42.

Lodgepole pine, p 44.

Western white pine, p 47.

Western red-cedar, p 52.

Alaska-cedar, p 54.

Incense-cedar, p 56.

**Conifers**

Mountain-ash, p 77.

Red elderberry, p 73.

Nootka rose, p 85.

Serviceberry, p 78.

Red-osier dogwood, p 72.

Indian-plum, p 77.

Bitter cherry, p 75.　　　　Salmonberry, p 82.

Shrubby cinquefoil, p 86.　　　　Thimbleberry, p 82.

Pacific ninebark, p 80.　　　　Ocean-spray, p 79.

**Trees and Shrubs:** deciduous　　　　*117*

Hardhack, p 85.

Subalpine spiraea, p 84.

Bitterbrush, p 81.

Orange honeysuckle, p 92.

Poison-oak, p 88.

**Shrubs:** deciduous, under 6′

White rhododendron, p 93.

Cascades blueberry, p 96.

Devil's club, p 90.

Wax currant, p 87.

Fool's huckleberry, p 94.   Red-flowering currant, p 86.

**Shrubs:** deciduous, under 6′

Pacific rhododendron, p 100.

Snowbrush, p 103.

Salal, p 101.

Green manzanita, p 102.

Oregon-grape, p 105.

Oregon-boxwood, p 104.

---

**Shrubs:** broadleaf evergreen, over 10"

White-veined pyrola, p 109.

Kinnickinnick, p 108.

Single pyrola, p 110.

Squawcarpet, p 113.

Oregon wintergreen, p 107.

Alpine laurel, p 112.

---

Yellow heather, p 111.

Twinflower, p 114.

White heather, p111.

Pipsissewa, p 108.

Pink heather, p 111.

Crowberry, p 113.

**Shrubs:** broadleaf evergreen, under 10″

Cottongrass, p 151.

Showy sedge, p 149.

Purple hairgrass, p 155.

Woodrush, p 153.

**Herbs:** grasslike

Avalanche lily, p 165.

Beargrass, p 171.

Skunk-cabbage, p 161.

May lily, p 162.

Trillium, p 162.

False-Solomon's-seal (*racemosa*)

(*stellata*), p 164.

Bead lily, p 164.

Fairy-bells, p 163.

Twisted-stalk, p 163.

Twisted-stalk, p 163.

---

**Herbs:** showy monocots (lilies)

Cascades lily, p 167.

Tiger lily, p 166.

Hooker's onion, p 169.

Chocolate lily, p 166.

Camas, p 170.

Death camas, p 171.

Tofieldia, p 168.

**Herbs:** showy monocots (lilies)

Oregon iris, p 173.

Purple-eyed-grass, p 173.

Glacier lily, p 165.

Cat's-ears, p 168.

Corn lily, p 167.

Twayblade, p 176.

Rattlesnake-plantain, p 177.

Round-leaved
bog orchid, p 174.

Lady's slipper, p 177.

Calypso orchid, p 177.

Phantom orchid, p 179.

Western
coralroot, p 178.

Striped coralroot, p 178.

Candystick, p 181.

Pinedrops, p 180.

Indian-pipe, p 180.

Pinesap, p 179.

---

**Herbs:** non-green dicots

Golden fleabane, p 183.

Aster, p 184.

Arnica, p 184.

Goldenrod, p 185.

Balsamroot, p 184.

Woolly-sunflower, p 184

Pearly
everlasting, p 187.

Yarrow, p 186.

Pussy-toes, p 187.

**Herbs:** composites

Coltsfoot, p 189.  Silvercrown, p 190.  Silverback, p 190.

Wormwood, p 188.  Thistle, p 186.  Saw-wort, p 189.

Pale agoseris, p 191.  Alpine microseris, p 191.  Hawkweed, p 191.

---

False-bugbane, p 193.

Snow willow, p 192.

Nettle, p 193.

Vanillaleaf, p 193.

Hedge-nettle, p 200.

Blue-eyed Mary, p 197.

Self-heal, p 200.

Broadleaf lupine, p 202, and Mtn.-daisy, p 183.

Monkshood, p 204.

Bleedingheart, p 203.

Corydalis, p 203.

**Herbs:** dicots, irregular flowers

Creeping penstemon, 197.    Elephant's head, p 196.

Sickletop, Birdbeak and Coiled-beak louseworts, p 196.

Menzies larkspur, p 203.    Common monkeyflower, p 195.

Crazyweed, p 201.          Dwarf lupine, p 202.

Monkeyflower, p 195.    Sweetpea, p 201.    Foxglove, p 198.

Pussypaws, p 202.          Violet, p 199.

---

**Herbs:** dicots, irregular flowers

Indian paintbrush, p 195.

Mountain-sorrel, p 199.

Youth-on-age, p 198.

Silky phacelia, p 209.

Sky-pilot, p 207.

Silverleaf phacelia, p 209.

Douglasia, p 208.

Alpine collomia, p 206.

Spreading phlox, p 205.

**Herbs: 5 petals**

Bluebells, p 210.

Bellflower, p 210.

Waterleaf, p 209.

Shooting star, p 208.

Skyrocket, p 206.

Buttercup, p 219, and Towhead baby, p 236.

Dogbane, p 207.

White catchfly, p 215.

Moss-campion, p 215.

Buckbean, p 210.

Wild strawberry, p 217.

Anemone, p 222.

Strawberry bramble, p 217.

Dotted saxifrage, p 213.   Fringecup, p 212.   Tiarella, p 212.

Alpine saxifrage, p 213.   Partridgefoot, p 216.

Grass-of-Parnassus, p 211.   Field chickweed, p 214.

Roseroot, p 214.    Columbine, p 219.    Purple avens, p 218.

Stonecrop, p 213.    St.-John's-wort, p 216.

Wood-sorrel, p 214.    Yerba de selva, p 216.

**Herbs: 5 petals**    *141*

Goatsbeard, p 219.

American bistort, p 223.

Martindale's desert-parsley, p 226.

Miner's-lettuce, p 221.

Chocolate-tips, p 226.

Valerian, p 227.

Springbeauty, p 220.

**Herbs:** 5 petals

Alpine willow-herb (open seed pods), p 228.

Broad-leaved willow-herb, p 228.

Yellow willow-herb, p 228.

Kittentails, p 231.

Veronica, p 231.

Toothwort, p 229.

Bunchberry, p 231.

Smelowskia, 229.

Meadow-rue, p 230.

Dirty socks, p 231.

Farewell-to-spring, p 228.

Wallflower, p 229.

Starflower, p 233.

Wild-ginger, p 232.

**Herbs:** dicots, 4, 3, or 6 petals

Yellow
pond-lily, p 235.

Oval-leaf buckwheat, p 233, and
Stonecrop, p 213.

Inside-out flower, p 233.

Towhead baby, p 236 (seed heads; flowers
shown on p 138).

Columbia lewisia, p 233.

Tweedy's lewisia, p 233.

**Herbs:** dicots, 6 or several petals

Baneberry, p 234.

Mountain bog gentian, p 236.

Marshmarigold, p 235.

Bitterroot, p 234.

**Herbs:** dicots, several petals

# 4

# Flowering Herbs

Defined simply as seed plants without woody stems, the flowering herbs include most plants thought of as "wildflowers," as well as grasses and similar plants. Some wildflowers, even small ones, are shrubs. Since the distinction between herbs and shrubs can be tricky, most borderline cases or "subshrubs" have been placed in Chapter 3 if they are plainly evergreen, and otherwise in Chapter 4.

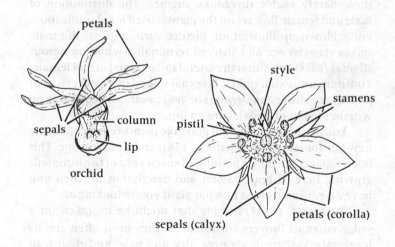

petals

sepals

column

lip

orchid

style

pistil

stamens

petals (corolla)

sepals (calyx)

Grasses, sedges and rushes are the three huge families of grasslike plants. Grasses proper are outnumbered by sedges above timberline here. To identify the family, roll a stem between thumb and forefinger; the rules of thumb:

**sedge** stems are triangular in cross-section, with V-shaped leaves in 3 ranks along the 3 edges ("Sedges have edges");

**grass** stems are round and hollow, with a swollen node at the base of each leaf; and

**rush** stems are round and pith-filled; their leaves are often tubular, and their pistils and seedpods 3-celled.

The reminder that "sedges have edges" is clearly true throughout genus *Carex*, which includes almost all or our sedges, but some sedges, including our cottongrasses (p 151), have somewhat rounded stems, often triangular only near the top.

When a *Carex* sedge blooms, straw-colored stamens adorn the sides of the (usually darker) flowering spikes, in conspicuous disarray. The units within the spike that bear stamens are male flowers, while female flowers each bear two or three barely visible threadlike stigmas. The distribution of male and female flowers on the plant is used for identification. For a blown-up illustration, picture corn, a grass: the male spikes (tassels) are all clustered terminally, while numerous stigmas (silks) show that the lateral spikes (ears) are all female. Those are unisexual spikes; bisexual spikes, in a clever twist of jargon, are either *androgyn*ous or *gynecandro*us, depending on whether males or females are on top.

Positive identification of grasslike plants requires a whole new vocabulary of grass parts (p 158), and a microscope. This book can include only a handful of species out of the hundreds growing here. If both habitat and description fit, then you have an educated guess of what plant you're looking at.

**Beargrass,** p 171, is a lily that might be mistaken for a sedge when its flowers are absent, as they most often are; its leaves are V-shaped, abrasive, dry and pale, and robust, in thick clumps. **Blue-eyed-grass**, p 173, is a slender iris.

# Black Alpine Sedge

*Carex nigricans* (cair-ex: the Roman term; nye-grik-enz: blackish).
Flower-spike single, quite dark, male flowers above females; leaves
4–9 per stalk, ± curling; plants 2–4", turf-forming. In small hollows,
alp/subalpine. Cyperaceae (Sedge family).

Dense, turfy concave beds of black alpine sedge underlie the
latest-melting patches of snow around timberline. They are a
Godsend—once they dry out in late summer, they offer the
perfect spot for basking, tumbling, or sleeping. What's good
for you is in this instance tolerated by the flora, too, as these
sedges are relatively resilent. Not even they are immune to
trampling damage, though; no subalpine community should
be camped on more than two nights in the same year. Camp
on sedge beds only when away from trails, and pick beds un-
marked by previous campers.

Late-lying snowbeds allow only an extremely short
growing season. Black alpine sedge is the speed demon among
grasslike plants, flowering five to nine days after its release
from snow. Even the quickest seed-setting is rarely quick
enough, so it spreads by rhizomes, producing a turfy (rather
than clumpy) growth habit. The only later-lying snowbed
communities—ones that survive many dormant years when
they don't melt clear at all—are largely mosses and lichens;
next to a black sedge bed you often see a strip of haircap moss,
indicating the direction of snowbank melting.

# Showy Sedge

*Carex spectabilis* (spec-tab-il-iss: showy). Flower-spikes ½–1¼",
several, terminal one male or androgynous, lateral ones female with
3 stigmas; plants 8–32". Alp/subalpine meadows. Cyperaceae (Sedge
family). Color p 123.

Our most abundant subalpine tall sedge grows in meadows
with moderate conditions—fairly early snowmelt, and deep
soil.

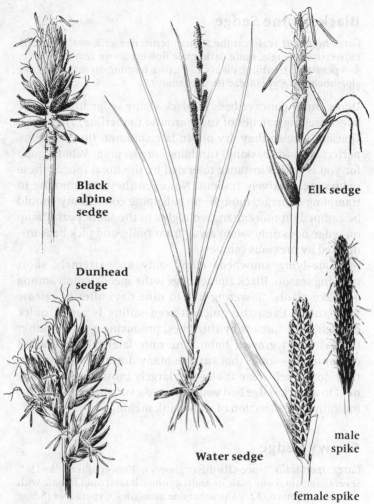

**Black alpine sedge**

**Elk sedge**

**Dunhead sedge**

**Water sedge**

**male spike**

**female spike**

# Dunhead Sedge

*Carex phaeocephala* (fee-o-**sef**-a-la: dun head). Flower-spikes pale, few, small, closely clustered; spikes with female flowers above males, or sometimes all female; plants 2–12". Abundant in alpine tundra and rock fields. Cyperaceae (Sedge family).

"Dun," "dusky" and *phaeo* are usually defined as medium-dark gray-brown in taxonomic usage; these straw-colored flower heads suggest instead a dun horse—pale, with washed-out pigments.

# Elk Sedge

*Carex geyeri* (guy-er-eye: after K. A. Geyer, below). Spike single, androgynous; female flowers exceptionally few (1–3), large (¼″ + 3¼″ stigmas), and set off from the close-packed males; leaves about as tall as stems; plants 6–20″, clumped. Open E-side forests. Cyperaceae (Sedge family.)

Elk sedge stands out among sedges for its form of inflorescence, its dry forest habitat, and its high forage value—equal to the better native grasses. Eastslope deer and elk rely on it briefly around snowmelt time. It grows with needlegrass and snowberry, under pine and Douglas-fir.

# Water Sedge

*Carex aquatilis* (a-qua-til-iss: of water). Spikes up to 2″, terminal one male or androgynous, laterals female, erect; plants 16–40″, rhizomatous. In water or wet soil. Cyperaceae (Sedge family).

The commonest sedge of Cascades marshes, water sedge is important waterfowl forage. Sedges are generally thought of as marsh plants, and a great many of them are; however, they grow in all but the driest moisture regimes. It seems that all their environments are in some way stressful; i.e., competition from grasses is discouraged.

# Cottongrass

*Eriophorum polystachion* (area-for-um: wool bearing; poly-stay-key-on: many spikes). Also **Alaska-cotton**. Spikelets 2–5, becoming white, cottony tufts ¾–1¾″ long (in seed); leaves triangular near tips; stems ± round to triangular, 8–36″. High bogs. Cyperaceae (Sedge family). Color p 123.

---

**Karl Andreas Geyer,** a German botanist, came to the Rocky Mountains with Sir William Drummond Stewart's 1843 expedition for science and pleasure. The party camped on Persian carpets under crimson canopies. Three of its botanists, perhaps succumbing to an unscientific surfeit of pleasure, collected little of note, but Geyer went on alone and made many new collections before sailing from Fort Vancouver.

---

The sight of mile after mile of cottongrass blowing in a breeze, up in Alaska, is hard to forget. Here, it is a little-known plant, though not uncommon in subalpine bogs.

---

**Herbs:** grasslike, round pith-filled stems

---

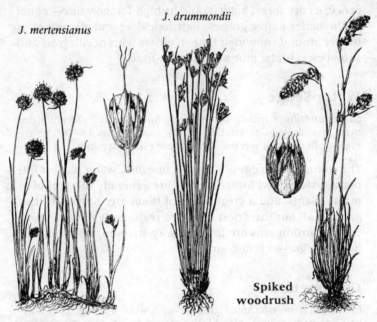

*J. mertensianus*

*J. drummondii*

**Spiked woodrush**

## Rushes

*Juncus* spp. (**junk**-us: the Roman term). Flowers of 6 dry tepals + 2 outer bracts; leaf blades tubular, resembling the stems; stems dark green, tubular, pith-filled, in dense clumps. These 3 species alp/subalpine. Juncaceae (Rush family).

*J. drummondii* (dra-mon-dee-eye: after Thomas Drummond, p 475). Tepals ¼"; flowers green, 1–4, in an apparently lateral cluster; stems 6–14", apparently leafless because the uppermost leaf, borne just below the inflorescence, looks like a continuation of the stem, and the lower leaves are reduced to sheaths.

*J. parryi* (**pair**-ee-eye: after Charles Parry). As above, exc with 1 leaf blade ¾–3" long from the uppermost basal sheath. Color p 123.

*J. mertensianus* (mer-ten-see-ay-nus: after K. H. Mertens, p 210).

Flowers dark brown, tiny (⅛" long), many, in a compact rounded terminal cluster; leaves 1–4, the uppermost one angled off just below the inflorescence; stems 4–12".

Rushes are tough, reedy, deep green, round-stemmed, round-leaved plants with chaffy tufts, often nearly black, for flowers. Ours are small; big "bulrushes" or "tules" are not rushes but sedges. Rushes generally grow in wet places, including bogs and marshes. These three grow in moist subalpine and (*parryi*) alpine soils, and in raw pumice or outwash gravel. Coast tribes made string out of larger coastal rush stems.

## Woodrushes

*Luzula piperi* (**luz**-you-la: light—; **pie**-per-eye: after Chas. Piper), *L. parviflora* (par-vif-**lor**-a: small-flowered), and *L. hitchcockii* (hitch-**cock**-ee-eye: after C. Leo Hitchcock). Flowers tiny, dry, green to brown, with 6 tepals and 2 sepallike bracts, in a very loose, often arching inflorescence; leaves grasslike, wide, flat, finely hair-fringed, from sheathing bases without swollen nodes; plants 6–20". Abundant alp/subalpine; also scattered in montane forest. Juncaceae (Rush family). Color p 123.

Like the glacier lily, the subalpine woodrush may melt its own hole to bloom through a few inches of dwindling snowpack. Its ancient name, *gramen luzulae* or "grass of light," observed the grace of an otherwise inconspicuous plant when bearing dewdrops in the morning light. More shade-tolerant than most grasslike plants, the woodrush can indeed grow in the woods, particularly near timberline, but it's more abundant in open meadows and on moraine gravels.

The Rush family here is divided into two genera, rushes and woodrushes. The tight inflorescences of the three preceding rushes above contrast with the open, delicate, spraylike ones of these three woodrush species, yet each genus includes other species with those characteristics reversed. (For example, the spiked woodrush, *L. spicata*, is an alpine plant 2–16" tall with a single, usually nodding, bristly flower spike ½–1¼" long.) A better across-the-board distinction is that a woodrush has a three-celled seed capsule with one seed in each cell, while a *Juncus* rush has a three-celled, many-seeded capsule. Grasses and sedges have one-celled, one-seeded fruits.

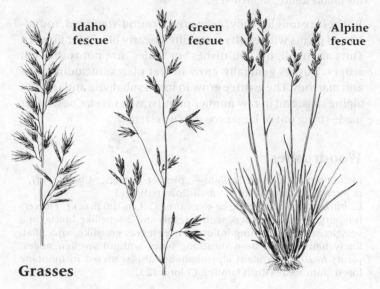

**Idaho fescue**　　**Green fescue**　　**Alpine fescue**

# Grasses

Family Poaceae.*

Grasses Primarily of the High Country

**Alpine fescue,** *Festuca ovina brevifolia* (fest-**you**-ca: the Roman term; oh-**vie**-na: of sheep; brev-if-**oh**-lia: short leaf). Spikelets of 3–4 florets forming a 1–3″ spikelike panicle; awns less than ⅛″; plants 2–8″. This variety alpine; the species ("sheep fescue") is very wide-ranging, and is planted in lawns.

**Green fescue,** *F. viridula* (vee-**rid**-you-la: green). Spikelets of 3–7 florets forming a 4–7″ spikelike panicle; awns lacking; leaves narrow, wiry, dark green; plants 16–32″. Subalpine and E-side in Cas.

**Idaho fescue,** *F. idahoensis* (Idaho-**en**-sis: of Idaho). Plants like green fescue (above), but with short (⅛″) awns. Subalpine meadows; abundant in Olys.

**Downy oatgrass,** *Trisetum spicatum* (try-**see**-tum: 3-awned; spic-**ay**-tum: spiked). Spikelets of 1–3 florets forming a 1–3″ spike; awns ¼″, bent outward; plants 4–16″, ± fuzzy all over. Common on alpine ridges, sporadic elsewhere; globally widespread.

*The grass family was previously known for centuries as the Gramineae—simply the Latin word meaning "grasses." See footnote, p 182.

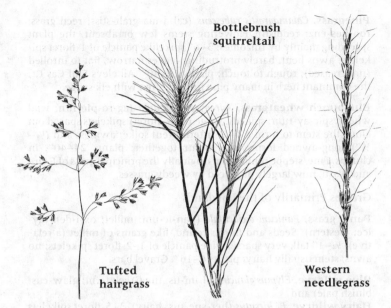

Bottlebrush
squirreltail

Tufted
hairgrass

Western
needlegrass

**Purple hairgrass,** *Deschampsia atropurpurea*\* (desh-**amp**-sia: after J. L. A. Loiseleur-Deslongchamps; at-ro-pur-**pew**-ria: black purple). Purplish spikelets of 2–3 florets in a wide, sparse panicle 2–4" tall; awns less than ⅛", bent inward, hidden within floret; leaves flat; plants 6–24". Abundant subalpine. Color p 122.

## Grasses Primarily of the East-Side

**Tufted hairgrass,** *D. cespitosa* (see-spit-oh-sa: growing in bunches). Spikelets of 2–3 florets in a wide, sparse panicle 4–10" tall; awns barely protruding from floret; leaves creased; plants 8–48". High open forests and meadows E of Cas Cr.

**Bottlebrush squirreltail,** *Sitanion hystrix* (sit-ay-nee-on: grain—; hiss-trix: porcupine). Spikelets of 2–6 florets in a 1½–6" spike which looks strikingly brushlike thanks to the many, ¾–4" long awns; plants 4–20". Pine forest with bitterbrush; also on alpine pumice.

**Western needlegrass,** *Stipa occidentalis* (sty-pa: tow of flax; ox-i-den-**tay**-lis: western). Also **needle-and-thread.** 1-floret spikelets held tightly erect against the main stem, in a 2–12" spike; awns ¾–1½" long, twice-bent, spreading at various angles; leaves often in-rolled; plants 8–40". All elevs.

\*Some texts split *Deschampsia,* resulting in *D. cespitosa* but *Vahlodea atropurpurea;* older texts placed both in genus *Aira.*

**Pinegrass,** *Calamagrostis rubescens* (cal-a-ma-**grah**-stiss: reed grass; roo-**bes**-enz: reddish). Flowering stems few or absent, the plant spreading mainly by rhizomes; 3–6" spikelike panicle of 1-floret spikelets; awns bent, barely protruding; leaves narrow, flat to inrolled (not creased), rough to touch; plants 16–40". All elevs E of Cas Cr; the dominant herb in many pine forests, often with elk sedge.

**Bluebunch wheatgrass,** *Agropyron spicatum* (ag-ro-**pie**-run: wild wheat; spic-**ay**-tum: spiked). Large (6–8- floret) spikelets spaced out along the stem to form a 3–6" intermittent spike; awnless and (¼–¾") long-awned forms may occur together; plants 24–40", in clumps. Pine/steppe timberline; originally the principal grass of E Ore and Wash, now largely displaced by weedy grasses.

## Grasses Primarily of the Lowlands

**Panic grass,** *Panicum occidentale* (**pan**-ic-um: millet; ox-i-den-**tay**-lee: western). Seeds and florets round, like grains of millet (a relative); ½–1" tall, very sparse, wide panicle of 1–2-floret spikelets; no awns; stem usually hairy; plants 6–16". Gravel bars.

**Blue wildrye,** *Elymus glaucus* (el-**im**-us: the Greek term; **glaw**-cus: bluish pale) and
**Hairy wildrye,** *E. hirsutus* (her-**sue**-tus: hairy). 3–5-floret spikelets in a 2–8", often nodding spike; awns vary from 1" long to absent; leaves flat, ⅛" wide; plants 16–48". W-side clearings, all elevs.

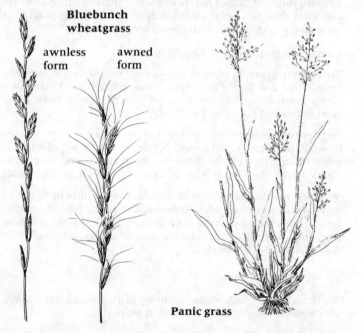

**Bluebunch wheatgrass**

awnless form    awned form

**Panic grass**

**Redtop,** *Agrostis alba alba* (ag-ros-tiss: Greek term for grass or fields; al-ba: white). Also **creeping bentgrass.** Purplish, awnless, 1-floret spikelets in a delicate, loose 2–10″ panicle; plants 8–40″. Lowlands, e.g., W-side Oly valleys; also in lawns and pastures.

Grasses are classically personified as humble, but in fact they are the hot new item on the paleobotanical scene—the most successful plant family, as measured by their rate and breadth of genetic diversification in recent geologic time. The ascendency of grasses has been magnified by the development of agriculture, which has always focused on grains (grass seeds) and on grazing (grass-eating) mammals. Of course, wild animals also coevolved with grasses, and make plenty of use of them, but it's we humans and our livestock that have really gone to town with them, literally and figuratively.

Few grasses like shade, so the Cascades and Olympics are not the best region for grasses. Our meadow communities are maintained typically by extreme snowiness, wetness, or infertility—and the first two conditions favor sedges and rushes more than grasses. New clearings here are mostly made by fire, and quickly reoccupied by stump-sprouting shrubs and plants with windborne or fireproof seeds. (Grasslands also excel at regenerating after fires, but the grasses have to be there

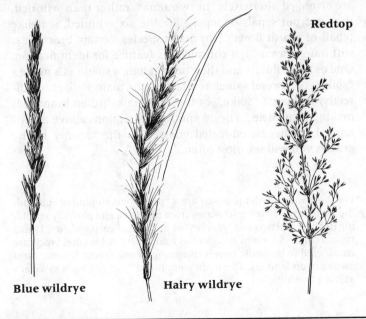

**Redtop**

**Blue wildrye**

**Hairy wildrye**

before the fire to spring up after it.) This is not to say there are no grasses here, just fewer than elsewhere. Our chief grass habitats are:

**open forests** and steppe margins on the lower East-slope;

**dry meadow** types at timberline and above; and

**gravel bars** and river terraces at early successional stages.

Periodic soil drought seems to be a factor in each.

Though some grasses are annuals, the 13 species described are perennial. Several are major forage plants. Tufted hairgrass is so sought out by sheep and cattle that its presence indicates a meadow in virgin condition; once grazed out from a meadow, it returns slowly, if ever. All of the 13 are native except perhaps creeping bent, which is so thoroughly naturalized that no one is quite sure whether it was always here.

Identification of grasses requires a special vocabulary. Keeping it to a minimum, grasses are flowering plants with simplified, undecorated, one-seeded, dry flowers ("florets"), each with three stamens and a usually two-styled pistil. (These sexual parts are short-lived and easily overlooked.) The florets are flanked by any number of scales or bracts. Since the bracts are arranged alternately, in two ranks, rather than whorled, they are not sepals or tepals like the six whorled, scalelike tepals of a rush flower.* In many species, certain bracts bear stiff hairs ("awns"), a conspicuous feature for identification. One or more florets and their bracts along a single axis make a "spikelet." Several spikelets attach to the main stalk either directly, making a "spike," or by small stalks (often branched) making a "panicle." The 13 species descriptions above are intended to suggest educated guesses on the identity of the grasses you will see most often.

---

*For those interested, the bracts are a "palea" enveloping or subtending each floret, a "lemma" across from the palea and partially enfolding its base, and a pair of "glumes" at the base of each spikelet. Thus a typical single-flowered spikelet has four bracts. Additional bracts are considered to be sterile florets consisting of one empty lemma. Most awns are on lemmas. The main stem is the "rachis"; each spikelet's axis is a "rachilla."

# Cattails

*Typha latifolia* (tie-fa: the Greek term; lat-if-**oh**-lia: broad leaf). Flowers minute, chaffy, in a dense, round, smooth spike of two distinct portions, the upper (male) thicker when in flower but withering as the lower (female) thickens and turns dark brown in fruit; stalks 3–10′; leaves half as tall by ¼–¾″ wide, smooth; from rhizomes in shallow water. Typhaceae (Cattail family).

Big lowland cattail marshes are avidly sought out by migrating waterfowl and by hunters thereof. Few such marshes lie among our mountains, though. The Quinault used to travel all the way to Gray's Harbor to pick cattails. Like all Northwest tribes, they wove the stalks (never the leaves) into thick, spongy mats for mattresses, kneeling pads in canoes, packsacks, baskets, rain capes, and temporary roofs in summer. Oddly, only a few of the tribes ate cattails, though the rhizomes and inner, basal stalk portions are pretty good baked, raw, or ground as flour.

---

**Herbs:** showy monocots

---

Monocots and dicots are named after and defined by their respectively single or paired seed leaves, or "cotyledons"—the first green part(s) to sprout from a newly germinated seed. Since those are an ephemeral feature, monocots are usually recognized by two less reliable traits:

**parallel-veined leaves**; and

**3-merous flowers** (generally 3 petals and 3 sepals, though these are sometimes nearly identical, in which case we call them 6 "tepals" in an inner and an outer whorl.)

Very few dicots have reliably 3-merous flowers. Only three such genera are in this book, and they are easily segregated since they don't have parallel leaf veins: **Inside-out Flower,** p 233; **Wild-ginger,** p 232; and **Wild-buckwheats,** p 232.

    The monocots on the following pages are distinguished from the grasslike monocots by (most of them) broader leaves and (all of them) moist, generally delicate flower parts evolved

---

for visual attractiveness—in a word, showy. Most fall into the Lily, Iris, and Orchid families, by having 6, 3, or 2 stamens, respectively. Three plants that don't fit those stamen numbers are described first.

**Wapato**

## Wapato

*Sagittaria* spp. (sadge-it-**air**-ia: arrow—). Also **arrowhead**. 3 petals white, spreading, nearly round, ⅜–¾"; 3 sepals green, pointed; many stamens; pistils on separate, ball-shaped female flowers, less showy and usually borne lower on the stem; flowers (both sexes) in whorls of 3; leaf blades narrowly 3-pointed or arrow-shaped, on long leafstalks all from the base, which is usually submerged in water. Alismataceae (Water-plantain family).

*S. latifolia* (lat-if-**oh**-lia: wide leaf). Leaf blades 4–12" long, almost as broad. W of Cas.

*S. cuneata* (cue-nee-**ay**-ta: wedge-shaped). Leaf blades 1–5" long × 1–3" broad. E of Cas.

In fall the slender rhizomes produce potatolike tubers about the size and shape of hens' eggs, with a flavor a little like roasted chestnuts. Traditionally retrieved from the mud with the toes, by humans, or with the bill, by ducks, wapato roots were widely traded among the tribes of the lower Columbia, and gratefully accepted by Lewis, Clark, & Co.

# Skunk-cabbage

*Lysichitum americanum* (lye-zic-eye-tum: loose tunic). Flowers greenish yellow, 4-lobed and 4-stamened, ⅛" diam, many, in a dense spike 2½–5", partly enclosed or hooded by a yellow, parallel-veined "spathe"; leaves all basal, ± net-veined, oval, eventually up to 3' × 1' or even bigger; from a fleshy, enlarged vertical root, in wet ground. Araceae (Calla-lily family). Color p 124.

Many plants evolved sweet aromas that attract sugar-loving pollinators, like bees. Others, such as various Eastern and European species of *Symplocarpus* (also called Skunk-cabbage) evolved putrid smells that attract pollinators, like some flies and beetles, that feed on, shall we say, decaying organic matter. Our skunk-cabbage lies in between, attracting bees as well as beetles, distinctly skunky to the human nose but not foul, merely rank, like long-faded skunk aroma. It releases different odors at different temperatures, each odor matched to a kind of pollinator likely to be out and about at that temperature.

The Calla-lily family has a characteristic inflorescence called a spadix (the fleshy spike of crowded flowers) and spathe (the large bract enfolding it). These voluptuous organs have long evoked sexual imagery in human imaginations; Whitman named a book of suggestive poetry after an Eastern species, Calamus.

In the case of our skunk-cabbage, spathe and spadix thrust up from wet forest or marsh ground in early spring. Leaves come later, and keep growing all summer to reach sizes unmatched north of the banana groves; they were universally used as "Indian wax paper" to wrap camas bulbs, salal berries, and other foods for steam pit baking or for storage. The leaf bases and the roots are themselves edible after prolonged cooking or storage breaks down the intensely irritating, "hot" oxalate crystals. Some tribes ate them during the perennial hard times of late winter. Origin myths told that in the bad old days before Salmon came, the people had nothing to eat but skunk-cabbage roots. Elk and bears eat them, with no complaints recorded. Food or not, skunk-cabbage was regarded as strong medicine—for example, to induce labor, either timely or abortive. Later, some white man patented and sold it under the name "Skookum."

---

# May Lily

*Maianthemum dilatatum* (my-anth-em-um: May flower; dil-a-tay-tum: widened). Also **false-lily-of-the-valley.** Flowers white, ¼" diam, with 4 stamens and 4 spreading tepals in a single whorl; berries ¼", ripening red, in a 1–3" raceme; 2 (occasionally 1 or 3) leaves, 2–4", heart-shaped but distinctly parallel-veined, ± shiny dark green; plants 4–14", slender, rhizomatous. W-side forest up to 3,500', often on nurse logs. Liliaceae (Lily family). Color p 125.

The insipid berries were eaten unenthusiastically by Northwest tribes, but in the woods you rarely see a full stalk of them even half-ripe, which shows that small mammals cherish them.

---

**Herbs:** showy monocots, 6 stamens (lilies)

---

# Trillium

*Trillium ovatum* (tril-ium: triple; oh-**vay**-tum: oval). Also **wake-robin.** Petals 1–3", white, aging through pink to maroon; sepals shorter, narrower, green; flowers single, on a 1–3" stem from the whorl of 3 leaves, 3–7", often equally broad, net-veined exc for the 5 or so ± parallel main veins; plants 6–16", rhizomatous. (Rarely with 4 or 5 leaves and/or petals and sepals, instead of 3 each.) Low forest. Liliaceae. Color p 125.

What a pleasure, seeing the year's first trilliums in March or April, just when the winter rains feel like Forever! Quinault elders used to warn their youngsters that picking trillium would bring rain—a safe bet in Quinault country at that time of year. From similar sentiments, whites spun the less reliable prediction that picking the flower will kill the plant. It merely discourages it; still, waiting to pick the seeds is a better choice. Then, with a little effort, you can have trilliums blooming year after year in a shady spot near your house. Trillium seeds require two winter chillings before producing a shoot; either use your refrigerator for one of them, before planting in fall, or be patient. The seeds come packed in a gummy oil which ants, the usual disseminators, find tasty.

# Fairy-bells

*Disporum* spp. (dis-por-um: two seed). Also **fairy-lanterns.** Flowers bell-shaped, pendent in pairs (or occasionally 1–4) from branchtips, white; berries red, egg-shaped, ¼–½", ± edible—juicy, sweetish but insipid; leaves 2–5", long-tapered oval, wavy-edged; stems 14–40", much branched. Liliaceae.

*D. hookeri* (**hook-er-eye**: after Sir Joseph Hooker, p 169). Tepals ¼–½", flaring; stamens exposed. Widespread in forests. Color p 125.

*D. smithii* (**smith-ee-eye**: after Sir James E. Smith). Tepals ½–1", nearly straight, hiding the stamens. W-side.

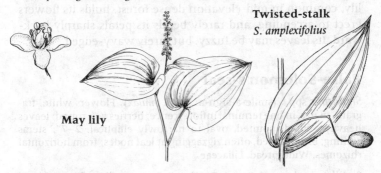

**Twisted-stalk**
*S. amplexifolius*

**May lily**

# Twisted-stalk

*Streptopus* spp. (strep-ta-pus: twisted foot). Flowers bell-shaped, 1 (sometimes 2) beneath each leaf axil; berries red, ¼–½", juicy, sweetish but insipid; leaves 2–5", tapered, elliptical. Liliaceae.

*S. roseus* (ro-zee-us: pink). Tepals ⅜", variably streaked rose with white, slightly reflexed at the tip; style 3-branched; flower-bearing stalklets straight to curved; stems 6–14", rarely branched, arching. Higher (3,000+') forests. Color p 125.

*S. amplexifolius* (am-plex-if-oh-lius: clasping leaf). Tepals ½", dull white, reflexed from near midlength; style unbranched; flower stalklets with a sharp kink; stems 12–40", much branched.

Twisted-stalk discretely hides its flowers under its leaves—quite a trick, since the the crotch *above* each leaf bears the flower. Look closely to see the flower stalk where, fused to the stem, it runs up it to the next leaf base. (The first leaf thus never has a flower under it.) The berries were called "snake-berries" by the Quileute, "frogberries" by the Kwakiutl, "owl-" or "witchberries" by the Haida—and inedible by all—but Makah women chewed the roots to help induce labor.

---

# Bead Lily

*Clintonia uniflora* (clin-toe-nia: after DeWitt Clinton; you-nif-lor-a: 1 flower). Also **Queen's-cup.** Flowers single, white, ± face-up, broadly bell-shaped to nearly flat-spreading; tepals ¾–1"; berry intense blue, ⅜", many-seeded, inedible; 2 or 3 leaves 3–6" × 1–2", heavy, smooth, ± shiny, basal, sheathing the 2–4" stalk; from slender rhizomes. Deep forest, 2,500–5,000'. Liliaceae. Color p 125.

The beady blue berry is more striking than the formal white blossom. Though they don't look much alike once you know them, you might confuse bead lily with avalanche lily. Bead lily, common in mid-elevation dense forest, holds its flowers erect to ascending, and rarely bends its petals sharply backward. Its leaves may be fuzzy, but rarely wavy-edged.

# False-Solomon's-seal

*Smilacina* spp.* (smile-a-sign-a: small *Smilax*). Flowers white, fragrant, many, in one terminal inflorescence; berries ¼", round; leaves heavily veined, pointed, oval to narrowly elliptical, 2–7"; stems arching, unbranched, often zigzagging at leaf nodes; from horizontal rhizomes. Widespread. Liliaceae.

*S. stellata* (stel-ay-ta: starry). Tepals ¼", flat-spreading; flowers 6–18; berries longitudinally striped at first, ripening dark red to blackish; stems 8–24". Color p 125.

*S. racemosa* (ras-em-oh-sa: bearing flowers in racemes, a misleading name for this plant). Tepals minute, stamens longer (⅛"); flowers in a ± conical fluffy panicle; berries speckled at first, ripening red; stems 1–3'. Color p 125.

A perplexing genus in many ways—for starters, take those sibyllant names. Neither Solomon's seal nor genus *Smilax* bear enough resemblance to this genus to justify insinuations of mimicry. And *stellata* is racemose, but *racemosa* is not—it's paniculate. Of more immediate concern, I have found false-Solomon's-seal berries in some places distinctive and delicious (and not purgative, as some authors report) but more often they are insipid. Both species are locally abundant in deep Westside woods, but also look quite happy in clearings, open forest, or even wide-open dry slopes. Form varies with envi-

---

*Some recent investigators lump *Smilacina* into genus *Maianthemum.*

ronment. In the silver fir zone, *stellata* spreads its leaves flat from a nearly horizontal stem; on the Eastside, where light is plentiful but moisture must be conserved, it holds its stem upright, grows narrower leaves, angles them upward, and folds them sharply at the midvein.

## Glacier and Avalanche Lilies

*Erythronium* spp. (air-ith-roe-nium: red—, the flower color of some species). Tepals 1–1½", spreading to reflexed either in an arc or from the base; flowers ± nodding, single or sometimes 2–6 on a 6–12" stalk; capsule erect, 1" tall, 3-celled and -sided; 2 leaves 4–8", basal, wavy-edged; from a scallionlike bulb. Abundant around timberline, scattered at lower elevs. Liliaceae.

**Glacier lily,** *E. grandiflorum* (gran-dif-lor-um: large-flowered). Flowers yellow. Color p 127.

**Avalanche lily,** *E. montanum* (mon-tay-num: of mtns). Flowers white with a yellow center, sometimes drying pinkish. Color p 124.

The snowy names refer to two striking traits of these lilies— they can generate enough heat to melt their way up and bloom through the last few inches of snow; and they prolifer- ate into overwhelming drifts of white, or yellow. Fluttering with illusory fragility in subalpine breezes, they seem ideal vehicles for those anthropomorphic virtues we love to foist on mountain wildflowers—innocence, bravery, simplicity, per- severance, patient suffering, etc. "They toil not, neither do they spin," and they don't taste half bad either. (Bulbs, leaves, and flowers are edible, in that order of preference, but not worth it except when far from trails.)

For some reason avalanche lilies are far more common in the Olympics than the Cascades, and vice versa for glacier lil- ies. In either case, lush subalpine meadows are the preferred habitat, though glacier lilies find their way down to near sea level. Below 2,500' occur two uncommon congeners called trout or fawn lilies. Pink fawn lily, *E. revolutum*, grows in Olympic and coastal lowlands. White fawn lily, *E. oregonum*, mainly of the Puget/Willamette lowlands, has white tepals that sometimes dry pinkish. Both have brown-mottled leaves, a clear distinction from glacier and avalanche lilies.

Common names in this genus are many. One English

name for the entire genus is "dogtooth-violets"; several European species are in fact violet, but resemble violets in no other way. "Adder's-tongue," another old name, is used in this genus and in unrelated genera. Immigrants to America, finding new species with no violet hues, but with mottled leaves, came up with "fawn lily" and "trout lily." But those sweet East Coast names didn't stick when settlers in the West met our unmottled montane species. Some texts call both colors "glacier lilies," while some others call them both "avalanche lilies." This book will shore up the plurality that calls our white ones "avalanche" and our yellow ones "glacier." But *E. grandiflorum*, the "glacier lily," also has a white-flowered variety in Idaho and Southeast Washington.

## Chocolate Lily

*Fritillaria lanceolata* (frit-il-**air**-ia: checkered; lan-see-oh-**lay**-ta: narrow-leafed). Also **fritillary, mission bells, rice-root lily.** Tepals ¾–1½", inward-curving, brownish purple, mottled with yellowish green; flowers pendent, 1–2+; capsule ¾", 6-winged; leaves 2–5", narrow, both whorled and single; stem 8–30"; from a bulb of a few large garliclike cloves with many tiny ricelike bulblets. Moist clearings up to 5,000'. Liliaceae. Color p 126.

These elegant-shaped flowers sell themselves short, in camouflage coloring and a fetid smell. The bulbs, universally gathered and eaten in the old days, are too rare and too bitter to justify digging up now. The bulblets suggest the common name "rice-root lily." Naturally, when the Haida, after eating lily bulblets for generations, were introduced to rice, they named it "fritillary-teeth."

## Tiger Lily

*Lilium columbianum* (lil-ium: the Roman term; co-lum-be-**ay**-num: near the Columbia River). Tepals orange with small maroon spots, 1½–2½" long, strongly recurved, making a ± full circle; capsule fleshy, 1½–2"; flowers/fruit nodding, several, on long stalks; leaves 2–4", narrow, the ones near midstalk longest and in the largest whorls; stem 2–4'; from a large, many-cloved bulb. Clearings and thickets up to timberline. Liliaceae. Color p 126.

The rather bitter bulbs were eaten by most tribes.

# Cascades Lily

*Lilium washingtonianum* (washing-tony-ay-num: after Martha Washington). Tepals white, often purple-tinged or -spotted, aging pink, 2½–3½"; flowers fragrant, bell-shaped, several, slightly nodding on short stalks all near the top of the 2–5' stem; capsules fleshy, 1"; leaves 2–4", narrow elliptical, often wavy, many of the upper ones in distinct whorls; bulbs large, many-cloved. Clearings and thickets, Ore High Cas. Liliaceae. Color p 126.

Our grandest lily grows only south of the Columbia, *washingtonianum* notwithstanding.

# Corn Lily

*Veratrum* spp. (ver-ay-trum: true black). Also **false-hellebore.** Flowers saucer-shaped, numerous; styles 3, persistent on the 1"-long 3-celled capsules, but usually lacking from the (staminate) lower flowers; leaves mostly 5–12", coarsely grooved along the veins, oval, pointed; stem 3–7', from a thick black rhizome. Wet meadows, mainly subalpine. Liliaceae.

*V. viride* (**veer**-id-ee: green). Flowers pale green, ½–¾" diam, in a loose panicle with drooping branches. Alaska to N Ore. Color p 127.

*V. californicum.* Flowers dull white, or only slightly greenish, ¾–1½", in a dense panicle with ascending branches. Wash Cas to S Calif.

Heavy beds of snow lying on steep meadows tend to creep downslope through the winter, scouring off all vegetation at the surface. Woody seedlings are frustrated year after year, while perennial herbs with fat storage roots and fast spring growth are favored; thus is the meadow community perpetuated. On many wet slopes, the fastest plant to thrust up from the mud as the snow recedes is the corn lily, whose clusters of blunt, wrapped shoots look so strange and rank as to suggest an imagined land like Venus. These were the only herbaceous shoots robust enough to push up through several inches of new ash on Mt. St. Helens; most herbs in the same meadows died. The big, lush leaves, startlingly clean and perfect when they unclasp from the stalk, often look ragged by the time the lily flowers. The young shoots and the roots are toxic, and used to be collected and ground up for a crop insecticide; Indians used them medicinally, in subliminal doses. The toxic alkaloids apparently degenerate or drain from the leaves by maturity, when they are eaten by elk and by insects.

---

# Cat's-ears

*Calochortus* spp. (cal-o-**cor**-tus: beautiful grass). Petals cream white, ¾–1" long, almost as broad, densely hairy ± all over their inner faces, usually with a fine purple arc at base of hairy part; sepals slightly shorter and much narrower than petals, pointed; flowers 1–5+; capsules nodding, 3-winged, with 3-branched tip; the only large leaf basal, flat, usually taller than the 3–10" stalk; from a small bulb. Liliaceae.

*C. subalpinus*. Sepals creamy yellow, with a purple dot near base; flowers usually all stemming from one point high on the stalk. Subalpine in Ore and S Wash Cas. Color p 127.

*C. tolmiei* (**tole**-me-eye: after William Tolmie, p 198). Sepals unspotted, cream to purplish; flowers often branching from several points along the stem. Uncommon; rocky slopes near Willamette Valley.

"Cat's-ears" is the most pictorial of several flattering names applied to these sensuous blossoms, perhaps all the more admired for being almost prohibitively hard to cultivate. Most members of the genus require bone-dry soil before the bulb can go into healthy winter retirement. Several are desert plants, with lavender to white flowers, and many reach just the east edge of our range. Called mariposa lilies (Spanish for "butterfly"), the desert species bear little resemblance to feline anatomy; their petals have few hairs, but these can be very long, as in *C. longebarbatus*, which ranges from Yakima County south.

# Tofieldia

*Tofieldia glutinosa* (toe-**field**-ia: after Thomas Tofield; gluten-oh-sa: sticky). Also **false-asphodel**. Tepals white, ⅛–¼", persistent while the 3-styled pistil grows out past them into a fat, reddish, 3-celled capsule; flowers/fruit in a dense cluster atop an 8–20", sticky, hairy stem; 1–3 grasslike basal leaves 2–6", and sometimes 1–2 smaller stem leaves. Boggy meadows, esp subalpine. Liliaceae. Color p 126.

# Wild Onions

*Allium* spp. (al-ium: the Roman term). Flowers ¼–⅜" long, several, on stalklets all from one spot between pointed, onionskinlike bracts (segments of the spathe or "onion-top" that encased the inflorescence in bud); stems often bunching; from small onions with the usual aroma. Liliaceae.

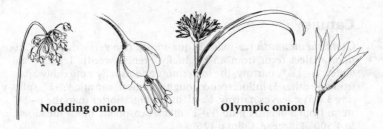

**Nodding onion**　　　**Olympic onion**

**Hooker's onion,** *A. acuminatum* (a-cue-min-ay-tum: pointed). Tepals pointed, purple or pink to (occasionally) white, the outer whorl ± spreading, bell-shaped, inner whorl smaller, narrowly jar-shaped; leaves 2–5, withering before flowers open, much shorter than the 4–12" tubular stem. Rocky openings, lower E Cas slopes. Color p 126.

**Nodding onion,** *A. cernuum* (sir-new-um: nodding). Inflorescence nodding (but often erect in fruit); tepals pink or white, oval, much shorter than stamens, all ± alike but inner and outer whorls separate; leaves several; stem 8–20". Dryish openings, W of Cas.

**Olympic onion,** *A. crenulatum* (cren-you-lay-tum: scalloped). Tepals pinkish, narrow, pointed, all ± alike; leaves 2, longer than the stem but not higher, being downcurved; stem 2-angled, 2–3". Alpine gravels, Wash Cas, Olys, and disjunct at Jefferson Park, Ore.

An onion is easy to recognize—indeed, often hard to miss—because it smells like an onion. If it doesn't, don't try a taste test; it might be death camas.

Sir William Jackson Hooker was a great British scientist of the nineteenth Century. He developed Kew Gardens into a major institution filled with plants sent from all over the world during the great Era of Plant Hunters. Earlier, while professor of botany in Glasgow, he had noticed the astonishing zeal of a teenaged gardener there, and taken this David Douglas as his protege on field trips; eventually he sent him out to explore North America. Hooker catalogued and named hundreds of new plants sent in by Menzies, Douglas, Drummond, Gairdner, Tolmie, and others. The plants included Hooker's onion and Smith's—not Hooker's—fairy-bells. (Rules forbid a namer honoring himself in a scientific name, though his name may turn up in the common name of the species. Hooker's fairy-bells were named after his son **Sir Joseph Hooker,** a noted plant hunter and colleague of Darwin.)

# Camas

*Camassia quamash* (ca-mass-ia qua-mosh: two versions of the Chinook dialect term from an originally French word). Tepals blue-violet, ¾–1½", narrow, the lowermost one usually noticeably apart from the other 5; inflorescence roughly conical; capsule ½–1", splitting 3 ways; leaves narrow, basal/sheathing, shorter than the 8–24" stem; from a deepset bulb, ½–1" diam. Seasonally moist meadows, to 4,500'. Liliaceae. Color p 126.

Camas bulbs were the prized vegetable food of most tribes in Oregon and Washington. The Nez Perce War was touched off by white settlers plowing up camas prairies for pastures. Camas cakes were second only to dried salmon in trade volume, especially north along the coast where no camas grows. In many tribes, a family would mark out, "own," and maintain a camas patch year-round for generations, weeding and burning it. This might suggest the beginnings of an agricultural economy, but it has a unique explanation. Year-round tending of a patch made it possible to weed out death camas when in flower and easy to recognize, so that camas digging would be safe when camas bulbs are best—in spring before flowering. Nevertheless, many people died from eating death camas. Camas-eating is **not recommended** to hikers.

David Douglas described the way he saw camas bulbs cooked: "A hole is scraped in the ground, in which are placed a number of flat stones on which the fire is placed and kept burning until sufficiently warm, when it is taken away. The cakes, which are formed by cutting or bruising the bricks and then compressing into small bricks, are placed on the stones and covered with leaves, moss, or dry grass, with a layer of earth on the outside, and left until baked or roasted, which generally takes a night. They are moist when newly taken off the stones, and are hung up to dry. Then they are placed on shelves or boxes for winter use. When warm they taste much like a baked pear. It is not improbable that a very palatable beverage might be made from them. Lewis observes that when eaten in a large quantity they occasion bowel complaints ... Assuredly they produce flatulence: when in the Indian hut I was almost blown out by strength of wind." The flatulence is attributable to inulin, an indigestible sugar in camas bulbs, Jerusalem artichokes, and a few other vegetables.

# Death Camas

*Zigadenus* spp. (zye-ga-dee-nus: paired glands). Also **zygadene**. Flowers white, ± saucer-shaped, in a tall raceme, withered tepals persistent; capsules ½–¾", splitting, 3-celled, 3-styled; most leaves basal, narrow, sheathing, but often 2 or more along the 8–30" stem; bulb 1". Liliaceae.

*Z. venenosus* (ven-en-oh-sus: poisonous). Tepals less than ¼", inner ones slightly longer than outer, all with an obscure oval dot near base; stamens longer than tepals. Grassy, ± open spots. Color p 126.

*Z. elegans* (el-eg-enz: elegant). Tepals all alike, ⅜", with a heart-shaped greenish spot near the base; leaves, stem often with whitish coating. Mainly alp/subalpine.

One of these two species has earned the name "death camas." Back during camas-digging days, *Z. venenosus* undoubtedly killed more people in the Northwest than any other plant ever will. Today it maintains its reputation with an occasional sheep death. No humans are known to have been killed by *Z. elegans,* which holds less toxin and sticks to high elevations where edible camas is rarely found. No one would mistake the small white flowers of death camas for the big blue ones of camas, but mistakes occur when populations of the two are intermixed and both have gone to seed or withered—and camas bulbs are ripest for eating.

**Beargrass**    **Death camas**

*Z. venenosus*

*Z. elegans*

# Beargrass

*Xerophyllum tenax* (zero-fill-um: dry leaf; ten-ax: holding fast). Stamens longer than tepals; flowers white, fragrant, saucer-shaped, ½" diam, numerous; inflorescence at first nippled, bulbous, 3–4" diam, slowly elongating up to 20", the lowest flowers setting seed before the highest bloom; capsules 3-celled, dry; leaves narrow, tough, dry, V-shaped, with minutely barbed edges, the basal ones largest, 8–30", in a large dense clump; stalk up to 60", covered with much smaller leaves; from rhizomes. Cas Cr and W-side, in ± open forests and clearings. Liliaceae. Color p 124.

---

Communities of beargrass may go years without one bloom—and then hundreds bloom at once. Like the related century plant, beargrass clumps grow slowly, accumulating photosynthates before venturing a flowering stalk. After flowering, the clump dies, having first siphoned its stored nutrients through a rhizome to a new offset clump.

The plants are slowest to flower in closed forest, but they are no less successful there; in fact beargrass makes a highly monopolistic understory community on ridgetops and high south and west aspects. Most ridges in the Oregon and Southern Washington Cascades are mantled with pumice that rained down from the skies during countless volcanic eruptions. Though the climate is wet, the water drains so fast through young pumice soils that herbs and shrubs have a hard time. Here beargrass excels, like a dry-leafed bunchgrass transplanted from the arid steppes, except that it's willing to settle for scantier light.

Beargrass may provide part of a bear's spring diet, but the neatly clipped leaf bases you see here and there are more likely the work of a smaller mammal, anything from deer to pika. The leaves are fairly tender early in the year, when there isn't much else to choose from. The wiry strong leaves made this a vital plant to Northwest Indians—certainly one of their strongest incentives for trips into the mountains. They wove beargrass into all kinds of baskets, also usable as hats.

As David Douglas recorded, "Pursh is correct as to their making watertight baskets of its leaves. Last night my Indian friend Cockqua arrived here from his tribe on the coast, and brought me three of the hats made on the English fashion, which I ordered when there in July; the fourth, which will have some initials wrought in it, is not finished, but will be sent by the other ship. I think them a good specimen of the ingenuity of the natives and particularly also being made by a little girl, twelve years old . . . I paid one blanket (value 7 shillings) for them." Douglas' overly imaginative biographer, William Norwood, later read between the lines of that and other journal entries to argue that Douglas, with his cavalier praise for a "little girl's" ingenuity, was repressing and disguising his romantic entwinement with a nubile "Chinook princess."

# Irises

*Iris* spp. (eye-ris: rainbow). 3 sepals spreading, 3 petals erect, + 3 smaller petallike parts (pistil branches) ± resting on the sepals and hiding the stamens; leaves grasslike, mostly basal; capsule 3-celled, splitting; from horizontal rhizomes. Iridaceae (Iris family).

**Oregon iris,** *I. tenax* (ten-ax: holding fast). Flower usually single, violet (or occasionally blue, white, yellow, or pinkish); stem 4–12", with 1–4 small stem leaves. Open woods, lower W-side. Color p 127.

**Western iris,** *I. missouriensis* (miz-oo-ree-en-sis: of the Missouri River). Flowers pale blue, 2 (rarely up to 4); stem 12–24", with no mid-stem leaves (or rarely 1). Open pine woods, lower E Cas.

David Douglas found Indians braiding iris leaves into snares for large game, even elk: "It will hold the strongest bullock and is not thicker than the little finger." Such tenacity in a slender leaf suggested the name *tenax.* Iris was the Greek goddess who flashed across the sky, bearing messages—the rainbow.

# Blue-eyed-grass and Purple-eyed-grass

*Sisyrinchium* spp. (sis-er-ink-ium: the Greek term). 6 tepals alike, ¼–¾"; flowers 1 or a few; leaves grasslike, ± basal (exc the bracts at base of flower stalks, the longer bract looking like a continuation of the stem above the flowers), shorter than the 6–14", 2-edged stem. Dry grassy sites that are briefly moist in spring. Iridaceae (Iris family).

**Blue-eyed-grass,** *S. angustifolium* (ang-gus-tif-oh-lium: narrow leaf). Tepals blue with yellow base; stamens usually fused into a single, 3-anthered column. Sporadic.

**Purple-eyed-grass,** *S. douglasii* (da-glass-ee-eye: after David Douglas, p 18). Also **grass widows, satinflower.** Tepals magenta; stamens fused less than half their length. Mainly (within our range) in and near the Columbia Gorge. Color p 127.

These delicate perennials complete their active season in wet soil in a few weeks of spring, then wither and die back for the rest of the year, going dormant through the long drought of summer. (See page 220 on spring ephemerals.)

---

The orchid's flower structure is a snap to recognize; one petal, lowermost and thrust forward, is always utterly unlike the others and usually much larger. It's called the "lip" and serves as a platform for insect pollinators. Above the lip is a combined stamen/pistil structure called the "column." Orchid flowers are among the most elaborate insect lures on earth; but while they evolved outlandishly, orchid roots mostly degenerated. A majority of the 500 or so genera of orchids are rootless lianas living on air and dripwater in tropical rain forest canopies. Our orchids are tenuously rooted, with vestigial root systems that tap into preexisting networks of fungal hyphae.

**Irregular dicots**, pp 194–204, have 5 petals or fewer.

## Bog Orchids

*Habenaria* spp.* (hab-en-air-ia: strap, referring to the spur). Also **rein orchids.** Flowers in a tall spike; lip with a long downcurved spur to the rear, 2 sepals horizontal, the other sepal and 2 petals erect, hooding the column. Orchidaceae (Orchid family).

**White bog orchid,** *H. dilatata* (dil-a-tay-ta: widened). Flowers white, spicy-fragrant, in a dense 4–12" spike; stem 8–40", with many clasping leaves, the lower ones up to 10" × 2", but much smaller upward. Wet ground, often subalpine.

**Round-leaved bog orchid,** *H. orbiculata* (or-bic-you-lay-ta: circular leaf). Flowers white to greenish or yellowish, in a loose spike; stem 8–24", leafless exc for a few tiny bracts; basal leaves typically 2, broadly oval to round, 2–6". Deep montane forest. Color p 128.

The distinguishing feature of bog orchids is a narrow nectar-filled pouch or "spur" projecting rearward from the lip. The spur is an element in the grand pattern of orchid evolution, which has allied each variety and species of orchid with one, or at most a very few, species of insect. The right insect is not

---

*Some authorities have recently split *Habenaria* into several genera, placing our species in genus *Platanthera*.

only powerfully attracted, but also physically unable to extract nectar from two successive blooms without picking up pollen from the first and leaving an adequate dose of it on the stigma of the second. Inadvertently, of course. Bog orchids' devices include:

**proboscis-entangling hairs** inside the spur to engage the insect for a little while;

**adhesive discs** that stick to the insect's forehead while instantly triggering the stamen sac to split open;

**little stalks** that each hold a cluster of pollen to the adhesive disc (now on the insect's head), at first in an erect position that keeps the pollen *away* from the stigma of that same flower, but then (when the insect flies on) drying out and deflating into the right position to push the pollen onto . . .

**the gluey stigma** of the next flower of the same species, where the insect must reassume the same position in order to extract more nectar.

All this to minimize waste of pollen in the wrong places.

The white bog orchid is our showiest but by no means our only member of the genus. Several species with small greenish

---

## Goals of Evolution?

The idea (expressed above in regard to orchid design) that species evolved certain traits *in order to* better accomplish certain functions is a timeworn fallacy about evolution. A few respected evolutionary scientists uphold subtle versions of this concept, but they are far outnumbered. A more accepted description is that the apparent present function of the trait may, over some evolutionary period, have conferred an advantage on individuals carrying it in terms of those individuals' reproductive success, thus preserving the trait in increasing numbers of descendents. (Other traits seem to appear and persist without any adaptive value whatsoever.) It's only because that sort of description is long, unwieldy, and abstract that this book succumbs to the charm of the teleological fallacy—an easy-to-grasp figure of speech.

---

flowers grow on wet ground, or in deep forest. The flowers are always spirally arranged on the spike, though less dramatically than in ladies-tresses (below). The inch-long apparent stalk between the flower and the stem is actually the flower's ovary; a 180° twist in it shows that the lip evolved from what was originally the uppermost petal.

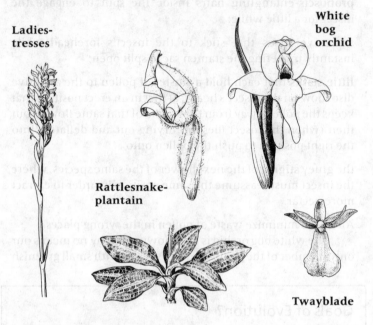

**Ladies-tresses**

**White bog orchid**

**Rattlesnake-plantain**

**Twayblade**

## Ladies-tresses

*Spiranthes romanzoffiana* (spy-ranth-eez: coil flower; roman-zof-ee-ay-na: after Count Nikolai Rumiantzev, p 219). Flowers ± white, ½" long, seemingly tubular, in (usually 3) ranks in a 2–6", dense, spiraling spike resembling a braid; leaves sheathing, mostly basal; stem up to 24"; from swollen roots in wet ground. Orchidaceae.

## Twayblade

*Listera caurina* (lis-ter-a: after Martin Lister; caw-rye-na: of the NW wind). Also **big-ears**. Flowers pale greenish, small (½"), with a flaring/rounded ± flat lip, several, in an open, short-stalked spike; leaves 2, both clasping the stem at mid-height, 1½–2½", pointed but usually broad; stem unbranched, 4–12". Various habitats, mostly forested, restricted to the PNW. Orchidaceae. Color p 127.

# Rattlesnake-plantain

*Goodyera oblongifolia* (good-yer-a: after John Goodyer; oblong-gif-oh-lia: oblong leaf). Flowers greenish white, many, in a one-sided spike up to 5"; lip shorter than and hooded by the fused, ¼"-long upper petals, all connected to the stalk by a twisted ovary; leaves 1½–3", in a basal rosette, thick, evergreen, very dark glossy green, mottled white along the veins; stem unbranched, 10–16". Widespread in dense forest. Orchidaceae. Color p 127.

"The Klallam informant, who is a devout Shaker, said that since she is a Christian she should not think of such matters, but formerly women rubbed this plant on their bodies to make their husbands like them better." (Guenther 1975)

The intensity of the snakeskin pattern on the leaves varies, even between side-by-side plants. Sometimes only the midvein is white, but usually there is enough white pattern to tell these leaves from those of white-veined pyrola (page 110). This is one orchid with enough leaf area to make it nearly independent, when mature, of its mycorrhizal partners.

# Lady's-slipper

*Cypripedium montanum* (sip-rip-ee-dium: Venus' slipper; mon-tay-num: of mtns). Flowers 1–3; lip 1" long, very bulbous, white with purplish veins and a yellow staminate structure at its base; upper sepal and lateral petals brownish purple, about 2" long, slender, often twisted; stem 6–24", with several broad, clasping-based, 2–6"-long leaves plus a smaller bract beneath each flower. ± open forest, E Cas slope, or rarely W in Ore. Orchidaceae. Color p 128.

# Calypso Orchid

*Calypso bulbosa* (ca-lip-so: hidden, the name of a Greek sea nymph; bulb-oh-sa: bulbous). Also **fairy slipper, deer's-head orchid**. Flower single, pink; lip slipper-shaped, white-spotted above, magenta-streaked beneath; other petals and sepals all much alike, narrow, ¾", leaf single, basal, growing in fall and withering by early summer; stem 3–7", from a small round corm. Moist, mature forest, mainly lowland. Orchidaceae. Color p 128.

A close look reveals these little orchids to be just as voluptuously overdesigned as their corsage cousins, which evolved bigger to seduce the oversized tropical cousins of our insects. Get down onto their level to see and smell them, using a

---

**Herbs:** showy monocots, 3 stamens (irises)          *177*

handlens if you like, but don't pick them. Their bulblike "corm" is so shallowly planted that it's almost impossible to pick them without ripping the corm's lifelines. A calypso is dependent on its fungal and plant hosts (page 262); its single leaf withers and disappears early in the growing season. Like other orchids, it produces huge numbers of minute seeds (3,770,000 seeds were found in one tropical orchid's pod) with virtually no built-in food supply, and an abysmal germination rate. They germinate only in the presence of particular species of fungi, which immediately supply nutrients. The black specks filling vanilla beans and classy vanilla ice creams are familiar examples of orchid seeds.

## Coralroot

*Corallorhiza* spp. (coral-o-rye-za: coral root). Entire plant (exc lip) dull pinkish brown, or rarely pale yellow (albino); flowers ¾–1¼" long (½ of that being the tubular ovary), 6–30 in a loose spike; leaves reduced to inconspicuous sheaths on the 6–20" stem. Forest. Orchidaceae.

**Western coralroot,** *C. mertensiana* (mer-ten-see-ay-na: after Karl Mertens, p 210). Lip redder than plant, often with 1 or 2 spots or blotches. Color p 128.

**Spotted coralroot,** *C. maculata* (mac-you-lay-ta: spotted). Lip usually white, with many magenta spots (sometimes also on petals).

**Striped coralroot,** *C. striata* (stry-ay-ta: striped). All petals and sepals brownish- to purplish-striped. Color p 128.

Coralroots usually grow in forest stands with few herbs or shrubs. They blend in with the duff and sticks until one of those few shafts of light hits, suddenly turning luminous their eerie, translucent russet flesh. Their rhizomes do resemble coral—curly, short, knobby, and entirely enveloped in soft fungal tissue. As in other orchid genera, a seed's embryo develops only if penetrated, nourished, and hormonally stimulated by a hypha from a fungus in the soil; fungal hormones suppress root hair growth and stimulate the orchid to produce mycorrhizae instead (page 262). In the case of coralroots, no true roots or root hairs ever form. The vestigial leaves do not photosynthesize, but oddly enough the flowers' ovaries do, in insignificant quantities.

# Phantom Orchid

*Eburophyton austiniae*\* (eb-ur-o-fie-ton: ivory plant; aus-tin-ee-ee: after Mrs. R. M. Austin). Entire plant ivory white (aging brown) exc for a yellow dot in the lip pouch; sepals, petals ½–¾", lip shorter; flowers 5–20; leaves mostly just sheaths, but 1 or 2 may have blades; stem 8–20". Forests; uncommon. Orchidaceae. Color p 128.

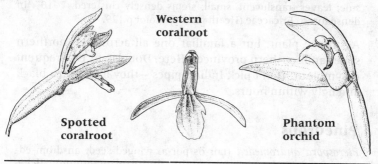

**Western coralroot**

**Spotted coralroot**

**Phantom orchid**

Herbs: non-green dicots

# Pinesap

*Hypopitys monotropa*† (hye-pop-it-iss: under pine; ma-not-ra-pa: turned one way, referring to the blooms). Entire plant fleshy, yellow (rarely red) to straw, tinged with pink, drying black; flowers narrowly bell-shaped, ⅜–¾", mostly 4-merous (rarely 5-), several, initially all downturned in one direction, but erect when fruit matures; seed capsules round, soft; leaves translucent, small; stems clustered, 2–10". In dense forest. Ericaceae (Heath family).‡ Color p 129.

Digging up a pinesap (Don't!) would reveal a soft mycorrhizal rootball only a couple of inches deep. Species of *Boletus* (page 273) are typical partners of this species, and a pine or other conifer is almost always hooked up to the same bolete.

\*Many texts place this species in *Cephalanthera*, an Old World genus of green-leafed orchids, but our authority disagrees.
†Curiously, another name—authored by Linnaeus and used still by some authorities—for *Hypopitys monotropa* is *Monotropa hypopithys*.
‡Many authorities separate this and the following three genera into either an exclusively non-green family, Monotropaceae, or a family Pyrolaceae which would also include *Pyrola* and *Chimaphila*. The Heath family (or families) is a moot point between splitters and lumpers, having been batted back and forth for generations.

# Indian-pipe

*Monotropa uniflora* (ma-not-ra-pa: a nonsensical name as applied to a one-flowered plant; you-nif-**lor**-a: one-flowered). Also **ghost-plant, corpse-plant.** Entire plant fleshy, white or pink-tinged, drying black; flower single, narrowly bell-shaped, ½–¾", mostly 5-merous (occasionally 4–6-), pendent, but erect in fruit—a soft round capsule; leaves translucent, small; stems densely clustered, 2–10". In dense forest. Ericaceae (Heath family). Color p 129.

A strange plant, but a familiar one all across the northern states and southern provinces. Here, Douglas-fir is a frequent cosymbiont. Don't pick Indian-pipes—they'll just turn black and ugly within hours.

# Pinedrops

*Pterospora andromedea* (tair-os-por-a: winged seed; an-drom-ed-ee-a: a name from Greek myth). Entire plant gummy/sticky, monochromatically brownish red exc for the amber, 5-lobed, jar-shaped corollas; flowers many, on downcurved stalks; capsules ± pumpkin-shaped; leaves brown, small and sparse on the 12–48" stem. Lower montane forests. Ericaceae (Heath family). Color p 129.

This year's translucent amber stalks of pinedrops, our tallest

## Non-Green Orchids and Heaths

Some of our most arresting herbs are those without chlorophyll. Some authors call these non-green plants saprophytes, a word long understood to mean that they live by extracting nutrients from dead organic material. But in fact, they obtain their nutrients from living fungi. Though some of the fungi are truly saprophytes, most fungi that feed non-green plants get their carbohydrate nutrients from living green plants they link up with. An accurate term for non-green plants is "epiparasites," or indirect parasites on green plants above them.

The fungi exchange nutrients with both green and non-green plants at the root level, through organs called "mycorrhizae." (Page 262.) The mycorrhizal symbiosis must have evolved because it lets fungi and plants take advantage of each other's strengths—the fungi's efficiency at getting water, minerals, and nitrogen from the substrate, and the plants' ability to make car-non-green species, shoot up alongside still-standing dry brown stalks from last year and even the year before. As the name implies, they may be found under (and mycorrhizally attached to) Ponderosa pines—but also just as often under Douglas-fir.

## Candystick

*Allotropa virgata* (a-lot-ra-pa: turned other ways, in contrast to *Monotropa*; veer-gay-ta: striped). Also **stickcandy.** Entire plant fleshy, bright red and white; no petals, 5 white sepals shorter than the 10 dark red stamens and pistil; flowers many, in a ± dense spike; stem thick, 4–16", leafy, sharply striped. Uncommon, in dense lowland forest. Ericaceae (Heath family). Color p 129.

This may be the most astonishing-looking plant you'll ever see around here, as it bursts, at once candylike and gothic, from the drab shady duff and litter. The genus *Allotropa* grows its only species only here. If you ever begin to get used to it, head south to find its Sierra Nevada cousin the snowplant, *Sarcodes sanguinea*, which is 100% scarlet.

bohydrates through photosynthesis. It would seem that some small mycorrhizal plants gradually contributed less and less to this exchange, and were able to get away with it so long as there were plenty of other plants plugged into the same fungus. As the free-loaders evolved, organs they no longer needed atrophied. They ended up as non-green plants with vestigial leaves, little or no chlorophyll, and no real roots—nothing but a stalk of flowers reproducing; the trees above provide the leaves, and the fungi below provide both the roots and lifelines to the trees.

It is possible, but by no means proved, that the non-green plant has some useful substance to offer on this underground market. Presumably the plant does something to induce the fungus to form mycorrhizae; secretions from non-green plants have been shown to stimulate growth in fungi. Some researchers call the secretions "vitaminlike," but others remind that it is common for parasites to stimulate growth in their hosts in ways that drain the host's energies, rather than augmenting them.

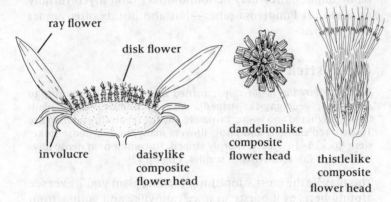

ray flower

disk flower

involucre

daisylike
composite
flower head

dandelionlike
composite
flower head

thistlelike
composite
flower head

Picture a daisy: seeming petals radiate from a cushiony "disk" in the center. On closer examination, the base of each petallike "ray" enwraps a small pistil. The ray and pistil together constitute a "ray flower," not a petal; the petallike length is an entire corolla of petals, fused, then split down one side and flattened. The central disk turns out to be lots of little flowers too—"disk flowers"; each has a (usually 2-branched) style poking out of a minute tube of five fused stamens, within a larger tube which is the (usually 5-lobed) corolla. What we saw at first as one flower is a "composite flower head," the inflorescence of the family Asteraceae,* or composites. "Involucral bracts" surround the head as sepals do the flowers of other families. True sepals may also be present in the form of a "pappus"—a brush of hairs, scales, or plumes that remain attached and grow as the seed forms, usually to provide mobility via wind or fur.

Some composites, dandelionlike, have only ray flowers and no disk. In these the ray flowers are bisexual, having stamen tubes as well as pistils. Other composites have only disk flowers, and some *appear* to have only disk flowers because their inner and outer flower types are equally unshowy; we

---

*This family was known for centuries as the Compositae. Recently, wide approval was given to a rule that all families should have names formed from the name of a member genus and the suffix *-aceae*. The term "composites" is still widely understood for the Asteraceae.

will call both kinds of rayless composites "thistlelike."

The tight fit of the stamen tube around the style is a mechanism for preventing self-pollination. As the pistil grows, it plunges all the pollen out of the tube; once its tip has grown free of the tube, it can split, exposing stigmas on the inner faces of its two branches.

Composite flowers excel at producing copious fat, well-nourished seeds. The family's strongest tactic here is rapid invasion; year-to-year survival in severe cold is not its forte, and in shade even less, so composites are fewer here than in many mountain regions. Still, they are conspicuous in meadows here, providing color in late summer when most flowers are gone and huckleberry foliage has yet to turn.

With somewhere between 15,000 and 20,000 species (several hundred in our range), the composites are even more diverse than grasses, though they don't cover quite as much ground or feed as many big stomachs. They are notoriously hard to identify at the species level. The larger genera are represented in this book by two or three species apiece.

---

Herbs: composites, disk and ray flowers (daisylike)

---

# Daisies

*Erigeron* spp. (er-**idge**-er-un: soon aged). Also **fleabanes.** Rays 40–80; disk yellow; heads usually single; leaves narrow, mostly basal; stem leaves much smaller farther up (few and tiny on alpine plants). Asteraceae (Aster family).

**Mountain daisy,** *E. peregrinus* (pair-eg-**rye**-nus: wandering). Rays violet to pale blue or pink, ⅜–1". Widespread.

**Golden fleabane,** *E. aureus* (aw-**ree**-us: golden). Rays bright yellow, ¼"; involucre woolly; stem 2–8". Alpine. Color p 130.

Daisies generally bloom during the spring flowering rush, earlier than most composites, which could be the implication of the Latin name. Insecticidal properties implied by "fleabane" are Medieval superstition, or confusion with *Pyrethrum* daisies. The name "daisy" (originally the English daisy, *Bellis perennis*) traces back to "day's eye" in Old English.

---

# Asters

*Aster* spp. (ast-er: star). Rays 6–25, lavender to blue, pink, or white, ¼–¾"; disk yellow. Asteraceae.

*A. ledophyllus* (leed-o-fill-us: rockrose leaf). Heads several; leaves ± elliptical, 1–3", all on the 8–30" stem. Mid to subalpine elevs. Color p 130.

*A. alpigenus* (al-pidge-en-us: alpine). Heads single; stems 1–3, 4–12", often oblique; leaves ± linear, basal ones 2–7", stem leaves very small. Alp/subalpine.

Asters look much like daisies, but tend to bloom later and to have fewer, broader rays.

# Arnicas

*Arnica* spp. (ar-nic-a: lambskin, referring to leaf texture). Heads entirely yellow, single to few; leaves opposite, plus some in stemless basal whorls from the rhizome; stems 4–24". Asteraceae.

*A. cordifolia* (cor-dif-oh-lia: heart leaf) and
*A. latifolia* (lat-if-oh-lia: broad leaf). Rays 8–15, ½–1¼"; leaves ± toothed, heart-shaped to broadly oval. Meadows and open forest, mostly E of Cas. Color p 130.

*A. rydbergii* (rid-berg-ee-eye: after Per Rydberg). Rays 7–10, ½"; leaves ± smooth-edged, narrow-elliptical, with 3–5 ± parallel main veins; stem 4–12". Alp/subalpine.

Arnicas bloom early to mid summer.

# Woolly-sunflower

*Eriophyllum lanatum* (area-fill-um: woolly leaf; lan-ay-tum: woolly). Leaves, stems, and involucres all thickly white-woolly; rays about 13, ¾"; heads all-yellow, single; leaves on stem often narrowly 3–7-lobed; stems 10–24", many, weak, in thick clumps or sprawling mats. Lower dry meadows. Asteraceae. Color p 130.

# Balsamroot

*Balsamorhiza sagittata* (balsam-o-rye-za: balsam root; sadge-it-ay-ta: arrow-shaped). Heads all-yellow, single, 2½–4" diam; leaves all basal (exc 1 or 2 small bracts on stem), triangular, 12" × 6", on long leafstalks; plant silver-velvety, esp under leaves when young, 10–30", in thick clumps. In sun, lower E-side. Asteraceae. Color p 130.

Balsamroots are spectacular in the spring, and they are also a good food plant. Deer and elk eat the leaves, and Indians ate the young shoots, the seeds, and the fat, fragrant, slightly woody taproot.

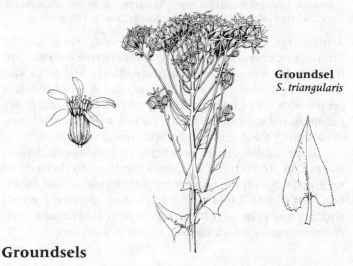

**Groundsel**
*S. triangularis*

# Groundsels

*Senecio* spp. (sen-**ee**-she-oh: old man). Rays ¼–½", few (4–10 or rarely none) sparse and disorderly; disk yellow; heads several; stem 1–6'. Asteraceae.

*S. triangularis*. Rays yellow; leaves narrow-triangular, 3–7", all on stem. In lush meadows.

*S. integerrimus* (in-te-**jer**-im-us: very smooth-edged). Rays yellow to white, or lacking; leaves ± elliptical, on long leafstalks, preponderantly basal, smaller upwards. Mainly E-sides.

# Goldenrods

*Solidago* spp. (so-lid-**ay**-go: healing, a misleading name). Heads yellow, numerous, in a large fluffy inflorescence; rays 8–13, minute. Asteraceae.

*S. multiradiata* (multi-ray-dee-**ay**-ta: many-rayed) and
*S. spathulata nana* (spath-you-**lay**-ta: spatula-shaped; **nay**-na: dwarf). Inflorescence ± round-topped, leaves basally crowded, tapering to leafstalks; stem 2–18". Alp/subalpine. Color p 130.

*S. canadensis*. Inflorescence pyramidal; stem 1½–6', very leafy. Widespread.

---

# Yarrow

*Achillea millefolium* (ak-il-ee-a: after Achilles; mil-ef-oh-lium: thousand leaf). Rays 3–5, white (rarely pink), ⅛" long and wide; disk yellow; heads many, in a flat to convex inflorescence; leaves narrow, extremely (though variably) finely dissected, fernlike, aromatic; to 3', or dwarfed (alpine). In sun. Asteraceae. Color p 130.

Achilles, the Greek hero of the Trojan war, was taught by Chiron the Centaur to dress the wounds of battle with yarrow. Northwest Indians also used yarrow poultices, and drank yarrow tea for myriad ailments. They steamed the homes of sick people with the pungent smell of yarrow leaves—rather like rosemary and sage. In China, the yarrow stalk oracle was systematized by Confucius and his followers in the *I Ching*.

Some *Achillea* species have perfectly smooth-edged linear leaves. Our "thousand-leaf" species was formerly divided into several species on the basis of varying degrees of leaf dissection, but it was found that transplanted specimens would within a few years alter their leaf shape to fit their new environment, nearly matching the yarrows around them.

---

**Herbs:** composites without rays (thistlelike)

---

# Thistles

*Cirsium edule* (sir-shium: swollen vein; ed-you-lee: edible) and *C. brevistylum* (brev-iss-tie-lum: short style). Flowers bright pink to purple; involucral bracts ending in long spines, cobwebby-haired at the base; leaves very spiny, pinnately lobed, up to 10" long, narrow, all on the 1½–7' stem. Widespread in sun, mainly W-side. Asteraceae. Color p 131.

These native thistles are herbaceous biennials, storing starches in their taproot their first summer, dying back to the ground, and then resprouting to flower, fruit, and die in their second year. The taproot and fat stems, peeled, are nutritious and tasty, highly rated in both ethnobotanical and survival-skills texts. Two European species are nasty weeds of Northwest lowland fields and lots.

---

# Pearly Everlasting

*Anaphalis margaritacea* (an-af-a-lis: the Greek term; margarite-ay-see-a: pearly). Disks ± yellow, surrounded by innumerable tiny papery white bracts (not ray flowers); heads ⅜" diam, many, in a convex cluster; leaves 2–5", linear, woolly (esp underneath), all on the 8–36" stem, the lowest ones withering. Widespread on roadsides, burns and clearcuts; also alpine. Asteraceae. Color p 130.

Pearly everlasting may appear daisylike, but in fact it has no ray flowers, only dozens of dry scaly white involucral bracts which persist everlastingly—either in the field or in a vase.

# Pussytoes

*Antennaria* spp. (an-ten-air-ia: antenna—). Disks ± white, soft-fuzzy, deep, surrounded by numerous scaly bracts (not ray flowers); heads several, ¼" diam; leaves woolly, mainly basal. Asteraceae.

*A. lanata** (lan-ay-ta: woolly). In clumps, occasionally broad but without runners; bracts white; basal leaves 1–6"; stems 4–12".

*A. alpina* (al-pie-na: alpine) and
*A. umbrinella* (um-brin-el-a: brown small). Mat-forming, with leafy runners; sepallike involucral bracts dark; few leaves over 1"; stems 1–7". Color p 130.

*A. microphylla* (micro-fill-a: small leaf). Mat-forming, with runners; bracts pink to white; few leaves over 1"; stems 6–18".

Alpine forms of many composites, including *umbrinella* and *alpina* pussytoes, usually reproduce asexually, the ovules maturing into seeds without being fertilized by pollen. Any given patch is likely a female clone, perhaps with interconnecting runners. These plants are adapted to a climate whose blooming season often zips by in weather too nasty for small insects (pollinators of tiny flowers like these) to be out and about.

---

*Some recent texts make *lanata* a subspecies under *A. carpatica* (carpat-ic-a: of the Carpathian Mtns).

---

# Wormwoods

*Artemisia* spp. (ar-tem-ee-zhia: the Greek term). Flower heads tiny, pale, within ⅛" diam cups of fuzzy bracts, forming tall to ± spreading inflorescences; spicy-aromatic and fuzzy-coated all over, the leaves silvery underneath, ± greener above. Asteraceae. Color p 131.

**Western wormwood,** *A. ludoviciana* (lu-doe-vis-ee-ay-na: of the Louisiana Purchase area) and
**Douglas' wormwood,** *A. douglasiana* (da-glass-ee-ay-na: after David Douglas, p 18). Plants 1–5' tall, leaves 1–4", most with a few irregular, fingerlike lobes. Locally abundant in dense meadows.

**Three-forked wormwood,** *A. trifurcata* (try-fur-cay-ta: 3 forked). Dwarfed plants, 2–10"; similar to above, exc leaves mostly basal, ½–1¼", with a few smaller leaves along the stalk. Alpine, in Wash.

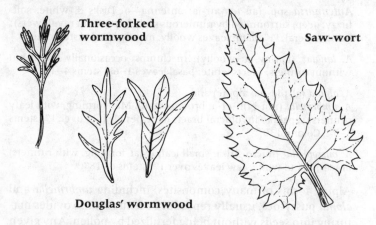

Three-forked wormwood

Saw-wort

Douglas' wormwood

Wormwood as a "bitter herb" figures in several dire biblical prophesies; they seemed to come true when apocalyptic disaster visited a town named Chernobyl ("wormwood" in Ukrainian). Bitter aromatic wormwoods of Europe include *vermouth* ("wormwood" in German) and the intoxicating absinth, both used to flavor liquors. A sweet herb, French tarragon or dragon wormwood, *A. dracunculus,* grows weedily in eastern Washington. Wormwood's flowers are nondescript but its silvery wool is striking, and its spicy fragrance more so; crush a leaf and you're perfumed for hours. Native shrubs of the same genus perfume about three-fourths of the American West. You guessed it—sagebrush. Our subalpine wormwoods are scattered late-summer dominants.

## Coltsfoot

*Petasites frigidus* (pet-a-**sigh**-teez: hat-shaped; **fridge**-id-us: of cold climates). Flower heads white or pink-tinged, occasionally with a few tiny rays, many, in a somewhat open round inflorescence, vanilla-fragrant; stems 6–24", with clasping, elliptical leafy bracts 1–3" long; true leaves basal on stout 16–24" leafstalks, palmately deeply 5–9-lobed, 6–16" broad by not quite as long. Wet places, mainly W of Cas Cr. Asteraceae. Color p 131.

Fast-growing coltsfoot shoots appear very early in lower forests. The flowers wither before the leaves reach full size in midsummer. A smaller, less common high-elevation variety has shallow-lobed leaf blades somewhat longer (up to 8") than wide.

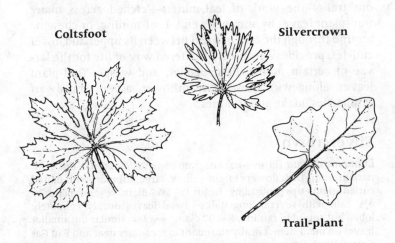

Coltsfoot          Silvercrown

Trail-plant

## Saw-wort

*Saussurea americana* (so-**sure**-ee-a: after Horace and Theodore Saussure). Disk flowers usually 13, dark purple, narrow-tubular, protruding haphazardly from a fluffy mass of pappus bristles; heads several, crowded; leaves ± triangular, sharply toothed, narrow, the lower ones up to 6", all on the 10–30" stem. Lower subalpine lush meadows, Wash. Asteraceae. Color p 131.

## Trail-plant

*Adenocaulon bicolor* (a-den-o-**caw**-lon: gland stem; **by**-color: two color). Also **pathfinder**. Flower heads tiny, white, several, in a very sparse panicle on a 10–32" stalk; leaves all on lowest part of stem, triangular, 4–6", dark green on top, white-fuzzy underneath. W-side forest up to mid elevs. Asteraceae.

This is an unusual composite in adapting primarily to deep shade. The leaves are suitably large, thin, and flat-lying—easy to recognize by their unique shape—while the flowers are rarely noticed. (They attract small flies by their smell, not their looks.) The combination of weak leafstalks and high contrast between upper and lower leaf surfaces led to the common name; a good woodsman tracking a large animal through the woods appreciated a conspicuous series of overturned trail-plant leaves. The sticky seeds are also said to "trail" us by adhering to our legs. And then there are the tiny circumlocuit-ous trails—the work of leaf-miners—etched across many trail-plant leaves by summer's end. Leaf-mining, or chewing around through the cells of a leaf between its upper and lower cuticles, provides a relatively sheltered way of life for the lar-vae of certain moths, flies, beetles, and wasps. Trail-plant leaves, along with those of foamflower, aspen, and dwarf raspberries, make good "mines."

## Silvercrown

*Luina nardosmia* (lu-eye-na: anagram of genus *Inula;* nar-**dos**-mia: nard aroma). Disk flowers bright yellow, large, tubular, with paired curled stigma tips protruding; heads ½–1¼" diam, several; plant 2–3½' tall, with several long-stalked basal leaves deeply palmately lobed but round in outline, 8–10" diam; + a few similar but smaller leaves on main stem. Locally abundant in meadows near and E of Cas Cr in S Wash Cas; uncommon elsewhere. Asteraceae. Color p 131.

On this handsome plant, the individual flowers in the compos-ite head are big enough for you to see them as flowers, with five petal lobes and two stigma branches.

## Silverback

*Luina hypoleuca* (hypo-**lew**-ca: white underneath). Flower heads distinctively cream-colored, several; leaves 1–2", oval, white-woolly on both sides or just the underside, attached directly all up and down the stalk; plants 8–24" tall, in ± dense patches. Rocky sites, widely scattered at high elevs. Asteraceae. Color p 131.

# Mountain-dandelions

*Agoseris* and *Microseris** spp. (a-gah-ser-iss: goat chicory; my-**crah**-ser-iss: small chicory). One 1"-diam flower head per stem; plants milky-juiced; leaves basal, ± linear or with a few widely spaced teeth. Asteraceae.

**Alpine microseris,** *M. alpestris* (al-pes-tris: alpine). Rays dandelion-yellow; plants 2–10". Alp/subalpine, Mt Rainier S. Color p 131.

**Pale agoseris,** *A. glauca* (**glaw**-ca: silvery). Rays pale waxy yellow, drying pinkish; dwarfed (4–10") on alpine slopes, where ± common N of Mt Rainier; or 10–30" in dry meadows E of Cas. Color p 131.

**Orange agoseris,** *A. aurantiaca* (aw-ran-**tie**-a-ca: orange). Rays red-orange, drying purplish; 6–12". Lush high meadows.

These are closely related to true dandelions (genus *Taraxacum*) and resemble them in their seed-parachutes as well as their flower heads, which on *Agoseris* tend to close up on hot days. True dandelions are also native, but uncommon here.

# Hawkweeds

*Hieracium* spp. (hi-er-ay-shium: hawk—). Flower heads ½" diam, in a sparse panicle; stems milky-juiced. Asteraceae.

*H. gracile* (**grass**-il-ee: slender). Heads yellow, few or single; leaves almost basal on long leafstalks, oval, smooth; stem 2–12", fuzzy. Alp/subalpine meadows.

*H. scouleri* (**scoo**-ler-eye: after John Scouler, p 67). Heads yellow, several; stems and leaf undersides covered with long white hairs; stem 10–30"; stems and leaf veins often deep magenta in spring. Low to mid-elev clearings, esp in Columbia Gorge. Color p 131.

*H. albiflorum* (al-bif-lor-um: white flower). Heads white to cream, several; leaves long-elliptical, hairy; stem 12–20". Open forest and clearings.

*Some texts include *Microseris* in *Agoseris;* others put it in *Nothocalais*.

"Without petals or sepals" applies differently to these four oddities. **Vanillaleaf** really does lack any petals or sepals, though the rest of its family has them. **False-bugbane** has four sepals but drops them as the flower opens; its family, the Buttercups, includes almost every number of flower parts. **Nettle** has four tiny, obscure sepals, but getting close enough to see them would likely prove painful. **Alpine willows** are our only catkin-flowered shrubs so tiny that the reader would probably look them up among the herbs.

Aside from stinging nettle, "without sepals" is applied pretty strictly, not to betray readers who look closely enough to see and count minute sepals. Barely countable sepals or petals are on **Mountain-sorrel**, p 199; **Pussypaws**, p 202; **Skunk-cabbage**, p 161; **Bunchberry**, p 231; **Desert-parsleys**, p 226; **Goatsbeard**, p 218; and **Baneberry**, p 234.

**Cascade willow**

# Alpine Willows

*Salix* spp. (**say**-lix: the Roman term). Female and male catkins on separate plants; leaves elliptical, ¼–¾"; prostrate shrubs forming cushiony mats to 4" tall. Alpine. Salicaceae (Willow family).

**Cascade willow,** *S. cascadensis* (cas-ca-**den**-sis: of the Cascades). Catkins on lateral spurs, 15–25-flowered, ½–¾" long.

**Snow willow,** *S. nivalis* (niv-**ay**-lis: of snow). Catkins on main branch tips, 8–12-flowered, ¼–½" long. Color p 132.

These are true willows reduced to alpine stature. They are fairly common in the North Cascades. Except for their catkins, you might mistake them for huckleberry, which becomes similarly carpetlike in the alpine environment.

## False-bugbane

*Trautvetteria caroliniensis* (trout-vet-ee-ria: for Ernest Rudolf van Trautvetter; carol-in-ee-en-sis: of the Carolinas). Flowers white, in broad rough clusters, consisting mainly of many stamens (up to ¼"), the 4 (3–7) sepals falling off as the flower opens; leaves predominantly basal, 4–10" wide, deeply 5–7-lobed; 20–36". Wet forest. Ranunculaceae (Buttercup family). Color p 132.

**Vanillaleaf**

## Vanillaleaf

*Achlys triphylla* (ay-klis: a night goddess; try-fill-a: 3 leaf). Also deerfoot. Flowers cream-white, minute, in a dense spike 1–2" × ¼", on a leafless stalk a little taller (8–16") than the one that bears the unique-shaped 3-compound, flat-lying leaf (6–10"). Abundant in W-side forest. Berberidaceae (Barberry family). Color p 132.

The vanilla fragrance is indetectable in growing leaves, but will be unmistakable and durable if you pick a few and let them dry. The odd leaf shape makes a good Rorschach pattern; where some see a deer foot, this writer sees a moose head.

## Nettle

*Urtica dioica* (ur-tic-a: burning; die-oy-ca: with male and female flowers on different plants, an untrue name for our nettles). Flowers 4-merous, tiny, pale green, many, in loose panicles dangling from the leaf axils, the panicles unisexual, with females higher on the plant (our varieties); leaves opposite, sawtoothed, pointed/oval, 2–6"; stem (2–6') and leaves lined with fine stinging bristles; from rhizomes. Moist, ± open lowlands, mainly W-side. Urticaceae (Nettle family). Color p 132.

Nettles are seen nowadays as something to avoid, but older traditions held them the most estimable of weeds. They're food: the young plants make delicious greens. They're fiber: the mature stems have been made into high-quality twine, cloth, and paper, substituting for flax in wartime as recently as World War II; and several tribes here made nettle twine nets for ducks and fish. They're medicine: the sting helped stoical Indian hunters stay awake through the night, and nettle tea has earned a medicinal reputation worldwide.

Nettle stingers are miniature hypodermic syringes, like bee and ant stingers, and employ the same irritant, formic acid. They evolved as a defense against browsing, but they don't save nettles from Milbert's tortoiseshell and other caterpillars. One folk remedy prescribes pulling up the nettle that bit you, to soothe the sting by crushing juice onto it from the nettle root. Wilting softens nettle stingers into harmlessness, and thorough steaming or drying renders them indetectable. If you ever camp in April or May near a bed of nettle shoots a few inches high, cook them up. Wear long sleeves and gloves (or socks) over your hands, lop them with a knife, and steam them limp in half an inch of water. (Cooking them only *al dente* leaves enough bristle to worry your lips and tongue.) Butter, sour cream, feta or bleu cheese, maybe crepes . . .

---

**Herbs:** dicots, strongly irregular flowers

---

Irregular flowers are those in which the petals (or the sepals, if they are the showy parts) are not all alike. You might think of them as bilaterally—as opposed to radially—symmetrical, but there are exceptions; louseworts are twisted and have no symmetry at all.

Irregular flowers display advanced specialization of form. Each species matches a particular form and size of insect for pollination. Irregular shapes run in families: you can probably think of the three familiar shapes—orchids, sweetpeas and snapdragons—which represent our three biggest families of irregular flowers, the Orchidaceae (pp 174–79), Fabaceae (pp 201–02), and Scrophulariaceae (pp 195–98).

Flowers are not included in this group if the irregularity consists of unequal-sized, but otherwise similar, petals; for example, **Veronica** and **Kittentails**, p 231, both with 4 small blue to lavendar petals. **Cow-parsnip**, p 224, has 5 unequal but similar 2-lobed white petals, and a Parsley-family look.

# Indian Paintbrush

*Castilleja* spp. (cas-til-ay-a: after Domingo Castillejo). Inflorescence most often red, varying to every shade of pink, magenta, orange, yellow, and greenish white; true flowers subtended and largely hidden by brightly colored bracts (lower bracts grading to green at their bases); calyx narrowly 4-lobed, same color as bracts; corolla a thin tube, dull green; leaves (incl the colored floral bracts) often narrowly 3-, 5-, or 7-pronged, elliptical, their main veins appearing parallel; stems usually several, unbranched, from a woody base; roots partially parasitic, esp on grasses and composites. Widespread, in sun. Scrophulariaceae (Figwort family). Color p 136.

Indian Paintbrush is easily and popularly recognized as a genus, but notoriously hard to identify to species; neither coloring nor hairiness are diagnostic, though within a small area the members of a species match up quite closely. Our commonest species is *miniata* ("cinnabar-red"), ranging from the coast to subalpine meadows and sometimes a little way into the forest, throughout the montane West. It grows 8–30" tall and is usually scarlet, varying to yellow-orange. The Makah used this paintbrush as hummingbird lure; they trapped hummers to use as charms for whaling.

# Monkeyflowers

*Mimulus* spp. (mim-you-lus: mime or clown). Flowers snapdragonlike—corolla with a long ¾–2") throat, hairy inside, and 2 upper and 3 lower lobes; calyx angularly 5-lobed; stamens 4, paired; leaves opposite; stems in dense clumps on stolons and/or rhizomes. Usually in small, often seasonal, streams. Scrophulariaceae (Figwort family).

**Purple monkeyflower,** *M. lewisii* (lew-iss-ee-eye: after Meriwether Lewis, p 234). Flowers deep pink to violet; leaves stalkless, pointed, 2–3"; plants 12–36", sticky-hairy. Mid to subalpine elevs. Color p 135.

**Common monkeyflower,** *M. guttatus* (ga-tay-tus: spotted). Flowers yellow, throat often red-spotted; 4–30". Common low to mid elevs. Color p 134.

**Mountain monkeyflower,** *M. tilingii* (til-ing-ee-eye: after Heinrich S. T. Tiling). Flowers yellow, throat often red-spotted; most leaves basal, roundish, ⅜–1"; plants 2–8". Subalpine. Cover photo.

Both "monkey" and the "mime" in *Mimulus* are impressions of this fat irregular blossom as a funny face. Some other monkeyflower species here don't depend on getting wet feet, but bloom very early to compensate.

# Louseworts

*Pedicularis* spp. (ped-ic-you-**lair**-iss: louse—). Corolla fused, with 2 main lips, the upper ± long-beaked, the lower usually 3-lobed; leaves (exc on sickletop) fernlike, pinnately compound; flowers many, on unbranched stems. Scrophulariaceae (Figwort family).

**Sickletop lousewort,** *P. racemosa* (ras-em-oh-sa: bearing racemes). Also **parrotbeak.** Flowers (our variety) pale bronze-pink, beak curled sideways, so inflorescence looks pinwheellike from above; calyx 2-lobed; leaves reddish, ± linear, fine-toothed, all on stem; 6–18". Openings near forest line, or high alpine. Color p 134.

**Elephant's head,** *P. groenlandica* (green-lan-dic-a: of Greenland). Flowers purplish pink, in a dense spike, elephantlike (upcurved beak as the trunk, lateral lower lobes as the ears); leaves preponderantly basal; 8–16". Subalpine rills and boggy meadows. Color p 134.

**Birdbeak lousewort,** *P. ornithorhynca* (or-nith-o-**rink**-a: bird beak). Flowers purplish pink, with a thin downturned beak, few, in a short round inflorescence; leaves almost all basal; 4–7". Moist alp/subalpine gravels in Wash Cas. Color p 134.

**Coiled-beak lousewort,** *P. contorta.* Flowers (our variety) pale yellow to white, in a loose spike; beak semicircular, arching back into lower lip; leaves mostly basal; 8–14". Alp/subalpine. Color p 134.

**Bracted lousewort,** *P. bracteosa* (brac-tee-oh-sa: with bracts). Flowers yellow (rarely purple to dark red), scarcely beaked, in a robust spike; stem leaves as big as basal ones, aging purplish; 24–36". Lush subalpine meadows.

**Mt. Rainier lousewort,** *P. rainierensis* (ra-near-en-sis: of Mt Rainier). Flowers yellow to cream, scarcely beaked, in a robust spike; stem leaves much smaller than basal ones; 6–16". Subalpine; only in Mt. Rainier vicinity.

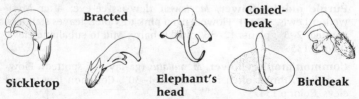

Bracted

Coiled-beak

Sickletop

Elephant's head

Birdbeak

Louseworts are as irregular and varied a genus of flowers as any. Their picturesque names are warranted; each flower shape suits the anatomy of one or more species of bumble bee—pollinators of the 100-odd species of lousewort. The undeserved name "lousewort" dates from an ancient superstition that cattle got lousy by browsing louseworts.

# Penstemons

*Penstemon* spp. (pen-stem-un: 5 stamen). Also **beardtongue**. Corolla typically blue or violet to pink or occasionally yellow-white, swollen-tubular, with 5 (2 upper, 3 lower) short rounded flaring lobes, and a broad ± hairy sterile stamen resting on the throat; fertile stamens 4, paired; sepals 5, hardly at all fused; leaves opposite, often basal-clustered in part. Rocky places and drier meadows; common. Scrophulariaceae (Figwort family).

**Creeping penstemon,** *P. davidsonii* (david-so-nee-eye: after Anstruther Davidson). Flowers blue to lavender, ¾–1½"; stems woody, forming dense mats 3–6" thick. Color p 134.

**Rock penstemon,** *P. rupicola* (roo-pic-a-la: rock dweller). As above exc more pink to red and leaves whitish-coated.

**Small-flowered penstemon,** *P. procerus* (pross-er-us: noble). Flowers usually blue/purple, ¼–½" long, in 1–3 apparent whorls; leaves narrowly elliptical, to 2" long; stem 4–24".

While most penstemons range between blue and pink, a white or yellow blossom isn't necessarily a rare species; several blue species, including *procerus,* may occur in pale form.

| *P. procerus* | *C. grandiflora* | *C. parviflora* |

# Blue-eyed Mary

*Collinsia* spp. (ca-lin-zia: after Zaccheus Collins). Also **innocence**. Corolla blue to violet, the upper 2 lobes fading to white; lobes seemingly 4, the fifth being the inconspicuous central lower lobe, creased shut to enclose the stamens and style; calyx 5-lobed, green; flowers in axils of upper, often whorled leaves; lower leaves opposite, ± linear; plants annual, 2–14". Sunny low to mid elevs. Scrophulariaceae (Figwort family).

*C. grandiflora* (gran-dif-lor-a: big flower). Corolla at least ½" long, bent over at right angles to the calyx. Color p 133.

*C. parviflora* (par-vif-lor-a: small flower). Corolla usually much less than ½", bent at an oblique angle to calyx.

# Foxglove

*Digitalis purpurea* (digit-ay-lis: finger—; pur-**pew**-ria: purple). Corolla pink (to purplish or occasionally white) tubular, 1½–2", with 5 very shallow lobes; sepals 5; stamens 4, paired; flowers many, in a very showy one-sided spike; leaves oval/pointed, up to 20 × 6"; stems fuzzy and ± sticky, 2–6'; biennial. Roadsides and scarified clearcuts. Scrophulariaceae (Figwort family). Color p 135.

Our most beautiful European weeds, foxgloves are the age-old source of digitalin, used as a heart stimulant in cases of cardiac arrest. Doctors prescribe it with care and precision; foxglove can be deadly to those with heart disorders.

# Youth-on-age

*Tolmiea menziesii* (tole-me-a: after William Tolmie, below; men-**zee**-zee-eye: after A. Menzies, p 94). Also **piggyback plant.** Calyx purplish to brownish green, bell-shaped but cut away on one side, ½", with 3 large and 2 small lobes; petals longer, threadlike, red-brown, 4 (or 5); stamens 3; flowers many, in a raceme; leaves shallowly palmate-lobed, 1½–3" broad, mostly basal on long leafstalks; stems several, 1–3', from rhizomes; entire plant hairy. Moist W-side forest. Saxifragaceae (Saxifrage family). Color p 136.

This anomalous saxifrage has several quaint names referring to its way of producing aerial offset plants from its leaf axils. These may take root when the mother stem reclines.

---

**William Frazer Tolmie** arrived at Fort Vancouver in 1833, a young Scot fresh out of medical school and into the employ of the Hudson's Bay Company. The same year, at his suggestion, the Company established there the Northwest's first lending library. In September while waiting for a ship north out of Fort Nisqually he set off to collect alpine plants from Mt. Rainier. Ankle-deep in fresh snow on a precipice directly facing the mountain (more likely Mt. Pleasant than the present Tolmie Peak) he dropped any notion of scaling Rainier. No white man had seen it so close.

He remained in the region all his life, a Company physician and later supervisor. His son became Premier of British Columbia. Like many early naturalists, he chose to be an M.D. because it was the standard way of being a biologist; botany and human physiology were not yet separated branches of natural science.

---

# Violets

*Viola* spp. (vie-oh-la: the Roman term). Also **Johnny-jump-ups**. Petals 5; lowest petal largest, ± different-colored than other 4, and with a rearward spur; sepals 5; stamens 5, short; flowers on long stalks; seed pods split explosively to propel seeds; leaves ± heart- to kidney-shaped. Violaceae (Violet family).

*V. sempervirens* (sem-per-**vee**-renz: ever green). Flowers yellow; leaves rather firm, evergreen, purple-flecked; flowers and leaves rising singly 1–4" from runners. Ubiquitous in W-side forest.

*V. glabella* (gla-**bel**-a: smooth). Flowers yellow; several flowers and leaves borne from the upper third of each 4–12" stem. Widespread, open forest and subalpine meadow.

*V. adunca* (a-**dunk**-a: hooked). Flowers lavender to blue; leaves heart-shaped to elliptical; stems clustered, 1–5". Scattered, esp in wet montane meadows.

*V. flettii* (flet-ee-eye: after J. B. Flett). Flowers lavender to blue; leaves kidney-shaped, often purplish; stems clustered, 1–6". Only in Olympics; subalpine, among rocks. Color p 135.

Our yellow violets bloom soon after the retreat of snow; often it's so early and chilly that pollinators are not yet on the wing. Violets respond by producing a second kind of flowers ("cleistogamous," meaning "closed marriage"), greenish, low, and inconspicuous, and able to pollinate themselves without ever opening, for sure-fire seed production.

# Mountain-sorrel

*Oxyria digyna* (ox-ee-ria: sharp—; **didge**-in-a: 2 ovaries). Flowers tiny, greenish, of 2 erect and 2 spreading lobes, 2 stigmas and 6 stamens, in rough spikelike panicles; fruit rust red, tiny, 2-winged; leaves kidney-shaped, 1–2" broad, on long basal leafstalks, coloring brilliantly in fall; stalks several, 4–18". Rocky streambanks and wet talus; alp/subalpine. Polygonaceae (Buckwheat family). Color p 136.

Several acids make mountain-sorrel leaves sourly tasty. The ascorbic acid (vitamin C) is good for you, but the oxalic acid—common to wood-sorrel (*Oxalis*, page 214), true sorrels (*Rumex*), and rhubarb—can make you sick if you eat too much.

# Hedge-nettle

*Stachys* spp. (stay-kis: ear of grain). Corolla purplish red to pink, ½–1¼", with a round upper lip and a longer, 3-lobed lower lip; calyx 5-pointed; stamens 4, paired; flowers in an open spike of several whorls; leaves opposite, elliptical, toothed; stem square, 1–4'; fetid when bruised. Marshy places, low to mid elevs. Lamiaceae* (Mint family). Color p 132.

Like the English, Northwest Indians named this plant after nettles, and sometimes mixed the two as greens or tea. Hedge-nettles often take stinging nettles for associates, but claim mints for family relations, as evidenced by their flowers, square stems, and foliage that looks like a nettle but "stings" only like a mint—sweetly, through the nose.

# Self-heal

*Prunella vulgaris* (pru-nel-a: brown small; vul-gair-iss: common). Also all-heal. Corolla blue-purple, ¼–¾", with one hooding upper lobe and 3 lower lobes, one liplike and fringed; calyx ½ as long; stamens 4 (rarely lacking); flowers blooming sequentially upward, in a crowded broad-bracted spike of opposite pairs neatly offset 90°; leaves opposite, elliptical, 1–3"; stem squarish, 4–16" tall, or sprawling. Introduced, now sporadic up to mid elevs. Lamiaceae (Mint family). Color p 133.

The mint family is full of medicinal, poisonous, and culinary aromatic herbs, including catnip, pennyroyal, horehound, oregano, sage, savory, thyme, and of course peppermint. However, neither Northwest tribal lore nor modern pharmacology agree with the old European belief in self-heal as a panacea.

The name *Prunella* suggests that Linnaeus somehow saw this plant as a "little plum;" more likely he just misspelled the old name *brunella*, from "brown," not "plum." Neither misspelling nor misinformation count against a scientific name's validity.

---

*The mints were long known as Labiatae, after their lip-shaped flowers. See footnote, p 182.

## Sweetpeas

*Lathyrus* spp. (lath-er-us: the Greek term). Petals 5, our varieties blue to violet (or sometomes albino), the upper one largest, the lower 2 partly fused, creased, enclosing the pistil and 10 stamens; calyx 5-toothed; stigma toothbrushlike—bristly on upper surface; fruit a pea pod; leaves pinnately compound, often tendril-tipped; leaflets 6–16; stems angled, climbing or (these species) ± erect, 4–32". Fabaceae* (Legume family).

*L. nevadensis* (nev-a-den-sis: of the Sierra Nevada). Flowers 2–10; leaflets 8–10. Clearings up to timberline. Color p 135.

*L. polyphyllus* (poly-fill-us: many leaves). Flowers 5–15; leaflets 12–16. Drier open woods, lower W-side.

Our two common sweetpeas rarely climb or clamber, unlike many members of their genus, so their tendrils are short, straight, and nonfunctional. Locoweeds and crazyweeds (below) lack tendrils altogether, the leaf terminating instead in a leaflet. Vetches (genus *Vicia*) are harder to tell from peas; their key difference is in the tiny pistil, which on a vetch is hairy near its tip on all sides—like a bottlebrush, as opposed to the toothbrush-shaped pistil on a pea.

## Crazyweed

*Oxytropis campestris* (ox-it-ra-pis: sharp keel; cam-pes-tris: of fields). Also **stemless locoweed**. Flowers pealike, pale yellow (our variety), 8–12; calyx black-hairy; pod thin-walled; leaves ± basal, densely silky-hairy, pinnately compound; leaflets 13–25; stalks several, 3–15". Alpine and arid meadows. Fabaceae (Legume family). Color p 135.

Crazyweeds and their close relatives the locoweeds, genus *Astragalus,* got their names and ill reputes from their several rangeland species that sabotage the muscular coordination and vision of cattle that graze them in quantity. However, the many mountain goats that graze this sweet alpine variety don't appear any crazier than other mountain goats, and you would think they wouldn't still be around if it made them uncoordinated.

*The legumes were long known as Leguminosae. See footnote, p 182.

# Lupines

*Lupinus* spp. (lu-**pie**-nus: the Roman term). Flowers blue to purple (sometimes white), pealike, small ½"), many, in several ± conical racemes; calyx 2-lobed; pods hairy; leaves mostly basal on long leafstalks, palmately compound; leaflets 5–9, center-folded. Fabaceae (Legume family).

**Broadleaf lupine,** *L. latifolius* (lat-if-oh-lius: broad leaf). Plant bushy, 1–3′, woody-based. Subalpine, abundant. Color p 133.

**Dwarf lupine,** *L. lepidus lobbii* (lep-id-us: charming; lobb-ee-eye: after William Lobb). Plant dwarfed, semiprostrate, the leafstalks and flower stems radiating horizontally but the racemes ± erect; leaves 1″ diam. Mostly on alpine gravels and tundra. Color p 135.

Lupine in bloom paints our bluest subalpine slopes—you can literally see it a mile off. On more intimate inspection, lupine leaves have a charming way of maintaining a little sphere of water all day on the centerpoint of each leaf. Other lupines, in the sagebrush, pine, and alpine zones where water is too scarce for such displays, are dramatically silvered with downy hair. This is not a useful trait for identifying species, but an environmental response for conserving water. The sheen reflects away some of the sun's radiant heat, while the fuzz keeps drying breezes off of the moist leaf surface.

We don't know for certain why the Romans named these flowers after wolves. ("Lupine" means "of wolves" or "wolflike" in English, too.) It may indicate an affinity they felt; remember their myth of a she-wolf as the mother of Rome, or at least its wetnurse. Or it may decry lupines so aggressively weedy as to seem predatory. Actually lupines, like other legumes, are nitrogen-fixing, and improve the soil.

# Pussypaws

*Spraguea umbellata*\* (sprayg-ia: after Isaac Sprague; um-bel-ay-ta: with flowers in umbels). Sepals 2, round, ⅛–⅜″ diam, sandwiching and nearly hiding the 4 much smaller petals and 3 stamens; flowers rust-pink to yellowish white, in fluffy, chaffy heads on prostrate stalks radiating well past the cushion of 1–3″, narrow, semisucculent basal leaves. Alp/subalpine, or dry E-side; rock crevices and ± barren gravels, esp pumice. Portulacaceae (Purslane family). Color p 135.

---

\*Some texts call this plant *Calyptridium umbellatum*.

---

# Bleedingheart

*Dicentra formosa* (di-sen-tra: 2 spur; for-mo-sa: beautiful). Corolla pink, ¾–1″ long, pendent, shaped like an elongate heart made of two fused petals, hiding a smaller pair of fused petals inside; sepals 2, falling off early; stamens 6; seed pod growing as long as 2″ out through the corolla mouth; stalks several-flowered, 12–20″; leaves fernlike, compound and incised. Open spots up to mid elevs, W-side forest. Fumariaceae (Fumitory family). Color p 133.

# Corydalis

*Corydalis scouleri* (cor-id-a-lis: the crested lark; scoo-ler-eye: after John Scouler, p 67). Flowers light pink, 15–35, angled at various tangents along a spike axis; petals 4, the upper 1 with a rearward spur at least as long as the forward part, the inner 2 ± hidden; sepals 2, small, falling off early; stamens 6; seed pod splitting explosively to propel seeds; leaves large, multiply compounded, round-lobed; plants 1½–4′ tall. Seeps and streambanks, lower W-side. Fumariaceae (Fumitory family). Color p 133.

There is also a very rare aquatic Corydalis, *C. aquae-gelidae,* limited to a few streams and marshes in the Cascades of extreme northern Oregon and southern Washington.

# Larkspurs

*Delphinium* spp. (del-fin-ium: the Greek term, derived from "dolphin"). Flowers deep blue to violet, ¾–1¼″ diam; sepals 5, petallike, spreading, the upper one with a long nectar-bearing spur behind; petals 4, the upper 2 spurred (within the sepal spur), often much paler; leaves 2–5″ diam, narrowly palmately lobed and/or -compound; stems often branched. Ranunculaceae (Buttercup family).

**Menzies larkspur,** *D. menziesii* (men-zee-zee-eye: after A. Menzies, p 94). 6–20″ tall; root tuberlike. Low W-side clearings. Color p 134.

**Rockslide larkspur,** *D. glareosum* (glare-ee-oh-sum: on gravel). 6–12″; leaves mostly basal, ± fleshy. Alp/subalpine rock crevices.

Larkspurs other than these two species may be much taller and may have smaller, narrower, or paler flowers, but all share the structure—a long-spurred upper sepal, and two unlike pairs of petals at the flower's center. All are poisonous to cattle when eaten in quantity. The alkaloid-bearing seeds have for millennia been ground up to poison lice.

# Monkshood

*Aconitum columbianum* (ac-o-**nigh**-tum: the Greek term; co-lum-be-**ay**-num: of the Columbia River). Sepals 5, petallike, blue-purple (rarely greenish or albino), the upper one hooding, helmetlike; true petals apparently two, smaller, hidden under hood; flowers ¾–1½" tall, in an open raceme atop the 1–6' stem; leaves 2–5" diam, palmately deeply incised. Moist sites, mainly E of Cas Crest. Ranunculaceae (Buttercup family). Color p 133.

Most blue irregular flowers are pollinated by bumble bees. Monkshood's odd-shaped flower excludes from its nectary all insects except highly motivated, intelligent bumble bees, whose advantage to the plant is their fidelity; having once been well rewarded with monkshood nectar, they will visit *only* monkshoods, whose pollen will not then be squandered haphazardly among several flower species. Unlike the nectar, monkshood foliage is toxic, and few animals eat it.

---

**Herbs:** regular flowers, 5 petals

---

These typically have 5 sepals and 5, 10, 15 or more stamens in addition to their 5 petals or corolla lobes (fused petals); the female parts aren't in fives. Also included (on pages 220–27) are flowers with just one set of conspicuous parts in fives—either petals or sepals but not both, and not the stamens. All these variables tend to reveal family relationships. For example, the Rose family mostly has 15 or 20 stamens; the Phlox family has only 5; the Primroses have 2 sepals and the Parsleys have none.

If you see a 5-petaled flower that you can't locate here, check for it among the following: **Shrubs** include many 5-petaled genera; very low plants barely recognizable as subshrubs are on pages 106–14 if their genus as a whole is evergreen, i.e., the leaves firm and glossy, each green through two or more winters. **Irregular dicots**, p 194, mostly have 5 corolla lobes, but these lobes are very unlike each other. **Composites**, p 182, have tight heads of tiny florets which are technically 5-parted, and sometimes visibly so. **Gentian,** p 236, and **Starflower**, p 233, have 5 petals in many blossoms, but usually have 6 or 7 on at least as many others in the vicinity.

---

# Spreading Phlox

*Phlox diffusa* (flocks: a Greek flower, meaning "flame"; dif-**you**-sa: spreading). Corolla white, pink, pale blue, or variously violet, ¾" diam, with 4–6 flat-spreading lobes at the end of a straight tube; pistil 3-tipped; flowers single on numerous leafy stems; leaves linear, pointed, ¾"; compact mats 2–4" thick. Dry ground, esp alpine but also at all elevs E Cas. Polemoniaceae (Phlox family). Color p 137.

High ridgecrests colonized by phlox fail to retain much soil or snow, which blow away. Without a snow blanket through winter, plants there are exposed to ferocious drying winds, and to hundreds of freezing/thawing cycles each year. The plants withstand frost action whose powers of pulverization are amply displayed on nearby rocks. Though the rocks continually break up, their particles tend to blow away as they reach soil size, preventing soil from accumulating.

Spreading phlox exemplifies the "cushion" form adaptive to this extreme environment. The smooth convex surface, like an airfoil, eases the wind on over with a minimum of resistance; often phlox contours itself in the lee of a larger rock or a crevice between rocks. The tiny, crammed leaves live in a pocket of calm partly of their own making, and there they trap windblown particles that slowly become a mound of nourishing soil. Mt. St. Helens' 1980 eruptions provided a striking demonstration: in the Goat Rocks area, the inch-thick blanket of ash was largely incorporated into the top soil layer within five years at most elevations, but from the alpine ridges it was completely removed by wind—except for what lay in deep crevices, or in conspicuous gray aureoles around the cushion plants.

Cushion plants also need very long (8–15') taproots. Though the ridgetops receive their share of our mountains' heavy snowfall, sweeping winds and a coarse rock substrate allow little of it to stay there. In effect, alpine habitats are arid despite generous precipitation.

# Skyrocket

*Gilia aggregata** (jil-ia: after Felipe Luis Gil: ag-reg-**ay**-ta: clustered).
Also **scarlet gilia.** Corolla scarlet, trumpet-shaped, the slightly flaring
tube twice as long (½–1¼″) as the ± recurved, pointed lobes; stamens borne near mouth of tube; leaves dissected, lobes linear; stems
to 3′, many-flowered, or dwarfed (4″) at high elevs. Dry clearings
near and E of Cas Cr. Polemoniaceae (Phlox family). Color p 138.

The long tubular corolla and bright red color are both clues
that this flower evolved with hummingbirds as pollinators.
Most insects cannot see red, but hummers crave it.

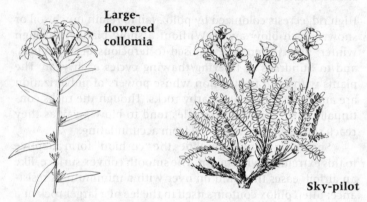

Large-
flowered
collomia

Sky-pilot

## Alpine Collomia

*Collomia debilis* (col-oh-mia: glue—; deb-il-iss: weak). Corolla
trumpet-shaped, ½–1″, lavender-streaked or white to pink; stamens
sometimes blue-tipped; each flower subtended by a green bract;
leaves of our varieties 3–7-lobed; typically forming dense cushions
2–3″ deep. Alpine. Polemoniaceae (Phlox family). Color p 137.

## Large-flowered Collomia

*Collomia grandiflora* (grand-if-**lor**-a: large flower). Corolla trumpet-
shaped, ¾–1¼″, light salmon-orange to ± white; stamens sometimes blue-tipped, often very unequal in length (split the corolla to
see); flowers subtended by green bracts, borne in a single head, plus
sometimes a few flowers in lower leaf axils; seeds gluey when moistened; leaves linear, 1–3″; plants annual, 8–40″ tall. Dry, ± sunny
spots at low to mid elevs. Polemoniaceae (Phlox family).

*Some authorities split *Gilia* and call this plant *Ipomopsis aggregata.*

# Sky-pilot

*Polemonium pulcherrimum* (pol-em-oh-nium: the Greek term; pool-ker-im-um: most beautiful). Also **Jacob's-ladder.** Corolla light blue with yellow center, ½" long; leaves pinnately compound, mostly basal; leaflets 9–21; stems several, 4–10" (subalpine) or 10–20" (mid-elev); plants often skunky-smelling. Around subalpine trees and outcrops. Polemoniaceae (Phlox family). Color p 137.

Like most flowers, sky-pilots produce a sweet flowery fragrance in their nectaries to attract nectar-feeding insects as pollinators. In many individuals, however, this is drowned out by a skunky aroma on sepals and bracts below the flowers. A researcher of a related Rocky-Mountain species proposed that the foul smell serves to repel ants, and tends to be produced only by plants at elevations where ants abound. Unlike bees and hover flies (good pollinators) nectar-feeding ants slip right past the pollen-bearing stamens to rob the nectary, and often destroy the pollen-receiving pistil in the process. See if you can find both skunky and sweet-smelling sky-pilots, and any pattern separating them, in our mountains.

# Dogbanes

*Apocynum* spp. (a-pos-in-um: away dog). Flowers bell-shaped, clustered; leaves opposite; stems milky-juiced, smooth, often reddish, much branched (resembling a shrub); seeds cottony-tufted, in paired long, slender pods. Scattered, commoner at low elevs. Apocynaceae (Dogbane family).

**Spreading dogbane,** *A. androsaemifolium* (an-dross-ee-mif-oh-lium: leaves resembling genus *Androsaemum*). Flowers pink, fragrant, ¼–⅜", with flaring lobes; seed pods 3–6"; leaves oval, ± glossy above, pale beneath, spreading flat or drooping; plants 1–2'. Dry but often ± shady sites. Color p 138.

**Indian-hemp,** *A. cannabinum* (can-a-by-num: hemp—). Flowers tiny, whitish, with ± straight lobes; seed pods 5–7"; leaves narrow elliptical, yellow-green, ascending; plants 1½–4½'. Usually in sun.

Indian-hemp was one of the best sources of plant fiber available to the Indians. Where none could be found, spreading dogbane was sometimes substituted. Dogbane attracts a huge variety of of butterflies and bees. To humans and perhaps to dogs, its foliage is "banefully" purgative and diuretic.

## Shooting Stars

*Dodecatheon* spp. (doh-de-**cayth**-ee-on: twelve gods). Corolla and calyx bent sharply back, ½–1" long, fused "collar" portion very short; stamens tightly clasping the pistil, or just slightly spreading; flowers several, many facing down (i.e., the petals up) but all erect when in fruit (a capsule); leaves basal on ± long leafstalks; stem 8–20". Subalpine streamsides and wet meadows. Primulaceae (Primrose family).

*D. pulchellum* (pool-**kel**-um: beautiful). Petals 5, pink with yellow collar and dark stamens.

*D. jeffreyi* (**jef**-ree-eye: after John Jeffrey, below). Petals usually 5, deep pink with white collar. Locally abundant. Color p 138.

*D. dentatum* (den-**tay**-tum: toothed). Petals 5, white with magenta collar and stamen base; leaves broadly oval, ± toothed. Uncommon.

*D. alpinum*. Petals 4, pink. Ore Cas, commoner southward.

## Douglasia

*Douglasia laevigata* (da-**glass**-ia: after David Douglas, p 18; lee-vig-ay-ta: smooth). Corolla crimson to deep pink; ½" diam, lobes round-tipped, stamens barely (or not quite) appearing at the narrow throat where the lobes spread at right angles from the narrow tube; flowers mostly 2–4 on leafless branched stalks; leaves ± linear, ½"; in compact mats 2–3" thick. Alpine, esp in Olys; or on Columbia Gorge bluffs. Primulaceae (Primrose Family). Color p 137.

Douglasia resembles and often grows near a commoner cushion plant, spreading phlox (page 205). To distinguish them, note that Douglasia's deeper-red flowers are on branched, leafless stalks, and each have a single, unparted stigma.

John Jeffrey was hired for plant hunting on the far edge of the New World by a club of Scottish gentlemen called the Oregon Association. He explored the Okanogan, Similkameen, Fraser, Willamette, and Umpqua drainages in 1851–53, arriving overland from Hudson's Bay. He sent his last shipment of plants from San Francisco at the height of the Gold Rush, and disappeared.

# Phacelias

*Phacelia* spp. (fa-**see**-lia: bundle, referring to the dense inflorescence). Flowers crowded, bristling with stamens about twice as long
as corollas; calyx hairy; leaves gray-silky-coated. Hydrophyllaceae
(Waterleaf family).

**Silky phacelia,** *P. sericea sericea* (ser-**iss**-ia: silky). Flowers purple (to
blue or white) in a dense round spike; stamens pale-tipped; leaves
deeply pinnately lobed, mostly basal, cushion-forming; stem 4–10".
In Wash, mainly alpine. Color p 137.

**Silverleaf phacelia,** *P. hastata compacta* (hass-**tay**-ta: halberd-
shaped, referring to the pair of leaf lobes). Flowers drab whitish (rarely lavender); leaves narrow elliptical, often with a pair of small lobes
or leaflets at the base; this variety rarely over 5" tall. Dry, rocky alpine
sites, Ore and S Wash Cas. Color p 137.

Each of our several *Phacelia* species has compact alpine and
leggy mid-elevation varieties. The silky is most striking, with
ornate leaves and a compact, colorful inflorescence.

**Shooting star**　　　　**Waterleaf**
*D. alpinum*　　　　　　*H. fendleri*

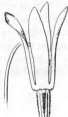

# Waterleaves

*Hydrophyllum* spp. (hydro-fill-um: water leaf). Flowers in compact
heads bristling with black-tipped stamens twice as long as the ⅜"
corollas; calyx hairy; leaves pinnately 5–11-compound. Hydrophyllaceae (Waterleaf family).

*H. capitatum* (cap-it-**ay**-tum: flowers in heads). Flowers lavender to
white, in spherical head; leaflets ± lobed, leaves basal, much taller
(4–10") than the flower stalk (1–3") exc in a Columbia Gorge variety
with 2–8" stalks and short leaves. Scattered E-side. Color p 138.

*H. fendleri* (fend-ler-eye: after August Fendler). Flowers white, occasionally purple-tinged; leaflets toothed and often cleft, leaves up to
12 × 6", with soft white hair underneath; plant 1–2". Lush thickets,
esp avalanche basins, mid elevs to subalpine.

---

Herbs: 5 fused petals, 5 stamens　　　　　　　*209*

# Bluebells

*Mertensia paniculata* (mer-ten-sia: after Franz K. Mertens, below; pa-nic-you-lay-ta: flowers in panicles). Also **lungwort**. Corolla blue, pink-tinged at first, short-lobed, narrowly bell-shaped, pendent, ½–¾"; stem leaves pointed-elliptical, basal leaves oval on long leaf-stalks, both often with bluish bloom; plants many-flowered, robust, 1½–5'. Streamsides or moist clearings, mainly E-sides. Boraginaceae (Borage family). Color p 138.

# Bellflower

*Campanula rotundifolia* (cam-pan-you-la: small bell; rotund-if-oh-lia: round leaf). Also **Scottish-bluebells, harebell.** Corolla pale blue, bell-shaped, ± pendent by maturity, ¾–1¼"; pistil 3-forked; stem leaves linear; basal leaves round to heart-shaped on long leafstalks, often withering by maturity; plants slender, 6–30". Open woods to alpine. Campanulaceae (Harebell family). Color p 138.

# Buckbean

*Menyanthes trifoliata* (men-**yanth**-eez: monthly flower; try-fo-lee-ay-ta: 3-leaved). Corolla white (often purple-tinged) with hairy-faced flat-spreading lobes (¼–½") on a straight tube (¼–½"); stamens purple-tipped; flowers in a ± erect, columnar raceme; 3 leaflets elliptical, 2–5"; stems and leafstalks usually ± prostrate and submerged; from rhizomes. Bogs and ponds. Menyanthaceae (Buckbean family). Color p 139.

---

**Karl Heinrich Mertens** accompanied the Russian **Count Fedor Lütke** on his globe-circling voyage of 1826–29, when both were in their early thirties. In London in 1829 they so impressed David Douglas (page 18) that he became obsessed with completing his second trip to the Northwest by sailing from the Russian colony at Sitka to the Siberian shore. He would then have walked the length of Siberia, collecting plants.

Mertens died the following year. His plant discoveries on Sitka Island (Southeast Alaska) include a half-dozen species named *mertensiana (-ianus, -ii)* and as many more named *sitchensis (-ense)*, such as Sitka spruce, as well as the partridgefoot, *Luetkea*, but not the *Mertensia* bluebells, which were already named after botany professor **Franz Karl Mertens.**

---

**Sibbaldia**

**Mitrewort**
*M. breweri*

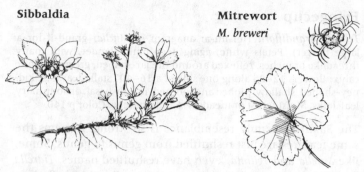

## Sibbaldia

*Sibbaldia procumbens* (sib-ahl-dia: after Sir Robert Sibbald; pro-cum-benz: prostrate). Petals tiny, yellow, sitting on top of 5 slightly longer green bracts alternating with 5 much larger (up to ¼") green calyx lobes; leaflets in 3s, ½–1½", 3(–5)-toothed at the tip, white-hairy; leafstalks and flower stems rising 2–4" from rhizomes or prostrate stems. Alpine, esp in sedge meadows. Rosaceae (Rose family).

## Mitreworts

*Mitella* spp. (my-tel-a: mitre—). Also **bishop's caps.** Petals threadlike, branched, sticking out between the white calyx lobes; flowers 10–20 along a 6–14" stalk; leaves ± kidney-shaped, scalloped to toothed, basal, on long hairy leafstalks. Moist subalpine forest or meadow. Saxifragaceae (Saxifrage family).

*M. breweri* (brew-er-eye: after William Brewer). Petals pinnately 5–9-branched; calyx saucer-shaped.

*M. trifida* (trif-id-a: 3-forked). Petals 3-branched; calyx bell-shaped.

It's the seed capsule that looks like a mitre—a bishop's tall, deeply cleft hat.

## Grass-of-Parnassus

*Parnassia fimbriata* (par-nas-ia: of Mt. Parnassus, i.e., montane habitat; fim-bree-ay-ta: fringed). Petals white, ¼–½", long-fringed near base; flower single on a 6–16" stalk; leaves heart-shaped, basal on long leafstalks, plus a small stalkless leaf halfway up stem. Subalpine wet meadows and streambanks. Saxifragaceae. Color p 140.

## Fringecup

*Tellima grandiflora* (tel-im-a: anagram of *Mitella;* grand-if-**lor**-a: large flower). Petals white, aging deep pink, slender, with many threadlike branches, reflexed around the jar-shaped, greenish white calyx; flowers 12–25 along one side of a 16–40" stalk; leaves ± kidney-shaped, shallowly lobed and toothed, mostly basal, on long hairy leafstalks. Moist forest or meadow. Saxifragaceae. Color p 140.

The Saxifrage family resemblance is so strong, it seems the same features are just reshuffled from genus to genus. Some, like *Mitella* and *Tellima,* even have reshuffled names. *Tiarella* and *Tolmiea* are typical—racemes of small, whitish flowers above toothed, roundish basal leaves in shady, damp habitats.

## Tiarella

*Tiarella trifoliata* (tee-ar-el-a: crownlet; try-fo-lee-ay-ta: 3-leaved). Flowers tiny, white, many, in a sparse raceme on an 8–16" stalk; petals threadlike, unbranched, less visible than the stamens; ovary and (later) seed pod have 2 sides very unequal; leaves mostly basal on short leafstalks, hairy, toothed, from 3-lobed to 3-compound and incised, 2–4". Abundant in dense forest. Saxifragaceae. Color p 140.

This inconspicuous but abundant forest flower has been saddled with more than its share of fluffy names—foamflower, laceflower, coolwort, false-mitrewort—without the consensus that would make any one name genuinely common. Why not stick with tiarella? It's a diminutive of "tiara"—more a household word than "mitre" or "wort." The original tiara was a mini-turban in ancient Persia.

For their part, taxonomists until recently saddled the plant with three different species names, distinguished by whether the leaves are lobed, compound, or both. But field research on the three types failed to turn up any clear genetic, geographic, or ecological boundaries among them, so they were reduced to varieties. Scientific Latin is an imperfect antidote to confusion over synonymous names, since Latin names are also changed when new research warrants.

## Dotted Saxifrage

*Saxifraga punctata* (sac-sif-ra-ga: rock breaker; punk-tay-ta: dotted). Petals white; flowers ⅜" diam, several, in a loose round-topped raceme; stem 6–12", slightly woolly toward the top; leaves nearly circular, 1–3", with large, even teeth, slightly fleshy, basal on 4–8" leafstalks. Wet places, esp subalpine. Saxifragaceae. Color p 140.

## Alpine Saxifrage

*Saxifraga tolmiei* (tole-me-eye: after W. Tolmie, p 198). Petals white; stamens white, flattened, red-tipped; flowers ⅜" diam, 1–4, on 2–4" stems; leaves fleshy (about as thick as wide), linear, ¼–½" long, many, densely matted. Among rocks, alp/subalpine. Color p 140.

This cute little bugger with tubby, succulent leaves resembles stonecrop. It earns its name "rock-breaker," pioneering in rock crevices of alpine talus where release from snow is too brief (annually) and too recent (geologically) for any real soil to have developed. It may be virtually the only plant in a nearly barren community, or it may be joined by woodrushes, sedges, and mosses, or sometimes lichens.

## Stonecrop

*Sedum* spp. (**see**-dum: the Roman term, derived from "sitting"). Petals (most species; see next species) bright yellow, pointed, ¼–⅜"; flowers several, in ± compact broad-topped clusters; seed pod 5-celled, starlike; leaves thick, fleshy, crammed together, often turning red; plants low (3–8"), spreading by rhizomes and/or runners. Dry rocky places. Crassulaceae (Stonecrop family). Color p 141.

Fat succulent leaves, like cactus stems, maximize the volume-to-surface-area ratio, and hence the ratio of water capacity to water loss through transpiration. Even with leaves for storage tanks, stonecrops grow only while water is available; they store water to subsidize flowering and fruiting. Water-filled leaves might be vulnerable to frost damage, but stonecrops (and the alpine saxifrage, above) resist freezing and do well in the alpine zone. You can squeeze liquid water out of stonecrop leaves at temperatures that will freeze it immediately.

Our commonest species is the spreading stonecrop, *S. divergens,* unique here in that its leaves are opposite; other distinctions among yellow-flowered species are technical.

---

Herbs: 5 petals, 10 stamens

# Roseroot

*Sedum roseum*\* (roe-zium: rose-red). Also **midsummer-men, king's crown.** Petals deep red, ⅛"; stamens protruding; seed capsules 5 per flower, deep red; or (rarely) flower parts in 4s; flowers several, crowded, male and female flowers on separate plants; leaves pale green, rubbery; stems 2–8", unbranched, in clumps, from rhizomes. Alp/subalpine. Crassulaceae (Stonecrop family). Color p 141.

# Wood-sorrel

*Oxalis oregana* (ox-al-iss: the Greek term, from "sharp"; or-eg-ay-na; of Oregon). Petals white, usually with fine red veins; flowers single, ¾–1½" diam; leaves cloverlike, of 3 leaflets, folding down at night and some other times; stems and leafstalks 4–7", from rhizomes. W-side up to 4,000'; characteristic of moistest forest type. Oxalidaceae (Wood-sorrel family). Color p 141.

Most of our forest-floor herbs hold their leaves horizontal to maximize interception of scarce sunlight. Wood-sorrel does this, but then at times it creases them sharply downward, taking about six minutes to fold up, and thirty minutes to flatten out again. This response follows a puzzling variety of stimuli. When a patch of intense sunlight comes along, folding up surely conserves moisture and avoids sunscorch. At night the reason may be similar; no photosynthesis will be lost by closing up shop, and at least a small amount of evaporation will be curtailed. However, folding in the rain just has to have a different reason—possibly to reduce raindrop impact stress.

Wood-sorrel's leaves make good mouth entertainment. It is close to European garden sorrels, genus *Rumex*, in flavor, but unrelated to them. The common ingredient, oxalic acid, is mildly toxic, so don't overdose.

# Field Chickweed

*Cerastium arvense* (ser-ast-ium: horn—; ar-ven-see: of fields). Petals white, each with 2 round lobes; styles usually 5; calyx 5-lobed; flowers 3–8 per stem; leaves ± linear, ½–1¼", opposite or basal; entire plant downy. Various habitats; ours mostly alp/subalpine with low compact foliage and stems shorter than 8". Caryophyllaceae (Pink family). Color p 140.

---

\*The name *Sedum roseum* reflects broad concepts of both that genus and that species; taxonomic splitters prefer the name *Rhodiola integrifolia* for our populations.

# Sandworts

*Arenaria* spp. (air-en-air-ee-a: sand—). Petals white, ¼–½"; sepals scarcely ¼", blunt (pointed in some other sandworts); styles usually 3; leaves linear, mat-forming; stem 2–6". Alp/subalpine gravels and crevices. Caryophyllaceae (Pink family).

**Thread-leaved sandwort,** *A. capillaris* (cap-il-air-iss: hair-leaved). Basal leaves ¾–1¼", dense; stem leaves ⅜–¾", opposite, 4–10 per stem; capsule splits 6 ways luminous. Also in steppes E of Cas.

**Arctic sandwort,** *A. obtusiloba* (ob-too-si-lo-ba: blunt-lobed). Leaves ¼", dense, on prostrate stems; capsule splits 3 ways.

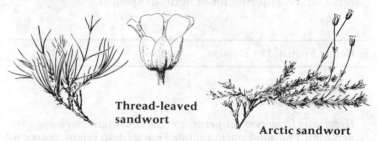

**Thread-leaved sandwort**

**Arctic sandwort**

# White Catchfly

*Silene parryi* (sigh-lee-nee: a Greek elf?; pair-ee-eye: after Charles Parry). Petals white or lavender-tinged, each deeply 4-lobed, with 2 more small lobes on the throat—a seeming inner whorl of 10 petals; calyx 5-lobed, hairy; leaves narrowly elliptical, 1–3", 2 or 3 pairs opposite, the rest basal; stem 6–15". Mainly subalpine, Olys and N Cas. Caryophyllaceae (Pink family). Color p 139.

# Moss-campion

*Silene acaulis* (ay-caw-lis: stalkless). Also **moss pink, cushion pink, carpet pink.** Petals pink (occasionally to white), separate though they form an apparent tube, bent 90° to spread flat; styles and/or stamens often protruding; styles usually 3; calyx shallowly 5-lobed; leaves linear, pointed, thick, crowded, to ½"; mosslike mats 2" thick. Alpine. Caryophyllaceae (Pink family). Color p 139.

This genus and the two preceding are in the Pink family, so called not because the petals are pink but because they are pinked, or notched at the tip. Pinked petals do run in the family, but other family traits such as ten stamens and unfused petals are more reliable, serving to distinguish moss-campion from phlox and douglasia. (See page 205 on cushion plants.)

# Yerba de Selva

*Whipplea modesta* (whip-lia: after A. W. Whipple). Also **whipplevine**. Flowers tiny (¼"), in a fluffy head; petals white to greenish, sticking out flat between the calyx lobes; leaves opposite, often persistent, ½–1", elliptical, vaguely toothed, 3-veined; stem rising 1–6" from leafy woody runners. Locally abundant in dry W-side forest, C and S Ore. Hydrangeaceae (Hydrangea family). Color p 141.

Yerba de selva means "forest herb" in Spanish.

---

Herbs: 5 petals, 15+ stamens

---

# St.-John's-wort

*Hypericum formosum* (hi-per-ic-um: the Greek term, meaning "under heath"; for-mo-sum: beautiful). Flowers deep yellow, orange in bud, many; stamens showy, as long (¾–1½") as the petals; capsule 3-celled; leaves opposite, oval, ± clasping-based; stems many, 4–30". Uncommon but striking; wet thickets and streamsides, esp low subalpine. Hypericaceae (St. John's wort family). Color p 141.

The original Old World St.-John's-wort was gathered to cast a spell for St. John'e Eve, June 23.

# Partridgefoot

*Luetkea pectinata* (loot-key-a: after Count Fedor Lütke, p 210; pectin-ay-ta: cockscomb—, referring to leaf shape). Flowers white to yellowish cream, ¼" diam, in a compact raceme on a 3–6" semishrubby stem; leaves ± persistent through winter, finely dissected into (usually 9) narrow lobes, mostly basal from runners or rhizomes; often carpet-forming. Abundant, subalpine and lower alpine. Rosaceae (Rose family). Color p 140.

Partridgefoot surrounds many of the sunken beds of black alpine sedge that emerge from late-melting snowbeds, indicating either a need for a slightly longer snow-free season than the sedge, or an intolerance for standing in water for a few days as the sedge does, or both. Similarly, on raw deglaciated gravel, it needs a slightly longer season than alpine saxifrage.

# Dwarf Raspberries

*Rubus* spp. (roo-bus: the Roman term). Petals ⅜" long, narrow; fruit a cluster of 1–5 glossy red 1-seeded drupelets ¼" long; flower stems and toothed leaves rising 1–2" from ± woody runners. Dense W-side forest, esp mid elevs. Rosaceae.

**Strawberry bramble,** *R. pedatus* (ped-ay-tus: 5-leafleted). Petals white; leaves compound, leaflets 5 (rarely 3, the lateral 2 not quite fully divided); runners smooth. Abundant in N Cas. Color p 139.

**Dwarf bramble,** *R. lasiococcus* (lazy-o-coc-us: shaggy berry). Leaves 3-lobed (sometimes 3-compound); runners smooth; ovary hairy.

**Snow bramble,** *R. nivalis* (niv-ay-lis: of snow). Petals usually pink; leaves ± heart-shaped, shallowly lobed (rarely 3-compound); runners fine-prickled. More common in Ore, and at lower elevs.

At their best, these berries miniaturize the essence of raspberry flavor as perfectly as wild strawberries do the essence of strawberry. More often, the ones we sample are flat-tasting and hard, since small rodents are close at hand to nab them as they ripen. It's hard to see them as raspberries at all, though they're the right color, flavor, and type of fruit—an "aggregate" of "drupelets"—because the big drupelets are simply too few to form the cup shape we expect of a raspberry.

**Snow bramble**

# Wild Strawberries

*Fragaria* spp. (fra-gair-ia: the Roman term). Petals white to pinkish, nearly circular, ¼–½"; sepals apparently 10; berry up to ½" long; leaflets in 3s, toothed coarsely and ± evenly, on hairy leafstalks; plant 3–8", from runners. Forest openings up to 4,000'. Rosaceae.

*F. vesca* (ves-ca: thin). Most leaves minutely hairy on top, ± bulging between veins. Color p 139.

*F. virginiana*. Most leaves smooth and flat on top.

Close kin of these precious morsels are proudly served (at champagne prices) by some of the most famous restaurants in France. Eat 'em up.

# Cinquefoils

*Potentilla* spp. (po-ten-til-a: small but mighty). Petals yellow, usually notch-tipped; 5 true sepals alternating with 5 shorter bracts; leaves compound, leaflets 1–1½"; stem 4–12". Rosaceae.

*P. flabellifolia* (fla-bel-if-oh-lia: fan leaf). Leaflets 3, thin, coarse-toothed. Subalpine meadow.

*P. diversifolia* (div-er-sif-oh-lia: varied leaf). Leaflets 5, deeply toothed. Alpine.

One cinquefoil is easily overlooked, but a few thousand of them in the high-country ultraviolet can take your breath away. Fan-leaf cinquefoils are among the earliest subalpine meadow flowers to bloom in quantity, following close upon springbeauties and glacier lilies, and preceding the lupine, valerian, and bistort that typically swamp them a few weeks later. Varied-leaf cinquefoil is equally common a bit higher, especially in the northeastern Cascades. There is also a shrubby cinquefoil (page 86). Both *cinquefoil* (French for "five-leaf") and *potentilla* (referring to medicinal potency) were names cut out to fit European species, and don't fit ours.

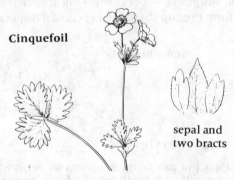

**Cinquefoil**

sepal and
two bracts

# Purple Avens

*Geum triflorum** (jee-um: the Roman term; try-flor-um: 3 flowers). Also **prairie smoke, old-man's-whiskers.** Flowers pendent, in 3s, dull reddish, 1' long, vase-shaped, scarcely opening; petals yellow to pink, mostly hidden by the 5 reddish sepals and 5 bracts; seeds long-plumed; leaves mostly basal, fernlike, hairy. Various habitats, esp alpine ridges, where 4–8" tall. Rosaceae. Color p 141.

*Some recent texts list this species as *Erythrocoma triflora.*

# Goatsbeard

*Aruncus sylvester* (a-**runk**-us: beard of goat; sil-**ves**-ter: of forests). Flowers cream white, minute, in large (to 14") stringy panicles, the males (on separate plants) much fluffier than the females, since the 15–20 stamens are the largest flower part; leaves twice- or thrice-compound, to 20", leaflets fine-toothed, long-pointed oval, 3–6"; plant 3–6', easily mistaken for a shrub. Moist thickets up to low subalpine. Rosaceae. Color p 146.

Goatsbeard roots and leaves made medicine for sores and sore throats among the Northwest tribes.

# Subalpine Buttercup

*Ranunculus eschscholtzii* (ra-**nun**-cue-lus: froglet; es-**sholt**-zee-eye: after J. F. Eschscholtz, below). Petals glossy yellow; seeds in a tight conical head; leaves generally 3-lobed to -compound, but extremely variable; stem 3–8", smooth. Subalpine meadow or scree, esp where wet. Ranunculaceae (Buttercup family). Color p 138.

Buttercups are called *Ranunculi* ("littlest frogs") for being small and green around the edges of ponds. They are called buttercups for the peculiar waxy (cutinous) sheen of their yellow petals—your first clue for telling buttercups from cinquefoils. Your second and surer clue is that buttercups lack the additional five sepallike bracts. (See opposite page.)

# Columbine

*Aquilegia formosa* (ak-wil-ee-jia: the Roman term, referring either to water-bearing or to eagle claws; for-**mo**-sa: beautiful). Flowers

---

**Johann Friedrich von Eschscholtz** (one of the most misspelled names in botanical Latin) and Adelbert von Chamisso were the naturalists on a Russian exploration of 1815–18, handsomely financed by **Count Romanzoff** and commanded by Captain Kotzebue. They collected extensively in Alaska and California, but apparently sailed right past Washington and Oregon. Chamisso, a poet from Berlin, was the primary botanist while Eschscholtz, an Estonian doctor, preferred insects.

---

nodding, several, the petal spurs and spreading ¾–1¼" petallike sepals red, while the stamens and short, cuplike petal blades are yellow; leaves compound with 9 round-lobed leaflets, mostly basal on tall leafstalks; stems branching, 1–3'. Lush subalpine meadows to lower clearings. Ranunculaceae (Buttercup family). Color p 141.

The shape of columbine flowers is unmistakable, while the colors change from species to species. The Colorado state flower is blue and white. A pale yellow Rocky Mountain columbine, *A. flavescens*, enters the northeast fringe of our range, where it is sometimes flushed with pink after messing around with our red species. Different columbines are interfertile and often grow together; they owe their separate identities to the fact that pollinators attracted to red flowers are uninterested in blue flowers, and so forth. The red/yellow combination is preferred by hummingbirds, many bees, and this writer. Bees nip the bulbous spur-tip to get the nectar, and may also go around to the front door for pollen.

---

Herbs: 5 petals, 2 sepals

---

# Springbeauty

*Claytonia lanceolata* (clay-toe-nia: after John Clayton; lan-see-o-lay-ta: narrowleaved). Petals ¼–½", white (rarely yellow, in N Cas) usually with fine pink stripes; stamens 5; stem leaves 2, opposite, ½–3", narrow-elliptical; stems 3–6", succulent, hollow, weak, several-flowered, often with 1–4 basal leaves; from a bulbous root. Meadows, esp subalpine. Portulacaceae (Purslane family). Color p 142.

Springbeauty bases its success on timing rather than brute size. It begins growing at its bulb tip in September, just when its neighbors are dying back. While snow covers it for the next eight months or so, holding the soil temperature close to freezing, the shoot inches up to the soil surface. Very few plants (all of them arctic/alpine specialists) are active at such low temperatures. Without the snow blanket it would be even colder and growth would be impossible.

As soon as the snow melts away from the shoot in spring, springbeauty bursts to its full height of three or four inches, expending in a few days its disproportionately large reserve of starches. It can push through the last inch or two of snow by combusting starch to melt itself a hole. It has two to four weeks to complete its life cycle—blooming, setting seed, and photo-synthesizing like mad to store up starches for the next spring. Then it withers, existing only underground for late July and August, the peak growing season for its associates.

Plants on this precocious schedule are called "spring ephemerals." Many, such as blue-eyed-grass, are common on semiarid land with just a few weeks of wet soil following snowmelt. Springbeauty does well on such sites on the lower East slope, but in subalpine meadows its timing has a different purpose—jumping the gun on bigger, leafier plants that will monopolize the light later in the season.

A remarkable adaptation in this and many early-bloom-ing high-country plants is a thin-fleshed hollow stem used as an internal greenhouse. When stored carbohydrates are burned off during the quick burst of growth, some of the heat produced is retained in the stem, making the internal air temperature warm enough for photosynthesis even when the outside air is not. Waste carbon dioxide from respiration also stays inside, available for synthesis into new carbohydrates.

With their concentrated starches, springbeauty bulbs are good survival forage. They taste radishy. Unfortunately, they are depleted when in bloom and hard to locate at other times, so don't bother them unless starvation impends.

## Miner's-lettuce

*Montia sibirica* (mont-ia: after Giuseppi Monti; sib-ee-ric-a: Siberi-an). Also **candyflower**. Petals ¼–½", notch-tipped, white with fine pink veins; sepals 2; stamens 5; stem leaves 2, opposite, pointed/oval, 2–4", ± clasping-based; basal leaves at least as large, several, on long leafstalks; stem 5–16", ± succulent, several-flowered. Moist forests. Portulacaceae (Purslane family). Color p 142.

Miner's-lettuce is one of our few common annuals—plants that grow from seed and die within a single growing season. Where it grows, you find it germinating and blooming at any time of spring or summer. To try "miner cuisine," find young unbloomed ones; they're mildest and tenderest.

The definition of petals and sepals states that if they're in just one whorl, they're sepals, no matter how colorful or tender. Technically, then, there are no flowers with 5 petals and no sepals, but in appearance there are quite a few. Parsley family flowers have a vestigial fleshy ring barely visible below the 5 petals; this ring is a calyx, but its lobes are so slight ("obsolete," in the official euphemism) as to be indetectable. Also described here is valerian, whose sepals open late, when the flower goes to seed, so they are not visible on the fresh flower.

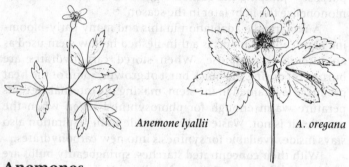

*Anemone lyallii*       *A. oregana*

## Anemones

*Anemone* spp. (a-nem-a-nee: wind—). Also **windflower**. Sepals 5 (sometimes 6), petallike; petals lacking; stamens many; stem leaves 3, in a whorl ¾ of the way up; basal leaves 1 to many; stems 3–12", from rhizomes. Ranunculaceae (Buttercup family).

*A. deltoidea* (del-toe-**eye**-dia: triangular). Flowers white (occasionally pinkish), 1–2" diam; stem leaves coarsely toothed, oval, attached directly to the stem without leafstalks. Common, W-side forests of Ore and S Wash Cas. Color p 139.

*A. oregana* (or-eg-**ay**-na: of Oregon). Flowers blue-violet (rarely pink or white), 1–2" diam; stem leaves 3–5-compound and lobed, on leaf-stalks. Columbia Gorge and E-side, Lake Chelan to C Ore.

*A. lyallii* (lye-ah-lee-**eye**: after David Lyall, p 51). Flowers white to pink or bluish, ¾" diam; stem leaves 3-compound and toothed, on leafstalks; stem 3–4". Uncommon, W-side up to timberline.

These woodland anemones bear a close resemblance to each other, but not to our subalpine anemones (page 236).

Newberry's
fleeceflower

## Newberry's Fleeceflower

*Polygonum newberryi* ( pa-lig-o-num: many "knees" or stem joints; new-bear-ee-eye: after John Newberry, below). Also **Newberry's knotweed.** Flowers greenish, in inconspicuous spikes wedged in leaf axils; calyx lobes 5, unequal; petals lacking; stamens 8; leaves gray-green, oval, ½–2"; stems to 16", often partly prostrate, with flaring papery sheaths above each leaf. Alpine gravels. Polygonaceae (Buckwheat family).

Fleeceflower shoots come up deep red in spring, often from semibarren pumice or serpentine gravel. The red may disappear, only to reappear by autumn. (See page 227.)

## American Bistort

*Polygonum bistortoides* (bis-tort-oh-eye-deez: resembling European bistort). Flowers white, small, chaffy, in a dense head about 1" × ½"; calyx lobes 5, unequal; petals lacking; stamens 8; stem leaves few, ± linear, sheathing; basal leaves much larger (3–6", on 3–6" leaf-stalks), elliptical; stem unbranched, 10–30". Subalpine. Polygonaceae (Buckwheat family). Color p 142.

Bistort typifies lush subalpine meadows, associating with lupine, fescue, or showy sedge. Few congeners look less alike than this bistort and the fleeceflower (above).

---

**John Strong Newberry** was surgeon-naturalist on an 1858 Army party looking for a railway route along the east flank of the Cascades from California to The Dalles. Specializing in geology and herpetology, he was probably prouder to be remembered for Newberry Crater—a great caldera comparable to Crater Lake, if not so famous—than for a fleeceflower.

---

# Cow-parsnip

*Heracleum lanatum* (hair-a-clee-um: the Greek term, after Hercules; lan-ay-tum: woolly). Flowers white; inflorescences 1 to several, nearly flat, 5–10" diam, petals near edge of inflorescence much enlarged and 2-lobed; stamens 5; leaflets only 3, huge (6–16"), palmately lobed and toothed; stems juicy, aromatic, hollow, 3–10'. Moist thickets up to low subalpine elevs. Apiaceae* (Parsley family).

Cow-parsnips, avidly eaten and widely eradicated by cows, are also popular with wild browsers. Northwest tribes ate the young stems, either raw or cooked but usually peeled first to remove a weak toxin. The stems taste milder and sweeter than the rank fragrance would suggest. Careless plant foragers in the Northwest have died from eating water-hemlock, *Cicuta douglasii*, and poison-hemlock, *Conium maculatum*, mistaking these plants for cow-parsnips. Stature, habitat, inflorescence, and purple-spotted stems contribute to this error; yet the poisonous parsleys have twice- or thrice-compound lacy leaves utterly unlike the three huge palmately lobed leaflets of cow-parsnip. To be safe, never eat any part of any wild "carrot-topped" plant with finely dissected or compounded leaves.

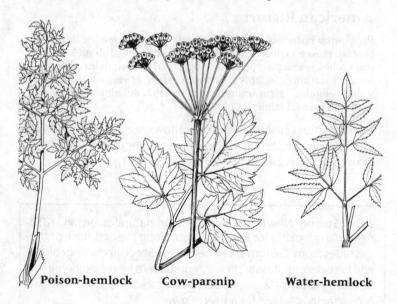

**Poison-hemlock**  **Cow-parsnip**  **Water-hemlock**

*The parsleys were long known as Umbelliferae. See footnote, p 182.

## Gray's Lovage

*Ligusticum grayi* (lig-us-tic-um: the Roman term, referring to the Liguria region; gray-eye: after Asa Gray). Also **licorice-root.** Flowers white (sometimes purple-tinged), in 1–3 slightly rounded inflorescences 2–4" diam; leaves like Italian parsley, nearly all basal; stem 8–24", from a thick fragrant taproot. Lush subalpine meadows. Apiaceae (Parsley family).

Lovage is the "Queen Anne's lace" look-alike common in subalpine meadows in late summer.

## Sweet Cicely

*Osmorhiza* spp. (os-mo-rye-za: odor root). Inflorescence loose; seeds slender, bristly; leaflets usually 9, oval, toothed; plant leggy, 1–3', licorice-fragrant. Apiaceae (Parsley family).

*O. chilensis* (chee-len-sis: of Chile). Flowers greenish white, tiny. Scattered, lower forests.

*O. purpurea* (pur-pew-ria: purple). Flowers pink to purple, tiny. Scattered, moist mid-elev openings.

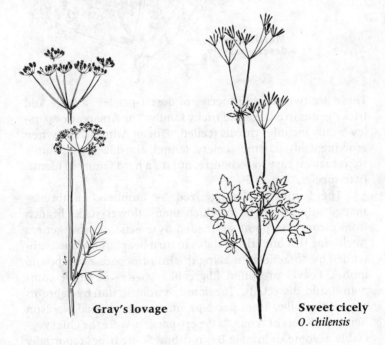

**Gray's lovage**

**Sweet cicely**
*O. chilensis*

---

# Desert-parsleys

*Lomatium* spp. (lo-may-shium: hemmed seeds). Apiaceae (Parsley family). Also **hog-fennel**.

**Martindale's desert-parsley,** *L. martindalei* (martin-day-lye: after I. C. Martindale). Flowers yellow, tiny, in several ½" broad-topped clusters on raylike stems from the root crown; leaves basal, pinnately twice-compound, in a nearly flat rosette. Dry, rocky (often pumice) ground, esp subalpine. Color p 142.

**Chocolate-tips,** *L. dissectum* (dis-ec-tum: finely cut). Flowers usually purple-brown, tiny, in many ½" balls on rays (like fireworks) from a common point on a 20–60", hollow, often purple stalk; leaves lacy, finely cut and twice-pinnate. In sun, esp lower E-side and Columbia Gorge. Color p 142.

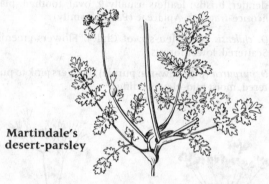

**Martindale's
desert-parsley**

These are two striking species of desert-parsley, a large and tricky genus in a large and tricky family. The Apiaceae or parsley family includes carrots (called "Queen Anne's lace" when growing wild), parsnips, celery, fennel, and dill. Strong family traits make it easy to recognize, but it's a hard family to identify to species.

The family is characterized by "umbels," umbrella-shaped inflorescences in which many flower-stalks branch from a common point subtended by bracts; in most genera (including this one) these stalks in turn bear "umbellets" subtended by "bractlets," making the inflorescence a compound umbel. Leaves are often filigreelike (twice- or thrice-compound and dissected). The family's robust starchy taproots range in edibility from parsnips and carrots to deadly poison hemlock. Roots of "cous," a desert-parsley, were the chief vegetable of some Columbia Basin tribes. Some tribes reportedly ate young chocolate-tips roots, but others poisoned fish and lice with them. The risk of deadly mistakes forbids recommending them (see cow-parsnip, page 224).

**Sitka valerian**

# Sitka Valerian

*Valeriana sitchensis* (va-lee-ree-ay-na: strong—; sit-**ken**-sis: of Sitka, Alaska). Corolla white (pink-edged at first), ⅜" long, with short lobes on a slightly asymmetrical tube; calyx appearing only as a "parachute" of plumes on the maturing seed; stamens 3, protruding; inflorescences round-topped, 1 at top, 2–8 smaller ones in opposite pairs below; leaves opposite, compound; leaflets usually 3 or 5, elliptical, pointed, vaguely toothed; plant 2–4', rankly fragrant. Abundant in subalpine meadows; also widely scattered. Valerianaceae (Valerian family). Color p 142.

As subalpine meadows are released by spring snowmelt (in June or July), deep reddish shoots of Sitka valerian soon shoot up abundantly. The redness disappears as the foliage matures, lingering longest in the budding flowers but disappearing as they mature to white. Most redness in plants comes from anthocyanin ("flower blue"), a complex carbohydrate pigment that may be red, blue, or anywhere in between; it may shift between them like litmus paper, depending on acidity.

Anthocyanin is suspected of several functions in high-elevation plants. First, it filters out ultraviolet radiation, which can be at least as hard on plant tissue as on human skin. Ultraviolet is especially intense at high altitude, where there is less atmosphere to screen it, and around the solstice in June, when sunlight is at its peak. In June the high country is still snowbank-chilled, and plant tissues young and tender, so that's where and when anthocyanin is brought out. Second, while reflecting ultraviolet radiation, it also seems to absorb and concentrate infrared, thus heating the plant. Third, anthocyanin is an interim form for carbohydrates on their way up from winter storage. In order to bloom and fruit early in their short growing season, high-country plants store carbohydrates in their roots, and then move them up fast after snowmelt, or even before (see page 220). In de-reddening, valerian stuffs itself with preserves from the root cellar.

Stonecrop and Roseroot, p 213–14, occasionally have flower parts in 4s, but usually in 5s in our region. One species of Shooting-star, p 208, has 4 petals. There are also 4-merous monocots: Skunk-cabbage and May lily, p 161–62.

# Fireweed and Willow-herbs

*Epilobium* spp. (ep-il-oh-bium: pod on top). 4 petals, 4 sepals and 4 stamens borne at the tip of an extremely slender ovary 1–5 times as long as the petals; dropping the flower, the ovary matures into a pod that splits into 4 spirally curling thin strips, releasing tiny downy seeds; leaves narrowly elliptical, all on stem, often opposite; from rhizomes. Onagraceae (Evening primrose family). Color p 143.

**Fireweed,** *E. angustifolium*\* (ang-gus-tif-oh-lium: narrow leaf). Flowers pink to purple, nearly flat, 1¼–2″, blooming progressively upward in a tall conical raceme; leaves 3–6″ × ¾″; plant 3–8′. Abundant in avalanche tracks and basins, recent burns, clearcuts, etc.

**Red willow-herb,** *E. latifolium*\* (lat-if-oh-lium: broad leaf). Flowers deep pink to purple, nearly flat, 1–1½″ diam; leaves 1–2″ × ½–¾″; 3–16″ tall. Mtn river bars to alp/subalpine talus.

**Alpine willow-herb,** *E. alpinum* (al-pie-num: alpine). Flowers deep or pale pink to white, nearly flat, ¼–¾″ diam, usually 2–4 per stem; petals 2-lobed; leaves ¾–2″; plant 2–12″. Alp/subalpine, esp in and along streams and seeps.

**Yellow willow-herb,** *E. luteum* (loot-ium: yellow). Flowers pale yellow, bell-shaped, ± erect; petals ½–¾″, deeply 2-lobed, crinkly; leaves 1–3″, fine-toothed; 8–24″. Streambanks and wet meadows.

# Farewell-to-spring

*Clarkia amoena* (clar-kia: after William Clark, p 236; a-me-na: delightful). 4 petals intense lavender to violet, with white streaking and often a carmine splotch at center; calyx shallowly 4-lobed, seemingly only 1 sepal; 8 yellow stamens; 4 cream stigmas; stem several-flowered, 4–32″, wiry, fuzzy, often reddish near base; leaves 1–2″ long, narrow. Sunny, well-drained slopes, low W-side; uncommon. Onagraceae (Evening-primrose family). Color p 143.

Growing annually from seeds, farewell-to-spring blooms while the other herbs on its summer-dry slope die back.

\*Some texts place the first two species in genus *Chamerion*.

# Toothwort

*Cardamine pulcherrima*\* (car-dam-in-ee: the Greek term; pool-ker-im-a: most beautiful). Flowers pink or lavender to white, ½–1″ diam, several; stem leaves 1–3, with 3 or 5 ± uneven, narrow lobes, purplish underneath; roundish basal leaves rising from the rhizomes separately; 4–14″ tall. W-side lowlands. Brassicaceae† (Mustard family). Color p 143.

Along with springbeauty, this is one of our earliest-blooming flowers. Toothwort is called springbeauty in some books, and the two are often confused; springbeauty (page 220) has five petals and grows in sunnier spots than toothwort.

# Wallflowers

*Erysimum* spp. (er-iss-im-um: the Greek term). Petals brilliant yellow, round, ½–¾″; flowers many, in a round-topped cluster; seed pods very narrow, splitting in 2; leaves 2–5″, narrow, often shallowly toothed, most in a basal rosette, a few on the stem. Brassicaceae (Mustard family).

*E. arenicola* (air-en-ic-a-la: sand dweller). Plant 10–20″ tall. Mostly alpine. Color p 143.

*E. asperum* (ass-per-um: rough). Plant grayish-hairy, 16–32″. Columbia Gorge and lower E slopes, in sun.

Some "wallflowers" may be modest, but not these beauties.

# Smelowskias

*Smelowskia* spp. (smel-ow-skia: after Timotheus Smelowsky). Flowers cream white or purple-tinged, ⅜–¾″ diam, several, in roundish clusters; leaves crowded, basal, 1–4″, pinnately compound or -lobed, gray-fuzzy; stems 3–8″. Alpine. Brassicaceae (Mustard family).

*S. calycina* (cay-lis-eye-na: cuplet). Sepals falling off when flower opens. Color p 144.

*S. ovalis* (o-vay-lis: oval pod). Open flowers retaining sepals.

---

\*Older texts list this as species *tenella* in a genus *Dentaria*.
†The mustard family was long known as Cruciferae. Nutritionists may be understandably irritated, at a time when they are campaigning for cruciferous vegetables in diets, to hear that botanists have renamed them "brassicaceous" vegetables. See footnote, p. 182.

# Bedstraw

*Galium triflorum* (gay-lium: milk—; try-flor-um: 3 flowered). Corolla white, 4-lobed, flat, up to ¼" diam; sepals lacking; flowers usually in sparse 3s branching from leaf axils; leaves in whorls of 5 or 6, narrow-elliptical; stems minutely barbed, 4-angled, generally sprawling or clambering, dense. Widespread in forest and thickets. Rubiaceae (Madder family).

Bedstraw grows clinging tangles of weak stems that are irresistibly easy to uproot by the fistful. This is the plant's way of attaching its seeds to passing animals. While other plants have developed barbed fruits for this purpose, bedstraw barbs its entire stem, and lets go at the roots. Several less common *Galium* species here differ in having four-leaved whorls, and often fuller inflorescences and a more erect stature.

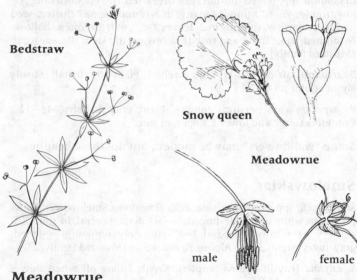

Bedstraw

Snow queen

Meadowrue

male          female

# Meadowrue

*Thalictrum occidentale* (tha-lic-trum: the Greek term; ox-i-den-tay-lee: western). Sepals 4 (or 5), greenish, ¼"; petals lacking; flowers many, in sparse racemes, male and female on separate plants, the males with numerous stamens of long yellow anthers dangling loosely by purple filaments; females less droopy, with several reddish pistils that mature into a starlike rosette of capsules; leaves twice- or thrice-compound, leaflets ¾–1½", round-lobed (similar to columbine); plants 20–40". Low subalpine lush meadows or open forest. Ranunculaceae (Buttercup family). Color p 139.

# Bunchberry

*Cornus canadensis* (cor-nus: dogwood). Also **ground-dogwood**. Inflorescence subtended by 4 showy (½–1") white bracts often mistak-
en for petals; true flowers tiny, 4-merous, in a dense head ½–¾"
diam; berries red-orange, several, 1-seeded, ¼" diam; leaves pointed,
oval, 1–3", in whorls of 6 beneath the inflorescence, and in whorls of
4 on flowerless stems (technically 4-leafleted compound basal
leaves) nearby; stems 2–8", from rhizomes. Dense mid-elev forests.
Cornaceae (Dogwood family). Color p 140.

How can this flower—a mere six-inch subshrub—look so
much like a dogwood flower? As a matter of fact, the dogwood
tree is the bunchberry's closest relative.


# Kittentails

*Synthyris* spp. (synth-er-iss: fused doors). Corolla pale blue to lavender, unequally 4-lobed, ⅜"; in small racemes; leaves basal, round to
heart-shaped, 1–2", with coarse blunt teeth. Scrophulariaceae
(Figwort family).

**Snow queen,** *S. reniformis* (ren-if-or-mis: kidney-shaped). Stems
and leaves sprawling. Warmer lowland forests, W Ore and SW Wash.

**Mountain kittentails,** *S. missuricus* (miz-oo-ric-us: of the Missouri
River). Stems erect, up to 12". Disjunct in Columbia Gorge; otherwise E of our range. Color p 143.

Snow queen has been known to bloom as early as December
in mild winters.


# Veronicas

*Veronica* spp. (ver-on-ic-a: from a Greek term). Also **speedwell**. Corolla blue-violet with yellow center, unequally 4-lobed, nearly flat,
⅜" diam; in a small raceme; leaves opposite, elliptical; stem 2–6".
Alp/subalpine, in drier meadows or with heather or conifer krummholz. Scrophulariaceae (Figwort family).

*V. cusickii* (cue-zik-ee-eye: after William Cusick). Style and stamens
longer than petals; leaves ± crowded.

*V. wormskjoldii* (vormsk-yol-dee-eye: after Morten Wormskjold).
Style and stamens shorter than petals; leaves scattered. Color p 139.

These flowers closely resemble the European veronica, *V. fili-
formis*, a weed in Northwest lawns.

---

Most 3- or 6-petaled flowers are monocots (pp 159–79) with 6 or fewer stamens and a total of 6 petals and sepals; most monocot leaves have conspicuously parallel veins. The dicot plants below differ in at least one of those characteristics.

"Several" petals in this heading means a variable number usually between 5 and 9. Most *apparently* several-petaled flowers are daisylike composites (pp 183–86); their seeming petals are ray flowers.

## Wild-ginger

*Asarum caudatum* (ass-a-rum: the Greek term; caw-day-tum: tailed). Calyx brownish purple, with 3 long-tailed lobes 1–3"; petals lacking; stamens 12, ± fused to the pistil; flower single on a prostrate short stalk between paired leafstalks; leaves heart-shaped, 2–5", finely hairy, spicy-aromatic, rather firm and often persistent, on hairy 2–8" leafstalks. Moist forests. Aristolochiaceae (Birthwort family). Color p 144.

This odd plant is unrelated to ginger, and even the tangy fragrance isn't really close; yet its stems as a seasoning won approval from cooks of trapping, pioneering, and wild-food stalking eras alike. The earthbound, camouflaged flowers are less fragrant. They attract creeping and crawling pollinators.

## Buckwheats

*Eriogonum* spp. (airy-og-a-num: woolly joints). Calyx 6-lobed; petals lacking; stamens 9; flowers small, many, above ± fused whorls of usually hairy bracts; leaves basal, oval, on short leafstalks, often woolly underneath. Rocky dryish places, alpine to steppes. Polygonaceae (Buckwheat family).

**Sulphur-flower,** *E. umbellatum* (um-bel-ay-tum: flowers in umbels). Flowers bright yellow to cream or reddish; stems 2–12".

**Dirty socks,** *E. pyrolifolium* (pi-roe-lif-oh-lium: pyrola leaf). Flowers dull pinkish or off-white, reddish-fuzzy, foul-smelling; only 2 bracts below inflorescence; stems often red, 2–5". Commonest on pumice. Color p 144.

**Oval-leaf buckwheat,** *E. ovalifolium nivale* (oh-val-if-oh-lium: oval leaf; niv-ay-lee: of snow). Flowers dull cream yellow to rose-tinged; leaves silvery white (scarcely at all green), tiny, densely matted; stems 1½–4" tall. Alpine. Color p 145.

## Inside-out Flower

*Vancouveria hexandra* (van-coo-vee-ria: after George Vancouver, p 94; hex-an-dra: 6 stamen). Also **duckfoot.** Apparent petals and sepals each 6, white, sharply reflexed, ¼"; another 6–9 outer bracts fall off as the flower opens; stamens 6; flowers many, in a very sparse panicle; leaves 9- (to 27-) compound, the leaflets ¾–2", vaguely 3-lobed; plant 8–20" tall. Moist forests in W Ore and SW Wash. Berberidaceae (Barberry family). Color p 145.

## Starflower

*Trientalis* spp. (try-en-tay-lis: one third, perhaps implying 4" height). Flowers white to pink, 1 to few, ½" diam; petals, sepals, and stamens each 5–8 (most often 6); leaves pointed-oval; capsule spherical; 3–8" tall. Primulaceae (Primrose family).

*T. latifolia* (lat-if-oh-lia: broad leaf). Leaves 1½–4", in a single whorl. Widespread, in forests. Color p 144.

*T. arctica*. Leaves up to 2", smaller downward along the stem. Bogs.

## Lewisias

*Lewisia* spp. (lew-iss-ia: after Meriwether Lewis, p 234). Petals 6–11; sepals 2; flowers several; leaves in a basal rosette; stem 4–8". Portulacaceae (Purslane family).

**Columbia lewisia,** *L. columbiana* (co-lum-be-ay-na: of the Columbia River.) Petals pink, red-veined, ¼–½"; stamens 5–6, red-tipped; leaves fleshy, linear, to 4". Alp/subalpine in Wash. Color p 145.

**Tweedy's lewisia,** *L. tweedyi* (twee-dee-eye: after Frank Tweedy). Petals apricot pink to cream, 1–1¾"; stamens many; leaves rather heavy, 4–8" × 1–4". Rocky slopes, limited to E-side Cas of C Wash. Color p 145.

Tweedy's lewisia is the most spectacular rare species in our mountains. If you should be lucky enough to see one, **don't touch it!** Please excuse my touchiness. The only specimens I've seen did well for two years between my visits to them, only to be uprooted by some wretch while I was up trail.

---

# Bitterroot

*Lewisia rediviva* (red-i-vee-va: reborn). Flowers pink, apricot, or white, 2–2¼" diam, borne very low to ground, 1 per stem in small clumps; 10–18 petals and 5–9 sepals both showy; leaves basal, linear, initially fleshy but withering often by time of flowering. Arid gravels E of Cas. Portulacaceae (Purslane family). Color p 146.

It is easy to imagine this dramatic dry-ground bloom becoming an instant favorite of Meriwether Lewis, who found it in the Bitterroot Mountains. It barely enters our range.

# Baneberry

*Actaea rubra* (ac-tee-a: elder, for the similar leaves; **roob**-ra: red). Numerous ¼" white stamens are the showiest part of the flower; 5–10 petals (occasionally lacking) white, shorter and scarcely wider than the stamens; 3–5 sepals petallike but falling as the flower opens; flowers (and berries) in a ± conical raceme; berries ⅜" diam, glossy bright red (or occasionally pure white); leaves 9 to 27-compound, leaflets pointed-oval, toothed and lobed, 1–3"; stem 16–40". Lower forests. Ranunculaceae (Buttercup family). Color p 146.

Baneberries, our most poisonous native berries, are less than deadly. A handful could render you violently ill, but even a small taste should start you spitting fast enough to save your stomach the troubles.

---

**Meriwether Lewis** and **William Clark**'s voyage of 1804–06 needs no lengthy description here; what you may not know is that biological discovery was its greatest distinction. After all, Alexander Mackenzie had crashed on through to the Pacific at Bella Coola in 1789, but he and others in the Northwest were interested in little but fur. Though not scientists by profession, Lewis and Clark were briefed intensively on natural history and cartography, respectively, before they set out. And they did their jobs well. Lewis ranks with Douglas and Menzies in the number of first collections of important Northwest plants credited to him. The trip's high point, both literally and figuratively, came in Idaho and Montana; here on the lower Columbia they were in already-discovered territory, and they spent a miserably wet winter.

---

## Yellow Pond-lily

*Nuphar polysepalum* (new-fer: from the Arabic term; poly-**see**-pa-lum: many sepals). Also **wokas**. Bright yellow, heavy, roundish, 1½–3" petallike sepals 4–8; smaller green outer sepals 4; true petals and stamens numerous, much alike, crowded around the large para-sol-shaped pistil; leaves heavy, waxy, elongated heart-shaped, 6–18" long, usually floating. Widespread in ponds and slow streams up to about 6' deep. Nymphaeaceae (Water-lily family). Color p 145.

Oregon and California tribes gather pond-lily seeds to eat. Northerly tribes reported broad-spectrum prescriptions using the roots; a sick person was steamed over them.

## Goldthread

*Coptis* spp. (cop-tiss: cut—). Petals and sepals similar, each 5–8, greenish white, threadlike, ⅛–⅜" long; petals shorter than sepals, with a tiny gland on a broad spot near the base; stamens many; leaves shiny, persistent, very fernlike, at least 3-compound, toothed, incised; roots bright yellow beneath their bark; stems 2–6". Low W-side forest. Ranunculaceae (Buttercup family).

*C. laciniata* (la-sin-ee-ay-ta: cutleaf). Ore and Columbia Gorge.

*C. asplenifolia* (a-splee-nif-oh-lia: spleenwort-fern leaf). Wet forest and bogs, from Stillaguamish drainage N in Wash.

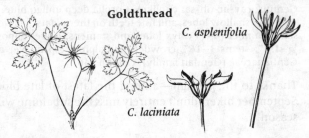

**Goldthread**

*C. asplenifolia*

*C. laciniata*

## Marshmarigold

*Caltha biflora\** (cal-tha: goblet; by-**flor**-a: 2 flower). Sepals 6–11, a few often 2-lobed, white, petallike, ½–¾"; petals lacking; stamens and pistils many; flowers usually 2 on a forked 3–10" stem; leaves basal on 2–3" leafstalks, kidney-shaped, 2–4" across, ± fleshy, edges ± scalloped, often curling. Wet places (often in streams) esp subalpine. Ranunculaceae (Buttercup family). Color p 146.

\*Some texts lump *biflora* under *C. leptosepala*.

# Towhead Baby

*Anemone occidentalis*\* (a-nem-a-nee: wind—; ox-i-den-**tay**-lis: western). Also **western pasqueflower.** Sepals 5–8, white, petallike, ½–1″ petals lacking; stamens and pistils many, styles growing to 1–2″ and feathery as seeds mature; stem 1–2′, hairy, 1-flowered, with a whorl of 3 leaves at mid height, plus larger basal leaves, all intricately twice- or thrice-compound, fernlike. Subalpine meadows. Ranunculaceae (Buttercup family). Color p 138, 145.

The most strangely lovely of subalpine "flowers" is actually the seed head of this Anemone. It looks like something Dr. Seuss would have dreamed up—or more traditionally, "the old man of the mountains," or a hirsute towhead. The flower attracts less attention, blooming early when the plant is only 2–6″ tall and there's still snow around to keep hikers away. After the petals fall, growth of the stem, leaves, and styles takes off; the styles become plumes on the seeds to catch wind. *A. drummondii* (after Thomas Drummond, page 475) has similar leaves and flowers (sometimes bluish), but no plumes, and grows at higher elevations.

## Mountain Bog Gentian

*Gentiana calycosa*† (jen-she-**ay**-na: the Greek term, honoring King Gentius; cay-lic-**oh**-sa: cuplike). Corolla deep indigo blue, 1½″ tall, with 4–7 shallow lobes, and fine teeth on the "pleats" between lobes; same number (4–7) calyx lobes and stamens; leaves opposite, oval, ½–1½″; stems 3–16″, crowded. Subalpine, esp in wet meadows. Gentianaceae (Gentian family). Color p 146.

Thanks to the gentian—among the latest of late bloomers—September hikers don't entirely miss the subalpine wildflower season.

---

\*Splitters of genus *Anemone* consider this species to be *Pulsatilla occidentalis*, while *drummondii* remains in *Anemone*.
†Some authorities split *Gentiana* into several genera, placing *calycosa* in genus *Pneumonanthe*.

---

# 5

# Ferns, Clubmosses, and Horsetails

The old term "vascular cryptogams" defines this informal group of plants. "Vascular" means having vessels, or veins—tiny tubes for conducting water and dissolved materials. Vessels are tiny, but their effects are conspicuous. Without them, a plant can't raise vital fluids more than a few inches from its moisture supply. *Non*vascular cryptogams (mostly mosses and liverworts) are therefore very low, while ferns and horsetails are taller, typically 6" to 48". Modern clubmosses are short enough to confuse with mosses, but they are closer to ferns. In the Mesozoic era, before seed plants took over, forests of fern and clubmoss "trees" covered much of the earth.

"Cryptogam" ("hidden mating") means that the sexual reproductive process in these plants is tiny and brief compared to the showy flowering and fruiting of seed plants, and doesn't produce seeds for extended dormancy or travel. The traveling function is left up to an asexual stage in the cryptogam life cycle—a one-celled spore. (Plant spores are more analogous to the one-celled pollen of seed plants than to seeds.)

In ferns, the dustlike spores are borne in and released from "sori"—tiny clusters appearing as dark spots, lines, or crescents on the leaf underside. In some ferns each sorus is shielded by a tiny membrane, in some others by a length of rolled-under leaf margin. Each fern frond and its stalk from the rhizome is one leaf; it is pinnately compounded or divided into "pinnae." The pinnae may be compounded an additional one to three times, but the word pinna(e) is reserved for those units branching directly from the central leaf stalk.

## Sword Fern

*Polystichum munitum* (pa-lis-tic-um: many rows; mew-**nigh**-tum: armed). Leaves 20–60", dark, leathery, once-compound, in huge clumps; each pinna asymmetrical at base, with an upward-pointing coarse tooth; stalks densely chaffy; sori round. Abundant on moist W-side forest sites; less so in drier forest, incl cool shady E-side spots. Polypodiaceae (Common Fern family).*

Sword ferns are not favored for food or forage, but florists prefer their fronds for funerals, gathering them in great numbers without apparent threat to their abundance. Indians sometimes bundled them up as mattresses. Makah children made a game of peeling off as many sword fern pinnae as they could on one breath, saying *"pila"* ("sword fern" in Makah tongue) once for each pinna.

## Deer Fern

*Blechnum spicant*† (blek-num: the Greek term; spik-ent: spiky). Leaves 12–50", dark, in clumps; pinnae slender, broadening toward the base and not separated all the way to the stalk; stalks dark brown, smooth. Moist W-side old-growth, esp in Wash. Polypodiaceae.

Deer fern leaves are plainly of two types—fertile spore-bearing leaves, and sterile, strictly vegetative leaves. The fertile leaves are taller, and stand segregated at the center of each clump; their pinnae are narrower and more widely separated than the sterile pinnae, rolling tightly in near-tubes around sori crowded on their undersides. The sterile leaves lack sori; they rise obliquely in a thick circle around the fertile center-piece. They are important winter forage for deer, elk, and cattle. The rhizomes were eaten, but not highly prized, by people.

*Some recent texts, including Lellinger (1985), split the large Common fern family into many families.
†Some texts keep *spicant* in a separate genus, *Struthiopteris*.

# Licorice Fern

*Polypodium glycyrrhiza* (poly-**poe**-dium: many foot; gly-sir-**eye**-za: licorice, from "sweet root"). Leaves 4–32", dark, smooth; pinnae broadening toward the base and not separated all the way to the stalk; sori exposed; stalk green. Typically growing among mosses upon rocks or trees; lower W-side. Polypodiaceae.

The flavor we call licorice occurs in several unrelated plants scattered around the globe, including star anise, fennel or sweet anise, this fern, and licorice, *Glycyrrhiza glabra.* The shared chemistry is called "glycyrrhizin." People worldwide have found it good for the appetite, the digestion, the spirits, the breath, and the dreams. In the Northwest, licorice fern rhizomes were sucked by hungry hunters or berry-pickers along the trail, or fed before meals to finicky young eaters. In quantity they may prove laxative, but most people find them too bitter to eat in quantity.

Licorice ferns grow on rocks, logs, and tree trunks like maples and alders, preferably in a good bed of mosses to keep their roots moist. On all but the moistest sites here the leaves are usually less than six inches tall, and die back in late summer when the moss mat dries out. New leaves sprout with the fall rains. Thus, though the plant is described as evergreen, most individuals here are summer-deciduous.

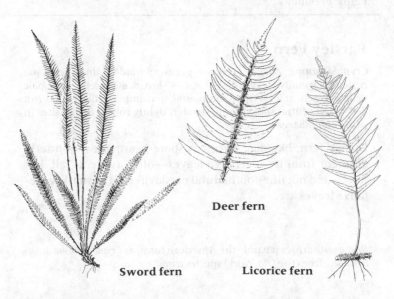

**Deer fern**

**Sword fern**

**Licorice fern**

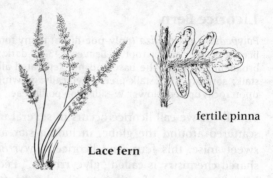

fertile pinna

Lace fern

# Lace Fern

*Cheilanthes gracillima* (kye-lanth-eez: margin flower; gra-sil-im-a: slenderest). Also **lip fern.** Leaves 3–10", slender, evergreen, pale, at least twice compound, in clumps; leaflets tiny, reddish-woolly underneath, with margins rolled under; upper part of stalk hairy. Rocky sites in sun. Polypodiaceae.

Not all ferns are particularly moisture-demanding. The little lace fern and parsley fern (below) are drought-tolerant, living almost exclusively in crevices of cliffs and rockpiles. Lace fern is partial to igneous rocks.

---

Ferns: deciduous

---

## Parsley Fern

*Cryptogramma crispa** (crypto-gram-a: hidden lines; **cris-pa:** curled). Vegetative leaves 4–10" × 2–4" broad, evergreen, firm, pale yellow-green, at least twice compound, in clumps; fertile leaflets (on 7–14" tall central stalks) long, slender, tightly rolled. Rocky sites in sun. Polypodiaceae.

Parsley fern, like deer fern, has spore-bearing leaves utterly different from its vegetative leaves—often twice as tall, but fewer, and not fine-toothed and parsleylike as the sterile vegetative leaves are.

*Some authorities separate the American form, as *Cryptogramma acrostichoides,* from the Old World species *crispa.*

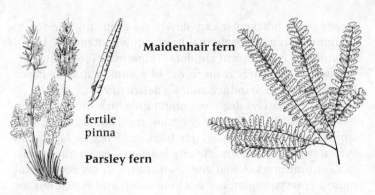

**Maidenhair fern**

fertile
pinna

**Parsley fern**

## Maidenhair Fern

*Adiantum pedatum* (ay-dee-an-tum: not wetted; ped-ay-tum: palmately compound). Leaf blades 4–16", fan-shaped, broader than long, twice compound; sori under rolled edges; stalks black, shiny, wiry. Saturated soil or rocks in shade. Polypodiaceae.

This is our easiest fern to identify. Its striking shiny black stalks are our only ones that split into two slightly unequal branches, the pinnae spreading fanlike. The stalks kept their dark shine well in decorative patterns in Makah and Quinault basketry. Not they but the masses of fine dark root hairs suggested the common name.

## Bracken

*Pteridium aquilinum* (teh-rid-ium: from the Greek term for fern, derived from "feather"; ak-wil-eye-num: eagle—). Also **brake fern**. Leaves 24–80" tall, ± triangular, twice to thrice compound, undersides fuzzy; sori under rolled edges. Widespread on ± sunny sites. Polypodiaceae.

Among Northwestern wild food gourmets, the fern "fiddleheads" picked for steaming or eating raw in salads are usually young bracken shoots. They taste like asparagus with a dash of almond extract and an unnervingly mucuslike interior. Most Northwest tribes ate bracken fiddleheads or rhizomes, or both. Stockmen, however, list bracken as a poisonous plant. It's true: 600 dry pounds of bracken consumed within a six-week period are enough to kill a horse. Cows are less sensitive, their lethal dose around a ton. They don't graze it willingly (only goats would be so omnivorous) but if their winter hay is weedy with bracken it can slowly do them in. The toxin, thiaminase, is an enzyme that breaks down vitamin B-1, and vitamin B-1 is a sufficent antidote. Thiaminase is equally toxic to humans, but there is no record of a human stuffing down enough bracken to induce vitamin deficiency.

Unfortunately, deadlier compounds lurk here. Bracken has been found to damage chromosomes, and a certain rare stomach cancer has a relatively high incidence in Japan and Wales, two far-flung lands where bracken is eaten traditionally. Consume bracken fiddleheads, if at all, only occasionally, as mouth entertainment, not a whole salad; and watch out for the vaguely similar unfurling shoots of monkshood, a very poisonous plant.

Bracken is doubtless the world's most widespread fern. Here, only sword ferns are more abundant, and no fern grows as tall or as fast; bracken has been measured at 16' in Washington, and clocked at several inches a day. Its best Cascade habitats are lower subalpine slope meadows annually scoured by avalanches or by snow creep.

## Lady Fern

*Athyrium filix-femina* (ath-ee-rium: no shield; fie-lix fem-eye-na: fern-woman). Leaves 16–80" tall, narrowing toward both ends, twice or thrice compound; sori exposed, or initially shielded on one edge; stalk base scaly. Wet ground, often with skunk-cabbage and devil's club. Polypodiaceae.

Medieval herbalists associated the female principle with a very large fern—larger than the male fern, *Dryopteris filix-mas*. They prescribed powdered roots and leaf infusions of lady fern for such diverse ailments as jaundice, gallstones, sores, hiccups, and worms; only since 1950 have male fern rhizomes fallen from pharmaceutical favor as a dewormer. Northwest Indians used lady and male ferns medicinally, and baked the rhizomes of these and other ferns for dinner.

The alpine lady fern, *A. distentifolium*,* is similar, but smaller (8–32" tall) and more finely incised. Common in open subalpine country, it can monopolize patches of wet talus.

*A. distentifolium* has been widely known as *A. alpestre;* our authority contends that name is invalid because the specimen originally named *alpestre* actually belongs to *filix-femina*.

## Oak Fern

*Gymnocarpium dryopteris* (gym-no-car-pium: naked fruit; dry-op-ter-iss: oak fern). Leaves 6–18" tall, broadly triangular, thrice compound, rising singly from runners; sori exposed; stalks pale, slightly scaly. W-side forest. Polypodiaceae.

Oak fern appears to have three similar leaves on each stalk. Technically, this is a single leaf with two basal pinnae, left and right, each nearly as big and as dissected as all the remaining pinnae put together.

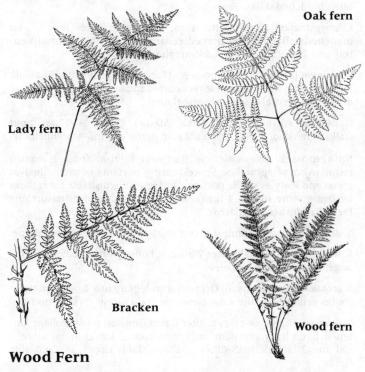

Oak fern

Lady fern

Bracken

Wood fern

## Wood Fern

*Dryopteris austriaca* (dry-op-ter-iss: oak fern; aus-try-a-ca: southern). Also **shield fern**. Leaves 8–36" tall, broadly triangular, thrice compound, in small clumps; sori round-shielded; stalk bases scaly. W-side forest. Polypodiaceae.

Indians said eating wood fern rhizomes cleans the system—after eating poisonous plants or red-tide shellfish, for example. Pharmacognosies call them laxative.

# Clubmosses and Spikemosses

**Clubmosses,** *Lycopodium* spp. (lye-co-**poe**-dium: wolf foot). Also **ground-pines.** Spore-bearing "cones" straw-colored, usually erect, often separated from green leafy stems by a slender stalk. Lycopodiaceae.

*L. clavatum* (cla-**vay**-tum: club-shaped). Cones ¾–3″ tall, on long, often branched stalks. Forest.

*L. complanatum* (com-pla-**nay**-tum: flattened). Cones ⅝–1¼″, on branched stalks; foliage flattened, cedarlike, leaves in the dorsal, ventral, and 2 lateral ranks of 3 distinct shapes. Forest.

*L. alpinum* (al-**pie**-num). Cones ⅜–1″, rising directly without a stalk from the leafy stem; lateral leaves curled and larger than the minute dorsal and ventral leaves. Alp/subalpine.

*L. sitchense* (sit-**ken**-zee: of Sitka, Alaska). Cones ⅜–1″, without stalks; 4 ranks of leaves much alike; in dense clumps. Alp/subalpine.

**Spikemosses,** *Selaginella* spp. (sel-adge-in-**el**-a: from a Roman term). Also **selaginellas.** Spore-bearing portions of stems just as green and leafy as sterile portions, but ± distinguishable from them by being more neatly 4-ranked, closer-packed, and often turning erect. Family Selaginellaceae.

*S. densa.* In dense clumps; spores ± orange. Rocky sites, often alpine.

*S. wallacei* (**wall**-a-sigh: after Wallace). Loosely branched; spores orange. Rocky sites, lower elevs.

*S. oregana* (or-eg-**ay**-na: of Oregon). Pendent up to 6′ long from trees; spores yellowish white. Low elevs, esp "rain forest" river bottoms.

*S. douglasii* (da-**glass**-ee-eye: after David Douglas, p 18). Foliage flattened, like a leafy liverwort, only more robust, leaves in dorsal/ventral and 2 lateral ranks differ in shape. Talus slopes, in Columbia Gorge.

These little fern relatives behave much like mosses, including their ability to resurrect, when wetted, from a dead-looking dried-up state. Their vessels (which mosses lack) enable their leaves to be thicker and more evergreen than moss leaves, however, so their closest semblance might be to heather or juniper. They bear spores at their leaf bases, but not on all their leaves; most species have visibly distinct, erect fertile portions loosely termed "cones" terminating some of their branchlets. In clubmosses these are more distinct than in spikemosses.

Of all plant and fungal spores, only those of clubmosses entered commerce, partly because they're easiest to collect. The cones are cut off, dried, pounded, rubbed, and finally sifted to collect the spores, used for centuries to dust wounds, pills, and babies' bottoms. No joke. Spores are extremely fine, smooth, slippery, nonreactive (except in the noses of allergy victims), water-repellent, and nonclumping. You may have noticed these qualities in pollen grains, which descended from spores through evolution. Pollen and spores need to be water-repellent and nonclumping to maximize air travel, their *raison d'etre,* in inclement weather.

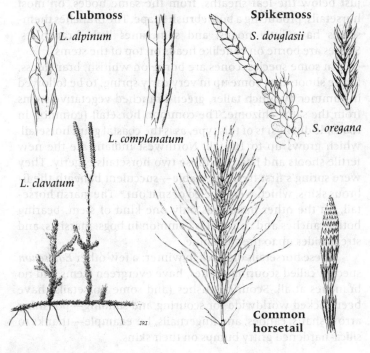

**Clubmoss**

*L. alpinum*

*L. complanatum*

*L. clavatum*

**Spikemoss**

*S. douglasii*

*S. oregana*

**Common horsetail**

## Horsetails

*Equisetum* spp. (ek-wis-ee-tum: horse tail). Equisetaceae.
**Common horsetail,** *E. arvense* (ar-ven-see: of fields).

**Giant horsetail,** *E. telmateia* (tel-ma-tie-a: of marshes).

**Marsh horsetail,** *E. palustre* (pa-lus-tree: of marshes).

Long ignored for being too primitive, common, and mono-chromatic, horsetails won their hour of media glory for sending the first green shoots up through Mt. St. Helens' debris of May, 1980. They can crack their way up through an inch of asphalt on highway shoulders. No wonder Quileute swimmers felt strong after scrubbing themselves with horsetails! And some Northwest gardeners feel weak after weeding them.

Leaves on *Equisetum* are reduced to sheaths made up of fused whorls of leaves, often straw-colored, growing from nodes at regular intervals along the stem. Additional whorls of slender green branches—easily mistaken for leaves—grow just below the leaf sheaths, from the same nodes, on most horsetails, producing a bottlebrush shape. The branches themselves have little nodes, and sometimes little branchlets. Spores are borne on conelike heads on top of the stems.

In some species, cones are borne on whitish, branchless, fertile shoots that come up in very early spring, to be followed in summer by much taller, green-branched vegetative stems from the same rhizome. The common horsetail (common in roadside ditches) is of this type, as is the coastal giant horsetail, which grows up to 10' tall. Northwest Indians ate the new fertile shoots and heads of those two horsetails eagerly. They were spring's first fresh vegetable—succulent beneath the fibrous skins, which were peeled or spat out.* The marsh horsetail, on the other hand, has only one kind of stem, bearing both branches and cones. It is common in bogs, marshes, and streamsides up to the subalpine.

These horsetails die back in winter; a few other *Equisetum* species, called scouring-rushes, have evergreen stems and no branches at all. Scouring-rushes (and some horsetails) have been picked worldwide for scouring and sanding—polishing arrow shafts, canoes, and fingernails, for example—thanks to silica-hardened gritty bumps on their skins.

---

*Like bracken (p 241), horsetails have caused thiaminase poisoning in livestock.

# 6

# Mosses and Liverworts

The nonvascular spore-bearing plants have little ability to conduct water and dissolved nutrients from the substrate up into their tissues; in many of them conduction takes place mainly along the outside of their stems, aided by surface tension. They compensate with an ability to pass water through their leaf surfaces almost instantly, both absorbing it and giving it up easily. After a few dry days in the sun, a bed of moss may be grayish, shriveled and brittle, and look quite dead. But let a little dew or drizzle fall on it, or a little water from your bottle, and see how the leaves revive before your very eyes, softening, stretching out and turning bright green. Again, when the temperature drops below freezing they give up their free moisture to crystallize on their surface rather than inside, where it would rupture cells.

Lichens share many of these characteristics with mosses and liverworts (though unrelated to them) and often lead similar lives. All three grow abundantly on trees and rocks in our area, undergoing countless alternations between their dried-out and their moist, photosynthetically active states. When they grow on trees or other plants they are called "epiphytes" ("upon-plants") meaning that they use the support plant to hold them up off the ground, but draw no substances out of it.

Mosses that grow on the forest floor stake out a seasonal niche as much as a spatial one, living their active season in spring and—where there is no snowpack—winter. Quick recovery from nighttime frost is essential. By late spring these mosses are shaded out by perennial herbs and deciduous shrubs, and go largely dormant until fall.

The sexual life cycles of spore-bearing plants evolved earlier than those of seed plants, and are more primitive but emphatically not simpler; they're so complex and varied that I won't even attempt to describe them. Many mosses and liverwort species propagate vegetatively from fragments much more often than from spores. Some produce multicelled asexual propagules, called "gemmae," just for this purpose. In mosses and seed plants both (see page 187) adaptation to arctic and alpine climates often entails virtually abandoning sexual reproduction in favor of cloning, either vegetative or by unfertilized spore or seed formation.

The "fruits" of mosses are spore capsules, usually borne on slender vertical fruiting stalks. To release spores, most open at the tip after shedding first an outer cap, the "calyptra," and later an inner lid, the "operculum." Keys to the mosses first separate two primitive families (Sphagnaceae and Andreaeaceae) with spore capsules that don't fit that description at all, and then divide the remainder into two growth forms based on where on the stem the fruiting stalks sprout:

**fruiting from the tip** of the leafy shoot, and typically growing **upright** in crowded masses or small tufts; or

**fruiting from midpoint(s)** along the year's new leafy shoot, which is typically **arching, trailing or pendent.**

Season of fruiting varies with species, but usually lasts several months. Positive identification of most mosses requires not only fruiting specimens but also a microscope and a highly technical key. The following pages offer tentative identifications of a few common species, with or without the help of a 10× or 12× handlens or monocular. A basic handlens is cheap, and fun to have along if you like plant forms.

# Haircap Mosses

*Polytrichum* spp. (pa-**lit**-ric-um: many hairs). Stems wiry, rarely branched, vertical, in dense colonies; leaves ⅜" average, narrow, inrolling when dry (exc *lyallii*), sheathing the stem at their bases; stem and stalk (or sometimes entire plant) rich reddish; capsule single, initially cloaked (exc *lyallii*) in a densely long-hairy cap. Polytrichaceae. Illustrated p 250.

*P. juniperum* (jew-**nip**-er-um: juniper). Leaves often bluish-coated, ending in a short reddish hair tip (may require 10× lens); leafy shoots 1–4" + 1–2½" fruiting stalk; capsule (after dropping the hairy cap) 4-angled. Sunnier sites at all elevs.

*P. piliferum* (pil-**if**-er-um: hair-bearing). Leaves end in a whitish translucent hair tip; shoots ¼–1¼" tall + ¾–1¼" stalk; capsule with 4 main angles. In sun, esp alpine.

*P. sexangulare* (sex-ang-you-**lair**-ee: six angled). Leaves tapered gradually (not to a hair-thin tip); shoots ½–¾" + ¾" stalk; capsule (underneath the hairy cap) 6- (or rarely 5-) angled. Mainly subalpine, in extreme late-snowbed sites.

*P. lyallii*\* (lye-**ah**-lee-eye: after David Lyall, p 51). Leaves tapered gradually, ± toothed near the tip (under 10× lens); shoots ½–2" + 1½–2½" stalk; capsule 4-angled, held horizontal, cap only sparsely short-hairy. Subalpine, abundant, near black alpine sedge.

Haircap mosses are palpably more substantial than other mosses, almost resembling small evergreen shrubs like heather or juniper. Their stems contain some woody tissue and primitive water vessels. They can even store carbohydrates in underground rhizomes, like higher plants. The leaves, too, are more complex and thicker than the translucent, one-cell-thick leaves of most mosses. Several species of haircaps have translucent leaf margins that, in drying, curl inward to protect the chlorophyllous cells. These traits generally adapt the haircaps to sunny sites, as well as to human use—the tough stems were plaited for baskets or twine, and the whole plants used for bedding. Linnaeus reported sleeping well on a haircap moss mattress on a trip to arctic Scandinavia.

\* Older texts place *lyallii* in a separate genus *Polytrichadelphus*.

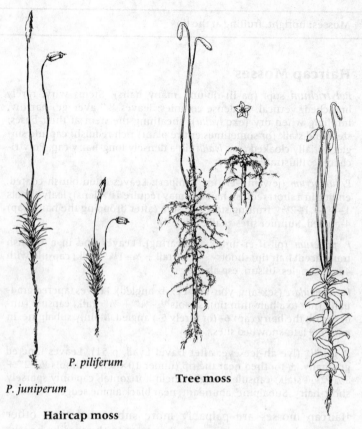

P. piliferum

P. juniperum

**Tree moss**

**Haircap moss**

**Badge moss**

# Bearded Moss

*Pogonatum alpinum* (po-gon-ay-tum: bearded). Stems wiry, reddish, ¾–4" tall + ¾–2" stalk, unbranched or slightly branched, vertical, in dense colonies; leaves heavy but fairly soft, narrow, tapering gradually, fine-toothed their entire length, the bases sheathing the stem; capsule cylindrical (not angled) underneath a densely hairy cap. All elevs W-side, esp subalpine near partridgefoot and black alpine sedge. Polytrichaceae.

To distinguish this from haircap mosses, which it often grows near, look for the nonangular capsules and, if you have a handlens, the teeth all along the leaf margins.

## Tree Moss

*Leucolepis menziesii* (lu-co-**leap**-iss: white scale; men-**zee**-zee-eye: after Archibald Menzies, p 94). Main stem 1½–3" tall, dark brown, lower portion with scattered large translucent white leaf-scales, upper portion with dense fine branches and branchlets bearing minute deep green leaves; fruiting stalks usually 2 or 3, 1½–2" tall, reddish, not twisted when dry; capsules nodding, ± pear-shaped, smooth, sometimes colorful; male plants terminating in conspicuous flower-like green rosettes (in place of fruiting stalks). Common on soil, W-side below 3,000'. Mniaceae.

This is by far the commonest moss here with the treelike growth form—vertical stems bearing leafy horizontal branches. It yields a yellow dye once used in Salish basketry.

## Badge Moss

*Plagiomnium insigne** (play-gee-ohm-**nye**-um: slant moss; in-**sig**-nee: badge). Leafy shoots 1¼–3" tall, unbranched; leaves bright green, large (up to ⅜") and broad, sheathing the stem at their bases, edges minutely toothed (under 10× lens), drastically shriveling when dry; fruiting stalks usually 3 to 5, 1–1¾" tall, reddish grading upward to yellowish; capsules nodding, yellow, smooth; male plant terminating in a green rosette around a fuzzy disk (the "badge"). Typically in ± pure colonies under maple and alder, or on wet lowlands. Mniaceae.

These are about the largest leaves you'll see on moss. At 12× magnification, held up against light, their cells can be just discerned.

## Peat Mosses

*Sphagnum* spp. (**sfag**-num: the Greek term). Robust mosses typically in massive spongy mats, the stems crowded, supporting each other; leaves tiny, mostly crowded on ¼–¾" branches, the branches either in groups of 2–5 along the stem or in a big tuft at the top; fruiting stalks short, several per shoot tip, each bearing one ± spherical blackish capsule which releases spores all at once, explosively. Typically floating in slow-moving water; sometimes terrestrial in wet climates; all elevs. Sphagnaceae.

---

*Older texts include *insigne* in a larger genus *Mnium*.

---

Peat mosses are the most primitive mosses, and at the same time the most important mosses both ecologically and economically. Estimates have them covering 1% of the earth's continents—making them one of the most extensive of all dominant plant types. They owe their success to their ability to change their environment to suit themselves and poison others. Specifically, they grow in slow-moving cold water; they draw oxygen and nutrients out of the water and replace them with hydrogen ions mainly in the form of uronic acids. The water, if too slow to replace itself frequently, is eventually too acidic and too poor in oxygen and nutrients to support the other plants growing there. At this point a whole new set of plants takes over, with peat moss the dominant. The new community is a "mire" or "peat bog" as opposed to a nonacid "marsh" or "fen." (Though peat mosses thrive in the acids they themselves create, they die in the sulphuric acids resulting from acid rain.)

Also suppressed—by cold, lack of oxygen, and certain antibiotics produced by peat mosses—are the bacteria that normally perform decomposition duties underwater. Very little decomposition takes place in a mire. The floating mass of peat moss lives and grows at the top, in the air, and dies bit by bit just below. The dead part, failing to decompose, gets thicker and thicker beneath the waterline. In some places (like western Ireland and northern Minnesota) this can go on indefinitely, the dead peat compressing and becoming a concentrated deposit of biomass suitable for fuel—the chief economic use of peat moss.

Other successions from peat moss are also well-known. The moss surface may rise high enough above waterline—either by flotation or by thickening to rest on the bottom—to become a seedbed for dry-land plants including conifers. This is a common pathway by which glacial cirque tarns (small lakes in high basins) are converted to forests. More typically in our mountains, though, tarns have plenty of streamflow to prevent their ever turning into mires. They silt up and turn into glorious meadows after a marshy (not boggy) transitional phase. Still, peat mires are not uncommon in Oregon's High Cascades, where there is mild subalpine topography; and peat mosses are fairly common here as minority members of marsh communities, and even as terrestrial mosses.

Peat moss was an invaluable material to some Northwest tribes, especially far north, where "muskeg" bogs abound. Its phenomenal water-absorbing capacity made it perfect for diapers, cradle lining, and sanitary napkins. Expectant mothers gathered quantities of peat moss, sometimes lining the entire lodge where the baby was to be born. Other mosses were sometimes given sponging, padding, and wiping tasks, mainly where peat moss was unavailable. Most tribes' languages didn't distinguish types of moss other than peat moss.

## Granite Moss

*Andreaea rupestris* (ahn-dray-ee-a: after G. R. Andreae; rue-**pes**-tris: on rocks). Plants brownish black even when wet (unlike grimmia, below; check with a few drops of water), in tight tufts usually less than 1" high; leaves minute; capsule hardly raised above the foliage, black, much less than ⅛" tall, opening by 4 lateral slits rather than at the tip. On rock (esp igneous) in full sun. Andreaeaceae.

The sooty pigmentation of this odd moss consists mainly of red anthocyanins on top of green chlorophylls; the former are there to protect the latter from the severe ultraviolet radiation on exposed high-altitude sites. The granite mosses are a primitive family set apart by capsule structure and leaf cells very different from other mosses.

## Alpine Grimmia

*Grimmia alpestris* (grim-ia: after Johann Grimm; al-**pes**-tris: alpine). Tiny cushions in rock crevices, deep green when wet, drying almost black, with a silvery surface sheen from whitish translucent hair tips on the leaves; stems about ½"; spore capsules usually numerous, much less than ⅛" tall, rarely protruding above the leaf tips. On high-elev rocks. Grimmiaceae.

Reflective whitish hair tips on moss leaves, like the silvery hairs all over some higher plants, conserve water on sunny sites by reducing radiant heat absorption. An experiment found them to be 35% effective on a similar *Grimmia* species; clumps with all their hair tips clipped off lost half again as much moisture in a day as their normal counterparts. Unshaded montane rock surfaces are ferociously hot habitats—exceeding 150° on some summer afternoons.

# Frayed-cap Moss

*Rhacomitrium canescens* (ray-co-mit-rium: ragged hat; cay-ness-enz: grayish-white). Stems 1–3", sprawling in large mats but typically erect near the tips, ± flattened because most of the leaves are on 2 opposite ranks of very short branchlets; spore capsules vertical, on ½" stalks that twist counterclockwise when dry; leaves tapering to whitish translucent hair tips that give the whole mat an ash-gray color when dry. On rocks in sun, abundant at lower elevs. Grimmiaceae. Color p 487 (with worm lichen).

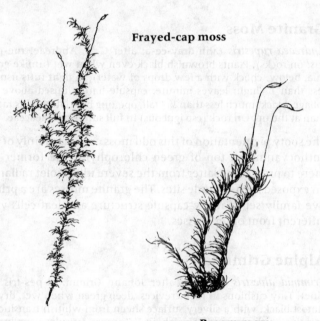

**Frayed-cap moss**

**Broom moss**

# Broom Moss

*Dicranum fuscescens* (die-cray-num: two head; fus-ess-enz: darkish) and

*D. pallidisetum* (pallid-iss-ee-tum: pale stalk). Also **heron's-bill moss**. Stems ½–2½", unbranched, reddish; leaves ⅜", all curving in arcs to the same side of the stem, or gracefully corkscrew-curled when dry; fruiting stalks single, ¾", yellow, twisting when dry; capsules angled upward, maturing in summer. Widespread on subalpine soil; *D. fuscescens* also abundant on trees at all elevs. Dicranaceae.

---

# Beaked Moss

*Stokesiella* spp.* (stoke-see-**el**-a: after Stokes). Leaves green to gold, in featherlike strands often 12" or longer, with closely spaced branchlets mostly on one plane; fruiting stalks ½–1" tall, appearing in late fall, the lower portion minutely roughened under 10× lens. Brachytheciaceae.

*S. oregana* (or-eg-ay-na: of Oregon). Branchlets hardly ever subbranched; forms luxuriant mats on ground, logs and tree bases below 4,000' on the W-side; the most abundant forest-floor moss in wetter W-side valley bottoms.

*S. praelonga* (pre-**long**-ga: elongated). Typically with short subbranches on some branchlets; thinner mats; common only on streamside rocks, all elevs.

# Rope Moss

*Rhytidiopsis robusta* (rye-tiddy-**op**-sis: wrinkled like, i.e., related to genus *Rhytidium*). Yellow-green to brownish moss in loose mats, the strands looking thick and ropy due to close-packed leaves and sparse branching; stems yellow-green; leaves ¼", irregularly deeply wrinkled (under lens), tending to curve all to one side of stem; fruiting stalks 1", red-brown; capsules often sharply crooked downward. The most abundant forest-floor moss of higher W-side elevs (rare below 2,000'). Hylocomiaceae.

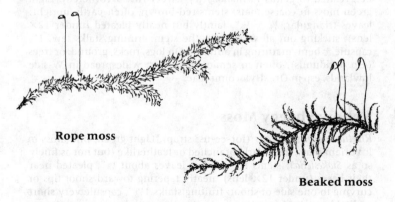

**Rope moss**

**Beaked moss**

*Both *Stokesiella* species were in genus *Eurhynchium* until recently.

# Fern Moss

*Hylocomium splendens* (hi-lo-coe-mium: forest hair; **splen**-denz: lustrous). Glossy gold to brownish green mosses in a distinctive stepwise growth form—each year's growth, shaped like a tiny (1½") fern with subbranched branchlets, rises from a midpoint on the previous year's stem, growing vertically at first and then arching into a horizontal position; a stem may show 10 or more such steps, but only the upper 1–3 look very alive; fruiting stalks few, ½–1" tall, red, not twisted; capsules ⅛", horizontal; leaves minute; forming luxuriant mats on rocks, logs, and earth. Locally dominant, common mainly in W-side lowlands. Hylocomiaceae.

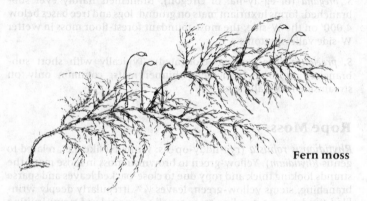

**Fern moss**

# Big Shaggy Moss

*Rhytidiadelphus triquetrus* (rye-tiddy-a-**del**-fus: wrinkled brother, i.e., related to genus *Rhytidium*; try-**kweet**-rus: 3-cornered). Light green moss in coarse mats; stems red-brown, often partly upright; leaves triangular, ¼ ± ⅛", faintly but neatly pleated (under 12× lens), sticking out all ways from the stem; fruiting stalks few, 1"; capsule ± bent, maturing in autumn. On logs, rocks, ground, or trees (esp deciduous), often in semiopen forests; widespread in W-side lowlands, esp in Ore. Hylocomiaceae.

# Yellow Shaggy Moss

*Rhytidiadelphus loreus* (**lor**-ee-us: strap). Light green, fine moss in luxuriant mats, irregularly branched or featherlike (but not as finely so as *Stokesiella*); stems red-brown; leaves about ⅛", pleated near their bases (under 12× lens), neatly tapering toward shoot tips or curving to one side of shoot; fruiting stalks 1½", capsule very short and tubby, ± horizontal, maturing in spring. On logs, trees, or sometimes ground; abundant in W-side valleys. Hylocomiaceae.

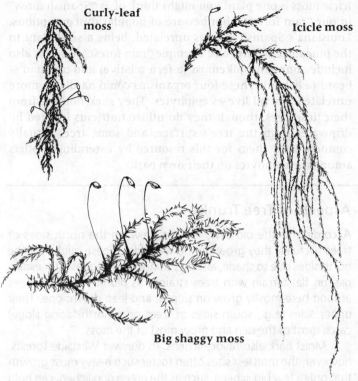

Curly-leaf moss

Icicle moss

Big shaggy moss

## Curly-leaf Moss

*Hypnum circinale* (hip-num: the Greek term; sir-sin-**ay**-lee: coiled). Very fine, delicate mosses in thin waterfall-like mats most often on conifers; stems reddish; leaves minute, narrowing (under 12× lens) to slender points in long arcs (to nearly complete circles) all to one side of stem, giving the shoot a braided look; fruiting stalks ¼–½", not twisted when dry; capsules very short (much less than ⅛"), maturing in winter. The most abundant moss on W-side conifers below 4,000'; also on other surfaces in forest. Hypnaceae.

## Icicle Moss

*Isothecium stoloniferum spiculiferum* (eye-so-**theece**-ium: equal capsules; sto-la-**nif**-er-um: many-stemmed; spic-you-**lif**-er-um: bearing spikelets). Glossy yellow-green hairlike growths festooning branches or sometimes rock faces; stems greenish brown; leaves minute, straight; fruiting stalks ¼–½", dark red; capsules maturing in winter. Mostly on trees in W-side lowlands. Brachytheciaceae.

Icicle moss is one plant you might think of as "Spanish moss" in our "rain forests," but beware of popular misconceptions. Louisiana's Spanish moss is unrelated, being a seed plant in the pineapple family; and Olympic "rain forest" festoons also include abundant spikemoss (a fern relative) and old-man's-beard (a lichen). These four organisms could hardly be more unrelated, but all live as epiphytes. They suck nothing from their host trees, though they do utilize nutrients leached by dripwater from the tree's surface, and some trees actually compete with them for this resource by extending rootlets among the epiphytes on their own bark.

## Aspects of Tree Trunks

According to the old saw, mosses grow on the north sides of trees. Actually they grow on the *wet* sides. These might well be north sides due to shade, all other things being equal—for example, on flat terrain with trees straight as plumb lines. But trees around here mostly grow on slopes, and lean downslope. Their upper sides (e.g., south sides of trees on a north-facing slope) catch most of the rain and grow most of the moss.

Moist bark also favors lichens. In our wet Westside forests, however, the moistest sites often foster such heavy moss growth that only a few big lichens, such as the green dog lichen, can hold their own. The most prolific lichen growth on these trees will likely be immediately adjacent—higher on the wet side of the trunk, and in a fringe bordering the moss carpet on both edges. The dry side may appear bare, but often it is covered by crust lichens which, being easily shaded out, are at the bottom of the pecking order for good microhabitats.

The very moistest microsites are flaring tree bases and the tops of larger limbs. Both accumulate litter, which breaks down into humus and eventually forms soil. The wet side of the trunk is a tall skinny triangle wrapping more than halfway around at the base, but tapering upward. The upper trunk, though dry all around, supports different epiphytes than the lower trunk's dry side. The upper trunk receives brighter light; wider temperature swings; and more wind, which causes fast drying, heavy wear and tear, and equal wetting by rain and mist on all sides.

# Thallose Liverworts

Order **Marchantiales.** Bright green, ± leathery flat lobes textured with close regular rows of bubblelike pale bumps; lobes regularly branching in 2-way equal splits; tall (¾–4") fruiting stalks present only briefly, in spring.

*Marchantia polymorpha* (mar-**shahn**-tia: after Nicolas Marchant; poly-**mor**-fa: many forms). Surface often bearing conspicuous cups holding gemmae (vegetative propagules); fruiting stalks umbrellalike, the female ones 9-lobed; on streambanks or burned or disturbed earth.

*Conocephalum conica* (co-no-**sef**-a-lum: cone head). No gemmae; surface bumps large and close-packed; fruiting stalks mushroomlike, the conical head fringed with spherical spore-capsules; aromatic when crushed; widespread on moist earth. Color p 489.

This type of liverwort bears no obvious resemblance to either leafy liverworts or mosses. It's more likely to be confused with foliose lichens, but the "liverlike" textural pattern distinguish it from the "lunglike" branching ridges of lungwort (page 289) or the patternless black speckles on green dog lichens (page 289). Coincidentally, liverwort, lungwort, and dog lichen are all names bestowed by medieval herbalists who looked for images of body parts—God's drug prescriptions—in the plant world. The bumps that give liverworts their alleged liver texture are air chambers, each opening to the outside by a tiny pore. They offer a favorable environment for photosynthesis.

*Conocephalum* is named for a female "cone head" that rests on the liverwort's surface for several weeks in spring, waiting for liverwort sperms to swim to it through a film of rain or dew or in the splash of a raindrop. When the spores are ripe and weather is favorable, it is raised a few inches into the air in the space of a few hours by special cells underneath it. These stem cells don't grow or multiply, they simply balloon up linearly. This unorthodox manner of growing a new stalk offers unique speed but little durability. You need to be sharp to spot one of these cute but short-lived fruitings.

Rarely are ecologists able to pinpoint the site requirements of a plant so confidently as in the following description of where *Marchantia* grows on Oregon Cascade streamsides:

"Occasionally a semiporous barrier will be deposited upon a slightly sloping, nonporous surface across which water seeps all year, such as a small log impeding drainage from an almost flat rock. If there is no disturbance, organic matter and extremely fine inorganic particles build up and form an aqueous muck." The researchers concluded that the *Marchantia* treacherously hides slick footing (Campbell and Franklin 1979).

## Leafy Liverworts

Order **Jungermanniales.** Mosslike growths of branched, flattened ribbonlike strips (usually less than ⅛" wide) of leaves overlapping in 2, 4, or 5 ranks.

*Scapania bolanderi* (sca-**pay**-nia: shovel; bo-**lan**-der-eye: after Henry Bolander). Leaves minutely toothed all around, (just visible under 10× lens), in 4 ranks all visible on top side, but only 2 apparent from underneath; a frequent dominant epiphyte species on wet sides of W-side conifers.

*Porella navicularis* (por-**el**-a: pore—; nav-ik-you-**lair**-iss: tiny boat-shaped). Leaves glossy olive green, smooth-margined; apparently 2-ranked in top view, but 5-ranked as seen from underneath. Common on broadleaf and conifer bark in lowlands.

Leafy liverworts are easily mistaken for mosses at first glance. The most dramatic difference is in the spore-bearing stalks. Those of mosses are sturdy and conspicuous several months out of the year, but the watery, insubstantial stalks of liverworts are rarely seen, since they last only for a few spring or summer days. The foliage is just as distinctive, at least from close up; contrasted with moss foliage, it reminds me of small plants flattened by winter snows.

*P. navicularis*          underside          *S. bolanderi* upper side

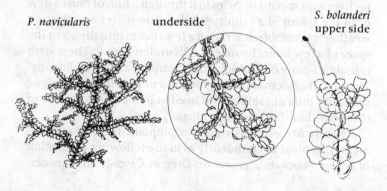

# 7

# Fungi

The mushroom is not the fungus—at least not by itself. Neither is a puffball, or a shelf fungus on a tree. Each is a fruiting body (what we might loosely call a fruit) of a fungus organism, which is something many times larger and older. As an apple tree produces apples to carry its seeds, the fungus produces mushrooms or other fruiting bodies to carry its spores.

A graphic illustration is provided by a "fairy ring"—a circle or partial circle of mushrooms that comes up year after year. These mushrooms are fruits of one continuous fungus body whose perimeter they mark; it expands year by year as the fungus grows within the soil. In some years the mushrooms may fail to appear, but the fungus is still alive and growing; it merely didn't get the moisture or other conditions for fruiting, that year. The largest fairy rings on record are over 600' in diameter. Divided by the observed rate of growth, that yields a likely age of five to seven centuries. By the same method, certain map lichens (page 285) are calculated to be 40 centuries old, rivaling the oldest trees.

The living, growing body of a fungus is a network of tiny tubes called "hyphae" (singular: "hypha"). Except when they aggregate in bundles, hyphae are too fine to readily see or handle; they range from 2 to 10 microns thick, and a cubic eighth-inch of soil or rotting wood can contain more than 300 linear feet of them. Fruiting bodies are also masses of hyphal cells. Two growing hyphal tips of any given species will normally fuse together if they happen to meet; the concept of an individual, as we know it from plants and animals, does not apply conveniently to fungi. The 500-year-old fairy ring is not exactly a 500-year-old individual, though that view would be more nearly correct than viewing each of its mushrooms as an individual. Some mycologists even argue that fungi are one-celled organisms in highly structured colonies.

A fungus is not a plant—at least not in my book. A growing consensus of upper-level taxonomists finds little reason to include fungi in either the Plant or Animal Kingdoms. (See the Five-Kingdom System on page 574.) Like animals, all fungi live by obtaining carbohydrates manufactured by chlorophyllous (green) organisms. Animals accomplish this by eating, whereas fungi employ four main nutritional modes:

**Mycorrhizal:** linking up with plant roots for two-way exchange of nutrients and water.

**Lichenized:** enclosing and "farming" algae and blue-green bacteria.

**Saprophytic:** decomposing (i.e., rotting) dead organic matter.

**Parasitic:** drawing nutrients from living plants or animals.

## The Fungus/Root Symbiosis

The word "mycorrhiza," coined in 1885 from Greek words meaning "fungus" and "root," names a special growth that connects fungi with plant roots. Over 90% of the vascular plant species of the world are known to form mycorrhizae, and over 50% of higher species of fungi feed almost exclusively by mycorrhizae. Yet this crucial arrangement of our living universe is little known, and barely acknowledged in popular writing.

Simply put, the functions of root hairs in plants are normally (well over 50% of the time) performed instead by symbiotic fungi. Soon after a plant germinates from seed, its rootlet is likely to meet up with a fungal hypha. Each contains hormones that will stimulate and alter the growth of the other, forming a joint fungus/root organ (the mycorrhiza) that immediately goes into use as a nutrient loading dock. Some plants start even earlier;

The first two modes, which cover a majority of the multicelled fungi, are often considered "mutualistic" symbioses, meaning that they may benefit both the fungal and the green partners in the relationship; the benefits are actually questionable (pages 181, 282) and variable, but at any rate all partners do survive. The third, fungal rotting, is essential to the health of plant communities in clearing away the dead members and recycling their nutrients. Only the fourth, parasitism, is usually an antagonistic relationship of fungi to plants and animals.

Very few mushrooms are seriously poisonous to eat. A greater number may make you sick in the stomach or uncomfortable somewhere else. Still more are considered edible by most who have tried them, but even among "good edibles" many reputations are tainted by reports of a few allergic reactions. Even the supermarket mushroom, *Agaricus bisporus*, upsets some tummies. Many other mushrooms go unrecommended on grounds of flavor or texture. One man's "edible and choice" is another man's "Bleccch!"

All in all, the odds favor mushroom eaters, but the risks are too extreme to forgive haphazard identifications. There are old mushroom hunters, the saying goes, and there are bold mushroom hunters, but there are no old bold mushroom

---

most orchid seeds must be penetrated and nourished by a hypha before they will germinate.

Different plant classes use different forms of mycorrhizae. The hyphae may grow into a net around the root tip, or they may digest cell walls, penetrate and inhabit the root cells, and eventually be digested back again by the cells. In any case the hyphae hormonally suppress the formation of plant root hairs; at the same time they provide a preestablished network of root hair surrogates finer and more efficient than plant roots. Water and soil minerals collected by the hyphal tubes pass into the root. After the plant grows and begins to photosynthesize, carbohydrates pass into the fungus to meet its energy needs. (Of course, the fungus wasn't going without carbohydrates up till now; it was already hooked up mycorrhizally with other plants.)

While the typical exchange includes water, minerals, and nitrogen from the fungus and carbohydrates from the plant,

---

hunters. (I'll bet there are really at least one or two!) Mushroom identification is more subtle and technical than most plant identification, often requiring chemical reagents and a microscope. There's much to be said for the unwritten rule that nontechnical guides (such as this one) not entirely devoted to mushrooms should forgo labeling edibles as such.

But fungal dinners in camp have been such pleasurable wilderness experiences for the writer that this book will share some favorable advice. For eating typical mushrooms—those with a gilled cap on a stem—I refer you to mushroom books (page 600); but I can recommend several fleshy fungi that either lack paper-thin gills or lack a distinct stem. Each of these species can be separated from poisonous species by carefully examining gross appearance alone. All the same, you must assume responsibility for your own results, gastronomic and gastrointestinal. If in doubt, consult a mushroom expert.

This chapter cannot offer a representative sampling of the hundreds of kinds of fungi, lichens and algae that grow here. Two huge and important groups of fungi that are omitted due to their poor visibility should be mentioned: microfungi, such as yeasts and molds; and underground-fruiting fungi, such as truffles. The latter live mycorrhizally just like many mush-

---

other materials may also be passed. Seedlings that rely on mycorrhizae receive carbohydrates indirectly from overstory plants until they are big enough to support themselves—a crucial benefit on our light-deprived forest floors; non-green plants (page 180) remain dependent all their lives. During dry seasons, deep-rooted trees may be the chief water suppliers to the network, since hyphae are largely confined to shallow organic soil.

Fungi cannot produce carbohydrates, and most mycorrhizal fungi are totally dependent on their plant partners. Technically, few plants are equally dependent on fungi, since most plants have the genetic information for making root hairs to obtain water and minerals. Planted in a rich, nutritious substrate, a seedling is likely to do just that, resisting mycorrhizal infection of its roots. But in the real world, nonmycorrhizal seedlings may not survive competition with fungus-assisted rivals, or they may die for want of water or free nitrogen; mycorrhizal fungi can break down soil

---

rooms, but instead of giving their spores to the wind to carry, they use ravishing aromas (as green plants may use sugar or fat) to entice mammals to eat them. (See page 32 .) Some of them also get by with tiny fruits that get moved around accidentally by burrowing animals.

Broad classification of fungi is so up in the air that it would serve no purpose to name families or orders in this book.

## Fly Amanita and Panther Amanita

*Amanita* spp. (am-a-**nigh**-ta: Greek term for some fungus on Mt. Amanus). Cap usually sprinkled with whitish warts; stem white, with skirtlike ring near the top and a bulbous (not quite cuplike) base; gills and spores white.

**Fly amanita,** *A. muscaria* (mus-**cair**-ia: of flies). Also **fly agaric.** Cap bright red to orange, or (uncommonly here) yellow to white; stem base has a series of slight rings, diminishing in size upward. Widespread, on the ground, mainly in fall. Color p 483.

nitrates, and reach deeper, faster, for water, enabling seedlings to thrive. This key relationship is appreciated by the Northwest's timber industry, which inoculates conifer seeds in the nursery with suitable fungal symbionts.

Much remains to be discovered about the connections, cooperative and competitive, in the mycorrhizosphere. Mycorrhizae help protect plants from disease; root-rot (parasitic fungi) is blocked by healthy mycorrhizae, which form both a physical and a chemical barrier around roots. Fungi regularly compete with other fungi by secreting selective toxins. They also secrete antibiotics to suppress some bacteria pathogenic to plants; on the other hand they appear supportive to nitrogen-fixing bacteria. And then again, fungi undoubtedly bring some kinds of trouble to their plant partners in the course of feeding countless soil-inhabiting animals. Mycorrhizal fungi are a literal embodiment of the "web of life" the natural community is sometimes called.

**Panther amanita,** *A. pantherina* (panther-eye-na: panther—). Cap tan or brown; bulbous base ± sharp-lipped above (where the universal veil broke away). Typically under Douglas-fir; spring, fall or rarely winter.

Long famous in Europe as the archetype of malevolently alluring toadstools,* the fly amanita has regained an older reputation as a recreational or spiritual drug. Siberian tribes and, according to some scholars of the Vedas, ancient Hindus used amanitas that way. Some people in the Northwest have used them repeatedly, taking precautions that apparently work for them. I suspect the variable reactions have less to do with technique than with luck (including genetic predisposition). A wild-eyed and euphoric stranger on the trail once handed a fly amanita cap to two friends of mine, who were impressed enough to down it on the spot. One friend had an enhanced afternoon while the other, from the other half of the same cap, was in abject misery. Cramps, spasms, sweating, vomiting, lethargy, stupefied sleep, "manic behavior," and subsequent amnesia are symptoms often reported, with death very unlikely, though that also has been recorded.

Problems of identification within this freely hybridizing genus are also unsettling. The panther amanita can be deadly, and its toxins are largely just higher concentrations of those (muscimol, ibotenic acid and a little muscarine) found in the fly amanita, as opposed to the liver-destroying cyclopeptides of the destroying angels. Panther and fly are distinguishable in the field mainly by color, and that's not much to go by, since the "white form" of the fly may be cream yellow and the panther ranges to pale tan. In any case, they do look fine; trailside specimens are a scenic resource to be left untouched.

Panther amanitas resemble panthers (i.e., leopards) only in being tannish and spotted, but fly amanitas have centuries of experience poisoning flies. They were traditionally left around the house broken up in saucers of sugared milk—children beware!

---

*A toadstool, in British usage, is any poisonous or unsavory mushroom. American usage broadened the word to mean any mushroom. Mycologists can be happy with the loss of meaning, if it entails the disappearance of dangerous old-wives'-tales about "how to tell a mushroom from a toadstool"—i.e., myths that edible and poisonous mushrooms are essentially different classes of beings. They aren't.

# Destroying Angels

*Amanita* spp. All-white mushroom, cap smooth, sticky or slimy when wet; stem with a substantial, tattered skirtlike ring (sometimes missing) and a ± bulbous base in a thin white cup (requires careful excavation to preserve); spores white; no odor; **do not taste.**

*A. verna* (ver-na: spring). In spring.

*A. virosa* (vir-oh-sa: smelly). Typically in late fall. Color p 483.

*A. bisporigera* (by-spor-**idge**-er-a: 2 spore bearing). Summer and early fall, typically under aspen or birch.

Pure white and lovely but monstrously poisonous, these are our most dangerous fungi, so it's important to know their characteristics even though they're rare here. All amanitas have white spores and more or less white gills. Most have a definite ring around the stem, the remnant of a "partial veil" that extended from the edges of the cap, sealing off the immature spores to help keep them moist. More distinctively, most also emerged from a "universal veil," an additional moisture barrier that wrapped the entire mushroom, from under its base to all over its cap. As a young "button," each amanita fruiting-body in its universal veil was egg- to pear-shaped, resembling a puffball but with the outline of cap and gills visible in cross-section. Remnants of this veil *usually* persist as a cup or lip around the base of the stem, and/or on top of the cap in the form of warts, crumbs or broad patches. But the absence of these remnants doesn't disprove any amanita, since the stem easily breaks off above the cup, and the bits on the cap may get washed off.

All white-spored, white-gilled mushrooms should be collected with great care and examined for ring and cup. The genus *Amanita* contains many good edible species, some of them long popular in Europe, but the chance of misidentifying a deadly one leads American guidebooks to disrecommend all amanitas. (By the same reasoning, this book doesn't recommend any stemmed, gilled mushrooms at all—it would take a complete mushroom book to identify them adequately to prevent mistakes.)

The destroying angels, together with the similar but tan or olive-capped death cap, *A. phalloides*, and the little-brown-mushroom genus *Galerina*, share one of the most insidious

---

poisons found in nature. It attacks the liver within minutes of ingestion, but symptoms don't appear for 10–14 hours (or up to 3–4 days) by which time the liver is seriously damaged. Over half of the poisonings recorded in America by these species have been fatal, though hospitals have developed intensive techniques that can improve the odds if the cause of poisoning is known before symptoms are too advanced.

## Autumn Galerina

*Galerina autumnalis* (gal-er-**eye**-na: helmeted; autumn-**nay**-lis: a misleading name for this species). Caps mostly ¾–1¾″ diam, sticky and deep yellow-brown when wet, dull tan when dry, radially striped near the edge; gills pale, becoming brown with spores; stem thin, brown, darkening toward base, with a thin whitish ring; spores rust brown. Typically clustered on rotting (often ± buried) wood, often among mosses, in late fall or sometimes spring. Color p 483.

This unprepossessing "little brown mushroom," or "LBM," contains the same cyclopeptide poisons as the deadly amanitas, is just as deadly, and is far commoner here. It is considered less of a danger simply because people don't normally bother with LBMs—unless they happen to be hunting psilocybin mushrooms. *Galerina* has a few traits in common with our native *Psilocybe* species, so you'd damn well better know what you're doing if you hunt *Psilocybe*. Galerinas have also been mistaken for the edible honey mushroom.

## Shoestring Root Rot / Honey Mushroom

*Armillariella mellea*\* (ar-mil-airy-**el**-a: banded, small; **mel**-ee-a: honey). Highly variable; cap cream to yellow-brown with ± conspicuous erect hairs or scales radiating from the center; often sticky or tacky; gills white, staining rusty with age; stem has a substantial, upflaring, whitish ring; clustered on wood or ground, caps often coated with spores where overlapped; spores white to yellow; coarse black threadlike rhizomorphs often visible around base, or netting across nearby wood, often under bark. Fall. Color p 483.

Mushroom-hunters know this as a long-popular if somewhat unreliable edible. (It is on this "not recommended" page solely

---

\*This genus was rather recently separated from *Armillaria;* further study may divide *mellea* into several species.

because of dangerous look-alikes.) Foresters know it as a pest, a killing pathogen on trees, and have devoted much study to eradicating it. Apparently it lives saprophytically much of the time, and switches to deadly parasitism for unknown reasons.

## Violet Cortinarius

*Cortinarius violaceus* (cor-tin-**air**-ius: curtained; vye-o-**lay**-see-us: violet). Cap and stem ± uniformly blackish purple; cap 2–5", shiny even when dry, covered with fine fibrous scales; cinnamon-brown spores soon color the gills. On ground in old-growth, esp in Olys, late summer and fall. Color p 483.

I admire this mushroom for its astonishing deep color, but I can't recommend it for eating because positive identifications within this huge and risky genus are beyond the scope of this book. Many of the 600-odd species of *Cortinarius* have rusty spores and varyingly deep shades of lavender in their caps. The genus is characterized by a type of veil called a "cortina," which stretches out like a filmy curtain as the cap expands and, after breaking, leaves cobwebby remnants rather than any substantial ring on the stem.

## Woolly Chanterelle

*Gomphus floccosus*\* (**gom**-fus: tooth, or cockscomb?; flock-**oh**-sus: woolly-tufted). Orange trumpet-shaped fungus with no differentiation of cap from stem; often deeply hollow down the center; inside surface roughened with big soft scales; outside surface paler and irregularly, shallowly wrinkled (no paper-thin gills); spores ochre. On ground, late summer and fall. Color p 483.

Mushroom guides, over the years, batted this species back and forth between the "Edible" and "Poisonous" lists. Chemical study seems to confirm it as "Not Recommended," but not so dangerous as to drag the true chanterelles down from the "Safe for Beginners" list on grounds of possible confusion with it. Unfortunately, this one is commoner than edible chanterelles in old-growth forest, at least in the Olympics, while the yellow chanterelle seems to prefer second-growth. The mechanism of these preferences is obscure, since both species accept Douglas-fir and other trees as mycorrhizal partners.

---

\* Older texts include this species as *Cantharellus floccosus.*

# Chanterelles

*Cantharellus* spp. (canth-a-rel-us: small vase). Vase-shaped mush-room (cap undifferentiated from stem), margin often irregular when mature; spore-bearing surface of rounded ridges (not paper-thin gills) sometimes with slight cross-wrinkles to make a netlike texture; slight peppery aftertaste when raw. On ground, in fall.

**Yellow chanterelle,** *C. cibarius* (sib-air-ius: for dinner). Cap yellow-orange, smooth, underside paler; spores ochre. Color p 484.

**White chanterelle,** *C. subalbidus* (sub-al-bid-us: somewhat white). Cap white, bruising yellow to (eventually) rusty orange; form as above but often extremely stout and short; spores white.

The yellow chanterelle is the mainstay of the burgeoning wild mushroom trade in the Northwest. Its popularity stems not so much from exceptional flavor as from its rich color, easy and safe identification, profuse local abundance, resistance to bugs, and established place next to boletes and morels in French cuisine. Our white chanterelle, being much rarer, is

---

## Cautious Mushroom-Eating

**A spore print** for observing spore color is made simply by laying a cap, without stem, flat on a piece of paper for an hour or so. If you put part of the mushroom over inked or tinted paper, then white spores will show up as well as dark ones.

**Nibbling by animals** is not evidence of safety.

**Try a nibble** before eating a quantity, and give your stomach at least two hours to test the species. Each member of your group should test his or her stomach. (Do this only after identifying the mushroom. Deadly amanitas and galerinas, pages 265–68, must be absolutely ruled out first, since even a nibble would be danger-ous. False morels, even the good ones, page 281, should not be eaten raw.)

---

regarded as even more choice. The only poisonous mushroom sometimes mistaken for a chanterelle (aside from the questionable woolly chanterelle, page 269) has true (paper-thin) gills. Thin gills, with a higher surface-to-volume ratio, are more efficient, and evolutionarily more advanced, than chanterelle-style wrinkles.

## Hedgehog Mushroom

*Dentinum repandum*\* (den-tie-num: tooth—; rep-and-um: wavy-edged). Also **sweet tooth mushroom.** Cap yellow-orange to buff or nearly white, vaguely differentiated from stem; underside covered with fine pale teeth of mixed lengths (average ¼") in lieu of gills; spores white. On ground; summer, fall and even winter. Color p 484.

Eyes scouring the ground for the soft gold of chanterelles may jump at this similar cap, but the fingers will be in for a surprise when they reach underneath to pluck it and find a curious soft, spiny texture. Don't be disappointed—this mushroom is also edible, and even harder to confuse with anything poisonous. Some ground-growing toothed relatives, however, are bitter, as even the tiniest taste-test will reveal.

\*Older texts list this species as *Hydnum repandum.*

---

**Small children** are much more susceptible to mushroom poisoning and allergies, and should not eat wild mushrooms.

**Eat only moderate quantities,** of only one species, in your first meal of that species.

Excessive **bugginess or worminess** occasionally causes stomach upset. If the stems but not the caps show larval bore holes, leave the stems behind so they won't infect the caps.

Carry mushrooms in **paper bags,** not plastic. (This helps store-bought mushrooms, too.) Keep them as cool as possible.

Don't forget to take **plenty of butter,** especially on fall trips. Gentle sauteing, without a lid, is rarely a bad way to cook a fungus. Scrambled eggs, toast, or crackers rarely fail to compliment.

---

# Oyster Mushroom and Angel-wings

*Pleurotus* spp. (ploor-oh-tus: side ear). Fan-shaped, usually stemless mushrooms growing in clusters to small groups from the side of fallen, dead, or sometimes just wounded trees; mildly fragrant and pleasant-tasting (edible raw).

**Oyster mushroom,** *P. ostreatus* (os-tree-ay-tus: oyster). On broadleaf wood, typically cottonwood or alder in our region; cap tan, cream or oyster-gray; spore print dries pale lilac. Color p 485.

**Angel-wings,** *P. porrigens** (por-i-jenz: spreading). On conifer wood, esp hemlock; all white; flesh thin; spores white. Color p 485.

The oyster mushroom may grow at any time of year, mainly fall but not uncommonly spring. It is one of the few temperate zone species that continue their growth through winter. This genus may also grow with a short, semihorizontal stem from one edge of the cap—never the center. But it's the stemless habit that warrants its place as the one genus of gilled mushrooms (Class Agaricales) that I can recommend for eating without a complete mushroom book for identification. There are several Not-Recommended species that share the stemless or offcenter-stemmed habit, but each undesirable is marked by at least one of these traits:

**tough,** leathery, thin flesh;

**saw-toothed** gills;

**yellow gills** contrasting with dark, drab cap;

**intense bitter** or peppery taste. (Nibble, and don't swallow.)

When you confirm oysters or angel-wings, search the entire log for more, because the fungus tends to exploit the entire tree and fruit here and there all over it. Slice the mushrooms off rather than ripping them out, and they may produce another crop in a couple of weeks. Tiny beetles living between the gills can be knocked loose by tapping briskly on the cap. Some people claim that oyster mushrooms, if cooked like

---

*Some authorities place *porrigens* in a separate genus *Pleurocybella*.

oysters, can pass for them. Angel-wings are so delicately tasty that much of the flavor may be the butter they're cooked in.

Carnivorous plants like Venus' flytrap are notorious, but carnivorous fungi are esoteric news. The oyster is one of several mushrooms recently found digesting tiny nematode worms and utilizing them nutritionally. Nematodes were long known to burrow into mushrooms for dinner, but the mushrooms were not previously known to turn the tables on them.

## Admirable Boletus

*Boletus mirabilis* (bo-lee-tus: clod; mir-ah-bi-lis: admirable). Cap 2–6" diam, with dark red-brown skin minutely fibrous-roughened, and cream-white flesh red-stained just under the skin; tubes yellow, bruising darker yellow (rarely bluish); stem streaked with same red-brown shade as cap; spores olive-brown. Usually on rotting wood, under hemlocks; in fall. Color p 484.

Boletes are generally mycorrhizal. Some serve as the normal host of non-green plants in the heath family, while also partnering with trees. The admirable is an unusual boletus in that it typically fruits on well-rotted wood, which might lead one to suppose it is saprophytic rather than mycorrhizal, but research has shown otherwise; this species lives by mycorrhizae with hemlocks, and associates with rot fungi for unknown reasons. An abundance of potentially mycorrhizal fungi in rotten logs might help explain the prevalence of nurse logs as a seedbed for hemlocks here. (See page 36.)

A choice edible boletus—but don't eat specimens that have been attacked by a white mold.

## King Boletus

*Boletus edulis* (ed-you-lis: edible). Also **cepe, steinpilz.** Cap 3–12" diam, skin tan to red-brown, often redder just under surface, ± bumpy but not fibrous, often sticky; flesh ± white, firm; pores white, becoming olive to tawny yellowish with age or bruising, but never blue; stem often very bulbous, white with ± brown skin, upper part finely net-surfaced; spores olive-brown. Usually near pines; late summer and fall. Color p 484.

## Suillus

*Suillus* spp. (sue-ill-us: pig). Also **slippery jack** (some species being slimy-surfaced).

*S. lakei* (lake-eye: after Lake). Cap rough with red-brown flat scales, yellow and ± sticky under the scales; flesh and tubes ± yellow, maturing orange, may bruise reddish; tube mouths angular and radially stretched when mature; stem has whitish ring when young; base bruises blue-green; spores brown to cinnamon. Always near Douglas-fir. Edible, tasty to some. Color p 484.

*S. cavipes* (cav-i-peez: hollow foot). Cap dry, with reddish to brown scales; flesh and tubes ± pale yellow, not bruising blue; tube mouths angular, strongly radial-arranged when mature; stem reddish-scaly below ring, smooth white and slightly narrower above ring, while the white-fibrous ring itself may disappear; stem becoming hollow; spores dark brown. Always near larches. Edible, choice to some.

---

### Cautious Bolete-Eating

Many of our better edible mushrooms are boletes, or members of the family Boletaceae. They look much like regular gilled mushrooms from above, but when picked reveal a spore-bearing undersurface consisting of a spongy mass of vertical tubes. If you tear the cap, you will see that the tubes are distinct from the smooth-textured flesh above them, and can usually be peeled away. Most boletes range between choice and merely edible, and none are commonly deadly. The relatively few nonrecommended species can be ruled out rather simply, so the rest can be eaten without necessarily identifying them to species. Boletes begin fruiting in midsummer, earlier than most edibles, and they are often large, supplying hearty eatings from few pickings. Their main drawback, as a group, is rapid susceptibility to insects, decay and moldy growths.

Here are the cautionary rules:

Never eat a bolete with **red-orange or pinkish-tinged** tube openings; this trait characterizes all the moderately toxic species.

Try a nibble of the cap for the **bitter, burning or peppery** tastes that make some other boletes undesirable.

Avoid boletes whose tubes **turn deep blue** (or "bruise") within a

---

*S. granulatus* (gran-you-**lay**-tus: granular, a misleading name). Cap with a tacky to slimy skin, tan to cinnamon; faintly mottled; flesh white to cream; tubes tan to yellow, small, "dewy" in youth, staining or speckling brown with age; no veil or ring; stem white, developing brown dots or smears with age; spores brown to ochre. Usually under lodgepole or ponderosa pines, often abundant. Good edible.

*S. brevipes* (brev-i-peez: short foot). Like *S. granulatus,* but cap starts deep red-brown and pales to ocher with age, and the short stem remains pure white except, rarely, for a few faint dots in great age; spores brown to cinnamon. Good edible.

Modern taxonomy places most boletes within one of two enormous genera, *Boletus* and *Suillus*. Both words were used by the Romans for certain mushrooms; one derives from the Greek for "a lump of earth," which boletes sometimes resemble, and the other from the Latin for "pigs," who are fond of

---

minute or two, where handled. While a few species mildly toxic to some people show this sharp reaction, as do some of the red-tubed toxics, many of the best edibles may also bruise bluish, at least slowly, at least sometimes or in some of their parts. Many mushroom eaters in this region are undeterred by blue bruising unless it is intense and quick, or the tube-mouths are at all reddish.

If the cap has a **gelatinous or sticky skin,** remove this. It upsets some people's bowels. Most people also prefer to peel off and discard the tubes from mature specimens.

Look closely for **larval pinholes** by slicing or snapping the stem from the cap; if there are few or none, the cap is in good condition. Then slice through the stem at mid height, and look again. If you still find few or no holes or discolorations, the stem can be eaten along with the cap flesh, especially if there is a shortage of good caps. If the stem is buggy, discard it immediately rather than give the larvae a chance to spread.

Don't expect boletes to keep longer than overnight in any case, except very clean young specimens in chilly weather or a refrigerator. Carry in paper bags, out of direct sun.

Also read the general mushroom-eating cautions, page 270.

---

eating them. Italians still name boletes *porcini*, after pigs. *Suillus* species almost all live mycorrhizally with conifers; many require pines or larches as partners, and are common east of the Cascades.

No American *Suillus* is poisonous, aside from numerous allergic reactions often blamed on the skin, which can be peeled before cooking. However, no single field trait distinguishes *Suillus* from *Boletus*, which includes poisonous species. Most *Suillus* have at least two of the following traits:

**Tube-mouths stretched out** and aligned in a direction radial from the center. (Probably transitional to gills in evolution.)

**A partial veil** that persists as a ring, at least for a while.

**A glutinous skin** on the cap, slimy when wet, tacky when dry.

**Glandular dots** on at least the upper stem, at maturity.


## Purple-tipped Coral

*Ramaria botrytis* (ra-mair-ia: branched; bo-try-tiss: bunch of grapes). Cauliflowerlike structure, the fat white bases (often clustered) much more massive than the branchings on top; the blunt branch tips brownish rose to red, dulling with age; base typically submerged in duff or moss, leaving only the tips exposed; delicately fragrant; spores ochre. Under mature conifers in early fall. Color p 485.

Like many worthwhile finds among the fungi, this species makes itself hard to spot, but once you have seen one and attuned your eyes to it you may find many more. It has provided me with fine camp dinners; the literature, however, reports an intensely bitter look-alike, *R. botrytoides*. The entire group of corallike branching fungi that grow on the ground or rotten wood is relatively harmless, but includes one stomach-upsetting species, *R. formosa*; it has the slim vertical branches (not thick like cauliflower) typical of the family, and is peachy-pink when fresh except for yellowish tips. Also not recommended are species that show translucent, gelatinous cores in cross-section. The remaining corals are nonpoisonous, but few are good eating.

# Cauliflower Mushroom

*Sparassis crispa** (spar-**ass**-iss: lacerated). Large (typically 5–24" diam) fragrant cream-white leafy structure resembling a bowlful of ribbon noodles; spores white. On conifer stumps or trunks quite near the ground, usually in late fall. Color p 486.

This looks less like cauliflower than the preceding species, and more like a great heap of fettucine; an Alfredo sauce (garlic, pepper and cream) would be perfect. Long slow cooking is recommended. Beauty, fragrance, size, and moderate rarity make this a prize for the mushroom hunter. Smart ones carefully note the location and return a year or two later.

# Bear's-head Coral

*Hericium* spp.† (her-**ish**-ium: hedgehog). Also **coral hydnum.** Large (6–12+") fungus consisting of many branches and sub-branches ending in fine teeth all pointing down; spores white. On dead wood or tree wounds, early fall.

*H. abietis* (ay-bih-**ee**-tiss: of fir trees). Initially pinkish to orange-buff, aging to white; sometimes huge (up to 30" high × 16"), ± evenly rounded, with short branches from a massive, solid base. On conifers only.

*H. coralloides* (coral-oh-**eye**-deez: corallike). All white, aging yellowish and brown-tinged. On hardwoods. Color p 486.

Spectacular appearance puts these fungi in both the "Safe for Beginners" and the "Scenic Resource" categories; trailside specimens should be left to feast the eyes. Whether they're really culinary prizes is debatable anyway. Fragrance and flavor are usually lovely, but the texture bothers some people; the larger branches can be tough and hard to digest. To prepare, cut into the mass discreetly, removing only modest quantities of parts that are pure white and have few or no bore holes from insect larvae. Soak them in a pot of water to float any beetles out from the crevices. Mince and saute gently.

---

*Older texts list this species as *S. radicata.*
†Nomenclature in this genus is confused. The species *weirii* and *americanum* of some authorities are included in *abietis* and *coralloides*, respectively, by our authority.

---

# Sulphur Shelf Fungus

*Laetiporus sulphureus*\* (lee-tip-or-us: bright pores; sul-**few**-rius: sulphur yellow). Also **chicken-of-the-woods**. Thick, fleshy shelflike growths 2–12″ wide, typically in large overlapping clusters; orange above, yellow below. On trees, summer through fall (long-lasting). Color p 485.

The shelf fungus family, Polyporaceae, is characterized by simple fruiting bodies that release spores through tiny pores on their smooth undersides. The pores are never separable from the cap flesh. Dense, often rather woody flesh makes these fungi much longer-lasting than mushrooms and hence more often seen than numbers alone warrant; but it doesn't do a thing for their edibility. The one relatively tender exception is the sulphur shelf, a good edible when young. To some tastes it's just like chicken; I find it lemony.

No serious toxins have been found in the shelf fungus family; occasional adverse reactions to this species have been blamed variously on eucalyptus as the host tree (no problem in the Northwest) or on allergies, excessive consumption, or overly mature specimens. To make sure of eating it young enough, take only the outer edges, and only if your knife finds them butter-tender, lemon-yellow rather than deep orange, and fat and wavy rather than thin and corrugated. New tender margins may grow in their place. Since this is both a gastronomic and a scenic resource to be shared with other hikers, limit yourself to a quantity you can easily eat, and don't even think about cutting up growths right next to a trail.

Most shelf fungi live as heart rot in trees. Since heartwood is dead tissue, they could be considered saprophytes rather than parasites. But they do kill trees by weakening them to wind breakage, and they spoil the lumber, so they are a serious economic pest in timber country.

\* Some texts include *sulphureus* in the large genus *Polyporus*.

# Warted Giant Puffball

*Calbovista subsculpta* (cal-bo-vis-ta: bald foxfart;* sub-sculpt-a: somewhat sculptured). Stalkless, slightly flattened ball 3–6″ diam, white patterned with brownish raised polygons. (Positive identification is technical, but the similar *Calvatia* species are all equally edible.) Typically subalpine among grass, or sometimes under conifers; in midsummer. Color p 484.

Most puffballs are smaller than golf balls, but *Calbovista* is baseball-sized, and some of its relatives in genus *Calvatia* can grow bigger than basketballs. Puffballs are a family of fungi whose round or pear-shaped fruiting bodies have neither cap nor stem nor gills. The maturing skin either splits open or opens a small hole at the top, and spores come puffing out by the millions, mostly when the puffball is struck by raindrops. You might think a rainy day a poor one for spore travel, but in fact spores, like pollen, are "hydrophobic" or resistant to wetting, so their flight is undamped by rain or fog. Puffball spores, a hundred times smaller than the majority of spores and pollen grains, are barely subject to gravity; squeeze a ripe puffball on a breezy day, and you may literally be sending a few spores on a trip around the world.

Puffballs have been used medicinally in many cultures, and some are thought to contain anticarcinogens. No true puffballs (family Lycoperdaceae) are poisonous, but young "buttons" of deadly amanitas can be mistaken for them, so you must **slice every puffball you pick through its center vertically,** and eat only those that are perfectly pure undifferentiated white inside, from tip to toe. Amanita buttons in cross-section reveal at least a faint outline of developing stem and gills. Other telltales for the discard pile include a yellow-, green-, or brown-stained center, or a punky center distinct from a smooth ⅛″ outer rind. The former indicates a maturing puffball, the latter a *Scleroderma* ("hard skin") fungus—both unpleasant though not really dangerous. Puffballs take a couple of weeks to mature, so you stand a sporting chance of

---

*This name is a compound of two other genera: *Calvatia* is Latin for "baldhead," while *Bovista* is old German vernacular for a puffball, with a literal meaning of "fox-fart." The same origin myth Latinized names genus *Lycoperdon* ("wolf fart"), our common puffball.

catching them young enough.

If it's your first taste of puffball, just pick a small sample. You may hate it. The flavor is mild, pleasantly fungal, but the texture is slippery where sliced, a bit rubbery within. Frying in batter, if you have the means, can mask the oddness. My own proudest trailside effort was a lunch of warted giant puffball sandwiches—half-inch slices fried in butter with garlic, on crackers thinly layered with anchovy paste and cheese.

## Myco-cuisines of the World

Moira Savonius, in the British *All Color Book of Mushrooms and Fungi,* attributes to the sulphur shelf fungus "an unpleasant sour smell as well as a nasty taste," while Vincent Marteka, in *Mushrooms Wild and Edible,* rates it as the first choice in a poll of American mushroom hunters, and mentions traditional use of it by Indians. There are many transatlantic differences of opinion over which mushrooms are delicious, and over which are poisonous. Some are based on the chemistry of geographic races which may eventually be reevaluated as distinct species.

Cultural attitudes also play a role. The English have a mycophobic tradition that regards only species of *Agaricus* (such as the supermarket mushroom) as deserving either the name "mushroom" or a place on the table. In contrast, Italians have a mania for amanitas and boletes; French go for truffles, morels, chanterelles, and boletes; and Slavs, the broadest-minded, are intimate with species in relatively nondescript genera like *Russula* and *Lactarius*. Chinese and Japanese traditions are bringing stranger fungal flavors and textures to American palates—leathery "tree-ears," resiny "pine mushrooms," leggy "straw mushrooms" grown in bottles, and "black fungus" cured in brine.

# Snow Morel

*Gyromitra gigas* (jye-ro-my-tra: round hat; jye-gus: giant). Also **snowbank false-morel**. Cap dull yellow-brown, convoluted, lacking gills, flesh in cross-section thin, white, brittle; stem white, often ± concealed by cap, nearly as big around as the cap and with similarly convoluted, thin, brittle flesh enfolding irregular hollow spaces. Under high-elev conifers immediately after snowmelt. Color p 486.

This bizarre-looking fungus is a decent edible, but unusual precautions must be observed. First, eat only specimens whose identity is corroborated by patches of melting snow-pack nearby. Second, boil the material after cleaning and chopping it; avoid breathing the steam; throw out the water; and rinse the fungus again before beginning your final cooking of it.

*Gyromitra* toxicity was long a mystery. Countless people, over hundreds of years, happily ate the so-called edible false-morel, *G. esculenta,* yet there were occasional fatalities. The toxin was recently isolated and found to be the same chemical as a rocket fuel manufactured for Apollo missions, so it is now a well-studied chemical—a deadly poison and suspected carcinogen. Its frequent failure to kill people results from at least two characteristics: highly volatile, it evaporates out during either drying or boiling; and it produces no symptoms until a victim exceeds a threshold level, for example by eating quantities of undercooked *Gyromitra* over a period of a few days.

Now that the explanation is known, some mushroom buffs eat *esculenta* after scrupulous parboilings, though mushroom guides say "POISONOUS" right under the name *esculenta,* which means "edible." Some guides extend the prohibition to the entire genus. A few other *Gyromitra* species contain the toxin, but none has yet been found in *gigas.* The several distinct, irregular hollows within the broad stem, together with time of fruiting immediately after snowmelt distinguish this from any toxic species. I recommend parboiling as an added precaution.

---

Lichens consist of fungi specialized to meet their carbohydrate needs from algae or bacteria they enclose. This symbiotic association is highly developed, producing organs, tissues and chemicals found only in lichens. Lichens differ conspicuously from other fungi in that they grow in the open and are tough, durable, and able to dry out and revive again. (Contrast this not with the mushroom but with the fungus' "body," which lives all year long in soil or wood beneath the mushroom site.) Most lichen fungi are so altered by their partnership that lichenologists can't firmly classify them in relation to free-living species. The algae, on the other hand, may have close nonlichenized relatives, and the bacteria may pass between free-living and lichenized states. (Cyanobacteria were formerly known as "blue-green algae," and popular writing on lichens often lumps together the photosynthesizing partners—properly "photobionts"—as "algae.")

Scholars today question whether the lichen symbiosis is mutually beneficial. Some (Ahmadjian 1982) call it "controlled parasitism" in which the fungi slowly kill their green partners. Plainly, the fungus gets its organic food from the alga, while the alga gets its water and minerals at least *through* the fungus. A cow is also fed by its farmer, and a slave by its master, but are they benefitted by these relationships? The alga can grow healthily on its own, after all, but the fungus cannot; the algal partner must be present to induce a fungus to even remotely resemble a lichen. Lichen algae, at about 7% of the tissue mass of a typical lichen, are in a minority, cultivated role not unlike that of intestinal flora that live symbiotically in mammals, and we don't think of ourselves as compound organisms—we're mammals, period!

Propagation of lichens is usually from fragments that include both partners together. These may be just any old fragment that happens to break off, or they may be specially evolved propagative bundles ("propagules") in the form of powders, grains or tiny protuberances. Most lichens also disseminate fungal spores from fruiting bodies analogous to those of other fungi. In many species, the spore must contact an appropriate alga to germinate. In others, spores can germinate and multiply into a tiny clot of fungus that can live saprophytically long enough for capture of a suitable alga to become reasonably likely. In either case, this route is far less common than growth from alga-containing propagules. Though infrequent, sexual propagation via spores is important in maintaining genetic diversity.

Lichen fungi are not absolutely particular about which green species they will partner; close relatives will usually do. The great majority of northern lichens contain the same genus of alga, *Trebouxia,* and a mere 37 other algal and bacterial genera partner the rest of the 400+ genera of lichen fungi.

Lichens are loosely (not taxonomically) fitted into three growth forms:

**Crustlike** ("crustose") lichens are thin coatings or stains on or in rock or bark surfaces—so thin that you would hardly even think of trying to pry them up.

**Leaflike** ("foliose") lichens are also thin, ranging from closely adhering, paintlike sheets with distinct, thick, more or less priable margins, to saladlike heaps attached to their substrate in only a few spots.

**Shrublike** ("fruticose") lichens consist of stems and branches, either erect or pendent.

Intermediate growth forms are also found.

Of all the familiar, visible organisms, lichens have perhaps the most modest requirements—a little moisture, sunlight either minimal or excessive, and solid materials carried in even the cleanest air. From minimal input comes meager output; growth rates measured in lichen species range from .0012" to about 4" per year.

Many lichens spend most of their lives dried out, all activity suspended. In this state they survive temperatures from 150° (typical of soil surfaces in the summer sun) down to near absolute zero (in laboratories). They have to dry out and suspend respiration to get through hot days in the sun, yet at 32° or a bit colder they remain active and unfrozen, thanks to their complex chemistry, including alcohols. Our alpine lichens get

---

some of their photosynthesizing done while buried in snow, which keeps them moist and lets plenty of sunlight in. Lichens on rocks can sustain activity as long only as they are supplied with atmospheric moisture—rain, snow, dew, fog, or mere humidity. As with mosses only more so, success in lichen niches is a matter of making hay while the rain falls, and then drying up. Among the best-known lichen habitats are alpine and recently deglaciated rocks. Far greater lichen biomass, in our region, grows in a seemingly opposite environment with moderate temperatures, low light levels and high rainfall— the bark of Westside forest trees. There, too, the lichens and mosses are subject to drying out many times a year.

Occasionally, lichens that colonize bare rock may, after centuries of life, death, and decomposition, alter the rock enough for plants to grow , initiating the long slow succession of biotic communities. Countless naturalists from Linnaeus to our high school biology teachers have been enamored of this concept, but in fact it is not a necessary or even a common path of primary succession from barren earth. In most climates, mosses can pioneer on rock, and trees and shrubs can pioneer in crevices holding a bit of fine gravel or dust. These plants advance succession much faster than their lichen neighbors. The ability of lichens to break down rock has been overrated; it is negligible compared to the effects of frost. Lichens secrete acids to dissolve needed minerals from rock, but the effect of this chemical etching on the rock mass is minuscule.

Slightly more substantial is lichens' physical crumbling of rock—effective on soft limestone and shale, and also on old stained glass of cathedrals. Accomplished by gelatinous gripping cups that expand and contract powerfully enough to crumble tiny bits of rock as they moisten and dry out, this effect has been experimentally replicated using plain gelatin. Contrary to the myth, lichens *obstruct* plant establishment and succession in some situations, including fresh Cascade mudflows.

A second appealing myth inspired by lichens' minimal needs proposes them as candidates for restarting succession, or even evolution, after a World War III. During the period of atmospheric hydrogen bomb tests, and again after the Chernobyl reactor disaster, arctic lichens accumulated fallout without apparent harm to themselves, while they passed on dangerous amounts of it into the bones of lichen-eating caribou and caribou-eating Inuit and Lapps. Lichens might be comparative survivors of radioactive contamination—but only in the unlikely event that it is not preceded or accompanied by other air contamination. Lichens' diet of airborne solids makes them extremely sensitive to airborne sulphur dioxide and other pollutants. No lichens survive in the worst urban air. Environmental scientists watch lichen species as a measure of pollution in an airshed over a period of years. In the Willamette Valley, for example, they devised a scale of lichen species found on white oak twigs at incremental levels of pollution (Denison and Carpenter, 1973).

Fungi: crustlike lichens

## Map Lichen

*Rhizocarpon geographicum* (rye-zo-car-pon: root fruit). Chartreuse yellow patches broken by fine black lines and surrounded by a wider black margin. On rocks; abundant in exposed situations. Color p 487.

Lichens grow at extremely slow but steady rates. Because of its ubiquity on arctic and alpine rocks, map lichen is an outstanding indicator of the number of centuries elapsed since glacial retreat from the rocks. Radial growth rates of 3 and 3⅓ millimeters per century—after a slightly faster initial "spurt"—were found for map lichens in Colorado and Wyoming; 3,000-year-old lichens were 4–4½" in diameter.

## Clot Lichen

*Mycoblastus sanguinarius* (my-co-blas-tus: fungus bud; sang-gwin-air-ius: blood red). Gray patches without sharp edges, on bark, texture grainy to warty, with clusters of ± shiny blackish fruiting disks .05–.1" diam.; under each black disk, if it's sliced off, is a bright red spot in the lichen cortex. Dry tree bark microsites (see p 258).

# Corkir

*Ochrolechia* spp. (oh-kro-lek-ia: ochre bed?). Also **cudbear**. Gray patches without sharp edges, on bark, with clusters of small white-rimmed, ± yellow-centered fruiting disks.

*O. pallescens* (pal-ess-enz: palish). Crust thin, often smooth; discs .05–.1″ diam. Abundant on broadleaf trees, esp alder, and sometimes conifers.

*O. oregonensis* (oregon-en-sis: of Oregon). Crust finely warty; discs .08–.15″. Abundant on conifers. Color p 487.

*O. tartarea* (tar-tar-ia: crumbly-rough). Crust thick, warty; discs often irregular, crowded in clumps of 2–4, quite concave, up to .25″ diam. On conifers only, commoner E of Cas.

Corkir's main significance to us is in lending its ghostly white-ness to the bark in young forests. Many whitish crusts grow on bark, especially red alder, but corkir is the easy genus to recognize with its distinctive white-rimmed discs.

Not whiteness, but brilliant shades of purple and red and rich browns have made corkir valuable. Mainly in Scotland, *O. tartarea* was scraped from rocks for use as a dye, going into international trade in such forms as Harris Tweeds and litmus paper. Lichen gathering rarely flourishes for long in any given locale, since lichens grow so slowly that they are quickly depleted. "Corkir" was an archaic Celtic word for this lichen, and "cudbear" was a more recent dialect name derived from "Cuthbert," a trade name. Neither term is common parlance in America, but then, few lichen names are. In North America, corkir grows on trees, hardly ever on rocks. This renders it useless for dye; the lichen is inseparable from the bark, which would muddy the dyebath with tannin. *O. pallescens*, in particular, permeates the bark, with only its spore-bearing disks sitting upon the surface. It looks more like a stain than a crust—you might think you see nothing there but bark.

A great many lichens yield yellow, olive or brown dyes when boiled in water. The transformation to purple and red comes when lichens containing gyrophoric acid, such as these three species and the rock tripes, are cured for several weeks in warm ammonia. Stale human urine was the form of ammonia used from the beginning (probably B.C. around the Mediter-ranean) through the nineteenth century. Yet the dyed fabric came out with a fine fragrance longer-lasting than the color.

# Imperfect Lichens

*Lepraria* spp. (lep-**rair**-ia: leprous, i.e., scurfy). Also **powdery paint lichens.** See text to distinguish species.

*L. neglecta* (neg-**lec**-ta: overlooked). Color p 487.

*L. membranacea* (mem-bra-**nay**-see-a: thin).

*L. candelaris* (candle-**air**-iss: luminous). Color p 487.

The most primitive (or perhaps degenerate) form of lichen growth is just a powdery layer of loosely associated fungi and algae. As long as they produce neither organized lichen tissues nor fruiting bodies, they are next to impossible to classify, describe, or identify positively. They are filed under *Lichenes Imperfecti*, a taxonomic dustbin indicating their way of growing more than their genetic relationships. Yet they appear to be proper symbiotic lichens, and in the field they are inescapable, albeit "overlooked," as one species name says.

Our three most inescapable types are filed under genus *Lepraria*. *L. neglecta* seems to fill in the bare spaces in almost any sort of alpine community, frosting old dead moss turf or bare soil with white, and characteristically producing a nubbly texture of 2" mounds.

*L. membranacea* probably covers more square inches of bark than any other lichen species. It specializes in twigs and in dry sides of trees, and looks simply like a powdery white coating with indistinct, fading margins and no specialized organs.

*L. candelaris* is less extensive but much more noticeable; it's the brilliant chartreuse green coating on wet cliffs such as the basalt beside Columbia Gorge waterfalls. It sometimes turns up on bark.

---

# Jewel Lichen

*Xanthoria elegans** (zan-**thor**-ia: yellow; **el**-eg-enz: elegant). Bright orange patches adhering tightly to rocks, thus easily mistaken for a crustlike lichen, but with slightly raised flakes and distinct, lobed edges; fruiting discs deeper orange, small, concentrated near center of patch. Widely scattered, esp alpine. Color p 487.

Jewel lichen favors sites fertilized by the nitrogen in urine; it marks habitual perches of rockpile-dwelling animals like pikas and marmots. More conspicuous than the pika itself, this orange splash may help us spot the source of an "eeeenk."

# Rock Tripe

*Umbilicaria* spp. (um-bil-ic-**air**-ia: navel—). Dark gray-brown leafy lichen attached to a rock by a single, central "umbilical" holdfast; tough and leathery wet, hard and brittle when dry; ± lobed and curling at the edges; ± smooth above, often with coarse black or pinkish tan hairs ("rhizines") beneath. Abundant on dry, exposed rock at all elevs.† Color p 487.

Indians boiled rock tripe for a fish roe adulterant—Indian hamburger helper, as it were. Chinese and Japanese fry it up to eat like potato chips, or relish it tender in salad or soup, drying and boiling it first to leach out dark, bitter flavors. Early *trappeurs* in Canada credited *tripes-des-roches* with saving them from starvation—but never with tasting good.

# Dog Lichens

*Peltigera* spp. (pel-**tidge**-er-a: shield bearer). Big sheets with lobed and curled-up margins; conspicuous dark red to chestnut, ± tooth-shaped fruiting bodies borne on margins; underside heavily netted with branching veins, raised from surface, bearing coarse hairlike rhizines. Moist forest soil or tree bases, often mingling with mosses.

---

*Some texts place *elegans* in genus *Caloplaca*.
†*U. virginis,* recognizable by extra-dense long rhizines underneath, is strictly limited to high elevations; indeed, it was first recorded in the Northwest from the 8200-foot level on 7965-foot Mt. Olympus!

---

**Dog lichen,** *P. canina* (ca-nye-na: of dogs). Greenish dark brown wet, gray when dry.

**Green dog lichen,** *P. aphthosa* (af-thoh-sa: blistered). Bright green above when wet, with many dark bumps; pale gray when dry; abundant on humus on high limbs of conifers. Color p 488.

Dark brown colors in lichens are a sign of cyanobacteria, formerly called "blue-green algae." *Nostoc*, a nitrogen-fixing blue-green genus, is the chief green partner in the darker dog lichen, whereas the green dog lichen mainly employs green algae, and has bits of *Nostoc* only here and there in dark superficial bumps cultivated where *Nostoc* colonies have fallen.

The medieval Doctrine of Signatures looked in nature for images of body parts, reading them as drug prescriptions in God's own handwriting. In the dog lichen's rows of red-brown fruiting bodies they perceived dog teeth, so they used dog lichen medicinally for rabies, the dogbite disease. The usual decoction for rabies, right up until the last century, was ground dog lichen and black pepper in milk. The Doctrine seems to have been a serious wrong turn in the history of herbal medicine; lichens do have medical value, and they were used more appropriately in pre-Christian Europe.

## Oregon Lungwort

*Lobaria oregana* (lo-bair-ia: lobed; or-eg-ay-na: of Oregon). Big "leaves" very pale green above when dry, bright green when wet, ridged in a branching pattern; the ridges make valleys on the lichen's underside, unlike dog lichen's veins; cream-colored underneath when dry, mottled with brown in the furrows; margins deeply lobed, curling; fruiting organs rare. On W-side trees or sometimes rocks. Color p 488.

Huge proliferations of lungwort grow in thick saladlike beds up on top of limbs in the conifer canopy. Lungwort litter on the forest floor gives a hint of the abundance above, but the true quantity was neither measured nor appreciated until rock-climbing hardware was brought to bear on the forest science (Denison 1973; Pike *et al.* 1975; Nadkarni 1985.) It was then calculated that this species supplies half the nitrogen in many mature Westside forests. Blue-green algal symbionts in the lichen pull nitrogen out of the air; the fungal cosymbionts appropriate it; and it moves on into the community nutrient cycle when the lichen is eaten by insects, red tree voles, or decomposing bacteria, or when it simply leaches out in rainwater. To intercept nutrients before they wash away, alders, maples, and cottonwoods actually extend rootlets among the lichens and mosses on their own bark.

The word "lungwort" derives from the medieval Doctrine of Signatures, which prescribed this lichen for lung ailments because its texture resembles lung tissue's. At least seven unrelated plants have also been called lungworts.

---

**Fungi: shrublike or twiglike lichens**

---

# Wolf Lichens

*Letharia* spp. (leth-**air**-ia: death—). Brilliant sulphur-yellow stiff tufts, the profuse branches cylindrical or ± flattened, often pitted, black-dotted. Drier tree-bark sites; abundant on pine and juniper.

*L. columbiana* (co-lum-be-**ay**-na: of the Columbia River). With dark, disklike fruiting cups ¼" in diam (up to ¾"). Color p 488.

*L. vulpina* (vul-**pie**-na: of foxes). Fruiting cups absent or rare. Color p 489.

Somewhere there was a tradition of collecting this intense-colored lichen to make a poison for wolves and foxes; my American sources say this used to be done in Europe, while a British source writes it off as American barbarism.

Wolf lichen imparts to fabrics a chartreuse dye close to its own color. Before they had cloth, Northwest tribes used the dye on moccasins, fur, feathers, wood, porcupine quills for basketry, and their faces.

# Puffed Lichen

*Hypogymnia enteromorpha* (hypo-**jim**-nia: naked underneath; enter-o-**mor**-fa: entrail shaped). Tufts of puffy hollow branches, sharply two-toned, pale greenish gray above with tiny black dots, blackish brown beneath, lighter brown at the tips; numerous yellowish fruiting cups on upper sides average ⅛" diam (up to ½"). Common on small limbs (and as forest litter) Color p 487.

---

# Globe Lichen

*Sphaerophorus globosus* (sphere-ah-for-us: sphere bearer; glo-bo-sus: round). Robust (not hairlike), brittle, stiffly bushy tufts 1–3", light red-brown to gray when dry, greenish wet; fruiting bodies, if present, are tiny (.08") globes full of sooty black spores, on branch tips; on trees, on moist microsites often with mosses. Color p 488.

# Horsehair Lichen

*Bryoria* spp.* (bry-or-ia: moss—). Blackish 2–16" festoons of extremely fine (less than .02" diam), weak, ± matted fibers, often well dusted with pale, powdery propagules; on high-elev trees, krummholz, and rarely rock surfaces. Color p 489.

These fibrous tendrils look ominously parasitic but, like other lichens, are actually as likely to contribute to the trees' nutrition as to steal from it. Even snow mold, *Herpotrichia nigra*, ("black creeping hair"), a fungus that turns many mountain hemlock branchlets into foul-looking mats of smutty needles, is not a serious pathogen. Black visual accents add drama to eccentric tree shapes, whose chief sculptor is the weather.

Indians eked some nutrition from horsehair lichens, baking them with camas bulbs or with meat to mask the soapy flavor. Some, too poor for leather footwear, bundled the lichen to make well-padded shoes.

# Old-man's-beard

*Alectoria sarmentosa* (alec-tor-ia: rooster; sar-men-toe-sa: twiggy). Pale gray-green festoons on trees; wispily pendulous, 3–30" long; when pulled, strands snap straight across. Mid elevs to timberline. Color p 489.

*Usnea* spp. (us-nee-a: Arabic term for a lichen). Pale greenish gray tufts on trees; our species range from densely bushy, 1¼–3", to pendent strands 20" long; when stretched gently, the thicker branches (unless extremely dry) reveal an elastic, pure white inner cord inside the brittle, pulpier skin—like wire in old cracked insulation. All elevs, but commoner low.

"Beard" lichens are abundant throughout boreal regions, and have figured in schemes to convert lichen starches into glucose for food, or alcohol for fuel or drink. Even the most abun-

---

*Bryoria* was recently created from part of *Alectoria*.

---

dant lichens are quickly depleted due to their slow growth; not enough can be gathered to supply starch to an industry.

In heavily browsed areas like Olympic valleys, old-man's-beard may show a browse line at the maximum height elk can reach. Deer and elk turn to this barely palatable browse not out of desperation but because it helps them absorb nutrients from the green plants of their winter diet. In high forests, an apparent "browse line" is more likely a marker of spring snowpack depth. *Alectoria*—easily our most conspicuous lichen on high-elevation trees—relies on spring as a growing season, when daylength and temperatures are rising but rain is plentiful.

## Iceland-moss

*Cetraria islandica* (set-rair-ia: shield; iss-land-ica: of Iceland) and
*C. ericetorum* (er-iss-e-tor-um: among heaths). Coarse shrublike clumps 1–3" tall by 2–8" wide, brown to olive when dry, consisting of narrow flattened lobes with sparsely fringed edges rolled into near-tubes. On soil, mostly alpine here. Color p 488.

The misleading but time-honored name "Iceland-moss" comes from Europe, where this is the best-known lichen in human consumption. It is sold as an herbal medicine in Sweden and made into teas and throat lozenges in Switzerland. Scandinavian sailors used to bake Iceland-moss flour in their bread to extend its shelf life at sea. The lichen had to be parboiled with soda before milling, or it would have been unspeakably bitter. I have seen no record of use in America.

## Worm Lichen

*Thamnolia subuliformis* (tham-no-lia: bushy, a misleading name; sub-you-li-for-mis: awl shaped) and
*T. vermicularis* (ver-mic-you-lair-iss: wormlike). White tapering tubes 1–2½" tall, without fruiting cups, typically in lackadaisical clumps of standing tubes with some reclining tubes and a few branched tubes. Alpine. Color p 487.

This pair of species, like the pair of Iceland-mosses above, are chemical variants, meaning they are distinguished at the species level solely on the basis of different "lichen acids" they contain. As a genus, worm lichens are distinctive. Several variable *Cladonia* species may be wormlike or awllike when they fail to fruit, but are gray-green or tan-brown in contrast to the bone white of worm lichens.

## Matchstick Lichen

*Pilophoron aciculare* (pil-ah-for-on: hair bearing; a-sic-you-lair-ee: needlelike). Pale gray stalks ⅜–1" long, rarely branched, with fat rounded black fruiting tips; standing or reclining in dense clumps, on rocks in forest. Color p 488.

## Reindeer Lichens

*Cladina* spp.* (cla-dye-na: branched). Profusely fine-branched shrublike lichens 2–4" tall, in dense patches, on soil but barely attached to it. Widespread; abundant among alpine sedges and heathers W of Cas Cr.

*C. rangiferina* (ran-ji-fer-eye-na: of reindeer). Tips mostly 4-branched, blunt; color ashy; uncommon.

*C. mitis* (my-tis: soft). Tips mostly 3-branched, sharper; yellow, sometimes with bluish bloom; common. Color p 489.

Though found at all elevations here, reindeer lichens are best known for covering vast areas of arctic tundra, where they are the winter staple food of caribou (known in Europe as reindeer). They aren't really very digestible or nutritious; even caribou prefer green leaves when they can get them, but to occupy that particular range they adapted to wintering on lichens. This requires ceaseless migration, since the slow-growing lichens are eliminated where grazed for long. Lapps even harvest lichens for their reindeer herds. Some Inuit prize a "saladlike" delicacy consisting of half-digested lichens from the stomachs of caribou killed in winter.

Closer to home we encounter them, dyed green and softened in glycerine, as fake trees and shrubs in architectural models. They also supply extracts for commercial uses including perfume bases and antibiotics. They are the chief source of the antibiotic "usnic acid" for German, Finnish, and Russian salves applied to ailments ranging from severe burns and plastic surgery scars to *Trichomonas* and bovine mastitis.

*Cladina* was formerly included in an enormous genus *Cladonia*.

# British Soldiers

*Cladonia* spp. (cla-**doe**-nia: branched).

*C. bellidiflora* (bel-id-if-**lor**-a: martial flower). Heavily flaky-coated greenish gray stalks ¾–2" tall, bearing 1 to several large, lobed, scarlet fruiting heads. On soil or logs. Color p 489.

*C. bacillaris* (bas-il-**air**-iss: club-shaped). Pale gray stalks ½–1" tall, ± roughened but not thick-coated, bearing 1 to several small, ± round scarlet heads, growing from a flaky mat. On rotting wood or sometimes tree bases.

The huge genus *Cladonia* has a two-part growth form. The primary growth is a mat of flakes, or "squamules," resembling some leaflike lichens only finer. After a while the fruiting growths—either stalks or bushy masses of branches—may rise from this mat. Sometimes the mats of squamules just persist and spread without fruiting; these sterile squamule mats can only be identified as *"Cladonia* spp." At the other extreme, species like *bellidiflora* have stalks covered with squamules, and the mat of primary squamules dies and disappears so early that it is not part of the lichen we know and describe. Many *Cladonias* fruit bright red; the name "British Soldiers" is borrowed from *C. cristatella*, the common red-capped species of the Northeast States.

# Pixie Goblets

*Cladonia pyxidata* (pix-id-**ay**-ta: goblet-shaped). Clustered greenish gray golf-tee-like fruiting stalks ½–1¼" tall, from a thick flaky mat; "baby tees" may sprout from rims of the main ones. Usually on soil, in open areas. Color p 489.

These fruiting stalks look like golf tees to me, but the image of goblets for pixies apparently suggests itself to some. The name *pyxidata* never saw a pixy; along with "box," it derives from *pyxis*, a Greek goblet-shaped container with a lid. The shape is found in many *Cladonia* species; this one is a relatively large, sturdy golf tee.

# 8

# Mammals

We scarcely need an introduction to mammals. We *are* mammals. We're well schooled in the salient characteristics of mammals, most definitive being the mammary glands which (on females) produce milk to nurse the young. Live birth is typical, though neither universal (egg-laying platypuses are mammals) nor unique (live-bearing snakes are not). Maintenance of an elevated body temperature ("warm-bloodedness") is universal among both mammals and birds, and has also been detected and studied in other animals and plants (pages 221, 453). In mammals only, a coat of keratinous hairs serves to insulate body temperature; hair is a second anatomical feature definitive of mammals. Even whales, armadillos, and other mammals less hairy than ourselves have at least a few true hairs at some stage in their development.

Mammal coats come mainly in shades of brown and gray. No fur is bright blue or green, as some feathers are—not even on exceptional mammals, like skunks, whose coloring is "showy" rather than camouflaging. For most mammals, it is crucial to be inconspicuous. They generally achieve this by wearing camouflage colors, by being elusive, and by being nocturnal. That explains why mammalwatching is a pastime less popular than birdwatching, and why knowing mammal tracks, scats, and other signs is key to being woods-wise to mammals. The mammals you are likely to see alive here each have particular reasons to be unafraid: porcupines have their defenses, tree-climbing squirrels and burrowing marmots and pikas have their alarm networks and refuges, and National Park-dwelling deer, elk, bears, and goats have big brains that have adjusted to their legally protected status. Watch out for them!

Small mammals are especially limited to the murky corner of the color spectrum. Experienced field naturalists learn to recognize quickly the species they are concerned with, relying on color differences that have become obvious to them through experience in that locale. They can even communicate these shades to each other in a jargon of browns and grays such as "dusky," "buffy," "tawny," "ochraceous," etc. These words are less useful to nonscientists. Photos in field guides can be helpful, but their color is often distorted by variation in lighting and printing. Worse, the animals themselves vary. In a given species, there may be different shades for different seasons, for juveniles than for adults, for intraspecific varieties and color phases, for the generally paler underside and darker back, and between the hair tips, underfur, and guard hairs. Among closely related species or varieties, animals of drier regions are generally paler than their counterparts of more humid terrain, each tending to match the color of the ground so as to be less visible to predators (especially owls) that hunt from above and in dim light. Dry-country (e.g., Eastside) creatures run around on pale, dry dirt, whereas moist (Westside) forests have dark floors of humus and vegetation.

Positive "identification" (as opposed to field "recognition") of small mammals utilizes the number and shape of molar teeth, and caliper measurements of skulls and penis bones (a feature of most male mammals) for which the creatures must first be reduced to skeletons. But don't worry—this book will not go into molar design or the meaning of "dusky," but will offer size, tail/body length ratio, form, habitat, and sometimes color, to facilitate what will admittedly be at best educated guesses as to small mammal identities. Most often, unless we trap them or find them dead, our glimpses of shrews, mice and voles are so fleeting and dark that we can make only a downright wild guess based mainly on habits and habitat.

# Shrews

*Sorex* spp. (sor-ex: the Roman term). Mouselike creatures with very long, pointed, wiggly, long-whiskered snouts, red-tipped teeth, and ± naked tails. Order Insectivora (Shrews and moles). Color p 490.

**Wandering shrew,** *S. vagrans* (vay-grenz: wandering). 2½" + 1½" tail; gray-brown. Ubiquitous in lowlands.

**Dusky shrew,** *S. monticola* (mon-tick-ola: mountain-dweller). 2¾" + 1¾" tail; dark brown. Common mid to high elevs.

**Trowbridge shrew,** *S. trowbridgii* (tro-bridge-ee-eye: after W. P. Trowbridge). 2½" + 2¼" tail; dark gray-brown exc white underside of tail. Lower W-side forests.

**Masked shrew,** *S. cinerea* (sin-ee-ria: ashen). 2¼" + 1¾" tail; brown with ± tan underside. Drier mtn forests in Wash; uncommon.

**Water shrew,** *S. palustris* (pa-lus-tris: of swamps). 3" + 3" tail; blackish above, whitish beneath. In and near mtn streams and lakes.

**Marsh shrew,** *S. bendirii* (ben-dear-ee-eye: after Charles E. Bendire, below). 3½" + 3" tail; blackish all over. Wet lowlands.

Shrews, our smallest and most primitive mammals, lead simple but exceedingly active lives. Day in and night out, they rush around groping with their little whiskers, sniffing, and eating most everything they can find. This goes on from weaning, between April and July of one year, until death, generally by August of the next.

Of course that's an oversimplification. They eat insects and other arthropods (often as larvae), earthworms, and a few conifer seeds and underground-fruiting fungi; they have been known to kill and eat other shrews, and mice. They have 24-

**Charles Bendire** (born Karl Emil Bender in Germany) watched birds and other creatures while stationed near Harney Lake in 1874–1877, and reported what he saw in copious letters to noted Eastern naturalists. To them he was a diamond in the rough, an army Major previously noted only as the intrepid Indian-fighter who dissuaded Cochise from returning to the warpath. His reports were published as *Birds of Southeastern Oregon* and later as a thick *Life Histories of North American Birds.* He was first to unmask the kokanee salmon as a landlocked form of sockeye.

hour cycles of greater and lesser activity based largely on when certain types of prey are easiest to get, but as a rule they can't go longer than three hours without eating, and the smaller species must eat their own weight equivalent daily. As with bats and hummingbirds, such a high caloric demand is dictated by the high rate of heat loss from small bodies: at 2 grams, masked shrews approximate the lower size limit for warm-blooded bodies. Baby shrews nurse their way up to this threshold while huddling together so that the combined mass of the litter of four to ten easily exceeds 2 grams. Whereas bats and hummingbirds take half of every day off for deep, torpid sleep (page 308), shrews never do. Nor do they hibernate. It's hard to imagine how our shrews meet their caloric needs during the long snowy season when insect populations are dormant, and heat loss all the more rapid. But they do—or at least enough of them do to maintain the population.

One cause of mortality seems to be a sort of Shrew Shock Syndrome triggered, for example, by capture or a sudden loud noise. Some scientists relate it to the shrew's extreme heart rate (1,200 beats per minute have been recorded) and others to low blood sugar caused by even the briefest shortage of food. At any rate, perhaps the most frequent sign of our abundant shrew population is their little corpses on the ground. Shrews are ill adapted to evade predators; their eyesight and hearing are poor. Their only defense is unsophisticated but effective: they are simply unappetizing. Owls, Steller's jays, and trout are among the small minority of predators known to have acquired a taste for shrews.

Marsh and water shrews, the types likely to tempt trout, spend much of their time in the water, going after tadpoles, snails, leeches, etc. They have such buoyancy, thanks to fur that traps an insulating air layer next to the skin, that they can skim or seemingly run across the water surface. When they dive and swim, they paddle even more frenetically to stay under; as soon as they stop, they bob to the surface. They have stiff, hairy fringes on the sides of their hind feet for efficient paddling. Terrestrial shrews are burrowers to some extent, and make runways through duff alongside a log or rock.

Shrews are ferociously solitary. To mate, they calm their usual mutual hostility with elaborate courtship displays and pheromone exchanges—a real-life "taming of the shrews."

# Shrew-mole

*Neurotrichus gibbsii,* (new-ro-try-kus, hairy wire, i.e. tail; gib-zee-eye: after George Gibbs, p 51). 3" + 1½" tail; gray-black; tail hairy—unlike any other shrews or moles; teeth not red-tipped; eyes tiny; no visible ears; forefeet and claws somewhat larger than rear ones, but not hugely so as in other moles; snout long, thin, whiskered. Mainly in W-side forest. Order Insectivora (Shrews and moles).

Though classed in the mole family, the shrew-mole certainly appears to be the odd half-breed its name suggests. (Shrews and moles are the two big families in the order Insectivora, meaning "insect-eaters"). It burrows so much less effectively than other moles that it is largely confined to the leaf-mold and loose-humus layer. On the other hand, it is the one mole that sometimes forages aboveground, even climbing bushes in search of bugs.

# Moles

*Scapanus* spp. (scap-an-us: digger). Burrowing animals, blackish with pink, ± naked tail and snout; eyes and ears barely visible; forefeet huge, turned-out, heavily clawed. Order Insectivora (Shrews and moles).

**Townsend mole,** *S. townsendii* (town-send-ee-eye: after J. K. Townsend, page 313). 7" + 1½" tail. W-side meadows.

**Coast mole,** *S. orarius* (or-air-ius: coastal). 5" + 1¼" tail. All elevs.

Of all our mammals, moles are most specialized for burrowing. They sort of swim through loose soil, either deeply or, more often, just beneath the surface, making long ridges. The forelimbs are heavily developed, while the pelvis and hindlimbs are small and weak. Eyes and ears are both almost entirely overgrown with skin and fur so as not to clog up with dirt. While the eyes barely function at all, the ears are quite sharp at receiving earthborn sounds, enabling the mole to detect and hunt down earthworms (its dietary staple) by sound.

These true moles are commonest in lowland pastures, but are also scattered in most parts of our range, even subalpine meadows. Their long ridges of turf pushed up are distinctive. The tunnels lack entrance holes; deeper ones may produce surface heaps of loose dirt, larger and more hemispherical than those made by gophers (page 317) or boomers.

---

# Bats

Order **Chiroptera** (kye-rop-ter-a: hand wing).

As evening gets too dark for swifts and nighthawks to continue their feeding flights, bats and owls begin to come out for theirs. Though the largest bats and smallest owls overlap in size, 5½" with a 16" wingspread, bats are easy to tell from owls by their fluttering, indirect flight. You are unlikely to see one well, since our species never venture out in the daytime. I won't discuss our ten or twelve species individually, but this doesn't mean bats are unimportant or uninteresting. They are

## Animal Sonar

You have probably heard that bats use a sort of ultrasonic radar to find their way around and to locate and catch prey. This is recent information. In 1793 it was first observed that bats get around just fine with their eyes blocked, but become helplessly "blind" with their ears blocked. The obvious deduction—that bats literally *hear* their way around as competently as other animals *see* theirs—was too outrageous to consider for more than a century. Not until 1938 were instruments able to detect bats' high-frequency squeaks, whose echoes bats hear in a sonarlike perceptual capacity called "echolocation."

A typical bat "blip" lasts a thousandth of a second, during which time it drops an octave and spreads out from a focussed sound to a nearly omnidirectional one. These precise shifts, and the very short wavelength, give the echoes such fine tuning that the bat not only locates objects but perceives their texture and their exact motion. It's strictly fast-food for a bat to nab a mosquito, for example, distinguishing it from a shower of cottonwood fluff amid an obstacle course of branches.

From a casual blip rate of several per second in the open air, the bat steps up to over fifty per second when objects of interest come within a yard or so. That's as far as a bat can echolocate, since high-pitched sounds don't carry far. (Contrast with the great carrying power of a grouse's low "booming.") To compensate for such "nearsightedness," the bat's reactions must be ex-

extremely abundant (exceeded among mammalian orders only by the rodents) and probably take a bigger slice out of the insect population than any other type of predator. Most bats, including all of ours, are insect-eaters, like their closest relatives the shrews, so calling them flying mice or flitter-mice (two colloquial terms probably descended from the German *fledermaus*) is a near miss—these are "flying shrews."

Bats catch flying prey either in the mouth or in a tuck of the small membrane stretched between the hind legs, from which the mouth then plucks it while the bat tumbles momentarily in mid-flight. Each wing is a much larger, transparently thin membrane stretched from the hindleg up to the

---

tremely quick, and its blip extremely loud; the decibel level an inch from a bat's mouth is several times that of a pneumatic drill at 20 feet. Those God-awful earsplitting nights in the country! (Well, they would be, if our hearing were sensitive to 80,000 cycles per second instead of its mere 15,000. "Concert A," the note orchestras tune by, is 440 cycles, roughly at the middle of our audible sound spectrum.) To protect the bat's own hearing from damage, its auditory canals vibrate open and shut alternately with the blips, admitting only the echoes. Bats also have lower-pitched (humanly audible) squeaks for communication.

The nocturnal aerial hunting made possible by echolocation is presumably a key to bats' success, since in the daytime they are at an overall competitive disadvantage with birds, whose swifter flight is made possible by feathers. Styles of echolocation in bats are highly diversified and specialized. (A few kinds of bats, though none in this part of the world, see pretty well, echolocate poorly, and shun nighttime activity.)

Aspects of echolocation have been found in many unrelated animals. Some moths evade bats by emitting similar blips to scramble the bats' radar. Porpoises and toothed whales echolocate as sophisticatedly as bats, and some shrews are not far behind. Cave-dwelling birds have learned to do it. The ability may be latent in most mammals; blind humans often learn to echolocate impressively, though rarely developing special calls for the purpose. The human auditory system, according to one theory, also vibrates shut to save us from the racket of our own voices.

---

forelimb and all around the four long "fingers." Since the wing has no thickness to speak of, it is less effective than a bird or airplane wing (page 368) at turning forward motion into lift. To compensate, bats have much greater wing area per weight than birds, and use a complex stroke resembling a human breast-stroke to pull themselves continually upward. Bats achieve only modest airspeeds, compared to birds, but they are much more maneuverable at close quarters. They actually chase flying insects rather than simply intercepting them.

Bats roost upside down, hanging from one or both feet. In this position, often in large groups, they sleep all day and hibernate all winter, except for a few winter-migrating species. Though our bats prefer to roost in caves—especially in winter, for insulation—they typically settle for tree cavities and well-shaded branches. After mating in autumn, most species' females reserve the sperm for delayed fertilization and bear a single young in spring. Except while out hunting, the mothers nurse the young almost constantly the first few weeks, hanging upside down in the roost.

Though bats may carry rabies, even rabid ones hardly ever bite people.

## Snowshoe Hare

*Lepus americanus* (lep-us: the Roman term). Also **varying hare.** 16" when stretched out, + 1½" tail; gray-brown to deep chestnut brown (incl tail) in summer; in winter, high Cas and E-slope race turns white, with dark ear-tips; W-slope race stays brown; ears slightly shorter than head. Nocturnal and secretive; widespread in forest. Order Lagomorpha (Rabbits). Color p 490.

Thanks to their "snowshoes"—large hindfeet with dense growth of stiff hair between the toes—these hares can be just as active in the winter, on the snow, as in summer. They neither hoard for winter nor hibernate, but molt from brown to white fur, and go from a diet of greens to one of conifer buds and shrub bark made all the more accessible by the rising platform of snow. They make the animal tracks we see most often while skiing. They also become the crucial staple in the winter diet of several predators—foxes, great horned owls, golden eagles, bobcats, and especially lynx. Though the hares' de-

fenses (camouflage, speed and alertness) are good, the predator pressure on them becomes ferocious when the other small
prey have retired beneath the ground or the snow. Hares can
support their huge winter losses only with even greater summer prodigies of reproduction, the proverbial "breeding like
rabbits." Several times a year, a mother hare can produce two
to four young ones, fully furred, eyes open, and ready to run
and eat green leaves within hours of birth—a key difference
between hares and rabbits. She mates immediately after each
litter, and gestates 36 to 40 days.

Drastic population swings, in 8 to 11-year cycles, are well
documented for showshoe hares in many parts of their range,
but not here. The other kind of "varying" the hare is known
for—from summer brown to winter white pelage—is true of
some hares here. A nonvarying subspecies lives in foothills
and mountains west of the Cascade Crest.

The semiannual molt is triggered by changing day length.
In years when autumn snowfall or spring snowmelt come ab-
normally early or late, the hares find themselves horribly conspicuous and have to lay low for a few weeks. In lowland areas
that fail year after year to develop a prolonged snowpack, the
hares have evolved permanently brown races, so they stand
out only during the occasional snows. The rest of the year they
make the most of their camouflage, foraging when they can
best see without being seen—by dawn and dusk, and sometimes on cloudy days in deep forest. They don't use burrows,
but retire to shallow depressions called "forms," under shrubs.

Though infrequently vocal, snowshoe hares have a fairly
loud aggressive/defensive growl, a powerful scream perhaps
expressing pain or shock, and ways of drumming their feet as
their chief mating call. A legendary courtship dance, in which
they literally somersault over each other for awhile, appears to
develop out of an ecstatic access of foot-drumming.

North America has one native genus, *Lepus*, of hares and
jackrabbits, and one of rabbits proper, *Sylvilagus* ("forest rabbit"). The mountain cottontail, *S. nuttallii*, a smallish rabbit
with pure white tail, may be seen along the forest-steppe margin east of the Cascades, and the brush rabbit, *S. bachmani*, in
Westside foothills.

---

# Pika

*Ochotona princeps* (ock-o-toe-na: the Mongolian term; **prin**-seps: a chief). Also **cony**. 8" long if stretched out, but appearing a thickset 5–6" long in typical postures; tailless; brown; ears round, ½" diam. On talus, in Cas. Order Lagomorpha (Rabbits). Color p 490.

A cryptoventriloquistic nasal "eeeenk" in the vicinity of coarse talus (rockpile) identifies the pika for you. Look carefully for it on the rocks. You would think of it as a rodent, but there's something definitely rabbitlike in its posture—perhaps the sharp nose-down head angle, the neck drawn back in an S curve. Pikas are in fact more rabbit than rodent;* they comprise a family in the rabbit order.

Pikas are thought of as subalpine creatures, and most of them do live up high, but others are just as happy on talus slopes down at the lowest elevations in our range—just above the highway in the Columbia Gorge. (On the other hand, they haven't migrated across the broad lowlands to reach the Olympics.) Their way of life depends on talus crevices for refuge; they make quick forays to harvest some of the surround-

---

*Rabbits and pikas were formerly classified as rodents, but are now a separate order, Lagomorpha. Members of the two orders share incisor teeth that grow throughout life as fast as they are worn down. While rodents have four such incisors, lagomorphs have eight—a second upper and lower pair being immediately behind the first; this is thought to be an early evolutionary branching. Lagomorphs are also nearly unique in having the testes in front of the penis.

---

## Coprophagy

An exclusive diet of vegetative parts of plants presents a severe challenge to mammalian digestive systems because of the high fiber content (and other disproportions: see page 330). As we learned in childhood, the large grazing mammals meet the challenge with cud-chewing and multiple stomachs, enabling them to take in lots of greens in a hurry, out in the open, and then retire to chew in a relatively safe hideout.

Pikas, most rabbits, and many herbivorous rodents employ a comparable method, but one that doesn't require squeezing ex-

---

ing vegetation, and run back carrying big mouthfuls crosswise. Pikas are known as "little haymakers." Each year, each pika stores several bushels of mixed greens for winter, under rocks selected for dryness.

The young stake out their own territory (usually toward the center of a rockpile, since choice outer sites near the meadow are taken by dominant individuals) and make their own hay their first summer, even if they're only about half grown at the time. The rockpile may appear to unite a colony, but pikas live solitarily within it except while mating.

The "eeeenk" call—amazingly loud for such a tiny creature—seems to serve as both territorial assertion and alarm. While no crevice large enough to admit a pika can keep out a weasel, the rockpile offers a maze where the pika may lose the weasel and lay low. The pika knows its rockpile, is superbly surefooted on it and, according to more than one report, may be aided by other pikas coming out from refuge to distract the weasel by running around like crazy. This kind of report sets some naturalists to arguing over whether it's proof of altruism (evolved traits endangering individual lives for the benefit of the genetic group) or merely foolish nervous agitation.

The name "pika" derives from the way Siberian Tungus tribespeople say "eeeenk," so no doubt we ought to pronounce it "peeeeka." Nevertheless, most Americans say "pike-a," perhaps with either "pica" or "piker" in mind. Others say "cony," but that name refers originally to several quite unrelated Old-World beasts.

---

tra stomachs into their tiny frames; they cycle their food through the whole digestive canal twice, eating their own (and sometimes others') "soft pellets" of partly digested material, and later excreting "hard pellets" of hard core waste. We don't know exactly what is going on here. In some cases the food is cultured with a bacterium that requires sunlight to complete its digestive work, but other animals eat their pellets immediately, allowing no time in the sun. Apparently, a longish spell in the caecum (a pocket at the start of the large intestine) releases essential vitamins which must go back to the stomach and small intestine to be absorbed —kind of like the cud being rechewed after initial digestion.

# Boomer

*Aplodontia rufa* (ap-lo-don-sha: simple teeth; roo-fa: red). Also mountain-beaver, mountain-boomer, sewellel, chehalis. 12–14" rotund body; tail vestigial, inconspicuous, 1–2"; dark brown above, slightly paler beneath; with blunt snout, long whiskers, small eyes and ears, long front claws. Scrubby moist W-side habitats. Order Rodentia (Rodents). Color p 490.

Boomers are big, slow, nearly tailless, partly arboreal, mostly burrowing rodents. They achieve several distinctions, all on the ludicrous side. They host the world's largest (⅜") species of flea. They are the only surviving species in the most primitive living family of rodents, having changed little since they first appeared, very early in the evolution of rodents. Naturalists never quite know what to call them, apart from "living fossils." Most manuals list them under Mountain beaver, but then immediately fall back on Aplodontia, apologizing that this is neither a beaver nor a mountain-dweller especially. Everyone in my end of the county that has boomers calls them "boomers," so I guess that's good enough for me—regardless of their inability to boom.

Boomers prefer wet, scrubby thickets and forests at all elevations. They are doubtless thriving in this century, with the proliferation of second-growth timber; Northwestern "stump farmers" know them well. They honeycomb their half-acre home ranges with shallow burrow systems, making many molehill-like dirt heaps. Mainly nocturnal, they are active year round, eating sword ferns and bracken, vine maple and salal bark, and Douglas-fir seedlings. The latter makes them unpopular with foresters, of course, but their digging does a lot for soil drainage and friability, and disseminates spores of certain desirable underground-fruiting fungi that have no other means of spore travel.

**Yellow-bellied marmot**

# Marmots

*Marmota* spp. (mar-moe-ta: the French term). Also **rockchuck, whistle-pig, whistler.** Heavy-bodied, thick-furred, large rodents of mountain meadows and talus, known for their piercing "whistles." Order Rodentia.

**Hoary marmot,** *M. caligata* (cali-gay-ta: booted). 20" long + 9" tail; grizzled gray-brown, with black feet and ± white belly and bridge of nose. alp/subalpine in Wash Cas.

**Olympic marmot,** *M. olympus*. Similar to hoary marmot (considered a subspecies by some) but face and feet markings less contrasty; back often yellowish by late summer, possibly bleached by urine-soaked burrow walls. Alp/subalpine in Olys. Color p 491.

**Yellow-bellied marmot,** *M. flaviventris* (flay-vi-ven-tris: yellow belly). 16" long + 6½" tail; yellowish brown, often gray-grizzled, with ± yellow throat and belly; feet darker brown. From Cas Cr E, mainly on basalt lowlands, but sometimes subalpine (as it is in Rockies and Sierra Nevada).

It's hard to feel you're really in the high country until you've been announced by a marmot, with a sudden shrill shriek. Sadly, subalpine habitat often lacks marmots for no apparent reason. The shriek (not a whistle, in that it's made with the vocal chords) is a warning that sends several other marmots lumping along to their various burrows. On the way they may pause, perhaps standing up like big milk bottles, to look

---

**Mammals**  *307*

around and see how threatening you actually appear. A more fearsome predator than yourself, such as a red-tailed hawk, would have elicited a shorter, descending whistle conveying greater urgency. In your case, they easily become nearly oblivious to you, or even quite forward and interested in your goods. Or you may get to watch them scuffle, box, and tumble, or hear more of their vocabulary of grunts, growls and chirps.

Marmots need their early warning system because they're slower than many other prey, and count all the large predators as enemies. They are rarely hunted by people any more, though Indians and Inuit used to think them worth hunting for both fur and flavor. To protect themselves from the phenomenal digging prowess of badgers, our yellow-bellied marmots locate their burrows in rockpiles. Olympic marmots, with no badgers to worry about, often burrow in loose meadow soils, but occasionally a whole hibernating family is dug out and eaten by a bear.

As befits the largest members of the squirrel family, marmots take their hibernating seriously. They put on enough fat to constitute as much as half their body weight, and then they bed down for more than half the year, the colony snuggling

## Torpor and Hibernation

Most warm-blooded species have normal body temperatures about as warm and about as precise as our 98.6°, but many of them spend much of their time at sharply reduced metabolic levels, which we lump together under the word "torpor." Torpor is a condition of deep sleep, with very slow breathing (one per minute) and heartbeat (four to eight per minute) at body temperatures close to the ambient temperature, down to a limit a few degrees above freezing. Its basic purpose is to conserve calories at times when they are hard to come by. There are several patterns of torpor:

**Daily torpor,** such as the daily sleep of bats and the nightly sleep of hummingbirds; ordinary sleep would waste too many heating calories from such tiny bodies, in temperate climates.

together to conserve heat. Resist the temptation to think of seven-month hibernation as a desperate response required by an extreme environment. It is just one of several strategies that work here; other small subalpine grazers like the pika and the water vole stay active beneath the snow at a comfortable constant 32°, while long-legged browsers forage above the snow, some (like mountain goats) staying subalpine, others (like elk) migrating downslope. These different wintering strategies go with tastes for different plants, so the grazing species rarely compete directly for one food resource.

Marmots concentrate a year's worth of eating into a brief green season. The season doesn't have to be summer. Olympic and hoary marmots hibernate through winter (late September to early May) as do yellow-bellied marmots of the high Rockies; but our yellow-bellies, low on the Cascades' east slope, fatten in April and May and go down for their seven-month slumber in midsummer, when the heat dries up spring's herbs and grasses. Summer torpor is "aestivation" as opposed to "hibernation," but these marmots perform the two consecutively without seeming to make any distinction.

Young yellow-bellies mature fast enough to disperse (i.e.,

---

**Seasonal torpor,** usually called either hibernation (from the Latin for "winter") or aestivation (from the Latin for "summer"); the animal may waken occasionally to excrete, stretch, eat stored food, or perhaps even go out and forage a bit.

**Occasional torpor,** a last-ditch response to food shortage, or even to a momentary shock, as in "playing possum."

Animals adapt to various habitats in their use of torpor. Some species of jerboas include subarctic races that hibernate, hot-desert races that aestivate, and in-between races active year round. Our chipmunks vary from year to year, as well as with elevation and latitude, as to whether they will hibernate or forage through the winter, and they may dehibernate if the weather turns better in midwinter. They also seem to include, within hibernating races, genetic minorities that never hibernate. Some

---

leave their maternal care and burrow) at the end of their first or, more often, second summer. Hoary and Olympic marmots, with an even shorter, colder active season, mature very slowly for rodents, dispersing only in their third summer when their mother's subsequent litter arrives. Young and yearling marmots suffer heavy casualties to both predation and winter starvation. Even more than a mother bear, a marmot mother is hard put to fatten enough for her nursing litter's hibernation as well as her own; in alternate years she is infertile, restoring her metabolic reserves while casually tending her yearlings. A dominant male may thus keep two mates, impregnating the fertile one and leaving her to run a nursery burrow while he shares a burrow with the infertile one and her yearlings—a social structure common to both hoary and Olympic marmot colonies. As summer wears on, parents increasingly work at chasing their grown two-year-olds away from the colony. Most of the marmot tussles you see are simply play between youngsters, but those that end in a one-sided chase are more likely adults making two-year-olds unwelcome.

ground squirrels and marmots go into aestivation in the summer and don't come out until they dehibernate in early spring. Most temperate-zone bats hibernate in addition to sleeping torpidly every day of their active season.

Hibernation is a strategy with great advantages, but also with severe costs and problems. First, the animal must put on a lot of weight—33% to 67% on top of its midsummer weight. Most species can't gain much weight on greens and fungi, so they have to do it all late in the season, after the carbohydrate-rich seeds and berries ripen. Most of the weight gain is in a special easily oxidized brown form of fat, similar to the brown fat in humans that magazines have publicized as a healthier form of fat for dieters and heart patients. However, considerable muscle tissue also has to be put on, since burning off fat draws water out of the animal's system, whereas burning off muscle adds water to the bloodstream, and dehydration is one of the worst problems hibernators have to deal with. Hibernation requires pituitary hormones to suppress urine formation in order to conserve water, and other special chemistry to ameliorate the toxic effects of the

# Douglas' Squirrel

*Tamiasciurus douglasii,* (tay-me-a-sigh-oo-rus: chipmunk squirrel; da-glass-ee-eye: after David Douglas, p 18). Also **chickaree, pine squirrel.** 7" + bushy 5" tail; gray-brown above, with reddish tinge; orange, variably grayed, below; the two color areas separated by a slight black line conspicuous only in summer. Dense conifer forest. Order Rodentia. Color p 491.

Noisy sputterings and scoldings from the tree canopy call our attention to this creature which, like other tree squirrels, can afford to be less shy and nocturnal than most mammals thanks to the easy escape offered by trees. "Scolding" lets all the neighborhood squirrels know there's a possibly dangerous animal nearby, but once they know about you they don't seem to consider you much of a threat. There *are* predators that take squirrels quite easily—martens, goshawks, and large owls—but apparently such predators were never common enough to put a dent in the squirrel population, and are scarcer than ever today, in retreat from civilization. Unlike squirrels.

Sometimes in late summer and fall we know Douglas'

---

urea that accumulates due to not urinating.

Exactly what triggers hibernation and dehibernation is a challenging problem. Scarcity of food, abundance of fat, outside temperature, day length, and absolute internal calendars are among the triggers for various species. Most hibernators gradually decrease their food consumption for some weeks in anticipation. Once hibernating, an animal is hard to rouse. To waken, most species spend several hours warming themselves by shivering violently. Marmots hibernate in heaps, and the shivering of one will trigger the others to join in a group shiver.

Black bears, probably the hibernators best-known to the public, are often said not to be "true hibernators" since they lower their body temperature only a little, and hence rouse into full activity quickly, when they rouse at all. But studies of bears show that their adaptation to "winter sleep" is at least as thorough and as successful as the "typical" hibernation of the squirrel family. Perhaps bears are simply so much better-insulated and bigger that to lower their body temperature while still sustaining life would be downright difficult, and serve no purpose.

---

squirrel by the repeated thud of green cones hitting the ground. Since cones are designed to open and drop their seeds while still on the tree, closed cones you see on the ground are likely a squirrel's harvest. The squirrel runs around in the branches nipping off cones, 12 per minute on Douglas-fir or up to 30 per minute for some smaller cones; then it runs around on the ground carrying them off to cold storage. True-fir cones are too heavy to drag or carry, so it gnaws away just enough of the outside of the cone to lighten it to a draggable weight, while leaving the seeds still well sealed in. Someday it will carry the cones back up to a habitual feeding-limb and tear them apart, eating the seeds and dropping the cone scales and cores, which form an eventually huge heap we call a "midden." Either the center of a midden or a hole dug in a streambank may be used to store cones for one to three years. The cool, dark, moist conditions keep the cone soft and the seeds fresh for the eating and also, incidentally, viable for germinating. Foresters learned to rob middens, and squirrels became the chief suppliers of conifer seed to Northwest nurseries by 1965. This scam is in decline now that nurseries are choosier about their genetic stock.

Mushrooms, which must dry out to keep well, are festooned in twig crotches all over a conifer, and later moved to a dry cache such as a tree hollow. With such an ambitious food storage industry, this squirrel has no need to hibernate. For winter, it moves from a twig and cedar-bark nest on a limb to a better-insulated spot, usually an old woodpecker hole. This-year's young winter in their parents' nest, unlike chipmunks and smaller rodents; they need most of a year to mature.

A huge proportion of all conifer seeds is consumed by rodents, birds, and insects—99% in some poor cone-crop years. Foresters used to regard seed eaters as enemies. But the proportion of conifer seeds that germinate and grow is infinitesimal anyway, and of those, the percentage that were able to succeed *because* they were harvested, moved, buried, and then neglected is substantial. Trees coevolved with seed eaters in this relationship, and some tree species depend on it (page 46). Most tree species are also more or less dependent on symbiotic fungi (page 262), many of which depend in turn on these same rodents to disseminate their spores.

# Chipmunks

*Eutamias* spp. (you-tay-me-us: true storer). Rich brown with 4 pale and 3 dark stripes conspicuous down the back and from nose through eyes to ears. Order Rodentia.

**Townsend's chipmunk,** *E. townsendii* (town-send-ee-eye: after John Kirk Townsend, below). 6" + bushy tail 5"; pale stripes ± gray. Dense forest.

**Yellow-pine chipmunk,** *E. amoenus* (a-me-nus: delightful). 5" + bushy tail 4" ; pale stripes yellowish brown. Open conifer forest and timberline areas. Color p 492.

Along with Douglas' squirrel, these small chipmunks are our most conspicuous forest mammals—diurnal, noisy, and abundant. They have a diverse vocabulary of chips, chirps, and tisks easily mistaken for bird calls. Townsend's is often heard without being seen, since it is a bit shy, in keeping with its predilection for heavy forest cover. Though sometimes seen together, the two species are more often separated by habitat if not by range; yellow-pine chipmunks require open forest, and tend to be either higher or farther east than Townsend's.

Our chipmunks are scarcely arboreal, nesting in burrows and finding most of their food on or in the ground—seeds, berries, a few insects and, increasingly toward winter, lots of

---

**John Kirk Townsend** came to Oregon with Nathaniel Wyeth and Thomas Nuttall (page 71) in 1834. Young (24) and enthusiastic, he wrote a charming popular account of their shared travels. Though best remembered among birders, he collected and described several new plants and animals (Townsend's mole, vole, chipmunk, etc.). He filled in as Fort Vancouver's physician between Gairdner's (page 440) and Tolmie's (page 198) tenures there, no mean feat during the epidemics that were slaughtering Northwest Indians at the time. Of portaging around the Cascades of the Columbia in heavy rain, Townsend wrote: "It was by far the most fatiguing, cheerless, and uncomfortable business in which I was ever engaged, and truly glad was I to lie down at night on the cold, wet ground, wrapped in my blankets, out of which I had just wrung the water. . . . I could not but recollect . . . the last injunction of my dear old grandmother, not to sleep in damp beds!!!"

---

underground fungi. To facilitate food-handling, they have an upright stance (like other squirrels and gophers) that frees the handlike forefeet. They store huge quantities of food, carrying it to their burrows in cheek pouches. Sometimes they hibernate and sometimes not, depending on their own genetics as well as on food supply and temperature. Typically, they might sleep in four-day spells at only moderately depressed body temperatures, then get up to forage a bit or perhaps only to excrete and eat from stores in the burrow. The young, though born naked, blind and helpless, mature fast enough to disperse and make their own nests for their first winter.

## Western Gray Squirrel

*Sciurus griseus* (sigh-oo-rus: the Greek term, derived from "shade tail"; **gris-ee-us:** gray). 12" + very bushy tail 10½"; gray frosted with silver-white hair tips; belly white; ears ± reddish. Mainly E-side pine forest; also W-side lowlands. Order Rodentia. Color p 492.

Scarcer than Douglas squirrels and scarcely ever vocal or bold, these gray beauties are infrequently seen or heard. With their huge tails they are gracefully athletic to watch. Large size and good meat make them popular game; this may explain their shyness, which they leave behind when they adapt to city life. Favorite foods are ponderosa and sugar pine nuts, and acorns, so you are likeliest to see them in pine or oak habitat. Fungi are eaten too.

## Golden-mantled Ground Squirrels

*Spermophilus* spp.* (sper-mah-fil-us: seed lover). Also **copperhead.** 7" + 4" bushy tail; medium gray-brown with 2 dark and 1 light stripe down each flank; no stripes on face, head or neck, unlike chipmunks; head and chest (the "mantle") rich yellow-brown; "milk-bottle" posture typical while looking around. E-side and subalpine. Order Rodentia. Color p 492.

*S. lateralis* (lat-er-ay-lis: sides, referring to stripes). In Ore.

*S. saturatus* (satch-er-ay-tus: dark). In Wash.

---

*Older texts include these species in the genus *Citellus;* some recent texts split the genus further, placing our species in *Callospermophilus.*

**Golden-mantled
ground squirrel**

S. Torvik

Golden-mantled ground squirrels occupy about the same open forest range as the yellow-pine chipmunk, but seem to avert out-and-out competition by being more arboreal and by hibernating for months, fattening grossly in the fall. To car campers they may appear more common than chipmunks, since they are campground scavengers. But if you're an Easterner exploring the West, forget what the ranger told you in Colorado or Wyoming about all the stripy critters being ground squirrels, not chipmunks; here, we have more chipmunks. Ground squirrels' cheek pouches—the mucus-lined mouth interior extending nearly to the shoulders, with a capacity of several hundred seeds—are similar to those of many squirrels and some mice.

# Northern Flying Squirrel

*Glaucomys sabrinus* (glawk-amiss: owl mouse, or perhaps blue-gray mouse; sa-**bry**-nus: of the Severn\*). 7" + 5½" tail (broad and flat); large flap of skin stretching from foreleg to hindleg on each side; eyes large; red-brown above, pale gray beneath. Ubiquitous, esp in ± open forest. Order Rodentia. Color p 490.

You aren't likely to see these pretty squirrels, since they are strictly nocturnal, active in the hours just after dark and before dawn. On a quiet night in the forest, you might hear a soft birdlike chirp and an occasional thump as they land low on a tree trunk. They can't really fly, but they glide far and very accurately, and land gently, by means of the lateral skin flaps that triple their undersurface. They can maneuver to dodge branches, and almost always land on a trunk and immediately run to the opposite side—a predator-evading dodge that includes a feint of the tail in the opposite direction. Large owls preying on them often pick off and drop the tail, making tails the most often-seen flying-squirrel parts.

Flying squirrels usually nest in old woodpecker holes, and have their young gliding competently at two months of age, around midsummer. Though omnivorous, they rely heavily on underground-fruiting fungi in summer and on horsehair lichens, which also insulate their nests, in winter. They don't hibernate, nor do they store such quantities of food as the nonhibernating Douglas' squirrel does. Meaty bait often draws them into winter traps set for furbearers.

\*Some texts identify Sabrinus as a river nymph in Greco-Roman myth, without stating why that name was applied to this species. The type, or first-described specimen, of the northern flying squirrel was found near the Severn River in Ontario. That river is named after England's Severn, originally named Sabrinus by the Romans.

# Pocket Gophers

*Thomomys* spp. (**tho**-mo-miss: heap mouse). 5½" + small ± naked 2½" tail; highly variable gray-brown tending to match the local soil; front claws very long, eyes and ears very small, incisors large and always showing. Open areas with loose soil; abundant on E-side steppes; also subalpine. Order Rodentia.

*T. mazama* (ma-**za**-ma: the Crater Lake volcano). In Olys and Ore Cas.

*T. talpoides* (tal-po-**eye**-deez: mole-like). In Wash Cas and E Wash and Ore.

Pocket gophers spend most of their lives underground, and hence have many adaptations in common with moles—powerful front claws, heavy shoulders, underdeveloped eyes, small hips for turning around in tight spaces, and short hair with reversible "grain" for backing up. But whereas moles are predators of worms and grubs, gophers are herbivores. They can suck a plant underground before your very eyes, making hardly a dent on the surface. They get enough moisture from the plants they eat that they don't even go to water to drink. In fact, only two occasions universally draw them into the open air. One is mating, in spring. That takes only a few minutes, and draws out only the males. Once mated, gophers return to mutually hostile solitude, plugging up burrow openings behind them. The second is the eviction of young gophers from their mothers' burrows. On the whole, though badgers and gopher snakes are well equipped to take gophers, the underground life is so safe that gophers limit themselves to one small litter per year.

One species, our *T. mazama*, takes slight exception to subterranean life, going out on occasion, mainly at night. Additionally, any gopher that lives where winters are snowy may, without exactly going out, exploit aboveground resources like bark and twigs by tunneling around through the snow. In spring (July or so in the high country) you see "gopher cores" —sinuous ridgelets about 3" wide of dirt and gravel that came to rest on the ground during snowmelt (color page 490). Shortly before snowmelt the gopher resumed earth burrowing, and naturally used the snow tunnels it was abandoning as dumps for newly excavated dirt. In nonsnow situations, dirt is dumped in a fan-shaped pattern around the plugged, oblique burrow entrance.

---

A pocket gopher's "pocket" is a cheek pouch used, like a squirrel's, for carrying food. Unlike a squirrel's pouch, it opens to the outside. Fur-lined and dry, it gets turned inside out for occasional emptying and cleaning. "Gopher teeth," or large protruding incisors, are used (at least by gophers) for digging, while the lips actually close behind them to keep out the dirt.

## Beaver

*Castor canadensis* (cas-tor: the Greek term). 25–32" + 10–16" tail; tail flat, naked, scaly; hind feet webbed; fur dark reddish brown. In and near slower-moving streams. Order Rodentia. Color p 491.

The beaver is by far the largest North American rodent today, and was all the more so 10,000 years ago when there was a giant beaver species the size of a black bear. Of all historical animals, the beaver has had the greatest effect on North American landforms, vegetation patterns, and Anglo-American settlement. If that's not enough, it's also Oregon's State Mammal.

Though beavers themselves are infrequently seen by hikers, they leave conspicuous signs—beaver dams, ponds, and especially beaver-chewed trees and saplings.

The sound of running water triggers a dam-building reflex in beavers. They cut down poles with their teeth and drag them into place to form, with mud, a messy but very solid structure as large as a few yards wide and several hundred yards long. Less conspicuous dams are typical in our range. A beaver colony may maintain its dam and pond for years, adding new poles and puddling fresh mud into the interstices.

Most beaver foods are on land; the adaptation to water is for safety's sake. The pond is a large foraging base on which few predators are nimble. Over most beaver range (though rarely here), pond ice in winter walls them off from predators. Living quarters, in streambank burrows or mid-pond lodge constructions, are above waterline, but their entrances are all below it, as are the winter food stores—hundreds of poles cut and hauled into the pond during summer, now submerged by waterlogging. Cottonwood, willows and aspen are favorites. A beaver can chew bark off the poles underwater without drowning, thanks to watertight closures right behind the incisors and at the epiglottis. Beavers maintain breathing space just under the ice by letting a little water out through the dam. They are well insulated by a thick fat layer just under the skin, plus an air layer just above it deadened by fine underfur and sealed in by a well-greased outer layer of guard hairs.

The typical beaver lodge or burrow houses a pair—life-long mates, typically—their young, and their yearlings. Territoriality is practiced with little apparent aggression. Two-year-olds dispersing in search of new watersheds may be found far from suitable streams.

When beavers dammed most of the small streams in a watershed, as they once did throughout much of the West, they stabilized river flow more thoroughly, subtly, and effectively than concrete dams are able to do today. Beaver ponds also have dramatic effects locally. At first they drown a lot of trees. Eventually, if maintained by successive generations, they may fill up with silt, becoming first a marsh and later a level meadow or "park" with a stream meandering through it. Such parks are abundant and much appreciated in the Rockies, but less common in our range.

Beavers were originally almost ubiquitous in the U.S. and Canada, aside from desert and tundra. Many Indian tribes felt kinship with beavers, showing special respect for them in lore and ritual, while also hunting them for fur and meat. Europeans, in contrast, trapped beavers to obtain a musky glandular secretion ("castoreum") for use as a perfume base. There was a busy castoreum trade for centuries, before, in the late 1700s, the beaver hat craze hit Europe, and the demand for dead beavers skyrocketed, not letting up until the supply became scarce in the mid 1800s. Strange as it seems today, the beaver pelt economy provided virtually the sole impetus for exploration of the North American West during that period, including the Lewis and Clark Expedition and the Louisiana Purchase itself. Beaver pelts are still on the market today, though changing fashion has allowed their value to fall far below that of many carnivore furs. Populations have recovered about as much as they can, reaching even into central Seattle and Portland, but will never foreseeably regain their natural numbers because so much beaver habitat has been lost, and the beaver's considerable ability to reconstruct it is not easily tolerated by an agricultural economy.

# Deer Mouse

*Peromyscus maniculatus* (per-o-**miss**-cus: boot mouse; ma-nic-you-lay-tus: tiny-handed). 3½" + 3–4" tail (length ratio near 1:1); brown to (juveniles) blue-gray above, pure white below incl white feet and bicolored, often white-tipped tail; ears large, thin, fully exposed; eyes large. Ubiquitous. Order Rodentia.

The deer mouse could be called the North American Mouse; it is far and away the most widespread and numerous mammal on the continent. Not the least bit shy of people, it makes itself at home in forest cabins, farmhouses, and many city houses in the Northwest. Don't confuse it with the unattractive, introduced Eurasian house mouse, common only in urban areas in the U.S.

**Deer mouse**

The deer mouse builds a cute but soon putrid nest out of whatever material is handiest—kleenex, insulation, underwear, moss, or lichens—in a protected place such as a drawer or hollow log. In winter it is active on top of the snow. It is omnivorous, with an emphasis on larvae in the spring and seeds and berries in the fall. Such a diverse diet adapts it well to recent burns and clearcuts; as the forest matures, deer mice are gradually displaced by the more specialized fungus-eating voles. Deer mice displayed an astonishingly stable response to the greatest disturbance our region has seen in this century—Mt. St. Helens' 1980 eruption. They were common in the blast zone so soon after the eruption that they are presumed to have survived the blast in their homes, and to have found sufficient food under the ash.

# Packrat

*Neotoma cinerea* (nee-ah-ta-ma: new cutter; sin-ee-ria: ashen). Properly **bushy-tailed woodrat**. 8–9" + 7" tail covered with inch-long fur, hence more squirrel- than ratlike; brown to (juveniles) gray above, whitish below; whiskers very long; ears large, thin. Ubiquitous. Order Rodentia.

"Packratting" is a behavioral offshoot of the industrious home-building known throughout this genus of oversized mice. Typically, packrats build huge (2–8') stickpiles with water-shedding roofs and several interconnected rooms—storerooms, nests and latrines. Other crevices may shelter casual roomers of various vertebrate species, as well as a rodent's usual complement of insect parasites.

For some strange reason, packrats also love to incorporate man-made objects, shiny ones especially, into their piles. Possibly some predators are actually spooked by old gumwrappers or gold watches, or perhaps the packrat's craving is purely aesthetic or spiritual. She is likely to already be on her way home with a mud pie or a fir cone when she comes across your Swiss Army knife, and she obviously can't carry both at once, so there may be the appearance of a trade, though hardly a fair one from your point of view. Packrats have other habits even worse than trading, often driving cabin dwellers to take up arms against them. They spend all night in the attic, the woodshed, or in the walls noisily dragging materials around—shreds of your fiberglass insulation, for example. Or they may mark unoccupied cabins copiously with a truly foul-smelling, red-staining musk.

Our common species, the bushy-tailed, is among the most active traders even though it has evolved away from the use of stickpiles as dwellings, preferring usually to nest in crevices of talus, cliff, mine or cabin. It may also build a modest stickpile and use it mainly as a food cache. If it does nest in one, this is likely to be high in a tree.

---

# Long-tailed Vole

*Microtus longicaudus* (my-**crow**-tus: small ear; lon-ji-**caw**-dus: long tail). 4¾" + 3" tail (length ratio 3:2); ears barely protruding; fur gray-grizzled, feet pale, tail bicolored. Brushy streamsides, clearcuts, etc. Order Rodentia.

Voles of this large genus, related to the arctic lemmings, are known for drastic population swings on a 3- or 4-year cycle, with each species synchronized over much or all of its range. Neither starvation nor predation is primarily to blame; mysterious hormonal/behavioral mechanisms seem to curb the population explosions somewhere short of mass starvation—though not always in time to prevent eating damage to seed or grain crops. The long-tailed vole is less social than the two *Microtus* species below, and rarely makes runways.

# Creeping Vole

*Microtus oregoni* (or-eh-go-nigh). 4½" + 1⅜" tail (length ratio 3:1); gray-brown fur exceptionally short, dense. Widespread, esp in clearcuts, drier meadows, and deciduous woods. Order Rodentia.

## Mice or Voles?

Out of the 4,000-plus species of mammals living today, almost 1,700 are rodents, and of those, almost 1,300, or 32% of all mammalian species, are myomorphs, or mouselike rodents. As with insects, songbirds, grasses, and composite flowers, such disproportionate diversification bespeaks competitive success in recent geologic times.

The three large groups within the Myomorpha are the Old World rats and mice; the New World rats and mice; and the voles or "field mice." Jumping mice are one of several smaller families. Mickey, Minnie and Mighty Mouse, with their huge ears, are all based on the house mouse, *Mus musculus,* which is in the Old World group along with the black rat and the Norway rat. This notorious scaly-tailed, un-American trio is adapted to life around humans, and has spread to every urban area in the world. They

This short-legged vole stays within an inch one way or the other of the ground surface, either burrowing shallowly in loose dirt or plowing little runways under the dead-turf mat, or under logs. Such habits offer protection from hawks, but little from weasels.

## Townsend's Vole

*Microtus townsendii* (town-**send**-ee-eye: after John Kirk Townsend, p 313). 5½" + 2¼" tail (ratio over 2:1); ears distinctly protruding. Moist to marshy meadows at all elevs W of Cas Cr. Order Rodentia.

This large vole feeds on the succulent stem-bases and root-crowns of lush sedge and grass meadows—a habitat also favored by northern harriers, which feed heavily on the succulent Townsend's vole. The voles make extensive runway complexes through the grass, swim well, and burrow to make their homes, often with underwater entrances.

---

are more closely related to each other than to any of our myomorphs, which all have more or less furred tails.

Aside from our abundant deer mouse and woodrat (both in the New World group) most of our "mice" are voles, distinguishable to the layman by their blunter snouts and smaller tails, eyes, and (often nearly invisible) ears. All voles are herbivores, whereas many rats and mice are omnivores.

Taxonomists agree as to the existence of the three big mouse groups, but have long debated their taxonomic rank. One extreme places them all in family Muridae, while this book's authority (Honacki, Kinman, and Koeppl, 1982), near the opposite extreme, considers them three families—Muridae, Cricetidae, and Arvicolidae. The latter (the voles) have traditionally been treated as either family Microtidae or subfamily Microtinae; "microtines" may persist as an informal term for voles among field biologists, who waste little love on taxonomic debate.

---

# Water Vole

*Microtus richardsoni** (richard-so-nigh: after Sir John Richardson, p 475). Also **water rat**. 6½" + 3" tail (length ratio 2:1); fur long and coarse, ± reddish dark brown above, paler beneath. In and near high bogs, streams and lakes. Order Rodentia.

Our largest vole might be thought of as a small muskrat, but it goes into water for refuge, rarely for forage. It also goes into water to defecate, and is a known carrier of giardia in watersheds too high for beaver. It dines on our favorite lush wildflowers—lupine, valerian, glacier lilies, and such—eschewing grasses and sedges. In winter it digs up bulbs and root crowns of the same flowers, or eats buds and bark of willows and heathers, while tunneling under the snow. Look for mud runways running straight to the water's edge from its burrow entrances, which are up to 5" in diameter.

# Muskrat

*Ondatra zibethicus* (ahn-dat-ra: Huron tribal term; zi-beth-ic-us: civet- or musk-bearing). 9–13" + 7–12" tail; tail scaly, pointed, flattened vertically; fur dark glossy brown, paler on belly, nearly white on throat; eyes and ears small; toes long, clawed, slightly webbed; voice an infrequent squeak. Largely nocturnal; scattered, in or near slow-moving water up to mid elevs. Order Rodentia.

We are tempted to think of the muskrat as an undersized beaver, and with good reason, even though its anatomy shows it to be actually an oversized vole, or "field mouse." Leading similar aquatic lives, beavers and muskrats grow similar fur, which was historically trapped, traded, marketed, and worn in similar ways. (The guard hairs are removed, leaving the dense, glossy underfur.) Several million muskrats are still trapped annually—more individuals and more dollar value than any other U.S. furbearer. In the South, the meat also finds a market as "marsh rabbit."

Like beavers, muskrats build either mudbank burrows or dome-shaped lodges with multiple underwater entrances. But they are smaller than beavers—half the length and rarely a

---

*Some texts place this species in a separate genus *Arvicola*.

---

tenth the weight—and their teeth and jaws are inadequate for cutting wood, so they build no dams or ponds. Soft vegetation like cattails, rushes, and water lilies makes up the bulk of their diets, their lodges, and the rafts they build for picnicking on. They also deviate from the general vegetarianism of the vole family, eating tadpoles, mussels, snails, or crayfish. Their interesting mouths remain shut to water while the incisors, out front, munch away at succulent underwater stems. They can take a big enough breath in a few seconds to last them 15 minutes underwater. Both sexes, especially when breeding, secrete musk on scent posts made of small grass cuttings. Neatly clipped sedge and cattail stems floating at marsh edges are a sign of muskrats.

The South American coypu or nutria, *Myocastor* ("mouse beaver") *coypus,* a similar rodent but about twice as big, may be seen at the Willamette Valley foot of the Cascades. Nutria fur farms were established in the early 1930s in response to an aggressive, deceptive promotion campaign. When profits failed to materialize, most of the hard-up farmers just turned their rodent herds loose, despite laws prohibiting this. The coypus have since multiplied to where they threaten some crops and native competitors, particularly muskrats. Today, their fur fetches a high enough price to warrant trapping and skinning them, but not enough to compensate for the ecological havoc they wreak.

## Red-backed Voles

*Clethrionomys* spp. (cleth-ri-**ah**-no-mis: keyhole mouse). 4" + 1¾" tail; gray-brown with distinct rust-red band down length of back; tail ± bicolored similarly; active day or night. Coniferous forests. Order Rodentia.

*C. gapperi* (gap-per-eye: after Gapper). In Wash.

*C. occidentalis* (ox-i-den-**tay**-lis: western). In Ore or rarely SW Wash.

Red-backed voles should be respected as gourmets, since they dine largely on fungi of the underground-fruiting types, which few of us are aware of except when, as "truffles," they are imported from France or Italy at $300 a pound. It's no accident that truffles and their kin are excruciatingly fragrant, delicious, and nutritious. These fungi have no way of disseminating their spores other than by attracting animals to dig them up, eat them, excrete the undigested spores elsewhere, and preferably thrive on them over countless generations. Many of our forest rodents evolved as avid participants in this scheme, but none are more dependent than red-backed voles. Since the fungi stop fruiting if their conifer associates die (see page 262), red-backed voles disappear after a clearcut or destructive burn, typically to be replaced there for a few decades by deer mice and creeping voles, who eat fewer fungi.

## Red Tree Vole

*Arborimus longicaudus*\* (ar-bor-im-us: tree mouse; lon-ji-caw-dus: long tail). 4¼" + 3" tail (ratio of lengths much less than 2:1); back light cinnamon, belly white with gray underfur; ears ± concealed; claws sharp, strongly curved. On conifers, in Ore. Order Rodentia. Color p 490.

\* Some texts include this species in genus *Phenacomys*.

## Rodents and Fungi

Underground-fruiting fungi—"truffles," loosely speaking—are the mainstay of rodent diets in Western conifer forests. (Though this finding dates back at least to 1951, and has been corroborated by countless stomach and fecal analyses, it remains poorly recognized in popular and even scientific literature.)

Compared to other foods of rodents, our wild fungi rate very high in protein (14–43% of dry weight), very low in fats (1–7%), moderate in carbohydrates (28–60%), and very high in some vitamins and minerals. High moisture content dilutes these percentages considerably, but the moisture itself is valuable; eating it is often safer and more efficient than making the trip to a stream or puddle to drink. To put that another way, moist food enlarges the small mammals' habitats by increasing the distance they can live from any year-round water source. On the other hand, some rodents like Douglas squirrels actually hang fungi up to dry, preserving them for winter when, as is widely observed,

Uncommon and rarely seen, the tree vole is one of the strikingly few animals that subsist on our most plentiful "green vegetable"—conifer needles. It is not an easy life. Tree voles are few and far-flung despite the ubiquity of the resource and the lack of competition for it. The hardship of eking nutrition out of needles is evidenced by prolonged gestation (28 to 48 days), small litters (of one to three), and slow infant development in tree voles, compared to similar-sized rodents.

Branchlets (Douglas-fir preferred) are cut at night and carried to the nest for painstaking nibbling. The vole can eat only the needle margins, leaving the two resin ducts whose pitchy contents would overwhelm its digestive system. A fraction of the resulting debris (100 needles may be consumed per hour) is used to line the nest, an airy edifice enlarged over the generations to include several rooms and escape tunnels. One escape tactic appears to consist of a leap and free fall from the conifer canopy, with legs spread wide like a flying squirrel. Experienced adults almost always land on their feet.

---

moisture is usually in good supply around here.

Not all fungal nutrients can be digested and used; a disproportionate share is contained in the spores. Passage of viable spores through the rodent's digestive tract and out the other end is the whole point of this partnership, as far as the fungus is concerned, so the mutualistic symbiosis requires indigestible spores. (See Red-backed Vole, page 325).

Fungal spores aren't the only potent stowaways in vole and squirrel droppings. Nitrogen-fixing bacteria, which live in the truffles, also pass unharmed through the rodents' bowels, as do yeasts which contribute nutrients the bacteria need in order to fix nitrogen. Since nitrogen fertility is often a limiting factor on conifer growth, the conifers may be as dependent on the bacteria and yeasts as on their mycorrhizal partners the truffles. Including the rodents that disseminate all four other partners, it's a five-way symbiosis—one that foresters would do well to heed. Currently, the timber industry poisons millions of small rodents to keep them from eating conifer seeds.

---

# Heather Vole

*Phenacomys intermedius* (fen-ack-o-mis: impostor mouse). 4¼" + 1⅜" tail (length ratio 3:1); gray-brown above, paler beneath; tail distinctly bicolored; feet ± white, even on top. Sporadic, mainly subalpine. Order Rodentia.

The heather vole lives around (and feeds on) the low heath shrub community of heathers, subalpine blueberries, kinnickinnick and manzanita. You may spot a heather vole's nest shortly after snowmelt—a 6–8" ball of shredded lichens, moss, and grass, typically with a big heap of dung nearby (color page 490). This winter nest was built in a snow burrow on the earth surface; the summer nest will be underground.

# Jumping Mice

*Zapus* spp. (zay-pus: big foot). 4" + 5–6" tail (longer than body); sides orange-brown flecked with black, back dark brown, belly buff-white; hindfeet several times longer than forefeet; ears small. Thickets and meadows near streams, May to Sept. Order Rodentia.

*Z. trinotatus** (try-no-tay-tus: 3-marked). Back nearly black. Mainly W-side. Color p 492.

*Z. princeps** (prin-seps: a chief). Paler and duller. Mainly E-side.

The jumping mouse normally runs on all fours, or hops along in tiny hops, or swims, but if you flush one it's likely to zigzag off in great bounding leaps of 3' to 5'. This unique gait gives you a good chance of recognizing them, even though they're largely nocturnal and not all that common. The oversized feet are needed for power, and the long tail for stability; jumping mice who have lost or broken their tails tumble head-over-heels when they land from long jumps. While none of our other mice or voles hibernate at all, this one hibernates deeply for more than half the year. It eats relatively rich food—grains, berries, and tiny (¼" maximum) underground fungi that grow on maple roots.

*Some researchers regard *Z. trinotatus* as a subspecies of *Z. princeps;* many specimens in our range appear to be intermediate, which argues against two distinct species.

# Porcupine

*Erethizon dorsatum* (er-a-**thigh**-zon: angering; dor-**say**-tum: back). 28–35″ long, incl the 9″ tapered tail; large girth; blackish with long coarse yellow-tinged guard-hairs and long whitish quills visible mainly on the rear half; incisors orange. Widespread, commoner E-side. Order Rodentia. Color p 491.

Porcupines' bristling defenses permit them to be slow and unwary. This is both good and bad news, for you. Though they're mainly nocturnal, you stand a good chance of seeing one some morning or evening, and possibly of hearing its low murmuring song. And you stand a good chance of having your equipment eaten by one during the night. Porcupines crave the salts in sweat and animal fat, and don't mind eating boot-leather, rubber, wood, nylon, or for that matter brake hoses, tires, or electrical insulation, to get them. (Car parts have also been consumed by marmots at Hurricane Ridge.) When camping in E-side valleys, always sleep by your boots, and make your packstraps hard to get at, too. Fair warning.

Quills and spines are modified hairs that have evolved separately in many kinds of rodents, including two separate families both called Porcupines. On our American porcupines they reach their very most effective form: hollow, very loosely attached at the base, and minutely, multiply barbed at the tip. The barbs engage instantly with enough grab to detach the quill from the porcupine; quickly swell in the heat and moisture of flesh; and work their way farther in (as fast as one inch per hour) with the unavoidable twitchings of the victim's muscles. Though strictly a defensive weapon, quills can eventually kill, either by perforating vital organs or, more often, by starving the poor beast that gets a noseful. But if you can keep your dog and your boots away from porcupines, don't worry too much about your own skin; just maintain a modest safe distance while the porky, most likely, retreats up a tree. A white-people's myth holds that porcupines throw their quills, but in fact the thrashing tail has only a slightly greater range than you would expect.

Since they have evolved no defenses other than their quills, porcupines have to be born fully quilled and active in order to survive infancy in adequate numbers. This requires very long gestation (seven months) and small litters (one or rarely two). But still . . . how to get the little spikers out of the womb? Answer: the newborn's soggy quills are soft and harmless, but harden in about half an hour as they air-dry. And then again . . . how to get mama and papa close enough to mate? Or baby close enough to nurse? Solving these problems has given porcupines their Achilles' heel, or rather, soft underbelly. With a quill-less underside including the tail, porcupines can safely mate in the same position as other mammals, provided she draws her tail scrupulously up over her back. (They mate in late autumn, but are otherwise solitary.) If it weren't for that unarmed belly, predators would have no place to begin eating a porcupine.

And predators there are, mainly fishers and cougars; rarely coyotes, bobcats, and even great horned owls. Many authors claim that the belly is the target of attack, but in fact it is well protected as long as the porky is alive enough to stay rightside up. Fishers know to attack via the head—and so should you, with a heavy stick, if you ever find yourself lost and starving and near a porcupine. Some states protect their porcupines on the grounds that they are easy edible prey for unarmed humans lost in the woods. They are not choice fare. They were eaten, though, by Indians, who turned the quills into an elegant art medium on clothing and baskets.

Porcupines eat green vegetation, new twigs and catkins in the summer, and tree cambium, preferably of pines, in winter. They select cambium near the top of the tree, where it contains the most sugar. Bright patches of stripped bark high up in pines are a good sign of porcupine use. Occasionally they kill a tree by girdling it, but more often they kill only the top. The tree can respond healthily by turning a branch upward to form a new main trunk, but this puts a permanent kink in the tree, which human economics regards as worse than death. Killing the tree would at least release its neighbors to grow faster, but kinking it doesn't, while greatly devaluing it at the sawmill. So foresters are alarmed at the increasing porcupine populations of this century, which resulted mainly from human persecution of predators.

Like the red tree vole (our other conifer-grazing rodent) the porcupine reveals the sheer nutritional stress of subsisting on this all-you-can-eat cafeteria. Porcupines inexorably lose weight in winter, even while keeping their grossly overdeveloped guts stuffed with bark. In summer, trees can't avoid providing more fattening fare in their new foliage, but have evolved to defend themselves with high levels of potassium, which in excess is toxic to mammals. For the porcupine kidney to keep up with the task of eliminating potassium, the porcupine is driven to find low-acid, low-potassium sources of sodium. Your sweaty boots and packstraps fit the bill tolerably; your car's wiring or hoses back in the parking lot might be even better. For similar reasons deer and elk eat the mud around soda springs, and mountain goats go after human urine.

## Coyote

*Canis latrans* (can-iss: dog; lay-trenz: barking). 32" + bushy tail 14"; medium sized, pointy-faced, erect-eared gray to tawny dog, grayer and thicker-furred in winter; runs with tail down, whereas wolf runs with tail horizontal. Ubiquitous. Order Carnivora, Canidae (Dog family). Color p 492.

In pioneer days, wolves and coyotes were known as "timber wolves" and "prairie wolves," respectively. Wolves ruled the forests, leaving coyotes to range over steppes, brushy mountains, and prairies. But during the nineteenth century, guns and traps tipped the scales in favor of the coyote by aiming at the bigger predators—cougar and grizzly bear as well as the wolf. Greater size made these animals more vulnerable than coyotes for at least three reasons: more fearless and unwary; more feared and hated by people; and fewer, because higher on the "food chain" pyramid. After the big predators were nearly extirpated from the lower 48, the next largest "varmints"—coyotes, bobcats, and eagles—inherited the brunt of predator-hatred, even though they are too small to significantly limit deer or sheep numbers. (They prey mainly on rodents, hares and insects.)

Coyotes are today America's most bountied, poisoned, and targeted predator—and they have proven uncannily adept at surviving and even increasing. Predator control appears to have done them more good than harm; they have been able to step in wherever the wolf has disappeared. That includes most of our range, though coyotes remain more abundant in brushy than in forested terrain.

Wolf sightings in our range are almost as rare and as hard to verify as Bigfoot sightings. It's likely that a few wolves, *Canis lupus*, roam near the international boundary, and it's conceivable that a family of them—escapees from captivity, perhaps—holes up in the Central Oregon Cascades, as is sometimes reported. But any given wolf sighting here is in all probability a misidentified coyote, a big German shepherd or something, who knows, maybe a sheep . . .

Considering how abundant they are, we rarely see coyotes either, thanks to the same wariness that enables them to survive human persecution. You may hear them howling at night, and can guess that hair-filled scats in the middle of the trail are likely theirs, especially when placed smack dab on a stump, hummock, footlog over a creek, in an intersection, on a ridgetop, or any combination of the above. Coyote feces and urine are not mere "waste," like yours, but more like graffiti signatures full of olfactory data which later canine passers-by, even other species, can "read." In Chinook myth, Coyote consulted his dung as an oracle! The male canine habit of fiercely scratching the ground after defecating probably deposits still more scents from glands between the toes. The long noses in the dog family really are "the better to smell you with, my dear"; the large olfactory chamber is arrayed with scent recep-

## Coyote and the Cedar

Coyote was traveling. He passed the mountains. He followed the trail through the deep woods. As he was traveling along, he saw an immense cedar. The inside was hollow. He could see it through a big gap which opened and closed. The gap opened and closed as the tree swayed in the wind. Coyote cried, "Open, Cedar Tree!" Then the tree opened. Coyote jumped inside. He said, "Shut, Cedar Tree!" Then the tree closed. Coyote was shut inside the tree.

Inside the tree, Coyote said, "Open, Cedar Tree!" The tree did not answer. Coyote was angry. He called to the tree. He kicked the tree. The tree did not answer. Then Coyote remembered that he was Coyote, the wisest and cunningest of all animals. Coyote began to think.

tors. Coyotes can detect the passage of other animals a mile or two away, or several days previous. No less important (in ways we puny-nosed ones have a hard time either imagining or measuring) is the ability to read "scent posts" for data about the condition and activities of fellow coyotes. "Asserting territory" does not well describe scent-marking by coyotes; that they are territorial at all is increasingly doubted.

As for the lovely coloratura howling at night, most people hearing it feel that it, too, conveys something above and beyond mere location—though helping a family group relocate each other is its best-understood function. Often it's hard to tell how many coyotes we are listening to; the Modoc used to say it's always just one, sounding like many.

Though preferring small mammals and birds, coyotes are prepared to subsist through hard times on grasshoppers, or on fruit, on winter-killed deer and elk, or occasionally on fawns. Stalking mice, they patiently "point" like a bird-dog, then pounce like a fox. Against hares they use the fastest running-speed of any American predator. To run down weakened deer, they work as a pack, like wolves, but this is very rare. Usually they hunt alone or pair up cleverly, one partner either decoying or flushing prey to where the other lurks. An unwitting badger, eagle, or raven, may be briefly employed as a partner,

---

After he thought, Coyote called the birds to help him. He told them to peck a hole through Cedar Tree. The first was Wren. Wren pecked and pecked at the great cedar until her bill was blunted. But Wren could not even make a dent. Therefore Coyote called her Wren. Then Coyote called the other birds. Sparrow came, Robin came, Finch came, but they could not even break the heavy bark. So Coyote gave each a name and sent them away. Then Owl came, and Raven, and Hawk, and Eagle. They could not make even a little hole. So Coyote gave each a name and sent them away. Then he called Downy Woodpecker. Finally Downy Woodpecker made a tiny hole. Then Pileated Woodpecker came and pecked a large hole. But the hole was too small for Coyote. So he saw there was no help from the birds. Then Coyote remembered again that he was Coyote, the wisest and cunningest of all the animals. Then Coyote began to think.

---

but most often it's the coyote's own mate.

The female tends to pick the same mate year after year, and the pair displays apparent affection as well as loyalty. To say they mate for life would be about as euphemistic as saying that Americans do. In years when coyotes are abundant or rodents are unusually scarce in a given area, as many as 85% of the mature females there may fail to go into heat, and those who do so will bear smaller-than-usual litters. On the other hand, they reproduce like crazy wherever their own populations have been depleted, such as where predator control men have been at work. This is another reason they're so hard for humans to get rid of—but an admirable example of the innate population control in many mammal species. Currently infertile females and the corresponding unattached males often spend the year with their parents, helping to raise the new litter. This extended family displays loyalty, but not of the ferocious sort typical of wolf packs, nor does it grow large enough to be called a pack.

Curiously, the fierce antagonism of wolf packs toward outsiders (including coyotes) may be crucial in maintaining wolves and coyotes as distinct species. Wolves, coyotes, jackals, and domestic dogs are interfertile and beget fertile off-

After he thought, Coyote began to take himself apart. He took himself apart and slipped each piece through Woodpecker's hole. First he slipped a leg through, then a paw, then his tail, then his ears, and his eyes, until he was through the hole, and outside the cedar tree. Then Coyote began to put himself together. He put his legs and paws together, then his tail, his nose, his ears, then his body. At last Coyote put himself together again except his eyes. He could not find his eyes. Raven had seen them on the ground. Raven had stolen them. So Coyote, the wisest and cunningest of all animals, was blind.

But Coyote did not want the animals to know he was blind. He smelled a wild rose. He found the bush and picked two rose leaves. He put the rose leaves in place of his eyes. Then Coyote traveled on, feeling his way along the trail.

Soon he met a squaw. Squaw began to jeer, "Oh ho, you seem to be very blind!"

spring. (Our authority regards wolves and domestic dogs as one species.) Yet they are dramatically different, physically and behaviorally, despite having long occupied overlapping ranges, so the frequency of their interbreeding must be very low. Wild "coy-dog" hybrids do occur, yet seem unable to establish reproducing populations, perhaps partly due to the dog half's maladaptation to the wilds and partly to a badly confused sense of a mating season—coyotes have one, domestic dogs don't.

Coyote the Trickster is a ubiquitous and complicated figure in all Western American Indian mythologies, possessing an unsurpassed, if devious, intelligence undermined by downright humanoid carelessness, greed, conniving lust, and vulgarity. Some myths tell that he brought the poor starving people rituals and techniques they needed for catching salmon. Some tribes, on learning about Christ, saw Him as a Coyote counterpart in that He came to Earth to improve people's lot. To other tribes, Coyote exemplified the bad, greedy ways of hunting that destroyed a long-gone, Edenlike abundance. In many origin myths Coyote is the Creator—which certainly suggests the mythmakers weren't fostering illusions that the world should always work perfectly.

---

"Oh no," said Coyote, "I am measuring the ground. I can see better than you can. I can see spirit rays." Squaw was greatly astonished. Coyote pretended to see wonderful things at a great distance.

Squaw said, "I wish I could see spirit rays!"

Coyote said, "Change eyes with me. Then you can see spirit rays."

So Coyote and Squaw traded eyes. Coyote took Squaw's eyes and gave her the rose leaves. Then Coyote could see as well as ever. Squaw could see nothing. Coyote said, "For your folly you must always be a snail. You must creep. You must feel your way on the ground."

Ever since that time snails have been blind. They creep.

—Clatsop tale,
slightly abridged from Katherine B. Judson (1910)

---

# Red Fox

*Vulpes vulpes** (**vul**-peez: the Roman term). 24" + 15" tail; shoulder height 16" (terrier size); usually red-orange with black legs and ears, white belly and tip of tail (the red phase). Uncommon; Cas Cr and E-side. Order Carnivora, Canidae (Dog family).

Foxes are absent from the Olympics and not much in evidence in the Cascades. They are largely nocturnal, shy, elusive, and alert, and though they can bark and "squall," they rarely make a pastime of it as their relatives do. Their tracks and scats are hard to tell from undersized coyote ones, and their dens are most often other animals' work taken over without distinctive remodeling. Rocky areas are preferred.

If you have the luck to see this fox, you should recognize it easily if it's a typical red one. Other color phases are rare in the Northwest, but do not strictly follow any geographic rules. Even littermates' may differ, like blonde, dark, and redheaded human siblings. All red foxes have the white tail-tip to distinguish them from the gray fox, *Urocyon cinereoargenteus* (you-rah-see-on: burnt dog; sin-ee-rio-ar-**jen**-tee-us: ashy silver). The gray fox has been seen in Central Oregon; elsewhere in our range, a gray-colored fox should be a silver phase red fox, unless it's a dog or coyote.

Red foxes mate in midwinter, bear their litters (of four to five) in early spring, and stay mated for the raising of them. In contrast to other carnivorous families, in which a mother has to chase the father away to prevent his eating his own offspring, canids make good fathers. Foxes eat lots of insects, fruit and seeds, in addition to the preferred mice, hares, frogs and squirrels. They hunt with devious opportunism and stealth, often culminating in a spectacular aerial pounce. They prefer timberline and other areas of broken forest cover.

Red foxes of the Puget-Willamette lowlands, like those of the Atlantic Coast, appear to be descendants of fur farm escapees—largely European stock. Red foxes' aversion to deep humid forest has sufficed, so far, to separate this introduced strain from our native Cascades red foxes, which range Westward only as far as timberline near the Crest.

---

*Older texts call American red foxes a distinct species, *Vulpes fulva*.

# Black Bear

*Ursus americanus* (ur-sus: the Roman term). 4–6" long (4" tail inconspicuous) 3–3½" at shoulder; jet black with a tan nose. (May be brown E of the Cas, or red-brown, "blue" gray, or white in a few parts of the West; but most of ours are so much blacker than anything else around as to call attention to them at a distance.) Ubiquitous; numerous in Olys. Order Carnivora, Ursidae (Bear family). Cover photo.

Along with sneaky Coyote, smart Raven and industrious Beaver, Bear has always been seen by humans as somehow kindred. Though Bear's reflection of human nature is at once darker and grander than Coyote's or Beaver's, it is hard to pin down the essential quality. Mammalogists would concede that bears are among the most humanlike of animals in terms of their feet and their diet. The feet are five-toed, plantigrade (putting weight on the heel as well as the ball and toes) and about as big as ours, so that the prints—especially the hind print—look disturbingly familiar. The diet includes almost anything, and varies enormously by season, region and individual. Plant foods predominate, starting in spring with tree sapwood or cambium, horsetails, grass, bulbs, and all kinds of new shoots, and working up to enormous berry gluttony in fall, the fattening-up season. The typical prey are small mammals, and insects or larvae where they can be lapped up in quantity, as from anthills, grubby old logs, wasp nests, or bee hives—preferably dripping with honey. An adult bear can chase virtually any predator from its kill, but is less adept at hunting for itself, so large animals are most often eaten as carrion. Bears are skilled at snatching fish from streams. Some develop predilections for robbing woodpecker nests, grain crops, fallen orchard fruit, garbage dumps, or hiker camps.

During heavy berry-eating, bear scats become semiliquid like cowpies, and show lots of fruit seeds, leaves (blueberry), or skins (apple). Earlier in the season, they are thick, untapered cylindrical chunks, perhaps showing animal hair, but often closely resembling horse manure. Fresh, they are usually as jet-black as the beast itself. Bears leave distinctive marks on trees from three activities. When they eat cambium, they strip away large swaths of bark with irregular incisor gashes, at 3–6' off the ground; this may kill the tree (color page 492). Second, they sometimes assert territory by marking selected trees with several long, parallel, often diagonal claw-slashes 5–9′ from the ground. Third, during the spring molt, a tree may show vague abrasion and lots of bear hair at 2–4′ from several bears rubbing their itchy backs against it. (Compare with cat scratch-marks, page 350).

Once the ripe fruits are all gone in the fall, there is little a bear can do to fatten up; if it isn't fat enough by then, it will likely die before spring, but that rarely happens. Activity begins to slow down even before hibernating; bears appear listless while preparing their dens. This may include building a substantial nest of fluffy stuff—ideally cedar bark, which stays resilient under the prolonged burden. The bear sleeps curled up in a ball with the crown of its head down. Between the insulative nest and the superlatively thick fur, the bear loses little heat to the air of its den, and maintains a body temperature of about 88° through winter, in contrast to the 40° hibernating temperature of many squirrels. But its heart rate may reach an astonishing low of just eight very weak beats per minute. And, unlike squirrels, it may go the full six or seven

## Hang Your Food!

There are two objectives to consider here. The first is to protect your precious body and equipment from serious accidental damage by bears going after your food. When in areas where "problem bears" have been reported (or, to be cautious, in any heavily camped area) set up an eating camp and a sleeping camp at least 150′ apart. In the former, you cook and eat, hang your food *and* garbage 10′ off the ground and 5′ out from the tree trunk, and store your cooking utensils *and* the clothes you ate dinner in *and* the shorts you spilled sardine oil on, day before yesterday. In the latter, you sleep peacefully away from previous campers' food smells as well as your own.

A good job of hanging will also reduce the odds of losing food to rodents, a much more frequent threat. But a smart "problem bear" may be able to retrieve your food from anywhere you can. They learn all about nylon cords and fabrics and flimsy little tin cans and foil packets—though some have suffered miserably from dehydrated food expanding in their stomachs. A few

months without eating, urinating, defecating or, presumably, waking up. (In other cases, it may wake easily and dehibernate briefly at any time of winter.) The urea waste that would ordinarily be excreted in urine is somehow recycled into new proteins, alleviating the problem of muscle atrophy during hibernation. Fecal accumulation is so reduced that the winter's worth can be saved until spring dehibernation.

Typically two or three cubs are born around January. The mother wakes up to give birth, then nurses them mostly in her sleep for the next few months, while they go between waking and plentiful sleep. A den of cubs nursing emits a hum like a beehive, only much deeper. Cubs are smaller at birth, in ratio to their adult weight, than almost any other mammals short of marsupials. In nursing them to viable size for the outside world, the mother may lose 40% of her weight during hibernation, as opposed to 15–30% for adult males. She does not undergo this stress in successive winters, but normally hibernates with her yearling cubs in alternate years.

Some speculate that the tree-climbing skills of black bears

---

seem to have learned to go after nylon without even having to smell food inside. And they can find their way to Camperland from virtually anywhere, so it's a waste of money to remove them with tranquillizers and helicopters. Most likely they'll end up shot by the rangers to protect human life and limb.

Clearly, the second objective—the surest way of protecting both human and ursine lives—is to avoid turning innocent bears into problem bears. Don't tempt them.

**Never:** **Discard** food or "leave it for the chipmunks."
**Cook** more than you will eat.

**Always:** **Pack** food in airtight, smelltight containers.
**Hang** food and garbage during day hikes and at night.
**Burn** your empty cans on your stove or in your fire.
**Pack out** all empty packets, bags, and burnt cans.

You owe it to the bears, and the hikers who will follow you here.

---

evolved partly as protection from the only animals on this continent big and mean enough to prey on them regularly—grizzly bears.* The well-known ferocity of black bear mothers in the company of their cubs more likely evolved to protect cubs from grown-up males of their own species; subadult bears may also wind up cannibalized if they are foolish enough to stand up to larger adults.

**Consider bears dangerous** even though black bears (as opposed to grizzlies) normally withdraw from any contact with people. Bears encountered by hikers often fit one of the two "abnormal" types—sows with cubs, and "problem bears" familiar with campers and camper food. Human injuries from black bears have been extremely rare historically, but are bound to increase if more and more naive campers leave more and more food around in areas where bears have been protected for generations.

Don't go backpacking in terror of bears, but go in knowledgeable wariness of them. If you follow game paths into thickets (where bears like to sleep through the day) make lots of cheerful noise as you approach. If a bear moseys into your camp vicinity, stand and bang pots and pans or blow whistles. If it still seems intent on approaching (i.e., a problem bear) by all means distance yourself from your food. If you meet one in the open, observe and let your intuition tell you if the bear is threatening. If so, retreat discreetly. Often, you can slowly and straightforwardly circle widely around a black bear, if it hasn't already run off. If a bear actually charges you, simply stand still; you have precious little chance of outrunning it. (Bears

---

*Grizzly bears, *Ursus arctos*, the world's largest terrestrial carnivores, are highly incompatible with human civilization. Though they specialize in mountain habitats more than black bears do, they have been gone from most of our range for decades. They are still sighted occasionally in the Pasayten Wilderness Area and a few remote parts of the North Cascades—presumably wandering from home ranges centered in British Columbia. The best identifying characters are a pronounced shoulder hump, and much longer claws than those on black bears.

are slow and bumbling only in comparison with deer and cougar.) The charge will probably be abandoned as soon as the bear can see you're a person. (Bears are very nearsighted.) If you're too scared to stand, or convinced you're being attacked, curl up face-down in the fetal position and play dead. That's the very very very last, unlikely resort; don't worry about it, worry about your ounce of prevention—for you and everyone else to hang your food.

## Raccoon

*Procyon lotor* (pro-see-on: a star named for its proximity to the Dog Star; low-tor: washer). 22″ + 12″ tail; gray with black "mask" across eyes and rings around tail; thickset and bushy-furred; all toes long and clawed. Lower elevs. Order Carnivora, Procyonidae (Raccoon family).

Northern outliers of a generally tropical family, raccoons, like bears, belong to order Carnivora but resemble the Primates, our own order, in two notable ways. First, they are omnivorous not merely by habit but by tooth structure, with plenty of blunt molars for grinding plants rather than cutting meat. Second, they are plantigrade—resting the heel down on the ground. With long toes on all four feet, raccoons rival primates in dexterity; they can turn doorknobs, and they often pick over their food manually before eating it. They raid garbage cans in Portland, and are actually commoner around civilization than in our mountains, where they concentrate around lowland lakes and streams. Their food includes berries, acorns, small mammals, frogs, bugs, fish, and crayfish. They climb trees for refuge, and are fond of large hollow trees for their dens, either at the base or in a crotch. Though not strictly territorial, they are generally solitary and mutually hostile except when mating.

# Weasels

*Mustela* spp. (mus-tee-la: the Roman term). Very fast, slinky, slender, short-legged, long-necked animals; in characteristic running gait, the back is arched; ears inconspicuous; rich medium brown above, white to orange-yellow beneath, incl feet and insides of legs and (long-tailed only) some of tail; most E-side and high-elev weasels turning pure white in winter, exc tip of tail always black; males almost twice as heavy as females. Ubiquitous. Order Carnivora, Mustelidae (Weasel family).

**Long-tailed weasel,** *M. frenata* (fren-ay-ta: bridled). 9–11″ + 5–6½″ tail (larger dimensions are male average; smaller are female); deep cream to yellow underneath (in summer).

**Short-tailed weasel,** *M. erminea* (er-min-ee-a: Armenian). Also ermine, stoat. 7–8″ + 2½–3½″ tail; white to light cream underneath. Color p 494.

Narrow, linear shapes like the weasel's are rare among the smaller warm-blooded animals because they are costly to heat; for weasels, there's no question of the shape being worth the price—a caloric intake requirement averaging perhaps 40% of body weight per day. Other small mammals share a roughly common shape which, during sleep or torpor, can be rolled up into an approximate sphere, the most efficient of all shapes for retaining heat. Weasels and their streamlined relatives roll up into, at best, a lumpy disk, which takes 50–100% more calories to maintain at a temperature than a spherically rolled rodent of similar weight. But when it needs to eat, the weasel can chase that rodent down any hole or through any crevice; a weasel is much thinner, faster, and fiercer of tooth and claw than anything close to weasel size. Weasels can also run down squirrels in trees and snowshoe hares on snow—prey several times their own weight. Mouse-sized prey are their staple.

Compared to large models within the weasel family line, they seem to make an easy living; they go after abundant prey and can be confident of catching it. The so-called food chain is really an extremely broad pyramid in which very few large predators can fit at the top.

Reports of weasel "killing sprees" in which they kill far more than they can eat are numerous and confirmed. It should be allowed, though, that human observation may have inhibited or overlooked the weasel's efforts to cache the leftovers for later use. There are also clear cases of weasel canni-balism, including juveniles eating their own litter mates, once they get carried away with the taste or smell of blood. They are undeniably among the most ferociously aggressive of preda-tors. They may themselves fall prey to owls, foxes, bobcats, or occasionally snakes. They nest in burrows of chipmunks, ground squirrels, moles, etc., often lining these with fur plucked from the body of the former occupant.

The term "ermine" is used by naturalists for the short-tailed weasel specifically; furriers and the general public call fur of either species "ermine" so long as it's in the white winter pelage. In fact, most "ermine coats" are made from long-tailed weasels because there are more of them out there to catch, and it takes fewer of them to make a coat. It still takes hundreds of them, though, so a single pelt commands a surprisingly low price. It has been proposed that the black tail-tip on the other-wise white winter fur serves as a decoy; the weasel can usually escape hawk or owl talons that strike this one body part that's conspicuous against snow.

## Mink

*Mustela vison* (vice-un: an archaic French term). 13–16″ + 6–8″ tail; long, narrow and short-legged; dark glossy brown except variable white patches on chin, chest, belly; ears inconspicuous. In and near streams, marshes, and sometimes lakes. Order Carnivora, Mustelidae (Weasel family).

The mink is an aquatic weasel, preying on fish, frogs, crayfish, ducks, water voles, and muskrats. Some populations become fully terrestrial for a while, subsisting on hares and voles. Oth-ers line the British Columbia and Alaska coasts, subsisting on crabs. The muskrat, a preferred prey, is also a mink's "most dangerous game" because it is much larger; a muskrat can drown a mink by dragging it under. In deep water, muskrats even attack minks fearlessly. On the other hand, a duck that thinks it can shake a mink by taking to the air may be in for a fatal surprise; cases are on record of minks hanging on for the flight until the duck weakens and drops. But whenever they can, minks, like weasels, kill quickly with a bite into the back of the neck or skull.

The foul discharges from under the tail that we associate with skunks are actually characteristic of the whole weasel family. Skunks alone have developed their marksmanship and range, maximizing the anal gland's potential as a defensive weapon, but minks have an even worse smell. They spray when angered, alarmed, or captured, when fighting each other (they are viciously antisocial) or to mark territory or repel raiders of their meat caches. It may seem ironic that there is so much blood, gore, and stench in the pedigree of a mink coat; but anyone lucky enough to watch a mink in its habitat is likely to admire it. Most mink coats these days come from commercial mink ranches. Newfangled colors of mink are bred just to keep fur fashions hopping.

# Marten

*Martes americana** (mar-teez: Old French term). Also **pine marten**. 16–18" + 8" tail; body narrow, legs short, tail fluffy, nose pointy; variably buff to cinnamon-brown to nearly black; (looks ± like a smaller red fox on shorter legs). In trees, in remote wilderness. Order Carnivora, Mustelidae (Weasel family). Color p 494.

Martens have evolved a striking resemblance in form, color, and habits, to their favorite prey—tree squirrels. They even eat conifer seeds sometimes, like Douglas squirrels, or a few berries, and occasionally raid birds' nests. Like weasels, they excel as predators because their prey is abundant and has only meager defenses against them. But unlike weasels, their populations, already self-limiting, have been further reduced by trapping and probably by a strong aversion to civilization. Today they are common only in mountain wilderness with conifers. Even there, we rarely see them because they are usually up in the branches, where they are fast, well camouflaged, and active mainly at dawn, dusk, or under heavy overcast. Still, curiosity and appetite sometimes lure one right into a hiker's camp. Winter forces them to forage on the ground more—often under the snow—hunting voles and hares.

Their musk secretions are milder than those of most weasel family members, and are used mainly to mark tree branches to ward off other martens. Except, of course, during a brief season when about 50% of other martens find the smell not repellent but, on the contrary, quite attractive.

---

*A few authorities lump American martens under European martens, *Martes martes*.

---

# Fisher

*Martes pennanti* (pen-an-tie: after Thomas Pennant). Also **wejack, pekan.** 20–25" + 13–15" tail; long, thin, and short-legged; glossy black-brown, occasionally with small white throat patch; ears slightly protruding. Rare; dense forest. Order Carnivora, Mustelidae (Weasel family). Color p 494.

Fishers scarcely ever fish—the name may derive from the Dutch *visse*, meaning nasty. Fishers eat mainly porcupines and snowshoe hares. In fact, they are the only predator that hunts porcupines by preference. Though porkies must undoubtedly be eaten via their soft underbellies, the myth that fishers attack there, by means of a flip with the paw or a fast burrow under the snow, is dubious. After all, those floppy quills lie on the ground and don't really leave a space for paw insertion. Darting, dodging attacks to the face, with either tooth or claw, have been observed, repeated for over 15 minutes until the porky is too dazed to turn or flail. Fisher stomachs soften broken quill bits enough to pass them safely through; scats containing pieces of quills are a good sign of fishers.

Fishers can rotate their hind feet almost 180° for running down tree trunks. Apparently they're fast enough to run down and kill martens. The only predators tough enough to overcome fishers rarely consider it worth the fight, and aren't fast enough to chase them either.

The only animals that threaten the fisher population are people. Fisher pelts closely resemble Siberian sable and usually rank as the highest-priced North American pelt. Trapping virtually eliminated fishers from the lower 48 states by 1940, but since then, foresters have been bringing them back to reduce the runaway porcupine populations that resulted largely from the lack of fishers. This seems to be working well in Wisconsin and Michigan. Oregon's fisher population may derive entirely from stock reintroduced in the Central Cascades and Wallowas. Washington hasn't reintroduced any yet, but probably has a few native fishers left. (Fishers are native to the Olympics, though porcupines are not.) Within our lifetimes there should be many more of these animals in our range.

Reproduction in the weasel family usually involves delayed implantation: the fertilized ovum undergoes its first few cell divisions and then goes dormant for a variable period before being implanted in the uterus and resuming its growth in time for springtime births. Increasing day length triggers implantation. The delay is especially long in the fisher, producing a total gestation of up to 370 days, only around 60 of them active. Thus, the female often goes into heat just a few days before or after giving birth, and mates before weaning her two or three helpless newborns.

## Wolverine

*Gulo gulo*\* (goo-low: gullet or glutton). Also **skunk-bear**. 26–30″ + 8–9″ tail; like a small bear but with ± distinct gray-brown to yellowish striping across the brow and down the sides to the tail; fur thick and long. Near timberline; very rare—sighted here only in N Cas and Three Sisters areas. Order Carnivora, Mustelidae (Weasel family). Color p 493.

Wolverines reached the verge of extirpation from the lower 48 states, but since 1965 seem to be coming back. They have raided a few campgrounds in Northern Idaho, so you may possibly have them to worry about in North Cascades camps before long. As they're the largest members of the weasel family, their reputation as the scrappiest, nastiest fighters on the continent should come as no surprise. Even cougars and bears will usually yield their kills to wolverines, who like nothing better than to polish off another predator's dinner. They're also known for raiding trappers' cabins and caches up North, trashing them thoroughly and spraying them up with truly execrable musk. The powerful scent repels other carnivores from the wolverine's meat caches, and is also crucial, along with a summer-long mating season, for enabling wolverines to locate compatible wolverines. ("SWF, attractive, into winter sports, for nonconfining relationship . . . ") They have always been extremely sparse and solitary, roaming continually over vast home ranges.

\*Some older texts use Linnaeus' original name, *Gulo luscus*.

# Otter

*Lutra canadensis* (loo-tra: the Roman term). Also **river otter**. 27–29"
+ thick, tapering, muscular tail 17–19"; dark brown with silvery
belly, pale whiskers, very small ears, webbed feet. In or near rivers,
lakes or ocean. Order Carnivora, Mustelidae (Weasel family). Color p
493.

Otters are among the unlucky species for whom people are
belatedly discovering fondness and admiration—only after
reducing them to a rarity. In both Europe and America they
have been trapped for fur or simply shot on sight as vermin,
largely because anglers suspect them of far greater damage to
trout populations than they actually inflict. Ancient Chinese
fishermen, in contrast, trained them to help by herding fish
into nets; a few European hunters trained them to retrieve
waterfowl.

Today, people invoke otters in arguments over the existence of nonhuman play and sensuous family fun. Reputable
observers report them running up snowy hills again and again
just to body-sled down, or body-surfing in river rapids for no
apparent reason; others claim these behaviors are mere trans-
portation. Otters would rather slide on their bellies than walk
anytime, even on level ground but preferably down a steep
otter slide with a big splash in the river at the bottom. They
frolic and tumble in the water, often in family groups after the
pups are six months old; up to that age the mother scrupulously keeps them away from the father, presumably for their own
safety. Oddly, the pups seem afraid of the water and have to be
taught to swim. Otters have a low, mumbly "chuckle" while
nuzzling or mating.

Fishermen here see otters regularly. Look on riverbanks
and lakeshores for otters' easily recognized slides, tracks or
"spraints." The latter are fecal scent-markers placed just out of
the water on rocks, mud banks or floating logs, and usually
showing fish bones, scales, or crayfish shell bits under a greenish, slimy (when fresh) coating which smells distinctive but
not unpleasant. Otter staples are crayfish and slow-moving
fish; they rarely compete with fishermen for game fish. Amphibians and voles round out their diet.

# Badger

*Taxidea taxus* (tax-eye-dee-a tax-us: both from the Roman term). 25" + 5" tail; very broad, low, flat animal with thick fur grizzled gray-brown, while ± yellowish, esp the tail; white stripe down face; fore-feet heavily clawed for digging. Visitor on E slope, up to Cas Cr in Ore. Order Carnivora, Mustelidae (Weasel family). Color p 494.

This squat, ungainly, but fantastically powerful burrowing creature lives mainly by digging ground squirrels, gophers, and snakes out of their holes. It is common in the drier country just east and south of our range.

# Skunks

**Spotted skunk,** *Spilogale putorius* (spil-og-a-lee: spot weasel; pew-tor-ius: putrid). 10–11" + 5" tail (kitten-sized); glossy black with many ± lengthwise intermittent white stripes; tail ends in a rosette of long white hairs.

**Striped skunk,** *Mephitis mephitis* (mef-it-iss: pestilential vapor). 18" + 11" tail (cat-sized); glossy black with 2 broad white stripes diverging at nape to run down sides of back, plus thin white stripe on fore-head. Both skunks widespread but much commoner in farmlands than in mtns here, and rare at high elevs. Order Carnivora, Musteli-dae (Weasel family). Color p 493.

Skunk coloration is the opposite of camouflage; it's to a skunk's advantage to be conspicuous and recognized, since its defenses are so good. The rare animal that fails to stay clear may receive additional warnings such as forefoot stamping, tail raising, or a handstand with tail displayed forward like a big white pom-pom. (The handstand, rare among striped skunks but well described among spotteds, has been explained as tempting an attacker to bite the tail, doing little damage to the skunk while fixing the attacker's face in the line of fire.) Only as a last resort does the skunk turn around and fire its notorious defensive weapon—up to six well-aimed rounds of N-butyl mercaptan in a musky vehicle secreted just above the anus. This substance burns the eyes, chokes the throat, and of course stinks like hell. It can be shot either in an atomizer-style mist or, more typically, in a water-pistol-style stream fanned across a 30–45° arc for greater coverage. Range is well over 12'. The skunk scrupulously avoids fouling its own tail. Folk remedies for skunk spray include tomato juice, ammonia, gasoline, and burning the affected clothes; juice is the least unpleasant, fire the most effective. Mere soap and water will hardly do. Paradoxically, the musk is extracted commercially and chemically stripped of scent to turn it into a vehicle for perfumes.

Only great horned owls seem to prey on skunks regularly. They may sometimes hit hard and stealthily enough to forestall the spray defense, but more likely they're just "thick-skinned," with the help of built-in goggles and a weak sense of smell. Many big owls smell skunky and have skunk-bitten feet. As far as the odds-makers of natural selection are concerned, skunk defenses are superlative. But like porcupines, skunks seem to be as prone to little parasitic animals as they are well-defended against big predatory ones.

Of our two skunk genera, the little spotted skunk is more weasellike—slimmer, speedier, and more completely carnivorous, though either skunk will eat some vegetation. Chief foods include insects and their larvae, mice, shrews, and occasionally ground-nesting birds and their eggs. Skunks' dens are most often burrows dug by other animals. They usually fatten up for winter and sleep in their dens—not torpidly—for days at a time during the coldest spells.

## Bobcat

*Lynx rufus*\* (links: the Greek term; roo-fus: red). Also **wildcat**. 28" long + 6" tail; tawny to gray cat, generally with visible darker spots, and bars on outside of legs and top (only) of tail; ears may show tufts, and cheeks ruffs, but these tend to be shorter than on lynx. Widespread, esp in brushy, broken, or logged terrain. Order Carnivora, Felidae (Cat family).

The bobcat is a lovely creature we have all too little chance of seeing, even though it inhabits every part of our range. You just might surprise one if you travel quietly, in unpeopled areas, but generally they keep well out of sight.

---

\*Some recent texts lump genus *Lynx* within *Felis;* they call the lynx *Felis lynx* and the bobcat *Felis rufus.*

**Bobcat**

Wild cats all like to work out their claws and clawing muscles on tree trunks, just like house cats scratching furniture. Bobcat or lynx scratchings will be 2′ to 5′ up the trunk (color page 493), cougar scratchings 5′ to 8′ up. These gashes may be deep, but rarely take off much bark; tree-clawing that strips big patches of bark is more likely bear work. Wild cats also often scratch dirt or leaves to cover their scats, at least partially. These scratchings may be accurately directed at the scats, unlike the random pawings of male dogs next to their "markers." Aside from that and the trail-center location typical of coyote scats, it is nearly impossible to tell bobcat from coyote scats.

Cat claws are kept retracted most of the time (thus not showing up in the footprint) so that their sharpness is preserved for slashing or gripping prey. The fifth toe (actually the first, or "thumb") has been lost from the hind foot, but on the forefoot has merely moved a short way up the paw, enlarging the grip. The hind legs are powerful, for long leaping pounces; but with the phenomenal exception of the cheetah, cats aren't especially fast runners. The cat jaw is shorter and thus "lower-geared" than most carnivore jaws, and bears fewer teeth. (Since mammals evolved from reptiles with many teeth, having fewer of them indicates further evolution. Teeth fewer in number are typically also more efficiently specialized. Humans are currently losing their rearmost molars, or "wisdom teeth.") Cats have relatively small and unimportant incisors, huge canines for gripping and tearing, and a quartet of enlarged, pointed molars called carnassials which, rather than meeting, shear past each other like scissor blades for cutting up meat. Cat tongues are raspy with tiny, recurved, horny papillae, which can clean meat from a bone, or hair from a hide. The cat nose is short, suggesting less reliance on smell than in the dog family. As in owls, the eyes are large, far apart, and aimed strictly forward to maximize three-dimensionality. Like owl eyes, they reflect fire or flashlight beams in the dark because they have a reflective layer right behind the receptor cells on the retina, to redouble light intensity at night. Except in cougars, who have round pupils, cat pupils narrow to vertical slits for maximum differentiation between night and day openness.

Bobcats live on rodents and hares, and in winter turn to deer somewhat, occasionally hunting young or even full-grown deer, but far more often seeking carrion.

# Lynx

*Lynx canadensis.** 31" long + 4" tail; gray cat ± tawny-tinged, never clearly spotted or barred exc the black tip of stubby tail; ears tufted, and cheeks ruffed, with long hairs; rarely recorded here, and mainly from the N Cas. (To confirm a lynx sighting, you would need to measure several footprints well over 2" long, and/or see a tail-tip black underneath as well as on top.) Order Carnivora, Felidae (Cat family).

The lynx is often thought of as a larger version of the bobcat. In fact, around here it averages about the same or a bit lighter than a bobcat, but may *seem* larger with its longer fur and legs and bigger feet—adaptations to deep snow and cold. Preying almost exclusively on the snowshoe hare, the lynx is perhaps the most single-minded of our predators. Lynx populations cyclically rise and plummet in response to fluctuations in the hare population.

# Cougar

*Felis concolor* (fee-lis: cat; con-color: all one color). Also **mountain lion, puma.** 4–5′ long + 2½′ tail; ours ± ruddy brown (deerlike), but species ranges in color from nearly black or slate gray to pale sand; no spots or ear tufts exc on kittens; our only cat with long thick tail. Vocalizations varied (purrs, chirps, yowls) but infrequent. Uncommon and elusive; widespread. Order Carnivora, Felidae (Cat family). Color p 495.

Cougars hunt alone, and take anything from grasshoppers and mice on up through porcupines and coyotes to elk—but their staple is deer. A male (the larger sex) can eat, at most, about one deer per week, up to 20 pounds at a time, burying the remains to come back to later. Buried meat, which may assault your nose, is a sign of cougar. He (or she) roams great distances, locates deer by smell or sound, stalks it very slowly, crouching, freezing for periods in a position a deer might mistake for a

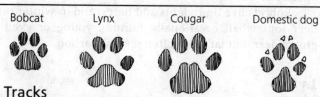

Bobcat    Lynx    Cougar    Domestic dog

## Cat Tracks

**Claws:** cat claws are normally retracted while walking, and hardly ever print. In contrast, claws show up clearly in *full* footprints of dogs and wild canids, not to mention otter, wolverine, badger, and bear prints, which all have five clawed toes rather than four.

**Pad:** the wide, central sole (behind the toes) of a cougar or bobcat print is indented or scalloped once in front and twice in back; dog pads are convex (not indented) in front, as are lynx pads though the fur usually obscures this feature in a lynx print.

**Proportion:** the two outside toes on a cat print are more nearly alongside the two central toes; on a dog print they're more nearly behind the front two.

**Size:** cougar's are 3–3½″ long and wide, bobcat's are 1¾–2″. Otherwise the two are much alike.

Also look nearby for scats and scratchings, described on p 350.

log, and then pounces the last 30 feet or so in a few bounding leaps. He bites the prey in the nape, where his canine teeth quickly grope for a space between vertebrae to cut the spinal cord. If that fails, he attempts to hang on until the prey suffocates. Hunting deer is not without risks; our cougars average only about 80% of the weight of a deer, much less of an elk. They are sometimes trampled or thrown hard enough to kill them, and one was recently found pinned until it starved under an elk it had killed.

The females may go into heat at any time of year. Cougars employ a mating "scream" that is said to be very loud, human, and bloodcurdling. The same could be said of a house cat, and we can imagine how much more true it would be of a cougar. One male travels and sleeps with the female for about two weeks, and no longer; if she let him stay around when the kittens are born, he might eat them. (This is true of several carnivores: the instinct to kill any easy prey encountered is inhibited in mothers, but not in fathers.) Though the home ranges of males overlap those of females, they live solitarily the rest of the year.

Originally the widest ranging of New World mammals, cougars have been persecuted as "varmints" until they are now essentially limited to mountainous wilderness. Though some misinformed hunters and ranchers demand their extermination, game managers now agree on the value of cougars as the only remaining major predator of deer. Their greatest benefit to deer and elk may lie in keeping them moving on their winter range, which helps avoid overgrazing of small areas. Unlike hunters, cougars select young, old, or diseased herd members—the easiest and safest to attack—minimizing their impact on deer and elk numbers. Oregon and Washington now classify cougars as game animals, and the long decline in population may be turning around.

Cougars hardly ever let hikers catch a glimpse of them, and attack people so rarely that the danger isn't worth worrying about. Really, don't lose any sleep over it—not even after learning that they commonly follow solitary hikers in our mountains, unseen, for days. Consider yourself lucky if you run across a clear set of cougar tracks.

# Mule Deer

*Odocoileus hemionus* (oh-doe-coe-ill-ee-us: hollow teeth; he-me-oh-nus: half ass, i.e., mule). 4¼' long + 6–8" tail (goat-sized or a bit larger); medium tawny brown or in winter grayish, with white patches on throat, inside of legs, and rump just under the tail; belly paler; males have antlers (see page 356); fawns are white-spotted; large (8–9") ears rotate independently of each other. Ubiquitous. Order Artiodactyla (Cloven-hooved mammals).

**Blacktail deer**, subspecies *columbianus*. Tail dark brown to black. W-side.

**Mule deer**, subspecies *hemionus*. Tail pale with black tip. Cas Cr and E-side. Color p 495.

The blacktail deer of our Westside, and the mule deer of our Eastside and the Rockies and Plains, were considered two distinct species until fairly recently. Now that they're lumped, authors differ as to which common name to lump them under. Both have tails black at least at the tip, and both have mulishly large ears in comparison to the Eastern whitetail deer.*

When fleeing, mule deer (unlike whitetails) break into a peculiar high-bounding gait called "stotting." This doesn't carry them as fast as a flat-out run could, but it enables them to respond to their typically rough terrain with abrupt, unpredictable changes of direction. This must be a hard act to follow for any predator giving chase, who has to respond at once both to the obstacles under its own feet and to the deer's changing course. Presumably, stotting evolved in steep and/or broken terrain, while sprinting is preferable on the more level Eastern prairies and deciduous woods.

Wolves and cougars were the chief natural predators of American deer, but since we've eliminated all of the wolves and most of the cougars from the lower 48, domestic dogs are probably a more frequent cause of violent death—along with

*Columbian whitetail deer, *Odocoileus virginianus leucurus*, were once widespread in the Puget-Willamette Trough, but have been reduced to small populations, one in the Umpqua valley and one on the lower Columbia. Oddly, while this NW subspecies dwindled (apparently incompatible with agriculture) the main subspecies of whitetails thrived and spread westward as far as NE Washington, from an original range in eastern North America.

hunters, cars, and trains. Even dogs that wouldn't know what to do with a deer if they caught up with one can run it to a death inflicted by barbed wire or a broken leg. Deer are transfixed by a strong beam of light at night, making them frequent victims of cars and trains. Coyotes, bears and bobcats rank as infrequent predators of deer, taking mainly fawns, or adults already close to death from other causes.

In the absence of wolves and cougars, deer tend to increase beyond the carrying capacity of their browse resource; then, during severe winters, huge numbers either starve or, weakened by malnutrition, fall prey to parasites and coyotes. Deer populations can sustain heavy loss of bucks to hunters, since it's normal for all the does to be impregnated by a small elite of bucks anyway. Even in the absence of hunting, deer have a higher mortality rate for growing males than females. State wildlife commissions issue doe hunting permits as needed to limit deer population to the capacity of each district. Overall, deer have thrived and increased with the spread of American civilization, thanks more to our war on forests than our war on predators; mature forests provide less browse than the second-growth and farmland replacing them.

Browsing seems to be a very sophisticated business. Try to spend a while sitting quietly near some deer (they are fairly approachable in Olympic National Park, where they've been protected for generations) and see exactly what they're eating. They often show an intense taste preference for a particular bush, which must contain high levels of some nutrient. They lap up springwater—no matter how muddy it has become from trampling hooves—that contains certain minerals or salts they crave. They strip the old-man's-beard lichen from tree limbs; it contains few nutrients, but enhances the deer's utilization of plant nutrients in the winter diet of twigs, evergreen needles, and leaves. Like other cudchewers, they are able to live on this high-roughage diet thanks to cellulose-digesting bacteria in their first (precudchewing) stomach. They have to browse for the nutritional demands of these bacteria; inadequate protein can kill the bacteria, leaving the browser literally starving to death with its belly full.

In winter, deer seek "thermal cover" on steep south aspects just above river bottoms. The low elevation and insulating conifer canopy offer warmer temperatures and shallower snow, while the southern exposure lets in a little of the low-angle sunlight, and tends to have more shrubs. In summer, our deer move upslope to meadows, clearcuts, and open woods, and fatten on herbaceous plants.

Deer are outstanding subjects for the study of phero-mones—chemical messages received olfactorily, mainly by others of the same species. Mule deer are rather antisocial. They use several pheromones apparently to repel each other (a function akin to territoriality, though mule deer are not territorial). Tarsal glands, for example, are buried in patches of dark hair on the inside of the ankle joints, midway up the rear legs; to activate tarsal pheromones, deer of any age and sex urinate on these patches and rub them together. While most pheromones are secreted in sweat or sebum (skin oil), deer urine is itself pheromonal—its chemistry reveals the animal's health and strength. A subordinate deer will sniff the tarsal patches of a dominant deer and then retreat to a respectful distance, showing that it got the message.

## Antlers

True horns are sheaths of keratin, like fingernails. They form from epidermal tissue at their bases; slowly slide outward over small bone cores; grow continuously throughout life; and never branch. They are found, generally on both sexes, in the cattle family, including sheep, goats, and antelope. Antlers, on the other hand, made of solid bone and usually branched, are a defining characteristic of the deer family./ They form inside living skin—complete with hair and blood vessels—and stretch the skin outward as they grow. This epidermal layer, or "velvet," must die and slough off before the antlers come into use in the fall. In late winter the antlers weaken at their bases and are soon knocked off. A new pair will begin to grow by early summer.

Cumbersome and easily entangled in brush, antlers probably affect survival adversely. Not they, but hooves, are a deer's defensive weapon against predators. (Caribou, or reindeer, are the one exception; most females have small antlers, and use them defensively.) Antlers exist to help establish dominance among males for

Other glands, on the forehead, are rubbed on shrub twigs to advertise the presence and condition of a dominant buck, or to mark possession of a sleeping bed. Interdigitate glands between the two toes of the hoof secrete a more attractive pheromone, marking a deer's trail for other deer to follow. Metatarsal glands, on the outside of the lower hind leg, secrete a fear- or alarm-signaling smell resembling garlic.

The only close social tie common among deer is between a doe and her fawns and yearlings. Males are solitary or form very loose small groups, except during rutting when dominant bucks follow single does in heat for a few days each. A doe seeks seclusion even from her yearlings before and after giving birth; it's up to the yearlings to reunite with her afterward. For their first few weeks she hides the new fawns—separately, if there are two or three—in nestlike depressions under brush. She browses in the vicinity, strives to repel other does, and comes back to nurse mainly at night. If a threatening large animal (like you) approaches, she will resolutely ignore the fawns, acting nonchalant. (She will meet a fox or bobcat,

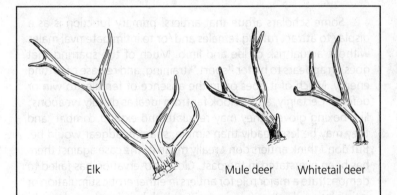

Elk          Mule deer     Whitetail deer

access to females during the rut. Big antlers, like bright plumages on small male birds, are an example of the kind of "fitness" evolved through sexual selection—the survival of the fittest is as a genetic line, not necessarily as long-lived individuals. In other words, highly competitive, large-antlered elk bulls tend to die younger than weaker ones who do not engage in many fights over cows, but the latter, leaving no offspring, are unfit.

however, with a bold counterattack.) Occasionally people come across the hidden fawn and not the foraging doe; sometimes they make the cruel mistake of "rescuing" the "abandoned" fawn. Unless you actually find the mother dead, assume a fawn is being properly cared for, and leave it in peace, untainted by human contact.

# Elk

*Cervus elaphus* (sir-vus: the Roman term for deer; el-a-fus: the Greek term for deer). Also **wapiti**. 7–8′ long + 4–6″ tail (large cow-sized); brown (in Olys) or tan (in Cas) with large, sharply defined tawny-pale patch on rump, and extensive darker tinges on neck, face, legs and belly; males have antlers; fawns are white-spotted. Widespread; abundant in Olys. Order Artiodactyla (Cloven-hooved mammals). Color p 496.

In this century, elk inhabit coniferous forests and high mountains of the West; before white settlement they were common all the way east to the Appalachians, even Vermont and South

---

Some scholars argue that antlers' primary function is as a display to attract rutting females and/or to intimidate rival males without actual risk of life and limb. Much of the sparring that goes on appears to be for "sport," training, and release of rutting energy. Much of it goes on in the absence of females to win or defend. Certainly, antlers look far from ideal as deadly weapons; like boxing gloves, they may regulate and extend combat, and they may be less deadly than simple, sharp headgear would be. But don't think antlers don't really mean it; the case against them has been overstated in the past. Closer observation has failed to demonstrate a major role for antlers in either erotic stimulation or intimidation displays (see page 361), while it *has* traced a substantial proportion of male mortality in protected natural situations to combat injuries.

Each antler on elk and whitetail deer has a single main beam from which all the other points branch. On mule (including blacktail) deer antlers, both branches off the first Y may be again branched. Average sexually mature elk have five-pointed antlers; older bulls may grow six, seven, or rarely eight. Mature mule deer

---

Carolina. Great herds of them on the plains were second only to buffalo both in sheer biomass and as a food and material resource for humans. Like the buffalo (properly "bison") they were shot in huge numbers, especially by ranchers wanting to eliminate grazing competition. They were able to hold on in the mountains and deep forests until an alarmed public, rallying around a famous hunter named Teddy Roosevelt, got refuges and hunting restrictions enacted to allow elk populations to recover. The very idea of conserving species was essentially new to (white) America at the end of the last century; even then the point, as popularly understood, was to conserve them for future generations of people to hunt.

The Olympic National Park (established in 1938) and Vancouver Island harbor the only pure stock of the native subspecies, Roosevelt elk. The Cascades have much sparser elk populations, and these seem to be descended mostly from Rocky Mountain elk (subspecies *nelsoni*), which were introduced here in the 1920s when Roosevelt elk seemed nearly extirpated from the Cascades.

---

typically have four or five points, rarely six or a record seven. Most yearling (18-month-old) bucks east of the Cascades grow "forks" of two points, while the less well-fed yearlings on the Westside, where the profuse greenery makes surprisingly poor deer browse, usually grow unbranched "spikes." The number tends to increase year by year, but also responds to nutrition, and thus serves, on top of the creature's sheer size, to advertise physical condition. Rich feeding in captivity has produced five-point antlers on yearlings, while meager range can limit even dominant bucks to forks.

*Deer (Cervidae) and cattle, etc. (Bovidae) are the two largest families of ruminants, or cud-chewers. The ruminants plus the pigs etc. (Suiformes) make up the cloven-hooved mammals (Order Artiodactyla). These and the much smaller Order Perissodactyla are the ungulates, or hooved animals. Perissodactyl feet stand on a single, symmetrical toenail (rarely with two lateral small nails for a total of three) rather than a "cloven" or split hoof of two equal toenails; most, such as the horse, lack any sort of horn, though the Perissodactyl rhinoceros has one or two pseudohorns of compacted hair tissue.

---

In Europe, a smaller, darker subspecies (*C. elaphus elaphus*) is the red deer or stag that Robin Hood hunted and commercial hunters still sell as venison. If you're beginning to suspect that this species known to the Greeks, Romans and English as a "deer" ought to be a deer, you're right. Quite a different genus exists which we likewise ought to call "elk," as Europeans do and had done for centuries before Americans ever dubbed it a "moose."* American races of both *Cervus* and *Alces* are much larger than their European counterparts, and the confusion must have begun when English colonists first met the American whitetail, about the size of a European red deer, *C. elaphus,* and called it a deer. When later generations, pushing Westward, met a race of *C. elaphus* about twice as big, they understandably reached for the name of the larger European cousin whose flattened antlers and massive, loose-fleshed muzzle they had doubtless never seen. The misnomer, "elk," stuck in common usage, even through generations of texts trying to replace it with the Shawnee name "wapiti." American elk and American moose were each formerly recognized at the species level (*Cervus canadensis* and *Alces americana*), but current taxonomic "lumpers" insist that our moose is merely a subspecies of European elk and our elk a subspecies of European red deer. Our deer are something else again.

The best routes for viewing elk are around timberline (especially off-trail) in the Olympics in early fall, though spring trips to Olympic river bottoms are also good. Most of the year, elk travel in herds segregated by sex, the mature bulls in bands of 10 or 15, and the females and young males in larger herds. In late summer, the bulls become mutually hostile, and the largest, most aggressive of them (called "primary bulls") divide out harems from the cow herds. They tolerate yearlings, but drive away two- and three-year-old males. The harem's

---

*Moose, *Alces alces*, have dramatically extended their range S through B.C. in recent decades. Though this thrust aims toward N Idaho and NE Washington, there have been reports of moose in the extreme N Cascades; there may be more before long. Swampy willow thickets in the NE Cascades are the most likely habitat.

movements, like those of the cow herd in winter, are subtly directed by a matriarch apparently respected for her maturity rather than her size or strength. The bull seems to tag along, rather than to lead. He may lose part or all of his harem if he lets himself be distracted. Other bulls invariably distract him, hoping to take over the harem by overcoming him in a clash of antlers staged at dusk or dawn. This is the rutting season.

Bulls have evolved curious behaviors for challenging each other and working up their sexual or combative frenzies:

**Bugling.** This unique vocalization includes a deep bellow and a farther-carrying whistle. (Elk cows also bugle, though less commonly, when calving in spring.)

**Antler-thrashing.** These attacks on small trees and brush (performed also by deer bucks) have been described as "polishing the antlers" or "rubbing off the itchy velvet" but they are now interpreted as making "visual challenges" or markers, as warm-up or practice for sparring, or as autoerotic stimulation.

**Pit-wallowing.** Shallow pits are dug and trampled out, and lined with urine and feces, for wallowing in. (If enough water collects there in the following winter, the wallow may turn into a pool lasting several years.) A bull may also use his antlers to toss urine-soaked sod onto his back.

The reek of urine advertises the bull's physical condition, presumably helping to avoid injuries by intimidating a challenger before combat begins; it reveals the degree to which the bull (or buck deer) has been metabolizing fat as opposed to fresh food or, worse yet, muscle. A bull metabolizing only fat is one so well fed that he can devote all of his energies to the rut without being weakened by hunger.

Keeping track of a harem, defending it from other bulls, and performing the sexual obligations (or reaping the rewards) is not only hard work, but so time-consuming that the bull can't eat or rest adequately. Almost inevitably in the course of the season, primary bulls succumb to lesser but better-conditioned rivals, and often these "secondary bulls" yield in turn to "tertiary bulls." After defeat, they wander off alone, catch up on sleep, and show no further interest in sex, at least that year. They may be weakened enough to reduce their chance of surviving winter. Large, sexually successful bulls have several years shorter life expectancy than bulls who never grow large enough to compete. The latter are called "opportunistic bulls" because, while spending the rutting season alone, they stand at least a slight chance of mating with a stray cow. All in all, elk have an unusually extravagant courtship system.

While Rocky-Mountain elk cows may bear young every year, starting at age two, Roosevelt elk only go into heat every other year, and also have poorer calf survival rates. This suggests that Westside vegetation is of low overall forage quality for elk and deer, no matter how luxuriant it may seem to us. Browsing techniques and care of the young among elk are much like those described for deer (pages 355–57) except that to an elk cow, humans are puny meddlers to be chased from the nursery. Biologists who tag elk calves are impressively quick tree-climbers. Barbed wire elk exclosures in the Hoh valley have demonstrated that selective browsing determines, in large part, the vegetation of Westside bottoms where elk herds winter. The effect is a pleasing one, including a relative scarcity of vine maples, and a preponderance of Sitka spruce, which are much too prickly for elk to eat.

## Mountain Goat

*Oreamnos americanus* (or-ee-am-nos: mountain lamb). 58" long + 6" tail, 36" shoulder height; all white, with "beard," shoulder "hump," and "pantaloons" formed of longer hair; hooves and sharp, curved horns black. Alp/subalpine in Olys and Wash Cas. Order Artiodactyla (Cloven-hooved mammals). Color p 496.

The mountain goat is a deservedly legendary creature that ought, it might seem, to be among the most sought-after wildlife encounters here. Unfortunately, most hiker-to-goat encounters in our range currently take place in the Northeast Olympics, where the goats are not native and are now pests. In 1981, the National Park finally braved tourist sentimentality and began flying them out, to alleviate damage to subalpine plant communities. Olympic National Park policy is ultimately to remove all goats. In the meantime, be careful to suspend your stashed gear and bestow your urine only upon the rocks.*

In the 1920s, before the Park was established, mountain goats were introduced for hunting, and they quickly overpopulated much of the range. Evidence suggests that mountain goats are nowhere primarily limited by predation. Eagles have been known to take kids and yearlings, dive-bombing to knock them off ledges, and cougars occasionally take young and adults both, but overall the goats are fairly predator-proof thanks to defensive weaponry (hooves and horns) as well as proverbial evasive skills on precipitous terrain. The hooves have strong, sharp outer edges and a hard, rubbery corrugated sole for superlative grip. Forage and climate seem to be the factors limiting goat populations. Winter starvation and disease are ranking causes of death, along with the inevitable attrition from falling. Mountain goats have been decimated wherever they were freely hunted. Permits to hunt them are currently issued on a lottery basis for some (nonpark) areas in Washington.

In the Olympics, goats have slowly radiated from the north corner where they were introduced, but seem reluctant to push west of the Bailey Range climatic crest because of too much wet snow. Apparently, cold wind doesn't penetrate their insulation, but wet snow does if they have to flounder in it all day and night, so they seek windswept, snow-bare ridges in winter. They need access to their winter staples—foliage, lichens, and mosses on subalpine fir—and when they can, they kick a hole down to the preferred grasses. They don't seem to need running water: they have been seen eating snow. The ridgecrest paths they beat are a help on hikers' high-routes in the Olympics, though they often betray us by leading out onto rock faces we think of as requiring hardware,

---

*Mountain goats crave the salts in human urine, as well as in certain salty springs; many Olympic goats are brazen enough to enter camps to eat freshly urine-soaked earth. In the process they demolish precious alpine turf, which is slow to heal such scars. To prevent this, urinate only on bare rock or gravel when in likely goat habitat—anywhere above tree line in the Olympics or Goat Rocks Wilderness.

climbing technique, or both. Tufts of white floss on branch tips remind us these are not paths for mere Vibram.

Mountain goats are classified in a tribe with the chamois, which lead a similar life in the Alps. Both differ from true goats and sheep in several respects, though all are in the cattle family. Mountain goat horns and skulls of males are not heavy enough to sustain bouts of butting. Male mountain goats are thickened at the other end instead; the skin of their rumps, where they are likely to be gored in their flank-to-flank style of fighting, has been known to reach 7/8" thick, and was used by Alaskan tribes for chest armor. More important, the males evolved a powerful inhibition against any real fighting or sparring at all; the occasional pair who get carried away and actually fight are usually both retired from further rutting competition, with broken horns if not severe wounds. The effective breeding males are those who manage to intimidate the others with visual and olfactory displays, without ever coming to blows. Pit wallowing is indulged in (as with elk) and males often spend the winter in filthy coats, looking rattier still in spring while these molt in big sheets, revealing immaculate new white for summer.

In a reversal of the usual mammal social order, the males generally let themselves be chased around by the females and immature males, who aren't as inhibited from using their horns. Females may viciously charge males who come too close, except during a brief sexually receptive period in fall when they allow the male a creeping, apparently submissive courtship. A single kid is born in May or June. Mothers are legendarily protective, walking, for example, against a kid's downslope flank to prevent a fall in steep terrain.

The bighorn or mountain sheep, *Ovis canadensis,* has been introduced in a few places on the eastern fringe of the North Cascades. Survival has been nip-and-tuck for these small bands. Even if they do establish themselves, they are not likely to spread far into our mountains, since their natural habitat is more arid grasslands. In the previous century they were widespread on the steppes of Eastern Washington and Oregon.

# 9

# Birds

Birds are winged, warm-blooded vertebrates with feathers. Wings first developed in some dinosaurs; birds evolved from winged dinosaurs while mammals grew from a line of humbler creatures that diverged from reptiles much earlier. Birds resemble reptiles in many ways that mammals don't. Both bird and mammal evolution produced keratinous skin growths to serve as insulation (corollary to warm-bloodedness) but while mammals got hair, birds got feathers, an optimally lightweight (corollary to flying) form of insulation. As you know from the sales pitch on down sleeping bags, bird plumage is still unequaled among resilient, deformable, durable substances, in its insulation value per weight.

In addition to insulating, feathers do much for the bird's shape, size, and color; plucked, a duck, a crow, a hawk, and a gull would look surprisingly alike in form, as well as pathetically small. Body plumage as much as doubles a bird's girth. The "fat" robins of winter lawns aren't fat, they're just fluffing out their plumage to maximize its insulation value, like what you are trying to do, unwittingly and ineffectually, when you raise goosebumps. Less fluffed out in flight (an intense activity that supplies heat in excess) body plumage serves the equally crucial function of streamlining. The long outer feathers of wings and tail, meanwhile, provide most of the bird's airfoil surface at very little cost in weight. They constitute typically 35% to 60% of a bird's wingspread and 10% to 40% of its length. Feathers make the bird.

Color in most plumages—nearly all female and juvenile plumages and a great many fall and winter male plumages—emphasizes camouflage. Since pale colors make good camouflage against the sky, darker colors against foliage or earth, most birds are paler underneath than on top. A mother and young in the nest need to be especially well camouflaged, since they can do little but sit tight when predators pass overhead. (The mother could fly, of course, but she stays to protect the flightless young.) Males are freer from the need for camouflage than females, and much freer than earthbound mammals. In them (and in the females of a very few species) sexual selection has been free to evolve gaudy plumage.

A showy plumage doesn't look equally good to all female birds; it looks better to females of the same species than to other females. For efficient pairing up, conspecifics need to recognize each other quickly and accurately; but no efficiency is lost if only the females have this ability. (Birdwatchers may be forgiven for supposing it's for their benefit that male birds carry specific markings.)

The same goes for bird songs. Females and juveniles in many genera are nearly impossible to identify to species when no males or songs are present. Songs, which by themselves suffice to identify many species, are male traits for the most part. Though most frequent during courtship season, songs may continue throughout the year, perhaps asserting territory. ("Songs" are distinguished from "calls" as being more extended and complex, and characteristic mainly of the perching birds, Order Passeriformes, also known as "songbirds.")

Birds molt, or replace their feathers, at least once a year even if they don't have different seasonal plumages. The latter are mostly a matter of making the male alternately showy for courtship and camouflaged for the nonbreeding seasons. Ptarmigan, however, simply alter their camouflages to suit the season; they may go through three seasonal plumages (to match snow, no snow, and patchy snow as depicted on page 499) and the females and juveniles go through these along with the males. With a few exceptions like the mallard, the large wing and tail feathers molt just a few at a time, making the bird look tattered but still permitting flight.

A second feather maintenance procedure is preening, done with the toes or the beak, and aided in most orders by oil from a preen gland. Much time is devoted to this crucial task of aligning and oiling plumage to keep it intact as an airfoil, as insulation and as waterproofing.

Along with wings and feathers, the evolution of bird flight entailed radically larger and more efficient respiratory systems. The stamina needed for a single day of flying, let alone for migrating across oceans, would be inconceivable in a mammal. It demands a lavish supply of oxygen to the blood, and of blood to the muscles. Cooling—direly needed during flight—is also done through the breath. Breathing capacity is augmented by several air sacs and, in many birds, by hollow interiors of the large upper leg and wing bones, all interconnected with little air tubes. Each breath passing through the lungs to the sacs and bones and back out through the lungs is efficiently scoured of its oxygen.

Hollow bones, their interiors crisscrossed with tiny strutlike bone fibers in accord with the best engineering principles, doubtless evolved to save weight, yet some birds lack them. Loons, for example, have solid bones—perfectly serviceable in a bird that dives for a living and doesn't fly much. But some diving birds have hollow bones, and a few soaring birds have solid bones.

The other large bone in a bird is the sternum, or breastbone, projecting keellike in front of the rib cage to provide a mechanically advantaged point of attachment for the flight muscles. The latter, known on the dinner table as "breast meat," are the largest muscles in flying birds. They power the flapping wings, and also guide flight by controlling the orientation of each and every feather along the wing edge through a system of tendons like ropes and pulleys. There is scant muscle in the wing and none in the foot, which is moved via tendons from muscles along the upper leg bone, held against the body. A flying animal ideally concentrates all its weight in a single aerodynamic "fuselage" close to its center of gravity.

The wing is analogous at once to both wing and propeller of an airplane. This is no accident; pioneers of mechanical flight studied and imitated bird flight for centuries before Wilbur and Orville finally, albeit crudely, got it right, using a design that separated the propelling and lifting functions of the bird wing. Like a propeller, wings provide forward thrust by slicing vertically through the air while held at a diagonal— the rear edge angled up on the downstroke and vice versa.

Once there is enough forward motion and/or headwind to provide strong airflow across the wings (bird or airplane), their shape provides vertical lift by creating a low-pressure pocket in the air deflected by their convex upper surface, while the lower surface is effectively flat. This upperside-convex principle is common to all flying birds except perhaps hummingbirds; but wing outlines have diverged into many specializations. Some of the most striking are water wings (Loons, page 370, and Dippers, page 408), soaring wings (page 374), speed wings (page 382), little-used wings (grouse, page 386), and hovering wings (hummingbirds, page 396).

Highly mobile and often migratory, birds wander from their usual ranges more than plants or mammals do. The 75 species described in this book are fewer than half of those reported for our range. Among those excluded is the American robin, which is common here but much commoner in cities. The great blue heron, on the other hand, though more characteristic of lowlands and even cities than of mountains, is included because it is sure to draw attention, admiration, and curiosity. I have mainly tried to include the birds most characteristic of our mountains, conspicuous or not.

Nomenclature and sequence of the species in this book are from the American Ornithologists' Union *Checklist* (1983 and supplements). Birds are the only major group for which decisions on names, both scientific and common, are made by committees and then accepted as "official" by just about everyone on the continent. Well-known common names that were once official are listed as "formerly" rather than "also." The 1983 *Checklist* added the modifiers "Northern," "American," and "Common" in several cases to distinguish American species from Eurasian ones. Where these are not needed to distinguish the species from any other North American bird, they are not repeated in the text.

The size figure that begins each description is the length from tip of bill to tip of tail of an average adult male; females are the same or more often a little smaller, except among raptors, where they are considerably bigger. Robbins' (1985) figures are used—measurements of "live birds hand-held in natural positions." These run about 10% shorter than those in other bird manuals which, following taxonomic tradition, are measurements of long-dead specimens or skins forcibly hand-stretched. Novice birders would do well to fix in their minds images of sparrow size (4–6"), robin size (8½"), jay size (9–12"), crow size (17"), and raven size (21") as a mental yardstick. Unfortunately, size isn't always helpful in the absence of other identifying characters; it's very difficult, for example, to tell the goshawk from the smaller Cooper's hawk against the open sky.

Habitat and behavior are useful clues, especially if you're content with a smart guess as opposed to a positive identification. Many small birds are faithful to plant communities they have adapted to for forage and/or cover. Certain species widespread in deciduous streamside thickets on the Westside, for example, are rarely seen in the surrounding conifer forest, and vice versa. Within a forest, one species prefers the canopy, another the tall shrub layer, and another the ground and low shrubs. But note that food and cover preferences of many birds are much altered during the nesting season.

# Common Loon

*Gavia immer* (gay-via: the Roman term; im-er: sooty). 24", ws 58" (variable size, generally much larger than ducks); bill heavy, tapered, ± ravenlike; breeding-season adults (both sexes) with iridescent green/black head, white collar, black/white checked back, white belly; winter plumage dark gray-brown above, white below; in flight, head is held lower than body, and feet trail behind tail. Lakes; rare. Order Gaviiformes (Loons).

The much-varied "laughs" and "yodels" of loons have been called beautiful, horrible, hair-raising, bloodcurdling, magical, and maniacal. Unequivocally they are loud.

Loons resemble diving ducks in their feeding and locomotion skills. They eat mainly fish, plus some frogs, reptiles, leeches, insects, and aquatic plants. Like ducks they use both wings and webbed feet to swim underwater. Diving either headfirst or submerging submarinewise, they can go deeper than any other birds (300 feet down!) thanks largely to their heavier bodies—only slightly less dense than water. Their heavy bones are a primitive trait that doubtless remains advantageous for diving. It is disadvantageous for flying; though loons can fly fast and far, they land gracelessly, with a big plop, and take off with great effort and splashing. Loons can become trapped for days or weeks on forest-lined lakes too small for their low-angle takeoff pattern—waiting to take off into a gale, when one arises. On their feet they're still more inept and cumbersome; the extreme rear placement of their legs is great for swimming but awful for walking. They go ashore (on an island) only to breed, and nest in soggy plant debris at the water's edge. She and he take turns on the eggs.

After wintering near the coast, some loons move to mountain lakes, arriving soon after the ice breaks up. Loons are sensitive to human disturbance during the breeding season, and have nearly disappeared from the Cascades as a breeding species. Ross Lake is our only lake known to have them still.

# Pied-billed Grebe

*Podilymbus podiceps* (pod-i-**lim**-bus: a name combining 2 other genera; pod-i-seps: rump feet). 9″ long, stocky; bill high, very stout with downcurved ridge, pale with a black band across it (on summer adults only); both sexes drab brown mottled white, broadly white under tail and (summer adults) marked with black on face and throat; in flight (rarely seen), head is slightly lower than body, and feet trail behind the very short tail. Lakes and marshes, uncommon. Order Podicipediformes (Grebes). Color p 497.

Like loons, grebes are a primitive order of birds poorly adapted for flying and worse for walking, but superlatively built for diving. Grebe toes—fat scaly lobes—paddle even more efficiently than webbed feet; they use a side by side stroke resembling the human butterfly stroke but without any help from the forelimbs. (Diving ducks, in contrast, paddle their feet alternately, and also use their wings.) By exhaling deeply to decrease their buoyancy, grebes can quietly submerge and skulk with only head or nostrils above water. Grebe hatchlings take up the submersible life before they can even swim, by clinging to a diving parent's back. The widespread pied-billed grebe is timid and rather antisocial, but capable of remarkable vocalizations sometimes compared to a braying donkey or a squealing pig. Crayfish are a favorite food.

# Great Blue Heron

*Ardea herodias* (ar-dia her-oh-dias: Roman and Greek versions of the term for it). 38″, ws 70″; gray (vaguely bluish) with some white, black, and dark red markings; bill, neck, and legs extremely long; neck held "goosenecked" in flight; huge birds seen in slow-flapping low flight over rivers and lakes are generally this species. Various loud, gutteral croaks. Order Ciconiiformes (Storks and herons). Color p 498.

The heron's way of life is to stand perfectly still in shallow water until some oblivious frog or small fish happens by, and then to pluck or spear it with a quick thrust of the beak. The prey see little of the heron but its legs and shadow, and perhaps mistake it for an odd reed or cattail.

# Mallard

*Anas platyrhynchos* (ay-nus: the Roman term; plat-i-**rink**-os: broad nose). 16"; males (Sept-June) have iridescent dark green head separated from red-brown breast and brown back by a crisp white neckband; bill yellow on males, black/orange on females; both sexes have a band of bright blue with black/white trim on upper rear edge of wing, and much white under wings; females (and July-Aug males) otherwise speckled drab. Loud quack. On marshy lakes, as in the C Ore Cas. Order Anseriformes (Waterfowl). Color p 497.

The mallard exemplifies the numerous surface-feeding, or "dabbling" ducks. It is conspicuous over much of the Northern Hemisphere—except during duck hunting season, when it makes itself perplexingly scarce. Tasty flesh contributes to its renown. Domestic ducks were bred largely from mallards, centuries ago, and breed with them when given the chance; city park ducks often include mallards and hybrids together.

Dabbling ducks feed by upending themselves in shallow water and plucking vegetation growing on the bottom. They also eat a few molluscs and insects, and a very few small fish. They take flight abruptly and steeply, unlike diving ducks which splash along the surface. Mallards are among the birds that molt their flight feathers all at once, becoming temporarily flightless; to escape predators they hide out for the duration in large groups in marshes. You may then see mallard mothers with chicks trailing behind, but no fathers anywhere. After the young are grown and independent, the females take their turn in molt as "sitting ducks."

# Harlequin Duck

*Histrionicus histrionicus* (hiss-tree-ah-nic-us: actor or jester, referring to clownlike facial markings). 12"; breeding males plumed in a clownlike patchwork of slaty blue-gray, rich brown on the flanks, and white splotches with black trim—but may appear merely dark from a distance; others dark brown with several small white patches on head, and whitish belly. Rough water. Order Anseriformes (Waterfowl). Color p 497.

Whether in whitewater rivers or heavy surf, harlequins display phenomenal pluck and strength as swimmers. They spend most of their time on rocky coastlines, where they dive for shellfish. After courting and pair-bonding, they move inland to nest along mountain streams, where they feed largely on insects. The grass-lined nest is built under streamside brush, or among boulders. The females stay to incubate and raise the chicks alone, the males returning to sea. Populations wintering on the Olympic Coast include some that nest on the Hoh in May along with others that nest in Alberta in July. Unlike most ducks, they rarely mix with other species.

Wags have compared harlequin plumage to unfinished paint-by-number art. The feathers were once collected to adorn women's hats.

## Common Merganser

*Mergus merganser* (mer-gus: diver; mer-gan-ser: diving goose). 18"; breeding males mostly white beneath with black back and head, and red bill, the head showing greenish iridescence in strong light, and becoming brown in non-breeding season; females with red-brown head sometimes showing slight crest on nape, and red bill, otherwise grayish, with darker back and ± white throat and breast. Lakes and streams. Order Anseriformes (Waterfowl). Color p 497.

Mergansers are carnivorous "diving ducks" as opposed to the more numerous "dabbling ducks" who don't dive, but merely dip, for their mostly vegetable foods. Instead of a Donald Duck-type broad bill, mergansers have a long, narrow bill with a hooked tip and serrated edges for gripping their slippery aquatic prey—amphibians, insects and other invertebrates, and fish. This diet makes mergansers less tasty, and less prized by hunters, than their herbivorous relatives. Of our three merganser species, the common is the largest, most montane, and certainly the commonest. Those who don't fly north to breed generally move up to mountain lakes, finding woodpecker holes or other cavities to nest in.

# Turkey Vulture

*Cathartes aura* (cath-ar-teez: purifier; aura: breeze). Also **turkey buzzard.*** 25"; ws 72"; plumage black exc whitish rear half of wing underside; head naked, wrinkled, pink (exc black when young); soars with wings in a shallow V, often tipping left and right, rarely flapping. Open country. Order Falconiformes (Hawks and allies).

> *bird of rebirth*
> *buzzard*
> *meat is rotten meat made*
> *sweet again and*
> *lean*
>
> —Lew Welch

Linnaeus was wise to name this creature "Purifier" to counter its unsavory reputation. Its bald head and neck are virtually self-cleaning (no feathers to foul) after mucking around in carrion. Its digestive tract is immune to disturbance by all the meat-rotting organisms that would do in the rest of us. Its beak

*"Buzzard," an old word tracing directly to the Roman *buteo*, has always been applied broadly by the English to include hawks (genus *Buteo*) as well as Old-World vultures, which are in the hawk family. When we Americans say "buzzard," we mean a vulture, not a hawk. American vultures and condors are their own family; they superficially resemble Old-World vultures due to convergent evolution. They include the largest U.S. bird, the endangered California condor, *Gymnogyps californianus* (**jim**-no-jips: naked vulture). Condors were common here in Lewis and Clark's day, but their range shrank inexorably, and they currently live only in zoos.

## Soaring Wings

The larger Falconiformes—eagles, vultures, ospreys, and buteo hawks—share a broad, spread-tipped wing (and tail) design specialized for soaring over land. These contrast with the short wings of accipiter hawks, the narrow wings of falcons, and the clean, linear wings of gulls and albatrosses—which also soar superlatively, but on entirely different air currents, over the ocean.

The big hawks stay aloft largely by seeking out "thermals," or columnar upwellings of warm air—the daytime convection

and talons are probably too weak to tear up freshly dead mammals, let alone kill live ones; predatory birds, in contrast, are equally ready to either kill or eat carrion. Vultures go long periods without food, and when they find it they gorge themselves, perhaps accounting for their apparent lethargy on foot and difficulty in taking flight. Once on the wing, they are the best soarers of all land birds, in the sense of least frequently needing to flap.

A congregation of vultures wheeling usually means carrion below, and other vultures seem able to read this sign, slowly gathering from miles around. Vultures are sharp-sighted, and locate their food primarily by sight. However, while most birds, including most vultures, have little sense of smell, this one species does respond to certain carrion smells—of mammals but not fish, reportedly, though they love fish, often gathering over salmon spawning streams in the Northwest.

pattern of low, warm air rising to trade places with higher, cooler air. Hawks don't try to travel across large lakes, or at night, or even in the early morning, since thermal air is warmed by land absorbing and reradiating sunlight. Dry, sparsely vegetated land does it best, so steppes and prairies are especially popular with soaring birds. To travel, they may climb one thermal spirally, then glide obliquely downward to the next thermal, and climb again. To migrate, they often follow long north-south ridges that produce wavelike updrafts by deflecting westerly winds. Tightly packed mountain topography, like ours, produces oblique thermals along creek headwaters; soarers can use these just as well.

# Osprey

*Pandion haliaetus* (pan-die-un: a mythic king; hal-ee-ay-et-us: sea eagle, the Greek term for osprey). Also **fish hawk.** Males 22", ws 54", females larger; blackish above, exc white crown; white beneath, with black markings most concentrated at wing tips and "elbows" (actually wrists); the "elbow"-break is sharper both rearward and downward than on similar birds, suggesting a shallow M while soaring. Frequent calls include loud whistles and squeals. Near rivers and lakes. Order Falconiformes (Hawks and allies). Color p 498.

In a hunting technique that wins respect and astonishment, the osprey dives into water from 50 to 100 feet up, plucks a fish from a depth of 1 to 3 feet, and bursts immediately back into flight, gripping a squirming fish sometimes as heavy as itself. The soles of its feet are toughened with minute barbules that help grasp wet fish.

Ospreys are a nearly worldwide species comprising a genus and subfamily by themselves. They are among several birds of prey that suffered heavy losses in some areas from toxic chemicals like DDT. (The pair in the photo, for example, has failed to raise offspring for several years now, despite valiant nesting efforts; pesticide-induced eggshell thinning can certainly be suspected as a culprit.) But in the Northwest, so far, ospreys coexist with civilization, perhaps even benefitting where rivers are dammed.

# Bald Eagle

*Haliaeetus leucocephalus* (hal-ee-ee-et-us: sea eagle; lew-co-sef-a-lus: white head). Males 32", ws 80", females larger; adults blackish with entire head and tail white; immatures (1–3+ years) brown; wings held flat while soaring. May be seen anywhere in our range, but common only on C Ore High Cascade lakes, the Nooksack and Skagit valleys, the Klamath Basin and the Olympic seacoast. Calls: various weak chips and squawks, or louder ± gulllike shrieks. Order Falconiformes (Hawks and allies). Color p 498.

The national symbol, bald eagles have been considered vermin to be shot on sight for most of America's history. Alaska even offered a bounty for shooting them until 1952, twelve years after it became illegal in the lower 48 states. Some Westerners still fear and hate eagles enough to break the law.

While they are now rare over most of North America, we

in the Northwest are lucky that they are fairly plentiful along the coast from the Aleutians to the Columbia, spilling inland along salmon streams and to high lakes where waterfowl concentrate. These areas (listed above) hold the largest breeding populations of bald eagles south of Canada. Fondness for fish and waterfowl keeps them usually near water, but they are sometimes sighted in various montane habitats here, and seem less averse to deep forest than other big soarers.

Despite a public image as a fearsome predator, the bald eagle far more often scavenges carrion or robs other birds—ospreys, gulls, kingfishers, and smaller bald eagles—of their prey. Most salmon it eats are spawned-out carcasses. At lambing time on sheep ranches, bald eagles eat afterbirths and stillborn lambs. Grossly exaggerated fears of lamb and salmon predation fueled America's animosity toward eagles for the last two centuries.

Our bald eagles breed in mid to late winter. That way the young, after an extraordinarily long time in the nest (six weeks incubating, under both parents taking turns, and ten to twelve weeks from hatching to flying) will be ready to fly up the coast in time for the summer coho salmon runs in Alaska. Those that migrate work their way back down the coast in conjunction with salmon runs that follow heavy fall rains; they reach Washington in November, and peak there in January. Many of our birds, especially those in Oregon, seem to be resident rather than migratory.

A spectacular courtship "dance" of bald eagles has been described many times: the male dive-bombs the female in midair, she rolls over to meet him, they lock talons and plummet earthward, breaking out of their embrace at the last possible instant. Juvenile bald eagles, like human juveniles in certain cultures, engage in every kind of courtship behavior short of mating, which waits for the fifth or sixth winter.

North American tribes almost universally associated the bald eagle with sickness, death, and healing, and hence as an ally or guardian of shamans. Eagle feathers were used in healing rituals, and it was variously said that an eagle would fly over a sick person, would eat a dead person, or would scream to a person soon to be killed by an arrow.

## Northern Harrier

*Circus cyaneus* (circus: Greek term for a circling hawk; sigh-ay-nee-us: blue). Formerly **marsh hawk**. Males 16¼", ws 42", females larger; conspicuous white rump patch on both sexes; males ash gray above, whitish underneath with black wing tips and tail bars; females speckled reddish brown, paler beneath; tail long and narrow; wings held well above horizontal while gliding. Marshes and lush meadows. Order Falconiformes (Hawks and allies).

The old British name "harrier" is perhaps more accurate than the traditional American name "marsh hawk"; this bird cruises dry grasslands and wet meadows alike. It nests among tall grass, and spends several hours a day cruising just inches above the tops of the reeds, grasses, or low shrubs. When a vole or similar creature takes off running underneath it, the agile hawk may harry it through many dashes and turns before dropping on it. It is an unusual hawk in hunting primarily by sound; in experiments, harriers precisely locate and attack tiny tape players concealed in meadow grass peeping and rustling like voles. In this respect harriers evolved convergently with owls, developing an owllike facial ruff of feathers thought to enhance hearing.

The harrier is also a unique hawk in mating polygamously. Some males manage to feed two or even three nests of about five young each; if they fail to keep up with the demand in the first two weeks after hatching, the mothers are forced to go out hunting, and some untended nestlings are likely to die of cold. The mother often leaves the nest momentarily to take midair delivery of food morsels from her mate.

Cooper's hawk

# Accipiters

*Accipiter* spp. (ak-**sip**-it-er: Roman term meaning a fast flier). Distinguished from other hawks by behavior, and by proportionally short, broad wings and long tails; the tail is often narrow (i.e., not fanned out), is broadly barred (± black/white) its full length, and has a pure white tuft under its base; 3 species hard to tell from one another at a distance—size ranges overlap, each species' females being about as big as next bigger species' males; both sexes slaty gray above, sometimes brownish; yearlings much alike, brown above, pale beneath with red-brown streaking. Calls cackling, infrequent. Order Falconiformes (Hawks and allies).

**Sharp-shinned hawk,** *A. striatus* (stry-**ay**-tus: striped). Males 10½", ws 21" (jay-sized); tail square-cornered, dark-tipped; underside pale with fine brownish barring. Often at forest/clearing edges, or alpine.

**Cooper's hawk,** *A. cooperii* (coo-per-ee-eye: after James G. Cooper). Males 15½", ws 28" (crow-sized); tail rounded, with slight white tip; underside pale with red-brown barring. Forest understory, edges, or streamside brush.

**Goshawk,** *A. gentilis* (jen-tie-lis: noble?). Males 21", ws 42" (± red-tailed hawk size) tail rounded, with slight white tip; underside pale with fine blue-gray barring. Deep wilderness forests. Color p 498.

Accipiter hawks evolved in the forests. Their long tails lend maneuverability in close quarters, while the short powerful wings minimize foul-ups with branches, and provide bursts of speed for overtaking quarry on the wing. Cooper's typically watches and bursts out from a concealed perch; the sharp-shinned flies around more or less constantly and randomly, and chases the birds it happens to surprise; the goshawk also cruises, flushing grouse and ptarmigan, its chief prey, and also taking some squirrels and hares.

Each day while young are in the nest, a sharp-shinned father brings home about two sparrow-sized birds apiece for his family of seven or so; that's a lot of hunting. He seems to stay in shape with playful harassment of larger birds, even ravens. Accipiter males select a limb not far from the nest and regularly take prey there to pluck it before eating it or delivering it to the nest. An indiscriminate scattering of thousands of small feathers may be a sign of one such pluckery in the branches above. Either parent may make a nasty fuss over an animal, such as yourself, happening near the nest.

Accipiters are often described as nonsoarers, with a typical flight rhythm of five flaps and a short glide—sometimes a useful identifying trait. This is a book about mountain habitats, however, and in the strong sustained updrafts of rugged mountains they appear perfectly capable of soaring till hell freezes over.

# Red-tailed Hawk

*Buteo jamaicensis* (bew-tee-o: the Roman term; ja-may-ik-en-sis: of Jamaica). Males 18", ws 48", females larger; tail (adults) red-brown above, pink below, broadly fanned out; brown above, highly variable beneath, from dark brown to (most often) white with delicate red-brown lines. Usually seen soaring. Call a short hoarse descending scream. Order Falconiformes (Hawks and allies). Color p 499.

The red-tail is easily America's oftenest-seen large bird of prey. For one thing, it is comfortable around freeways, often perching on roadside fenceposts. Its preferred natural habitat is grassland mixed with some trees, such as our Eastside ponderosa pine timberline.

Adaptability, opportunism, and economy are key to the red-tail's success. Lacking the speed that makes falcons and accipiters perfect at their specialties, it adopts a variety of hunting ploys based on acute eyesight and effortless soaring. Hares and ground squirrels, the preferred prey, may get an aerial swoop following a patient, stealthy approach. Or the hawk may brazenly land between a rodent and its refuge, forcing the rodent into an end-run; the hawk is sometimes quick enough to grab it. Band-tailed pigeons may be swooped upon while feeding in brush. Lizards, snakes, and slugs demand no particular technique, and form a large part of the diet in some seasons. Red-tails are also pirates of smaller hawks, robbing them of their kills. In winter, individuals remaining in the snowy North have shown a remarkable ability to conserve calories and subsist on a skimpy and irregular diet.

## Golden Eagle

*Aquila chrysaetos* (ak-will-a: Roman term for eagle; cris-ay-et-os: gold eagle). Males 32″, ws 78″, females larger (up to twice as heavy); adults dark brown all over; juveniles have white bases to their main wing and tail feathers (but never an all-white tail, as in bald eagle, nor a white rear edge of wing, as in turkey vulture); wings held flat while soaring. Call (rarely used) is rapid chipping. Open country; wary of humans. Order Falconiformes (Hawks and allies). Color p 499.

Golden eagles inhabit any North American terrain that has plenty of vertical relief, few trees, and populations of both hares and large diurnal rodents like marmots or ground squirrels; the hares provide winter fare while the rodents hibernate. Winter may also impel an eagle to attempt larger prey such as fox, or rarely coyote or deer. This is reportedly accomplished with a plummeting, falconlike dive; momentum multiplies the eagle's 10 pounds into enough force to overpower heavier prey.

Normally, the golden eagle hunts from a low, fast-soaring cruise, using angular topography both for visual cover and for updrafts. It also robs hawks and falcons of their kills; hawks rarely venture, much less nest, within half a mile of an eagle's aerie. The latter is a stick structure 4′ to 6′ in diameter, and growing from 1′ deep to as much as 5′ with many years of reuse—yet somehow hard to spot against its cliff.

# Prairie and Peregrine Falcons

*Falco* spp. (fahl-co: the Roman term). Order Falconiformes (Hawks and allies). Males 16", ws 40", females larger; wings pointed, tail narrow.

**Prairie falcon,** *F. mexicanus.* Dusty to (less often) slaty brown above, strongly mottled cream white beneath, with blackish wingpit (analogous to armpit) patches; may have a vertical streak below eye, but narrower and less distinct than on peregrine. E-side steppe year-round, wandering in summer to timberline meadows near Cas Cr. Color p 497.

**Peregrine falcon,** *F. peregrinus* (pair-eg-rye-nus: wandering). Also **duck hawk.** Color variable, ours most likely slate blue to (immatures) dark brown above, white beneath with dark mottling and ± reddish wash except on pure white throat and upper breast; face has a distinctive high-contrast rounded dark bar descending across and below eye. Very rare; current reintroduction habitats are Crater Lake and both sides of the Columbia Gorge.

The prairie falcon can take birds, as other falcons do, but its chief prey are ground squirrels; when it hunts them around Cascade timberline after its spring nesting season in the East-side dry country, it is our most conspicuous falcon larger than a kestrel. A parent with a fledgling may defend it by means of dive-bombing attacks at the heads of intruders, veering away at the last instant—and provide a lucky hiker with one of the great adrenalin-stimulating wildlife encounters.

## Speed Wings

Falcons in flight are told from other hawks by their pointed, swept-back wings and straight, or even slightly tapering, tails. (They also soar, though infrequently, with tail fanned out like other hawks.) This must be the optimal wing design for sustained speed, since so many of the fastest flyers—falcons, swifts, swallows, and nighthawks—evolved it independently.

The fastest of all falcons in level flight is said to be the largest falcon, the gyrfalcon, *Falco rusticolus,* of the Arctic (rarely reported seen here). In ancient falconry, the gyr was so highly esteemed that kings reserved it for their exclusive use. Almost as highly rated was the slightly smaller peregrine. The peregrine has often

The prairie falcon favors alpine areas less than the other medium to large falcons—merlin, peregrine falcon, and gyrfalcon—but the latter are all rare here. Peregrines were heavily politicized during the 1960s; use of DDT (as well as the capture of peregrines for falconry) was banned in the U.S. after it was established that the severe worldwide decline of peregrines was caused chiefly by breakage of thin-shelled eggs from mother falcons contaminated with DDT and related insecticides. Programs for captive breeding and release of peregrines were funded and a recovery process begun. Success is coming slowly, hindered partly by poaching (peregrines fetch astronomical prices from falconers, especially in the Middle East), but mainly still by pesticide residues.

In addition to reintroduced pairs, one or two pairs are rumored to have survived in the Washington Cascades. A distinct coastal subspecies resides on the Northwest Coast; their shorebird prey carry much less pesticide contamination than inland food chains do. Inland peregrines are recently exploiting a new habitat, large cities, where two of their favorite things are in generous supply—spectacular high ledges to perch and nest on; and pigeons to eat.

been called "world's fastest animal" on the basis of its "stoop," not exactly flight but more like skydiving targeted at airborne prey. The wings are held close to the sides to provide a modicum of control and a minimum of drag. Most hawks and eagles will stoop occasionally, when presented with an irresistible target, but only the peregrine virtually limits itself to midair prey, and has perfected the stoop. The figure 180 miles per hour has been bandied about on the basis of one 1930 report by a small-plane pilot who actually claimed that while he was diving at 175 miles per hour he was passed by a stooping peregrine doing about double the plane's speed. The figure has never been corroborated, and is discredited by many ornithologists who believe that, among birds, the race may yet belong to the swifts.

# American Kestrel

*Falco sparverius* (spar-ver-ius: sparrow). Formerly **sparrow hawk**. Males 9", ws 21", females slightly larger; red-brown above, exc wings blue-gray on adult males only, and tail tipped with heavy black band and slight white fringe; brown-flecked white beneath (exc dark tail). Call a sharp, fast "killy-killy-killy." ± open country. Order Falconiformes (Hawks and allies).

**American kestrel**

The kestrel is one of the smallest, most successful, least shy, and most often seen raptors. When not perched on a limb or telephone wire, it often hovers in place, wings fluttering, body tipped about 45°, facing upwind, 10–20' above a field or roadside. From this vantage it can drop and strike prey quickly, as a larger, broader-winged hawk might do from a low soar. Grass-hoppers and other big insects are staples; mice are also taken. If all these are scarce, the kestrel may fly down sparrow-sized birds.

The 1982 decision to change the name from "sparrow hawk" to American kestrel conforms to global nomenclature; the Eurasian kestrel is a similar falcon, whereas the Eurasian sparrow hawk is an *Accipiter* and a frequent hunter of spar-rows. The change incidentally answers a century of demands to rename the sparrow hawk "grasshopper hawk" as a kind of pardon. Earlier Americans tended to associate any predator with some of the largest prey it was known to take, especially if the prey was valued by humans. This bias fueled the organ-ized violence with which predatory "vermin" were hunted down. Naturalists tried to counter it by dividing predators into "harmful" and "beneficial" species. They were sure farmers could learn to appreciate predation on grasshoppers and field mice, while for their own part they were willing, as bird-lovers and often hunters or fishermen themselves, to join in vilifying predators of birds and game animals. As more observations and stomach content analyses were published, more and more predators turned out to feed mainly on insects and rodents, and could be moved to the "beneficial" list.

In short, it made a difference whether this falcon was a sparrow hawk or a grasshopper hawk. But today, with the misleading name abolished at last, wildlife biologists have al-ready moved on to the next phase of reforming public atti-tudes—to appreciate predators for what they are. Even the ones from the old "harmful" list are beneficial from a broader perspective; eliminating a predator usually does its prey popu-lation more harm than good. Virtually all species evolved, and function normally, *within* predator/prey relationships. Preda-tory culling speeds and hones the evolution of prey popula-tions, and helps keep them healthy.

# Spruce Grouse

*Dendragapus canadensis*\* (den-**drag**-a-pus: tree lover; can-a-**den**-sis: of Canada). Also **Franklin's grouse, fool-hen.** 13"; males slaty to brownish black above, with red-orange eyebrow-comb, black beneath, with white bars or large flecks on flanks and upperside of tail; others mottled red-brown (darker than ruffed grouse, redder than sooty grouse); tails of both sexes dark with ± distinct reddish tip. Courting males hoot like sooty grouse but even lower, and beat their wings in a noisy flutter. Near timberline, uncommon, mainly in N Cas. Order Galliformes (Turkeys and allies).

The spruce grouse earns its nickname "fool-hen" by being preposterously fearless around humans—vulnerable even to sticks and stones. Combine that unfortunate naivete with tender, juicy, tasty flesh (at least in summer when the bird hasn't been eating conifer needles) and you get severely reduced numbers of spruce grouse, concentrated today where protected in National Parks.

Birds of the chickenlike order, Galliformes, feed and nest on the ground, and fly only in infrequent short bursts. They have undersized wings and pale breast meat, revealing a scanty supply of blood to the flying muscles. After the fowl has been flushed a few times in quick succession, these muscles are too oxygen-short to fly again until they are rested. Tirelessness in flight is more the norm among other kinds of birds which, being infrequent walkers, have dark breast meat and pale and scanty leg meat.

# Blue Grouse

*Dendragapus obscurus*† (ob-**skew**-rus: dark). Also **sooty grouse,**† **hooter.** 17"; adult males mottled dark gray above, pale gray beneath, with yellow eyebrow-comb; others mottled gray-brown; both sexes have blackish tail. Male's courtship call a series of 5 or 6 low hoots; hen with chicks clucks. All elevs in montane forest. Order Galliformes (Turkeys and allies). Color p 499.

---

\*Formerly in a separate genus as *Canachites canadensis.*
†Rumors have it that the sooty grouse, currently our variety of Blue Grouse, may be raised to species rank. It would then be *Dendragapus fuliginosus* (foo-li-jin-**oh**-sus: sooty).

---

Some blue grouse populations migrate vertically in an unorthodox direction—upslope for winter. They relish the berries and insects of the valleys during the season of abundance, but during winter they have to subsist on conifer needles anyway, so they may as well go up to timberline where there's little pressure from competitors or predators. They find adequate protection from the elements in the dense branches of timberline conifers.

Both the courting "hoot" and the chief visual display of the males—bare yellow patches on the neck—are performed by inflating a pair of air sacs in the throat.

## White-tailed Ptarmigan

*Lagopus leucurus* (la-go-pus: hare foot; loo-**cue**-rus: white tail). Also **snow grouse.** 10"; underparts white; upper parts white in winter, mottled brown in summer, patchwork of brown and white in spring and fall; feet feathered; bright red eyebrow-comb on males in spring. Soft clucks and hoots. Alpine, to subalpine in winter; Wash Cas. Order Galliformes (Turkeys and allies). Color p 499.

Though smaller than their grouse relatives, ptarmigan are the largest creatures that make our alpine zone their exclusive home. Dwarf willows are their staple food, with help from crowberries. Like other grouse, they rely on camouflage to protect them from predators; considering the visual acuity of hawks, this seems an especially dangerous life in the absence of trees. In winter they switch to pure white plumage and stay on pure white snow as much as possible, digging into it for shelter or to reach willow buds. Once their summer plumage grows in, they stay off the snowfields.

The "p" in ptarmigan is not only silent but silly. It must have found its way into print long ago in the work of some pedant who assumed "ptarmigan" to be Greek. It's Gaelic.

# Ruffed Grouse

*Bonasa umbellus* (bo-**nay**-sa: bull; um-**bel**-us: umbrella). Also **drummer**. 14"; mottled gray-buff; tail red-brown or gray (two color phases) with heavy black band at tip and faint ones above; males have slight crest; black neck "ruff" is erected only in courtship display. Distinctive "drumming" in lieu of a vocal mating call is common in late spring, occasional (territorial?) at other seasons; also an owllike hoot is sometimes heard. Lower forests. Order Galliformes (Turkeys and allies).

The male ruffed grouse has a mysterious noise to contribute to your wilderness experience. (Even Press Expedition men—seasoned frontiersmen who made the first recorded trip across the Olympics—seem to have mistaken it for, of all things, unseen geysers bubbling.) Traditionally known as drumming or booming, it's an accelerating series of muffled thumps he makes with sharp downstrokes of his wings while perched on a log. You probably can't tell what direction or distance it's coming from. Very low-pitched sounds carry farthest in forests, but they sound nondirectional. An attracted female must have a tantalizing search in store for her.

At the end of it, she can watch his fantailed, ruff-necked dance, and then mate with him, but that's all he has to offer her; she will incubate and raise the young by herself. She nests on the ground, like other grouse, and trusts her excellent camouflage up until the last second, when you're about to step on her unaware. Then she flushes explosively, right under your nose. She may actually try to scare you away, if there are young to protect, or to draw you away from them with her famous broken wing routine. You are touched. She knows a thing or two about psychology.

Though they occupy a range congruent with the conifer-dominated forests, ruffed grouse are usually found near deciduous trees.

# Spotted Sandpiper

*Actitis macularia* (ak-tie-tiss: shore dweller; mac-you-**lair**-ia: spotted). Also **teeter-tail**. 6¼"; light brown above with white eye stripe and wing bars, white below; summer adults dark-spotted white below, with ± yellowish legs and bill; dips and teeters constantly when on the ground; flies with wings stiffly downcurved. Call a high clear "peep-peep," usually in flight or when landing. Single or in pairs, along streams and lakes, esp subalpine. Order Charadriiformes (Shorebirds). Color p 500.

Like the dipper, which also feeds in frothy streams, the spotted sandpiper dips from the knees; it distinguishes itself in tipping forward and back, hence its nickname "teeter-tail." When threatened by a hawk it can dive like a dipper as well. In feeding it doesn't dive, but plucks prey from shallow water or the bank nearby, or sometimes from midair. Prey range up to trout fry, though insects predominate. Spotted sandpipers are widespread, breeding in almost any mountains north of Mexico and then wintering on seacoasts where they can count on mild weather. That just barely includes the Washington Coast.

# Band-tailed Pigeon

*Columba fasciata* (co-**lum**-ba: the Roman term; fas-ee-**ay**-ta: banded). 13½"; a large gray pigeon with purplish and whitish hues, broadly fanned gray tail, yellow legs and black-tipped yellow bill, thin white band across nape (adults only). Call a low, owllike "whoo-whoo." Order Columbiformes (Pigeons).

Flocks of band-tailed pigeons rove our Westside forests, and timberline and croplands as well, from April through October, then winter far to the south. Here they gorge on our tasty berries (salal, madroño, elder-, salmon- and blackberries, etc.) and, not coincidentally, become tasty and desirable game themselves. A federal ban was placed on hunting them in 1916, perhaps in fear of repeating the passenger pigeon tragedy, but their numbers recovered somewhat and a short season for them was reinstated. Their habit of laying only one egg per nest kept recovery slow. As many as 50 pairs may nest in one large conifer, with as many again in the next tree over.

---

# Owls

Order Strigiformes.

**Great horned owl,** *Bubo virginianus* (bew-bo: the Roman term; vir-jin-ee-ay-nus: of Virginia). Males 20", ws 55", females larger; large "ear" tufts or "horns," yellow eyes, and reddish tan facial ruff; white throat patch, otherwise finely barred and mottled gray-brown. Identifiable by long, low hoots (4–8 in series) heard year-round, but oftenest in Jan-Feb breeding season. Nocturnal; widespread and common in forests.

**Spotted owl,** *Strix occidentalis* (strix: as in "strident," Greek term imitating screech owl; ox-i-den-tay-lis: western). Males 16", ws 42", females larger; no "ear" tufts; dark eyes, yellow bill; dull brown, white-spotted above and -barred beneath. Strident hoots, somewhat doglike, in series of 3–4. Nocturnal; uncommon, in mature dense conifers.

**Barred owl,** *Strix varia* (vair-ia: variegated). Males 17", ws 44", females larger; very similar to spotted owl, but white flecking on back suggests horizontal bars, and on chest is more a white field with vertical brown flecks. Strident hoots in rhythmic series of 6–9, ± like a dog yelping at the moon. Nocturnal; in forest.

**Northern saw-whet owl,** *Aegolius acadicus* (ee-jo-lius: another Greek term for owl; a-kay-di-cus: of E Canada). Males 7", ws 17", females larger; reddish brown above, white/brown smeared (adults) or golden (juveniles) beneath; yellow eyes, V-shaped white eyebrows. Rarely-heard call is short, monotonous squeaks, like filing on saw teeth, i.e., saw-whetting. Nocturnal, rarely seen; in conifers.

**Northern pygmy owl,** *Glaucidium gnoma* (glaw-sid-ium: small owl, derived from "gleaming"; no-ma: gnome). Males 6"; ws 15" females larger (i.e. barely robin-sized, but clearly owl-shaped); brown with slight pale barring, dark eyelike pair of spots on nape; eyes yellow; longish tail often held cocked; flight swift, darting, audible, with rapid beats—qualities atypical of owl flight. Call like whistling over a bottle (± like a squirrel). Primarily diurnal; ± open forest.

Owls have universally evoked human dread, superstition, and tall tales with their ghostly voices and silent, nocturnal predatory flight. Various parts of their anatomies found their way into countless talismans and potions both medical and magical. Owl pellets seem almost ready-made talismans, while they provide naturalists with clear information on owl diets and distribution. Pellets are strikingly neat oblong bundles coughed up by owls to rid them of indigestible parts of small prey (and anything else) they swallow; hard, angular parts are

**Great horned owl**

smoothly coated with fur. A spot with many pellets suggests an owl's roost on a limb above. (Hawks make similar but smaller and fewer pellets, eating their prey in smaller pieces and digesting it more completely.)

While all owls are formidable hunters, evolution has specialized them for a broad spectrum of habitats and roles, roughly paralleling the specializations of hawks and eagles. The ferocious little pygmy owl darts about catching insects and also birds up to and greater than its own size. Like some 40% of owl species, it hunts mainly by day. The nocturnal great horned owl can hunt mammals much heavier than itself, such as porcupines and large skunks, as well as almost any sort of creature down to beetles and worms. Where common, cottontails are its chief prey.

The most striking things about owls are their sensory adaptations. They have the broadest skulls of any birds, separating their eyes and their ears as widely as possible to maximize three-dimensionality of vision and directionality of hearing. They hunt about as much by hearing as by sight; in experiments, barn owls easily locate and catch mice by hearing alone, in absolute darkness. The facial ruff of feathers, plus ear flaps hidden under its outer rim, funnel sound to the highly developed inner ears. (The "ear tufts" or "horns" on top of some species' heads are unconnected with hearing, but serve expressive and decorative functions.) Owls' eyes are the most frontally directed of any bird's; this narrows the field of vision but makes nearly all of it 3-D. The bill is squashed down out of the way for the same reason. (Try straining your eyes left, then right, to see the translucent profile of your nose on either side; only that portion—about a third—of your field of view lying between the "two noses" is seen in 3-D, by both eyes.) The owl's adaptive tradeoff—narrower field of view than other birds', but all of it three-dimensional—favors zeroing in on prey, not watching out for predators. To look around, an owl can twist its neck in a split second to anywhere within·a 270° arc. The eyeballs, which don't rotate in their sockets (the neck does it all), evolved an optimal light-gathering shape—somewhat conical, like a deep television tube, with a thick powerful lens. The retina has a reflective backing (the kind that makes nocturnal mammals' eyes gleam in your headlights) behind the photoreceptor cells. These are almost all rods (high-sensitivity vision) and few cones (color vision). Owls' light perception threshold is between a thirty-fifth and a hundredth of ours. They have much sharper acuity for detail than we do, even by day, and a modicum of color perception as well.

Most owls practice utterly silent flight, a magical thing to witness. Their feathers are literally muffled, or damped, with a velvety surface and soft-fringed edges, costing some efficiency and speed as a tradeoff. Their flapping is slow and easy, thanks to low body weight per wing area. (With extra fluffy body plumage, owls are far slimmer and lighter than they look.) Silent flight enables owls to hear the movements of small rodents while in flight, while in turn it keeps the sharp-eared prey in the dark over someone coming for dinner. This tempts one to fancy that mice know owls only as invisible agents of disappearances from the family.

Saw-whet and spotted owls depend on camouflage rather than escape when approached by large creatures; people alert enough to spot them can approach quite close. Spotted and great horned owlets seem to attempt flight from the nest before they are ready; landing on the forest floor, there they must reside until fully fledged. The parents continue to feed and guard them—aggressively in the case of the great horned.

If you find an adorable, fluffy owlet on the ground, **back off.**

The spotted owl, though never commonly seen any-where, is currently a media star and a pawn in political chess, because it needs extensive old-growth forest for habitat. (This is doubtless true of many species, but best-proven of this one.) The species will almost certainly die out if all old-growth for-ests outside of present parks and Wilderness Areas are cut be-fore others mature to replace them. The Endangered Species Act enables environmentalists to use the spotted owl as a legal tool to protect old-growth. Much data suggests that the acre-age of old-growth needed for the species survival is equal to or even greater than what now exists. High government officials have replied that, at $300,000 a year per bird in lumber value foregone, the Forest Service cannot afford spotted owls. A wide range of estimates are quoted by partisans of one side or the other as to how many spotted owls survive today, how

## Raptors

The hawk and owl orders (Falconiformes and Strigiformes) are spoken of collectively as "raptors." Coming from a Latin root meaning "snatcher," raptor could perhaps mean any predator, but ornithologists don't apply it to the many predatory birds— mergansers, herons, kingfishers, swifts, shrikes, ravens, etc.— classed in other orders, while they do apply it to vultures, who aren't predators at all. "Raptor" is just useful jargon, not a tax-onomic grouping; i.e., the two orders are not closely related. Their common traits evolved "convergently" in two evolutionary lines separately specializing in predation.

The dramatically convergent traits of raptors are heavy, hooked bills; large, heavily muscled feet; and females larger than males. The connection of the first two to the predatory life is clear, but that of the third is conjectural. The fact that hawks and owls separately evolved larger females, while equal or larger males are almost universal in the other bird orders, argues strongly that the trait must offer a significant advantage to birds of prey. It does not prevail among mammals, which have larger males with the exception of hyenas (predators), nor among lower animals which, predatory or not, more often have larger females.

many are needed for the species to survive, how many hundreds of acres each pair requires, how much use they can make of second-growth as part of their habitat, how widely separated their parcels of old-growth can be, and so on. Further research should provide better data on these issues and, most important, better establish that it's not just spotted owls, but a whole, almost inconceivably intricate ecosystem, and the productivity of the lumber and fishery industries themselves, that are endangered by disregard for the value of extensive old-growth.

Larger owls are involved in the threat to the spotted owl. Its range in the Northwest is being invaded by its close cousin, the barred owl, which is more aggressive, bears larger broods, and seems to require less territory. Barred owls were virtually unknown in our range until the 1980s, but are spreading with astonishing speed. One hypothesis is that the replacement of old-growth with clearings and second-growth enlarges the range of great horned owls, which prey upon spotted owls, while barred owls are able to occupy nearly the same niche as spotted owls, and are just large enough to stand their ground against great horned owls.

## Common Nighthawk

*Chordeiles minor* (cor-die-leez: evening dance; minor: lesser, a false name since this is now the larger species of nighthawk). 9″, ws 23″; wings long, bent backward, pointed, falconlike (see p 382); mottled brown/black, with white wing-bar; males also have white throat and a narrow white bar across tail. Marshes and ± open areas, all elevs. Order Caprimulgiformes (Nightjars). Color p 500.

Nighthawks are most often seen at dusk; no self-respecting nighthawk would be on the ground at that hour, when insects are on the wing. Like swifts and swallows, they prey on insects by flying around with their mouths open. Their flight is wild and erratic like a bat's, but swift like a falcon's—though they aren't really any kind of hawk at all. Males may interrupt their erratic feeding flights with long steep dives that bottom out abruptly with a terrific raspy, farting noise of air rushing through the wing feathers. This is their courtship. The vocal call is a softer, nasal beep repeated while feeding. Nighthawks are our briefest breeding visitors, here only from June through early September. They nest on gravelly ground with very little construction work, and they have an odd style of perching, lying lengthwise along a branch.

## Swifts

Family Apodidae, Order Apodiformes (Swifts and hummingbirds).

**Vaux's swift** ("Voh's"), *Chaetura vauxi* (key-too-ra: hair-thin tail; voh-zigh: after William S. Vaux). 4½"; slaty to brownish gray, somewhat paler beneath; wings long, pointed, gently curved; tail short, rounded. Rapid chipping. Widespread. Color p 499.

**Black swift,** *Cypseloides niger* (sip-sel-oh-eye-deez: like *Cypselus*, the Old-world swift; nye-jer: black). 7"; black all over; wings long, pointed, gently curved; tail slightly forked. Call rarely heard. Subalpine; uncommon (only one breeding site known in Ore Cas, a handful in Wash, the Newhalem vicinity being best known).

True to their name, swifts are probably the world's muscle-power speed champions. One *Chaetura* species was clocked at 106 miles per hour. (Compare that with a cruising speed of 20 for the fastest similar-sized songbirds.) They often fly 600 miles in a day—and make it look playful, interspersing short glides between spurts of flapping, often appearing to flap left and right wings alternately. Unfortunately they do most of this, especially in fair weather, too high up for us to see any charming details.

Swifts can only maintain their high metabolisms on a steady supply of flying insects, and few insects fly during cold weather. So, when it's cold and gray, swifts may fly off for a few days, as far as they need to go to find sun, or else stay home but go cold and torpid (see page 308). If young are in the nest, they remain in torpor while the parents vacation in the sun. People sometimes mistake torpid swifts and nighthawks for dead, and pick them up, only to be startled when they fly off in perfect health as soon as they're warmed up.

In Southeast Asia, swifts' nests are gathered for the famous curative delicacy "birdsnest soup," featuring filmy masses like boiled eggwhites—actually the special gluey saliva swifts work up for constructing nests. Vaux's swifts glue their nests to the insides of hollow trees or chimneys. Their stiff, spine-tipped tail-feathers help them perch on sheer surfaces. Black swifts nest on cliffs, often behind waterfalls.

---

# Hummingbirds

Family Trochilidae, Order Apodiformes (Swifts and hummingbirds).

**Rufous hummingbird,** *Selasphorus rufus* (se-lass-for-us: light-bearer; roo-fus: red). 3½"; adult males red-brown with iridescent red throat patch, some white on belly, green on wings and crown; others mostly green with some reddish near base of tail. Flight produces a deep dragonflylike hum. Our common hummingbird. Color p 500.

**Calliope hummingbird,** *Stellula calliope* (stell-you-la: starlet; ca-lie-o-pee: a Greek muse). 2¾"; bronze-green above, red-tinged white below; throat has long deep purple streaks on adult males, small dark speckles on others. Uncommon; timberline and E-side.

The redeeming virtue, for me, of the garish reds and yellows the camping suppliers sell us is their attractiveness to hummingbirds. My red bootlaces alone draw several hummers a day in the alpine country in July; they pause only an instant to discover my boots are no bed of columbine. I can only hope the pleasure they bring me doesn't cost them much.

The smallest of all birds, hummers have frantic metabolisms in common with shrews, the smallest mammals.* Ounce for ounce, a hummingbird flying has 10 times the caloric requirement of a person running—and we don't have to spend all day running, while hummingbirds must spend most of their waking hours flying around in search of nectar for those calories. Daily, they consume up to half their body weight in sugar. Their whirring flight, suggestive of a huge dragonfly, does in fact work more like insect flight than like that of other birds, and allows stationary and backward hovering, but no gliding. The wings beat many times faster than other birds' thanks to an extremely shortened wing with long feathers; there's little mass to flap, other than the feathers themselves.

*The smaller the warm-blooded body, the more energy is required to keep it warm. Heat is produced in proportion to size (or volume) but is dissipated into the air in proportion to surface area. Reducing the length and width of a body by half, for example, divides its surface area by four (2 × 2) and its volume by eight (2 × 2 × 2), thus doubling the rate of heat loss.

Hummers' hyperkinetic days are complemented by energy-conserving nights, with temperature and metabolism sharply lowered (see page 308).

Hummingbirds are belligerent toward their own and larger species, and the males' courtship displays are assertive in the extreme. A rufous male swoops around and around in an ellipse hundreds of feet high, at the lowest point passing at high speed only inches from the demure object of his attentions, simultaneously eliciting a shrieking noise from his wings. Both sexes may use this sort of display to assert feeding territory. Males flash their "gorgets" (iridescent, erectile throat patches) both in courtship and in aggression.

## Belted Kingfisher

*Ceryle alcyon* (ser-i-lee al-see-on: two Greek terms for kingfishers). 12"; head looks oversized because of large bill and extensive crest of feathers; blue-gray with white neck and underparts and (females only) reddish breast band. Call a long, peculiar rattle. Along streams. Order Coraciiformes (Kingfishers and allies).

The Greeks had a myth that the Halcyon, a kind of bird we presume was a kingfisher, floated its nest on the waves of the sea while incubating and hatching the young. Hence "halcyon days" are a fortuitous respite from the storms of life.

Our kingfishers, in real life, raise their young amid a heap of regurgitated fish bones at the end of a hole in a mud bank. Is "nest" too sweet a term for such debris? They look for their prey—fish, crayfish, waterbugs and larvae—from a perch over a stream or occasionally a lake. (They can also hunt from a hover where branches are in short supply, as is rarely the case on our streams.) After diving and catching a fish, they often return to thrash it to death against their branch before swallowing it headfirst. Fishermen have long resented kingfishers' success rate, but statistically the birds are unlikely to significantly reduce trout numbers. The kingfisher population has plummeted during the advance of civilization; there was once a pair of kingfishers for virtually every creek in the U.S.

# Woodpeckers

Family Picidae, order Piciformes.

*Pileated woodpecker, Dryocopus pileatus* (dry-oc-o-pus: tree sword; pie-lee-ay-tus: crested). 15"; all black exc bright red, large, pointed crest, black/white-streaked head with (males only) red moustache, and white underwing markings visible only in flight; drumming very loud, slow, irregular; call a loud rattling shriek with a slight initial rise in pitch. Deep forest with many standing dead trees or snags.

**Northern flicker,** *Colaptes auratus* (co-lap-teez: pecker; aw-ray-tus: golden). 11"; gray-brown with black-spotted paler belly, black-barred back, black "bib," and (on our subspecies, the red-shafted flicker) reddish crown, moustache (males only) and underside of wings and tail. Varied calls include a flat-pitched rattle. Semiopen habitats. Color p 500.

**Hairy woodpecker,** *Picoides villosus* (pic-o-eye-deez: from the Roman term; vil-oh-sus: woolly). 7½"; black and white exc (males only) a small red patch on peak of head; wings strongly barred. Widespread.

**Downy woodpecker,** *Picoides pubescens* (pew-bes-enz: fuzzy). 5¾"; like Hairy only smaller. Less common: ± restricted to lowlands with some deciduous trees. Color p 499.

**White-headed woodpecker,** *Picoides albolarvatus* (al-bo-lar-vay-tus: white mask). 7¾"; black all over exc white head, throat, and patches near tip of wings; male has small red patch on nape. Ponderosa pine forest, where locally common though inconspicuous; walks head-down or sideways on tree trunks.

Woodpeckers are the one kind of "outsiders" that often flocks with birds of the perching bird order (Passeriformes); woodpeckers resemble perching birds, but differ in a number of specialized traits. Most have two front and two rear toes on each foot, rather than the 3-and-1 arrangement of perching toes. Strong sharp claws on 2-and-2-toed feet, plus short stiff tail feathers, give them the grip and the bracing they need for hammering the full force of their bodies into a tree. Naturally, they also have adamantine chisellike beaks, and thick shock-absorbing skulls to prevent boxers' dementia. Before diving into work on a tree, they listen for the minute rustlings of their insect prey boring around under the bark. Insect larvae and adults are 90% of their diet, seeds and berries the remainder.

For snatching grubs out of their tunnels, the woodpecker has a barb-tipped tongue much longer than its head. The tongue shoots out and then pulls back into a tiny tubular cavity looping around the circumference of the skull. Our one woodpecker exception is the white-headed; it subsists mainly on pine seeds, and forages for insects merely by prying bark up—reserving its drilling skills for excavating a nest.

Females are smaller and thinner-billed than males. They excavate less than males, foraging often by prying bark up, and so exploit a slightly different resource than their mates. The males apply their prodigious pecking to two additional vital tasks. First, they chop large squarish holes for their nests, padding them with a few of the chips. Such nests are dry and easy to defend, allowing prolonged rearing—advantages that have led several other species to depend on abandoned woodpecker holes for their nests. Second (white-headed excepted), they drum on resonant trees as a mating and territorial "call." Not that either sex lacks a voice—flickers vocalize diversely and often; downies and white-headeds rarely.

The pileated, our largest woodpecker, has an especially strident and impressive call like a harsh, maniacal laugh, which it seems to use when disturbed—such as whenever you or I come around. Once you know the pileated's "laugh" you'll know, when you hear it, to pause and look for a glimpse of one of our flashiest birds. Each pileated requires a great many standing dead trees, so the species dwindled as the old-growth forests vanished from most of the continent. Our lower Westside forests are among their best remaining habitats. Even our cities are acceptable, wherever there are enough snags; a few pileateds have become pests, chopping away at house and phone-pole timbers.

# Olive-sided Flycatcher

*Contopus borealis* (cont-o-pus: short foot; bor-ee-ay-lis: of Northern forests). 6¼"; olive-gray with pearly white smear down throat and breast; white downy tufts on lower back visible esp in flight. Order Passeriformes (Perching birds).

The olive-sided's song, usually asserted from some conifer pinnacle, has been written down as "THREE cheers," "Quick! ... FREE beer," or "Tuck ... THREE bears." These flycatchers summer throughout our forests, clearcuts, and timberlines. They eat winged insects, spotting individuals from their perch and darting out to snatch them. In contrast to the constantly flying grab-bag method of swifts and swallows, which nets small insects, flycatchers eat medium-sized insects. They are reportedly fond of bees, but do not nest near beehives.

# Western Flycatcher

*Empidonax difficilis* (em-pid-o-nax: gnat king; dif-iss-il-iss: difficult). 5"; olive brown above with 2 whitish wing-bars; pale yellowish olive beneath; white eye-ring, two-toned bill. Usually in deciduous understory. Order Passeriformes (Perching birds). Color p 500.

The specific name *difficilis* means that it's next to impossible to identify *Empidonax* birds in the bush. They are hard to see, staying camouflaged among foliage. This one has a recognizable call, though—a very high, rising "pseeeet." The song is similar but much repeated, in threes, with slight variations.

# Swallows

Family Hirundinidae, Order Passeriformes (Perching birds).

**Violet-green swallow,** *Tachycineta thalassina* (tacky-sin-ee-ta: fast moving; tha-lass-in-a: sea green). 4¾"; dark with green/violet iridescence above; white below, extending around eyes and up sides of rump to show as two white spots when seen from above; wings back-swept and pointed, tail shallowly forked. Various high tweets. Open areas; in flocks.

**Tree swallow,** *T. bicolor** (by-color: 2-colored). 5"; iridescent blue-black above, white below (not extending above eye or rump); otherwise like violet-green swallow. Color p 501.

*Formerly in a separate genus as *Iridoprocne bicolor*.

**Cliff swallow,** *Hirundo pyrrhonota*\* (her-un-doe: the Roman term; per-o-no-ta: red back). 5"; blackish wings and tail, whitish below, with cinnamon-red cheeks and paler rust-red rump and forehead; wings backswept and pointed, tail square to barely notched.

A swallow's flight is graceful, slick, and fast, though not quite as swift as a swift's. A swallow's swallow, or more precisely its gape, is striking, and important to the bird's sustenance. The wide, weak jaws are held open almost 180° while the swallow knifes back and forth through the air intercepting insects. The colonial nesting habit is suited to the swarming habit of the insect prey. Cliff swallows build striking gourd-shaped mud nests on cliffs or under bridges or eaves; *Tachycineta* swallows use ready-made crevices and tree holes.

Gray jay

## Gray Jay

*Perisoreus canadensis* (pair-i-sor-ee-us: "I-heap-up"). Also **Canada jay, camp-robber, whiskey-jack.** 10"; fluffy pale gray with dark brownish gray wings, tail, rump and (variably) nape and crown; short, ± black bill; juveniles all dark gray exc pale cheek-streak. Calls a whistled "Whee-oh," and various others. Deep forest and timberline. Order Passeriformes (Perching Birds).

Like their close relatives the crows, jays seem coarse and vulgar but are intelligent, versatile and successful. Their voices are harsh and noisy, but capable of extreme variety and accurate mimicry of other birds. Equally versatile in feeding, ours eat mainly conifer seeds, berries and insects, but also relish meat when they can scrounge some up or kill a small bird or rodent. Gray jays, or "camp-robbers," were known for thronging around logging-camp mess halls; a trapper would sometimes find them nipping at a carcass while he was still skinning it. You will find them just as interested in your lunch,

\*Formerly in a separate genus as *Petrochelidon pyrrhonota*.

and bold enough to snatch food from your fingertips. Their nonchalance reminds skiers that winter here is perfectly livable for the well-adapted. Gray jays and Clark's nutcrackers are so comfortable in the cold that they nest and incubate their young with plenty of snow still on the ground—or even on their heads. To store food for winter, the jays glue little seed bundles with sticky saliva, and leave them in bark crevices or foliage.

**Steller's jay**

**Clark's nutcracker**

## Steller's Jay

*Cyanocitta stelleri* (sigh-an-o-sit-a: blue jay; stell-er-eye: after Georg Steller, see below). 11"; deep ultramarine blue with ± black shoulders and dramatically crested head. Widespread. Order Passeriformes (Perching birds).

Our Western "blue jay's" repertoire of noisy calls includes a near-perfect imitation of a red-tailed hawk's scream. This presumably deceives other birds into clearing out while the jay feeds, but perhaps it sometimes warns of an actual hawk. Jays

**Georg Steller** was the first European naturalist-explorer on northeast Pacific shores. A German, he crossed Siberia to accompany Danish Captain Vitus Bering on a Russian ship built and launched from Kamchatka in 1741. They were just east of Cordova, Alaska when they turned around late that year, and they didn't quite make it back to Kamchatka. Marooned for a terrible scurvy-ridden winter on small, rocky Bering Island, the Captain and many of the crew died. Steller survived, but took to drink and died in Siberia without ever reaching Europe again.

and crows are the birds most often seen harassing or "mob-bing" birds of prey—an aggressive group defense against larg-er birds. The smaller, paler Blue Jay proper is a rare visitor here from the Eastern States. Westerners sometimes say "blue jay" either for Steller's or for the scrub jay, *Aphelocoma coerulescens*, a crestless jay, blue above and white beneath, seen increasing-ly on dry slopes in and near the Willamette Valley.

## Clark's Nutcracker

*Nucifraga columbiana* (new-sif-ra-ga: nut break; co-lum-be-ay-na: after the river). 11"; pale gray (incl crown and back) with white-marked black wings and tail (white outer tail-feathers and rear wing-patch); long thin grayish bill. Call a harsh "kraaa, kraaa." Mostly in subalpine habitats with whitebark or lodgepole pines. Order Passeri-formes (Perching birds).

Flashy black/white wings and tail distinguish Clark's nut-cracker from the more numerous gray jay, or "camp robber." The nutcracker has also been known as "camp robber," but mainly in California where there are none of the nervy jays that better deserve the name.

Clark's nutcrackers and whitebark pines provide a fine example of coevolution. While many conifers hide and protect their seeds from seed-eaters as long as possible, the whitebark pine exposes them early and conspicuously, and baits them with rich oils and proteins. It also fails to provide them with membranous wings (found on most pine seeds) to help the wind carry them. Clark's nutcracker obliges by collecting the seeds compulsively, selecting the best ones (most nutritious and most viable), eating some on the spot and burying the rest, a few at a time, for retrieval in winter.

Though canny enough to relocate caches even after snow has come and gone, the birds are so industrious that they store two or three times as many as they can consume. The remain-der are thus planted, often many miles from their source—a great advantage to the pines. The strong-flying nutcrackers have a typical foraging range of 14 miles. They can carry 150 whitebark pine seeds in sublingual pouches that bulge con-spicuously at the throat. Lodgepole pine seeds are also taken. Poor pine seed crop years cause irruptions of nutcrackers into far-flung lowlands in search of substitute fare.

## American Crow

*Corvus brachyrhynchos* (cor-vus: raven; brak-i-**rink**-os: short beak). 17"; black with blue-purple-green iridescence; tail squared-off; rarely glides more than 2–3 seconds at a time. Lowlands. Order Passeriformes (Perching birds).

Most of the crowlike birds you see in our mountain wilderness are ravens; crows characteristically inhabit farmlands, and avoid both deep forest and cliffy terrain. You are most likely to see them here along lowland streams and meadows, or in part-deciduous forest. They scavenge carrion, garbage, fruit, snails, grubs, insects, frogs, and eggs and nestlings from birds' nests. Species as large as ducks, gulls, and falcons may be victims, and some songbirds count crows as major enemies.

Common raven

## Common Raven

*Corvus corax* (cor-vus: the Roman term for raven, from cor-ax: the Greek term, imitating raven's voice). 21", ws 48"; black with purple-green iridescence, and shaggy grayish ruff at throat; bill heavy; tail long, flared then tapered; alternating periods of flapping and flat-winged soaring. Call a throaty croak. Subalpine and E-side (lower) timberlines. Order Passeriformes (Perching birds).

You can tell Raven from his little brother Crow by his larger size, heavier bill, ruffed throat, more prolonged soaring, and by his voice—an outrageously hoarse and guttural croak rather than a familiar "caw." While both species stick to open country, at timberline in our mountains and near the Eastside steppes you are more likely to see ravens. Their courtship flights in spring are a sight to remember; they do barrel rolls or chaotic tumbles while plummeting, then swoop and hang motionless, all the while exercizing their vocabularies. The ensuing family group—four or five young are typically flying by June—often stays together through the following winter, and may flock with other families. The nest is high, solid, and cozy, preferably lined with deer hair, but filthy and smelly by our standards, and often flea-ridden. Ravens eat anything crows eat and then some; they can take live prey as large as hares by grouping up to harry them. I once watched a raven rob a gadwall (duck) nest to bring an egg home and stuff it, whole, down the throat of a fledgling.

Ravens were once abundant throughout the West, but white men's westward expansion pushed them back to the remote deserts and mountains. Just as the wolf was more vulnerable than the smaller coyote, the raven was more vulnerable than the crow. In raven's case there were especially feeble grounds for supposing a conflict with agriculture.

Ravens were accepted members of tribal life in Northwest Coast villages. Raven was seen as at once a trickster and a powerful, aggressive, chiefly figure—much the equivalent of Coyote among intermountain tribes. He was a Creator in origin myths. His croaks were prophetic, and a person who could interpret them would become a great seer. To inculcate prophetic powers in a chosen newborn, the Kwakiutl would feed its afterbirth to the ravens. A mythic view of ravens and crows as powerful, knowing, and somewhat sinister is virtually universal worldwide. French peasants thought that bad priests became ravens, and bad nuns, crows.

The raven is the largest of all Perching birds* (Passeriformes). They are the most recently evolved avian order, with by far the greatest number of species, an indicator of current success and rapidity of evolution. Whether crows or sparrows are the newest family within the order is a matter of long

---

*Perching birds are also commonly known as "songbirds." Both perching and singing are characteristic of this order but can be found, depending on how you define them, in many other birds as well. Many naturalists prefer to use "perching bird" because the term refers to an anatomical structure—the foot with one long rear toe opposing three front toes—that is more clearly defined than "song."

Anglo-American dispute. The rapid evolution and success of seed plants was accompanied by a huge development of specialized seed-eaters—mainly perching birds and rodents—and a wave of predators specializing in eating the seed-eaters. The raven rides the crest of both waves—eating rodents, nestlings, and eggs, as well as seeds. By some accounts it is the most advanced product of bird evolution.

**Black-capped chickadee**

## Chickadees

*Parus* spp. (pair-us: the Roman term). Order Passeriformes (Perching birds).

**Chestnut-backed chickadee,** *P. rufescens* (roo-fes-enz: reddish). 4¼"; black crown and throat, white cheeks and belly, rich red-brown back and sides. Abundant year-round in W-side forests.

**Mountain chickadee,** *P. gambeli* (gam-bel-eye: after William Gambel). 4¼"; black crown, eye-streak and throat; white eyebrow and cheeks, grayish belly, gray-brown upperparts. Subalpine and E Cas slope.

**Black-capped chickadee,** *P. atricapillus* (ay-tri-ca-pill-us: black cap). 4½"; black crown and throat, white cheeks; grayish belly, gray-brown upper parts. Lower thickets, streamsides.

The various chickadees, gleaners of insects from the branches, reside here year-round, nesting in fur-lined holes dug rather low in tree trunks either by woodpeckers or by themselves, in soft punky wood. When, rarely, their respective habitats overlap, they tend to be separated vertically—chestnut-backed feeding in the canopy, black-capped in the shrubs. "Chick-a dee-dee-dee" transcribes the most characteristic of their calls; their song (mountain and black-capped only) is more relaxed, on two notes descending. Chickadees seem to epitomize the chipper dispositions people want to see in songbirds, and they let us see it, being tamer than most.

# Nuthatches

*Sitta* spp. (sit-a: the Roman term). Order Passeriformes (Perching birds).

**Red-breasted nuthatch,** *S. canadensis*. 4"; blue-gray above, pale reddish below, with white throat and eyebrow, black (males) or dark gray crown and eye-streak. Widespread in forest, esp W-side. Color p 501.

**White-breasted nuthatch,** *S. carolinensis* (carol-in-en-sis: of the Carolinas). 5"; blue-gray above, white below (may have a little red under tail), with black (males) or gray crown, but all-white face. Subalpine and E-side pine forest.

Nuthatches are known for walking headfirst down tree trunks, apparently finding that way just as rightside up as the other. They glean insects from the bark, and eat seeds. They nest in dead snags, quite inconspicuous but identifiable by their odd habit of smearing pitch around their hole. Even in deep wilderness they aren't shy, and draw our attention with their penetrating little call, a tinny "ank" or "nyank." They migrate downslope some winters, down south for others—generally when cone crops here are poor.

# Brown Creeper

*Certhia americana* (serth-ia: the Greek term). 4¾"; mottled brown above, white beneath; long downcurved bill. Call a single very high, soft sibilant note. On tree trunks, year-round, widespread and common but very inconspicuous. Order Passeriformes (Perching birds). Color p 502.

You probably won't notice a brown creeper unless you happen to catch its faint, high call, and then patiently let your eyes scour nearby bark. This is a well-camouflaged full-time bark dweller, gleaning insect prey there and nesting behind loose bark. In contrast to nuthatches, which usually walk down tree trunks, the creeper spirals up them, propping itself with stiff tail feathers like a woodpecker. These opposite approaches to gleaning suggest that different food resources are exploited.

---

# Winter Wren

*Troglodytes troglodytes* (tra-glod-i-teez: cave dweller, a misleading name). 3¼"; finely barred reddish-brown all over; tail rounded, very short, often (as in all wrens) held upturned at 90° to line of back. Near the ground in forest with dense herb or low shrub layer; year-round. Order Passeriformes (Perching birds).

The winter wren is conspicuous mainly by its song, a prolonged, varied, often-repeated, virtuoso sequence of high trills and chatters. It moves in a darting, mouselike manner, eats insects, maintains a low profile among the brush, and goes to great lengths to keep its nest a secret. Several extra nests are often built just as decoys, and the real occupied nest has a decoy entrance, much larger than the real entrance but strictly dead-end. Real and decoy nests are camouflaged to boot.

In Europe, this species (the only wren native there) has long been familiar around cities and towns, in sharp contrast to its insistence on undisturbed habitat here. Apparently, adaptation to civilization is possible even for species that resist it for dozens of generations.

# American Dipper

*Cinclus mexicanus* (sink-lus: the Greek term). Also **water ouzel**. 5¾"; slate gray all over, scarcely paler beneath, often with white eye-ring; tail short; feet yellow. In or very close to cold mtn streams. Order Passeriformes (Perching birds). Color p 500.

It's no wonder that dippers make such an impression on campers throughout America's Western mountains, considering how much trouble campers have keeping warm. Winter and summer, snow, rain, or shine, dippers spend most of their time plunging in and out of frigid, frothing torrents, plucking out invisible objects—actually aquatic insects such as dragonfly and caddis fly larvae, and sometimes tiny fish. Somehow they walk on the bottom, gripping with their big feet; they can also swim with their wings, quickly reverting to flight if they get swept out of control downstream. They are known to dive to considerable depths in mountain lakes, and occasionally they forage on snowfields. They show little interest in drying off. Even in flight they are usually in the spray zone over a stream, and they often nest behind waterfalls. They never really get soaked to the skin, thanks to extremely dense body plumage and extra glands to keep it well oiled.

No, "dipper" implies no shyness toward water, not in these birds. The name refers to their odd, jerky genuflections repeated as often as once a second while standing, and accompanied by blinking of their flashy white eyelids. Their call, "dipit dipit," is forceful enough to carry over the din of the creek. In midwinter they often break into lyrical, extended song.

## Kinglets

*Regulus* spp. (reg-you-lus: small king). Order Passeriformes (Perching birds).

**Golden-crowned kinglet,** *R. satrapa* (sat-ra-pa: ruler). 3½"; gray-green above, whitish below, with 2 white wing-bars; central yellow (female) or orange (male) stripe on head is flanked by black and then white stripes at eyebrow. Very high, lisping "chee, chee" call. Widespread, in conifer canopy.

**Ruby-crowned kinglet,** *R. calendula* (ca-lend-you-la: small glow). 3¾"; gray-green above, whitish below, with white eye-ring and 2 white wing-bars; rarely-visible scarlet spot on crown is displayed only by excited males. Scolding "jit-it" calls, and long, variable song of chatters, warbles, and rising triplets. Mainly winter and spring, in brush and lower canopy of W-side clearings.

Constant movement—wings twitching even when perched—characterizes kinglets. They catch insects, sometimes in flight but mostly on bark and foliage, where their tiny size enables them to forage on twigs too weak to support other gleaning birds. Gleaning is a full-time job; insects are a less concentrated energy source than seeds, and occur less predictably. Traveling in flocks seems to help gleaners locate insect populations. Mixed flocks of kinglets, chickadees and woodpeckers are often seen during the nonbreeding seasons—by cross-country skiers, for example.

# Bluebirds

*Sialia* spp. (sigh-ay-lia: the Greek term). Order Passeriformes (Perching birds).

**Mountain bluebird, *S. currucoides*** (cue-roo-co-eye-deez: warbler-like). 6"; summer males turquoise above, shading through pale blue beneath to whitish on throat; females and winter males gray-brown with varying amounts of blue on tail, rump, and wings. Soft warbling song at dawn; "phew" call. Alp/subalpine in summer, and in open pine forest year round. Color p 502.

**Western bluebird, *S. mexicana*.** 5½"; summer males deep blue above, rich red-brown breast and backband, whitish belly; females similar but much duller, with dark bluish-gray head. Single, soft "phew" calls. Semiopen lowlands; uncommon.

Bluebirds eat 10% berries, 90% terrestrial insects which they drop on from a low hover or perch. They nest in woodpecker holes; the severe decline of the Western bluebird is partly caused by introduced sparrows and starlings competing for nest holes.

# Townsend's Solitaire

*Myadestes townsendi* (my-a-des-teez: fly eater; town-send-eye: after J. K. Townsend, p 313). 6¾"; gray with white eye-ring; dark tail has white feathers on sides (like the more abundant junco, p 414; solitaire is longer, slenderer, more upright, more arboreal); dark wings have buff patches, visible underneath in flight. Call a high, ringing "eep"; song a long melodious warble heard at any season. Conifer forest and timberline. Order Passeriformes (Perching birds). Color p 500.

The solitaire returns to the mountains early, searching the first snow-free areas for a nesting cavity in a stump or rotting log. It likes the edges of clearings, and seems to be on the increase in the Cascades as this habitat type proliferates in the form of patch clearcuts. After the breeding season it may gather in large flocks, belying its name.

# Swainson's and Hermit Thrushes

*Catharus* spp. (cath-a-rus: pure, referring to the songs). 6¼"; gray-brown above; pale eye-ring; belly white, breast spotted. Order Passeriformes (Perching birds).

**Swainson's thrush,** *C. ustulatus* (ust-you-lay-tus: singed). Back and head ± reddish; tail less so. In various partly open lowland habitats.

**Hermit thrush,** *C. guttatus* (ga-tay-tus: spotted). Tail, but not back, is rusty red; tail is "nervously" raised and lowered every few seconds, while wings may twitch. Mainly subalpine. Color p 501.

Thrushes of this genus have some of the most lyrical and virtuosic of songs, spiraling flutelike up and down through the scale, and often repeated at different pitches. Hermit prefaces the fast phrase with a single long clear note; Swainson's with a slow phrase of two to four notes. Both forage on the ground for earthworms, insects, and berries.

**Varied thrush**

## Varied Thrush

*Ixoreus naevius* (ix-or-ius: mistletoe mountain, referring to food and habitat; neev-ius: spotted). 8"; breast, throat, eyebrow and wingbars rich rusty-orange (males) or yellow-buff (female), contrasting with slate-gray breastband, cheeks, crown, back, etc; whitish belly; perches or walks with body more horizontal than the similar robin. W-side forest. Order Passeriformes (Perching birds).

The varied thrush sings a single note with odd, rough overtones, like two slightly dissonant notes at once; after several seconds' rest, it sings another tone, similar but higher or lower by some irrational interval. Prolonged early or late in the day, in deep forest or fog, this minimal music acquires powers of enchantment over people. The thrush itself is close cousin to a robin, living on worms, larvae, and a few berries. It migrates vertically quite early in the season—upslope typically in late April, and down in late August.

# Hutton's and Solitary Vireos

*Vireo* spp. (veer-ee-oh: green). Greenish-gray above with 2 white wing-bars, paler beneath. Order Passeriformes (Perching birds).

**Hutton's vireo,** *V. huttoni* (hut-un-eye: after William Hutton). 4"; slight eye-ring and white area in front of eye; underside pale olive. 2-note call repeated metronomically, the second note higher and accented. Year-round, W-side part-deciduous forest.

**Solitary vireo,** *V. solitarius*. 4¾"; heavy white eye-ring and line in front of eye; flanks yellowish, belly white. Varied, low 2–3-note phrases at 2–4-second intervals, second note accented. Forest understory, in summer. Color p 500.

Female and male vireos take turns warming the eggs, and won't leave them even if approached and handled; they're oddly sluggish birds. The nest is a small cup of grasses lined with lichens, moss, or feathers, often decorated outside with bark, petals or catkins held on with spiderwebbing. It dangles from a fork in horizontal branches not too high up in a conifer.

# Warblers

Subfamily Parulinae, Order Passeriformes (Perching birds).

**Yellow-rumped warbler,*** *Dendroica coronata* (den-droy-ca: tree house; cor-o-**nay**-ta: crowned). 4¾"; yellow in 5 small patches (ranging to mere tinges on females): crown, rump, throat, sides; mostly gray/black (breeding males) to soft gray-brown (others), with white eye-ring and 1 ± vague wing bar. Song a long trill. Widespread. Color p 501.

**Townsend's warbler,** *D. townsendi* (town-send-eye: after J. K. Townsend, p 313). 4¼"; bright yellow breast and sides of face (surrounding a dark cheek patch); whitish beneath, ± greenish gray above, with 2 white wing-bars. Song a series of wheezes rising to 1 or 2 clear notes. Mid-elev to high forest.

**Hermit warbler,** *D. occidentalis* (ox-i-den-tay-lis: western). 4¼"; bright yellow over almost entire head; black throat "bib" on breeding males; very white beneath, ± greenish gray above with 2 white wing bars. Song of 5–7 slurred notes, ± wheezy but bright. Low to mid-elev forest.

*The 1983 O.A.U. checklist lumped our common race, Audubon's Warbler, hitherto *D. auduboni*, with the generally eastern Myrtle Warbler, *D. coronata*, whose throat is white, to comprise the Yellow-rumped Warbler.

**MacGillivray's warbler,** *Oporornis tolmiei* (op-or-or-nis: autumn bird; tole-me-eye: after William Tolmie, p 198). 4½"; yellow beneath, grayish olive above; solid gray "hood" (head and throat) with incomplete white eye-ring. Short song of about 3 rising and 2 falling notes. All elevs, typically in dense shrub growth.

Sparrow-sized birds around here with some yellow on them are generally some kind of warbler. This large subfamily is known for long winter migrations to the tropics and for distinctive (but not always warbling) songs. Most warblers also have "chip chip" calls. Birders concentrate on learning the songs, since there are so many kinds of warblers and they all tend to keep themselves inconspicuous among foliage. Most are gleaners of insects; many also hawk at larger insects, or vary their diets with fruits and seeds.

## Western Tanager

*Piranga ludoviciana* (pir-ang-ga: Tupi tribal term; loo-do-viss-i-ay-na: of the Louisiana Purchase area). 6¼"; summer males have bright red to orange head, yellow breast, belly and rump, black backband, tail and wings, and white wing-bars; others yellowish to greenish gray above, yellow beneath. Generally near treetops in ± open forest, in summer. Order Passeriformes (Perching birds).

Lewis and Clark described many new plant species, but only three new birds: Lewis' woodpecker, Clark's nutcracker, and the western tanager, whose scientific name honors their voyage in exploration of the Louisiana Purchase. "Tanager" and "Piranga" are native names for the birds from deep in the Amazon rain forest, where some tanagers winter. This particular species travels only as far as Central America. Its breeding plumage here rivals that of gaudy jungle birds, but when in the jungle it wears dull winter plumage.

Tanager beaks are intermediate between the insect-picking thin beaks of the preceding birds (pages 406–413) and the heavy, seed-crushing beaks of birds to follow; tanagers switch from an insect diet to one of ripe berries in late summer.

# Song Sparrow

*Melospiza melodia* (mel-o-spy-za: song finch). 5½"; brown with blackish streaking above, white below with brown streaking convergent at throat, above a mid-breast brown spot; pumps its tail in flight. Widespread year-round, esp in thickets. Order Passeriformes (Perching birds). Color p 501.

"Sparrow" is a catchall term for a lot of common, drab brown birds which few beginners care to identify. They aren't really a taxonomic group—less so with each new revision of bird taxonomy. This species, as its name suggests, is a melodious sparrow, and is one of the most widely distributed birds in America. Its typical song—heard on spring and early summer mornings—is a few clear piping notes, then a lower, raspy buzz or series, ending with around three quick, clear but unemphatic notes.

# Dark-eyed Junco

*Junco hyemalis* (junk-oh: rush, a plant with no obvious connection; hi-em-ay-lis: of winter). Formerly **Oregon junco.*** 5¼"; tail dark gray-brown except for white feathers at sides; belly white, throat and above variably gray and brown. Song a simple, hard trill. Ground-foraging and -nesting; probably the most abundant bird in our mtns. Order Passeriformes (Perching birds). Color p 501.

Though juncos migrate, we have them year-round. Many migrate upslope from northwest cities and farms in early summer while others arrive from California, and still others, having wintered here, leave for the Yukon to breed. They are primarily seed-eaters, turning to insects in summer and feeding insects and larvae to their young. After the young leave the nest, juncos travel in loose flocks until the next summer.

*The 1983 O.A.U. Checklist lumped our commoner form, hitherto *J. oreganus,* with the Slate-colored Junco, *J. hyemalis,* to comprise the Dark-eyed Junco.

# Rosy Finch

*Leucosticte arctoa* (lew-co-**stick**-tee: white patch; arc-**toe**-a: arctic). Formerly **gray-crowned rosy finch.*** 6¼"; brown with blackish face, gray crown or entire head, and (males esp) reddish tinges on rump, shoulders, belly. Either hoarse or high chips and chipping chatters. Arctic fellfields, snow, etc.; not common. Order Passeriformes (Perching birds).

The finch family name, Fringillidae, comes from the same Latin root as "frigid." Rosy finches are the most purely alpine birds of the West, at least in summer; they have been seen as high as 11,000' on Mt. Rainier. They nest in high rock crevices, reportedly even in ice crevasses, and often forage on glaciers, utilizing a resource all climbers have noticed—insects that collapse on the snow, numbed by cold after being carried astray by diurnal upvalley winds. The bulk of their diet is vegetable, including seeds, white heather flowers, and succulent alpine saxifrage leaves. Most of our rosy finches winter east of the Cascades, where they prove congenial with such lowland conveniences as window feeding-boxes, grain elevator yards, railroad beds where grain has spilled, and the company of dozens of other rosy finches.

# Purple and Cassin's Finches

*Carpodacus* spp. (car-**pod**-a-cus: fruit eater). Older males ± reddish, esp toward late summer, with dark gray wings and white belly; females (and males up to breeding yearlings) a sparrowlike gray-brown with streaked to spotted white breast and belly, white eyebrow-streak. Order Passeriformes (Perching birds).

**Purple finch,** *C. purpureus* (pur-**pew**-rius: purple). 5½"; older males broadly dusky rose to wine red; undertail feathers white. Song a long rolling warble. W-side forests, esp where broken.

**Cassin's finch,** *C. cassinii* (ca-sin-ee-**eye**: after John Cassin). 6"; males have crimson crown and pale red breast; females' eyebrow streak scarcely visible; dark-streaked white under tail. Warble full of breaks and squeaks. Subalpine and E-side. Color p 501.

When the red-flowering currant blooms, the purple finch is sometimes seen eating the flowers for their nectar-rich ovaries. Cassin's finch is more likely eating conifer buds. Both are primarily seed-eaters later in the year.

*The 1983 O.A.U. checklist lumped our gray-crowned form, hitherto *L. tephrocotis*, into a species with the other Rosy Finches.

# Red Crossbill

*Loxia curvirostra* (lox-ia: oblique; cur-vi-**ros**-tra: curve bill). 5½";
older males red; females yellowish beneath, greenish gray above;
young mature males often orange while grading from yellowish to
red; wings (all) solid dark gray. "Chip, chip" call; warbling song. In
flocks, in the coniferous canopy. Order Passeriformes (Perching
birds). Color p 502.

You have to be close to see the crossed bill, but it's a fact—the
lower mandible hooks upward almost as much as the upper
one hooks downward. This design is for prying cone scales
apart to extract the seeds. The jaws also open wide for the
bird's tongue to glean aphids and terrestrial insects. While
bugs are important in its diet, the crossbill migrates wherever
some kind of conifer is bearing a good cone crop, so it's occasionally abundant in any given locale. A shower of conifer
seed coats and seed wings often means a crossbill flock is
above.

# Pine Siskin

*Carduelis pinus* (card-you-ee-lis: from "thistle"; **pie**-nus: pine).
4¼"; gray-brown with subtle lengthwise streaking; yellow in wings
and tail may show in flight. Various distinctive scratchy twitters and
sucking wheezes. In large flocks in treetops; abundant subalpine
year-round. Order Passeriformes (Perching birds). Color p 501.

Pine siskins look nondescript, but call and fly quite distinctively. Often a flock's twitterings seem to match its breathtaking
undulations and pulselike contractions and expansions in
flight. I once saw one of these aerial ballets soon after a display
by the Navy's Blue Angel jets, and thought, "Why do people
bother to watch that clumsy beginner stuff?"

Siskins' narrow, sharp bills (in contrast to the heavy conical ones typical of their family) limit them to lighter seeds—
thistle, foxglove, and birch seeds, for example—along with
insects and buds. Siskins hang upside down from catkins while
extracting the seeds.

---

# 10

## Reptiles

Though we may habitually say "reptiles and amphibians" in one breath, or lump them together as "herptiles," they have important differences. Amphibians originally evolved from fish; they pioneered onto land with the use of two key innovations—air-breathing apparatus (lungs, in some groups) and legs. Certain lunged amphibians later evolved into reptiles with the innovation of two key moisture barriers—scaly skin and hard eggshells—which enabled them to exploit sunny and arid terrains. Reptile evolution led in turn to birds and mammals. All five of these evolutionary stepping-stones, from fish to mammals, are of equal, distinct status as "Classes" on the taxonomic family tree. The jump from amphibians to reptiles is, if anything, broader than that from fish to amphibians or from reptiles to birds.

### Northern Alligator Lizard

*Gerrhonotus coeruleus** (jer-o-no-tus: wicker back; see-rue-lius: blue). 8½–10" long; body thick, legs and toes slender; scales heavy, rough; dark olive back with light stripe; belly gray to bluish. Diurnal; mainly W-side to subalpine. Anguidae (Anguid lizard family).

Scaly skin distinguishes lizards from salamanders. The scales on alligator lizards are heavy enough to serve as armor, but at considerable cost to flexibility; breathing is facilitated by a pair

*Some recent texts place this species in a separate genus *Elgaria*.

**Northern alligator lizard**

of large pleats along the lizard's sides. If you pick up an alligator lizard you may encounter its other defenses—it can bite quickly and hard, albeit harmlessly, defecate all over your hand, or snap off its tail. Many lizards share this famous trait of fracture planes in the tail. It lets a lizard escape while a predator holds and likely consumes the still-writhing tail. The lizard grows a new tail within weeks. Northern alligator lizards are themselves predators of millipedes, snails, and other crawling things. They walk stiffly, somewhat snakewise, on legs too weak to get their bellies off the ground. The young are born live, fully formed, in litters of 2 to 13 in late summer.

**Western fence lizard**

## Western Fence Lizard

*Sceloporus occidentalis* (sel-op-or-us: leg pores; ox-i-den-tay-lis: western). Also **bluebelly**. 6–8½″ long; belly relatively wide, legs big, with very long toes on rear; scales spiny, abrasive; back olive brownish; belly pale, with yellowish under legs, with many light blue flecks and patches; males have 2 blue throat patches. Diurnal; in warm sunny habitats from C Wash South. Iguanidae (Iguana family).

With their long toe-claws, fence lizards are agile climbers, fond of fenceposts. In winter they hibernate, perhaps in an old log. In spring, females lay clutches of 5 to 15 eggs in loose dirt. For courtship, territory assertion, and other communication they have a precise gestural language including pushups, head-bobbing, teeth-baring, throat-puffing, and side-flattening, much of it tailored to display their blue patches. (Each iguanid species seems to have its own language.) Contact with you may alarm one enough to provoke several of these gestures—or a quick scurry to the far side of the tree or rock.

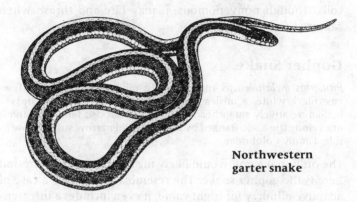

**Northwestern garter snake**

# Northwestern Garter Snake

*Thamnophis ordinoides* (tham-no-fis: shrub snake; or-din-o-eye-de-ez: patterned—). 15–24" long; black, brown, tan, gray or greenish, usually with 1 or 3 yellow, orange or red stripes down back; belly yellow to gray, sometimes red-spotted. Common in ± open areas, W-side below 4,000'. Family Colubridae.

Our range is not prime snake country; these pretty little snakes are easily our commonest, and even they avoid deep forest. They like to bask in the sun, sometimes intertwined in groups. They eat worms, slugs, and small amphibians. They give birth in summer to 3 to 15 live, worm-sized snakelets. (Most reptiles lay eggs.)

Our other two commonest snake species are larger (18–50") garter snakes, the western terrestrial, *T. elegans*, and the common, *T. sirtalis*. Unlike the northwestern, these two species try to stay near water, which they use for refuge.

## Racer

*Coluber constrictor mormon* (col-ub-er: Roman term for some snake). Our subspecies the western yellowbellied racer. 22–36+"; grayish olive above, yellowish beneath, adults unpatterned but young may be ± blotched. Any open lowlands. Family Colubridae.

This agile, slender snake is true to its name "racer," but not to its name *constrictor* since it kills its prey—mostly insects and some small rodents and reptiles—with its mouth, not with its coils. Though nonvenomous, it may bite and thrash when captured.

## Gopher Snake

*Pituophis melanoleucus* (pih-too-o-fis: pine snake; mel-an-o-lew-cus: black/white, a misleading description.) 36–72" long, heavy-bodied, relatively small-headed; tan with blackish blotches, largest ones along the back, flanked by 2 nearly solid narrow stripes. Low E-side. Family Colubridae.

The only snake you would likely mistake for a rattler around here is the gopher snake. The resemblance may be a case of adaptive mimicry for fright value; it even includes a threatening vibration of the tale which, when it rustles dry leaves, can sound much like a rattler. This snake is nonvenomous, killing its prey, including gophers, by constriction. It can also climb trees and prey on nestling birds. The constrictive technique characterizes one small branch of the huge family Colubridae. Containing some 68% of all snakes, it is probably recognized as a single family only because there is little agreement on how exactly to split it.

Gopher snake

---

Cascade-Olympic Natural History

# Western Rattlesnake

*Crotalus viridis oreganus* (crot-a-lus: rattle; **veer**-id-iss: green; or-eg-ay-nus: of Oregon). Our subspecies the northern Pacific rattlesnake. 16–60" long, heavily built, with large triangular head; tail terminates in rattle segments; variable brown, black, gray, and tan colors in a pattern of regularly spaced large ± geometric dark blotches against paler crossbars. Lowest E-side and S Willamette Valley elevs, typically on talus or rock outcrops. Viperidae (Pit viper family).

We're lucky in having no poisonous snakes except along the eastern and southern low fringes of our range where the rattler may be found. Avoid it. This species can inject a dose of venom lethal to small children and many mammals, though almost always merely painful to adult humans. (A solitary victim might possibly be incapacitated long enough to die of hypothermia if warm clothes and food are far away.) Rattlesnakes hunt at dusk, dawn, and night. A light, dry rustling sound from their rattles is the typical warning to large intruders. Far from being aggressive toward humans, they will almost always flee if given the chance.

**Western rattlesnake**

# 11

# Amphibians

Though considered terrestrial vertebrates, amphibians are only marginally terrestrial; they lack an effective moisture barrier in either their skins or their eggs, so to avoid deadly drying
they must return to water frequently, venturing from it mainly at night or in the shade, and never far. Most of them hatch in
water as gilled, water-breathing, legless, swimming larvae
(e.g., tadpoles) later metamorphose into terrestrial adults, and
return to water to breed. Hence the word "amphibious," from
"life on both sides." A few amphibians actually manage to be
completely terrestrial in moist habitats; several others have
regressed to a fully aquatic life. Western Oregon and Washington, abounding as they do in moist habitats, are very strong on
amphibians, and weak on reptiles.

Among vertebrate animals, mammals and birds are
"warm-blooded," whereas reptiles, amphibians and fish are
"cold-blooded." This doesn't mean they're self-refrigerating,
but they're never much warmer than their environment.
They need some ambient heat to help them be active, yet they
can sustain activity in astonishing cold—long-toed salamanders in our high country typically breed in 32° water with
winter's ice still on it. Nevertheless they hibernate through
most of the freezing season. At the other extreme, amphibians
rarely survive heat over 100°. Ironically, their intolerance of
heat won salamanders and newts a superstitious reputation as
fire*proof*. (No doubt it was because they know how to survive

a ground fire; they take refuge in a familiar wet crevice or burrow, just as they do from the midday sun.) This mistaken reputation may have got some newts into hot water of the black magic variety—

> Fire burn, and cauldron bubble.
> Fillet of a fenny snake,
> In the cauldron boil and bake
> Eye of newt and toe of frog . . .

## Rough-skinned Newt

*Taricha granulosa* (tair-ic-a: mummy; gran-you-low-sa: grainy). 6–7" long; back warty-textured (exc on breeding-season males) green-black to almost translucent brown; underside orange to yellow; ribs not visible. W-side. Salamandridae (Newt family). Color p 503.

Newts are a family of salamanders with relatively bright color displays and toxic skin secretions. Salamanders in general protect themselves with a combination of skin toxins and nocturnality, but this newt is so strong in the poison department (it won't hurt your fingers, but it can sicken or even kill a smallish animal that eats it) that it doesn't need to be nocturnal. This makes it our most often-seen salamander, commonly foraging on gray summer days in forest not far from its retreat in a pond or marsh. When threatened, it displays its bright underside by curling up until its tail is near or in front of its upturned chin. This probably deters many predators by reminding them of the newt's poisonous skin.

At breeding time (winter or spring) the male's skin smooths out, his tail flattens, his genital region swells, and his underside turns a brighter orange. The female lays eggs one by one on submerged vegetation.

## Olympic Salamander

*Rhyacotriton olympicus* (rye-a-co-try-ton: brook sea-god). 3–4½" long, slender, bug-eyed; olive to chocolate brown above, variably flecked yellow beneath; 14–15 rib grooves; males have squarish anal lobes behind rear legs. W-side. Dicamptodontidae. Color p 503.

This delicate, pretty little salamander generally stays in mountain creeks or within splash range of them.

# Long-toed Salamander

*Ambystoma macrodactylum* (am-bis-ta-ma: blunt mouth; macro-dac-til-um: big toes). 4–6" long; wide, blunt head, rather long legs; dark gray-brown with a thin, irregular, full-length back stripe bright yellow to greenish or tan; 12–13 rib grooves. Widespread. Ambystomidae (Mole salamander family). Color p 503.

This is our most widespread salamander, the only one to range over both sagebrush country and alpine meadows. Like other amphibians it has to stay close to water, especially in arid terrain. In the high country it breeds even before the ice is gone from its lake. High-country larvae rarely metamorphose into adults before their second summer, since their active season is short and frigid.

# Pacific Giant Salamander

*Dicamptodon ensatus* (die-camp-ta-don: twice-curved teeth; en-say-tus: sword). 7–12" long, stocky; brown to purplish with black splotches; belly light brown; ribs indistinct. W-side forest. Family Dicamptodontidae. Color p 503.

While other salamanders are limited to eating insects and other small invertebrates as food, this one—the largest terrestrial salamander—can catch and eat mice, garter snakes, and small salamanders. It is also the only salamander with a real voice, variously described as a "yelp" and a "rattle." Salamanders have no eardrums or external openings to receive communications, but they do have inner-ear organs which are presumed sensitive to vibrations transmitted up through the legs. Larvae have intricate plumelike red (blood-filled) structures where we might expect ears; these are external gills that "breathe" or absorb oxygen from suspension in the water. Some Pacific giant larvae metamorphose into terrestrial adults at about 3", in their second summer. Others never do metamorphose, but instead mature sexually as aquatic larvae and sometimes grow as long as 12". This latter variation ("neoteny") is common in both the mole salamanders and this family. The Dicamptodontids were only recently given family rank, and are known only from the Pacific Northwest.

## Western Red-backed Salamander

*Plethodon vehiculum* (pleth-o-don: many teeth; ve-hic-you-lum: carrier). 3–5" long, slender; brown or black exc for a ± full-length broad, even-edged back stripe red to orange or yellow, tan or even greenish; 16 rib grooves, or 15 in C Ore. W-side forest. Plethodontidae (Lungless salamander family). Color p 502.

Salamanders of this family lack lungs, and breathe through their skins. This species lives mostly under moss and logs in the forest, or under stones in wet talus. The eggs are laid on the ground and guarded by the mother. Lungless salamanders of the West all hatch into miniature "adults," having breezed through the larval stage within the egg, and are thus completely terrestrial. They are known for elaborate mating rituals, often ending with her mounting him while, like other salamanders, he walks along dropping gelatinous sperm-cases for her to retrieve with her cloacal lips for internal fertilization.

## Larch Mountain Salamander

*Plethodon larselli* (lar-sel-eye: of Larch Mtn). 3–4" long, slender; gold-flecked blackish above, with a ± broken reddish tan back stripe; red to salmon pink-orange beneath. Found only in the Columbia Gorge (both states). Plethondontidae (Lungless salamander family). Color p 503.

This salamander does not adhere to streams and seepage as many of its relatives do, but it finds cool, moist, deep refuges in the basalt talus substrate of its Columbia gorge habitat. It employs a curious defense against large creatures, coiling and uncoiling its body so rapidly as to fling itself about, or sometimes simply coiling up, perhaps in mimicry of poisonous millipedes.

## Oregon Slender Salamander

*Batrachoseps wrighti* (ba-tray-co-seps: frog lizard; right-eye: after Margaret and A. H. Wright). 3¼–4" long, extremely slender in adulthood; dark brown above, with a ± vague, reddish to greenish back stripe; white-blotched black beneath; 16 or 17 rib grooves. Found only in the Ore Cas, at low to mid elevs. Plethodontidae (Lungless salamander family). Color p 502.

Salamanders of wormlike slenderness crawl around in termite or beetle tunnels in rotten wood, preying there on springtails, mites, beetle larvae and worms. They are characteristic inhabitants of big old logs, which they depend on for refuge from summer heat. Slender salamanders reportedly employ both the coiling defense (page 425) and the tail-breaking defense.

## Western Toad

*Bufo boreas boreas* (bew-foe: the Greek term; bor-ius: northern). Also (our subspecies) **boreal toad.** 2–5″ long; thick; sluggish; skin has large bumps, the largest being 2 oval glands behind the eyes; olive to grayish, with a narrow pale stripe down back, and blotches on belly. Widespread. Bufonidae (Toad family). Color p 502.

Toads are distinguished from frogs by their warty skin, toothless upper jaw, sluggish movement (generally walking rather than hopping), and parotoid glands behind the eyes. These bulbous protrusions exude a thick, white, nauseating, burning poison similar in part to digitalin, effectively deterring predators. (It does not cause warts.) Toads' slow pace limits them to slow, creeping sorts of invertebrate prey.

Toads resist drying better than most frogs and salamanders. Our only toad, the western, inhabits animal burrows and rock crevices, and is often seen in mountain meadows and woodlands well away from watercourses. It is less strictly nocturnal than most toads, especially at elevations where nights are too cold for much toad activity. Lacking the inflatable vocal sac many of its relatives boast, it has a weak, peeping voice.

## Red-legged Frog

*Rana aurora* (ray-na: the Roman term; aurora: dawn, referring to redness). 2–4″ long; grayish to reddish brown with small dark blotches; yellow underneath, ± reddish on rear legs and lower belly; toes only slightly webbed. Lower W-side forest. Ranidae (True frog family). Color p 503.

In moist woods, this diurnal frog can venture away from watercourses. Though preferring low elevations, it ranges farther north and seems to require a colder macroclimate than the Cascades frog. Its croak is feeble, rough, and prolonged.

## Cascades Frog

*Rana cascadae* (cas-cay-dee). 1¾–2¼" long; brown to yellow-olive with black spots above, yellow underneath; yellow jawline; toes only slightly webbed. In Cascades, mid elev to timberline, always near water. Ranidae (True frog family). Color p 502.

The diurnal Cascades frog is often seen basking on high-country shores and marsh meadows. It croaks in a rapid, raspy chuckling manner, several blips per second. Escaping into water, it swims off across the surface rather than diving as its relatives do. Countless tadpoles (the larval stage) of this species inhabit shallow seasonal ponds.

Northwest tribes associated frogs with wisdom. Certain barbaric cultures make them into good luck amulets or sacrificial victims to bring rain.

**Pacific treefrog**

## Pacific Treefrog

*Hyla regilla* (hi-la: of forests; ra-jil-a: queenlet). 1–2¾" long; skin bumpy; toes bulbous-tipped; color changes with background: green to black or tan, with large irregular black splotches; males' throats gray. Ubiquitous. Hylidae (Tree frog family). Color p 503.

Tree frogs are distinguished by their bulbous toe pads, which offer amazingly good grip on vertical surfaces such as trees. Our species probably spends more time in water and on the ground than on shrubs and trees. It has a sticky tongue for catching insects. Active and vocal day and night, it has a rather musical, high-pitched call. Voiced frogs employ a variety of calls, including alarm, warning, territorial, and male and female release calls. The familiar pond frog choruses are likely (at least around here in spring) to be mating-call duets and trios of male treefrogs. The male amplifies his voice with a resonating throat sac he blows up to three times the size of his head.

---

# Tailed Frog

*Ascaphus truei* (ask-a-fus: not digging; true-eye: after F. W. True). 1–2″ long; skin has sparse small warts; olive to dark brown with large irregular black splotches and black eye-stripe. Widespread in mtn streams. Ascaphidae (Tailed frog family).

Tailed frogs spend most of their time in fast, cold creeks. They attach their eggs like strings of beads to the downstream side of rocks. The tadpoles suck firmly onto rocks, or perhaps to your leg or boot when you wade a creek, but don't worry, they aren't bloodsuckers. The silent, nocturnal adults are less easily encountered. They don't have real tails; those little soft protuberances are male cloacas, and they fertilize the females internally. You might think a seeming penis prototype to be an advanced item on an amphibian, but actually this is considered the most primitive family of frogs. Its other members are all in New Zealand.

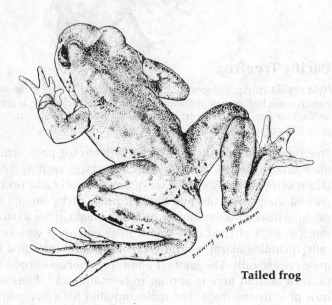

Drawing by Pat Hansen

**Tailed frog**

# 12

# Fishes

Fishes in our mountains have recovered from the last Ice Age less completely than other life forms. When the glaciers advanced, aquatic life necessarily retreated before them; when the glaciers retreated, fish were able to return only so far as they could swim, so each stream is generally fishless above a certain impassable waterfall. Only strong-swimming, cold-loving species, mostly salmon and trout, make it into our mountains at all, and many of these fight their way to their upper distributional limit just once in their lives, when at their peak, and at a cost of total and terminal exhaustion. They may be phenomenal waterfall-leapers, but still, there are limits. Above the critical waterfall live perfectly healthy aquatic communities whose animal members—invertebrates, amphibians, and small mammals—all got there overland or airborne.

Higher still, in mountain lakes, similar communities have recently been rejoined by trout who rode up the trail in saddlebags or flew there in airplanes. Fish population of our high lakes is almost entirely a product of human management— a second major factor in fish distribution here that favors salmon and trout. Sometimes trout competitors have been introduced accidentally (as live bait that escaped) and whole lakes have been temporarily poisoned with rotenone, and restocked with trout.

Some of the most popular, accessible fishing lakes are stocked with catchable-size trout before and during the fishing season; such trout are made of dry fishfood, just like super-market trout, and only freshness makes them taste any better. More often the stocking is done with fingerlings soon after the fishing season. Still, the lake may get nearly fished out each season, so that almost all the catch is two-year-old hatchery trout. Relatively inaccessible lakes may be stocked only occasionally and have near-natural, self-sustaining populations. Or they may support healthy trout for years at a time but lack reproducing populations, for want of a proper spawning bed. The bed must be clean gravel of the right size, at the right depth, with a moderate current to keep it aerated and silt-free, since silt suffocation is the chief cause of egg and infant mortality. Check a lakeshore near an inlet or outlet stream and follow the stream to its first waterfall (rarely very far) and if you find a shallow gravelly spot in early summer, you may well see a trout busy swimming back and forth over it.

Another limiting factor on fish in high lakes, strange as it may sound, is high temperature. Except in tiny pools, the water is never what we would call warm, but trout may suffer or die of heat in water that could chill and "freeze" a person in short order. Fish are sensitive to small differences in the 32° to 65° range. Being "cold-blooded," their body temperatures drop with that of the water, but their health isn't endangered; only their activity and growth rates are reduced. Under ice for nine months, high-country trout spend a very slow winter without danger or discomfort, but also without growing. In the smallest and highest lakes that support trout at all, they never grow very big, maturing and spawning at three to six years of age while only 4–5" long and still displaying the parr marks typical of juvenile trout elsewhere.

Some lakes are poor fish habitat by very reason of their cleanness. The aquatic food pyramid rests upon algae, which in turn depend on minerals not present in rain or snow; water has to pick these up in its passage over or through the earth. Small drainage basins, high snowfall, barren or impermeable terrain, and rapid turnover of lake water often combine, in our region, to severely limit the nutrient quality of lakes. Fish and Wildlife staff may stock such lakes, but the trout grow poorly.

All the fishes described in this book are carnivores, a few steps up the food pyramid from algae. Each species in its tiny

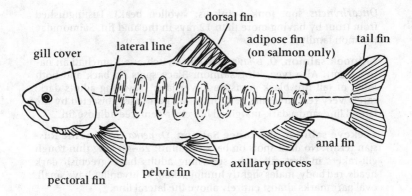

gill cover
lateral line
dorsal fin
adipose fin
(on salmon only)
tail fin
anal fin
axillary process
pelvic fin
pectoral fin

"fry" stage eats zooplankton until it grows large enough to subsist on larger crustaceans and insects of all life stages. The young fish soon add snails, worms, isopods, freshwater shrimp, amphibian larvae, and fish eggs and fry to their diets. Any fish species, even their own, may be fair game, so long as it's currently smaller. For example, Chinook salmon eggs and fry are seasonal staples for the little sculpin, which at the same time is a major food item for one- to two-year-old Chinook. Salmon and trout that reach large size in fresh water even eat swimming bird chicks, adult amphibians, and water shrews. In salt water they become voracious predators on fish after first growing up on a diet of tiny krill.

Upon their return to freshwater for spawning, most eat little or nothing, but metabolize stored fat and muscle tissue. Spawning, they often look like they're at death's door, which they are. With pale, crumbling flesh, they make very poor eating—though bald eagles don't seem to mind. The horrifying rate of decomposition during and after spawning may have evolved because of the nutrition the young derive later from the residue. Some family members prolong their freshwater stay; sockeye salmon take a last summer vacation feeding in a lake before completing their spawning run, and a small minority of smelt and sea-run trout individuals survive spawning to return to sea and spawn again the following year.

# Pacific Salmon

*Oncorhyncus* spp. (onk-o-**rink**-us: swollen beak). Distinguished from trout by having more than 12 rays in the anal fin. Salmonidae (Salmon family).

**Chinook salmon,** *O. tshawytscha* (cha-**witch**-a: a Kamchatkan native term). Also **tyee, king salmon.** Black spots on back and both lobes of tail fin; black gums on lower jaw; spawning adults dark, rarely very red; juveniles at 4″ have tall parr marks bisected by the lateral line, and a dark-margined but clear-centered adipose fin.

**Sockeye salmon / Kokanee Salmon,** *O. nerka* (**ner**-ka: the Russian term). No dark spots on back or tailfin; 28–40 long thin rough gill-rakers in first gill arch; spawning adults have greenish dark heads, red body, males slightly humpbacked; 4″ juveniles have small, oval parr marks almost entirely above the lateral line.

**Coho salmon,** *O. kisutch* (**kiss**-utch: a Kamchatkan native term). Also **silver salmon.** Black spots on back and upper half of tail fin; white gums on lower jaw; spawning males have brilliant red sides; 4″ juveniles have tall parr marks bisecting the lateral line, an all-dark adipose fin and some dark pigment on leading edge of anal fin.

**Chum salmon,** *O. keta* (**ket**-a: Russian term). Also **dog salmon.** No dark spots on back or tailfin; 19–26 short smooth gill-rakers on first gill arch; spawning adults develop vertical streak patterns with variable tinges of red, esp on males; the iridescent silvery green fry with small, high, ± faint parr marks rarely exceed 2¼″ before reaching sea.

**Pink salmon,** *O. gorbuscha* (gor-**boo**-sha: Russian term, meaning humpback). Also **humpback salmon.** Large oblong dark spots on back and both lobes of tail; spawning males develop high hump on back ridge; fry have no parr marks and rarely exceed 1¾″ before reaching sea.

Pacific salmon range the Pacific Rim from California to Korea, and some also range the Arctic from the Mackenzie to the Lena River; they swam up most river systems nearly to their headwaters until they were recently prevented from it, in many cases, by high dams. They were originally described to science by Steller in Eastern Siberia, which is why their specific names are in Russian and Kamchatkan. They were called salmon after the Atlantic salmon, *Salmo salar,* a close relative of our rainbow/steelhead trout.

Salmon are "anadromous" ("up-running") fishes; they migrate upstream from the sea to spawn. Their relatives the

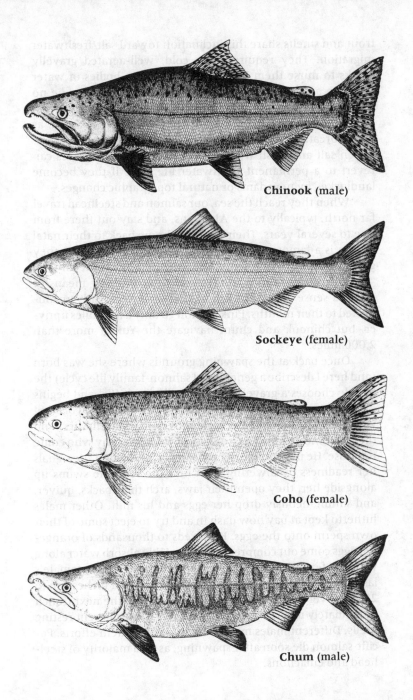

**Chinook** (male)

**Sockeye** (female)

**Coho** (female)

**Chum** (male)

Fishes 433

trout and smelts share this inclination toward salt/freshwater migration. They require clear, cold, well-aerated gravelly creeks to nurse them in infancy, and larger bodies of water richer in animal life to rear them to maturity. It may be no surprise that they find the ocean ideal for their maturation phase, but it's remarkable that they're perfectly adapted, as water-breathing creatures, to handling the chemical shock between salt and fresh water. Apparently all these species can revert to a permanent freshwater life cycle if they become landlocked by new dams or natural topographic changes.

When they reach the sea, our salmon and steelhead travel far north, typically to the Aleutians, and stay out there from one to several years. Their ability to home back to their natal stream is a famous wonder of animal navigation. Once they get close to the right river mouth, they zero in on the mineral "recipe" of the precise tributary of their birth by means of smell, a sense located in fish in two shallow nostrils unconnected to their mouths. Pink salmon go only a few miles upriver, but chinook and chum navigate the Yukon more than 2,000 miles.

Once back at the spawning grounds where she was born (and here I describe a generalized salmon-family life cycle) the female chooses a gravelly spot, usually in a riffle, and begins digging a trough, or "redd," by turning on her side and beating with her tail. A dominant male moves in to attend her, and they become aggressive to outsiders, attacking any who come too close. He may nudge her repeatedly; and when she signals her readiness by lowering a fin into the redd, he swims up alongside her, they open their jaws, arch their backs, quiver, and simultaneously drop her eggs and his milt. Other males hitherto kept at bay now dash in and try to eject some of their own sperm onto the eggs. Hundreds to thousands of orange-red eggs come out compressed, but quickly absorb water along with the fertilizing milt and swell to roundness. The female's last maternal act is to cover them under several inches (up to 2') of gravel as she extends her redd, digging the next trench immediately upstream from the last, with intermittent resting spells. Different males may attend her subsequent efforts. Pacific salmon die soon after spawning, as do a majority of steelhead and eulachons.

The eggs take a few months to hatch—the colder the water, the longer they take. The newly hatched "alevin" remains in the gravel layer several weeks, still nutritionally dependent on the yolk-sac suspended from its belly. As the yolk is depleted, the fish adapts to a diet of zooplankton and emerges from the gravel as a "fry." Pink and chum fry migrate down to sea almost immediately, during the nights, but the other species remain longer in freshwater. In a year or two they grow to "parr" size, displaying the "parr marks" that make them easier to identify. We call them "smolts" at whatever age they go to sea—from month-old, ¾" pink salmon fry to 10-year-old, 12" cutthroat trout in severe Alaskan habitats.

Salmon were incalculably important to the economies of all the Northwest Coast and Columbia River tribes. The Indians' gratitude to the salmon was kept alive in countless myths that spoke of dreadful times before the salmon came, or before someone (usually Coyote) taught the people how to catch them. The people sure did learn how to catch them: with hook and line, with spears, from platforms over waterfalls, with one- or two-man dipnets of spruce-root twine, with larger basketry nets or long seine nets or wicker traps or wooden weirs. Fishing was all done during the spawning runs; smoking and drying kept the bounty "in season" as long as possible, long beyond what we might consider palatable today. Tradition also maintained a painful awareness that the runs were not entirely reliable. Elaborate taboos, different for each species, were observed in order not to offend the Salmon. Yet the overall salmon population was so large in proportion to the human market, even including heavy trade to desert tribes, that there was really no call for restraint in harvesting it. Salmon fishing camp was Fat City.

Nineteenth century white men who promoted a worldwide market for canned Pacific salmon liked to reassure each other that they were developing an inexhaustible resource and sparing scarcer resources on land; by the end of the century they had to face the fact that the salmon were seriously overfished. Salmon populations today are a small fraction of what they once were. It's uncertain whether even a complete halt to fishing could restore them. Loss of spawning habitat associated with dams, logging and irrigation is a more permanent problem than overfishing. For example, clearcutting can aggravate erosion and seasonal streamflow fluctuation, ruining spawning beds for years if not forever.

Of the five species, Chinook is the least abundant and the most highly prized in both sport and subsistence (Indian) fisheries. Since it favors large rivers, it has suffered the most from dam building. A landlocked population of Chinook survives above Cushman Dam in the Olympics.

Sockeye are abundant and commercially valued, especially for deep red flesh richest in oil. Sea-run sockeye are rarely caught for sport, but the small, landlocked form (Kokanee salmon) rivals trout as a sport fish in many lakes of Northeast Washington and the Rockies. The sea-run form (unlike other salmon) spends parts of its life cycle in a lake—one or two juvenile years, and then a final summer before spawning.

Coho are a popular game fish. In Northwest markets they are sold cheap as "silver salmon," fine in flavor but less rich. They are noted for their tolerance of tiny urban creeks.

Chum were called "dog salmon" by Alaskan tribes who fed them to their dogs, favoring the richer Chinook for themselves. (Never let a dog eat salmon or trout raw. Many dogs are lethally susceptible to a bacterium in some salmon; wild canids and bears are generally immune following a sublethal infection in youth.) Chum appear to be the salmon best suited for intensive rearing, including "salmon ranching."

Pink are the most abundant salmon in our part of the Pacific, supplying about half the catch, but they rarely reach our mountains since they swim only a few miles up rivers. They are on a uniquely strict two-year life cycle; on any given river they run only in odd-numbered or only in even-numbered years. Most North American runs, including all of Washington and Oregon's, are in odd years.

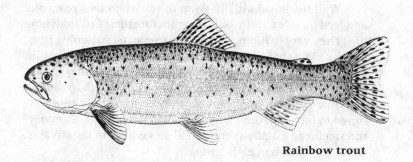

**Rainbow trout**

## Rainbow Trout / Steelhead

*Salmo gairdneri* (sal-mo: the Roman term; gaird-ner-eye: after Meredith Gairdner, p 438). Coloring extremely variable; flesh bright red to white; spawning adults usually have a red to pink streak (the "rainbow") full length on each side, much deeper on males; returning sea-run fish (steelhead) are silvery all over with guanine (a protein coating actually common to all salmonids fresh from the sea) that obscures any coloring underneath; dorsal fin rays typically 11–12, pelvic fin rays 10, anal fin rays 10 (range 8–12); juveniles have a distinct row of 5–10 small dark spots along the back straight in front of the dorsal fin, plus 8–13 oval parr marks along the lateral line. The common trout of high lakes. Salmonidae.

Native up and down the Pacific Slope from Mexico to the Alaska Peninsula and inland throughout the Columbia River System, the rainbow trout is now raised and stocked throughout the world's temperate zones, and is the common trout of dinner tables. Trout culture was developed on the European trout, *Salmo trutta*, (called brown trout where introduced in America) but the rainbow has proven even more amenable to intensive rearing. Its dominance of contemporary trout fishing draws few complaints, since it is highly rated for sportiness as well as flavor and color. As with any species, though, a young hatchery specimen will neither challenge nor reward the angler the way a big, experienced wild fish can.

Sea-run rainbow have fabulous mystique among fishermen who, with good reason, think of steelhead as altogether different fish. At sea the trout grow faster and reach larger sizes than those who stay in fresh water. Big and strong, "steelies" are notoriously hard to land, and harder still to locate. A favorite riffle for steelhead is a fiercely guarded secret.

Wild steelhead usually swim to sea when two years old, and feed for one to four "salt years" before their first spawning run. They vary in timing of both the smolt and spawning runs, but the main spawning run is in midwinter—another reason for the popularity of steelhead fishing, since few other species offer good fishing then. Steelhead also eat more than salmon do while running upstream, so they're somewhat more inclined to bite, and their flesh is in better condition. Summer-run steelhead usually spend all fall and winter in the streams, waiting to spawn in early spring.

State hatcheries developed methods and feeds to produce sea-ready smolts in one year; they used to release mostly one- and two-salt-year stock which came back as six- to nine-pounders, but recently they've responded to pressure from fishermen for a broader, more natural range of size and age classes. Certain Washington rivers are stocked with smolts from parents selected for exceptional size, which correlates with a three- or four-salt-year trait that produces many 15- to 25-pound steelhead.

---

**Meredith Gairdner** arrived at Fort Vancouver in 1833 expecting to make a big name for himself as naturalist, but fortune was not so kind to him as to William Tolmie (page 198), the other Scottish doctor on the same ship. (Five years later, another two brave young Scottish botanists were unluckier still, drowning near The Dalles even as they arrived in the territory.) The Hudson's Bay Company had Gairdner mind the trading post when he wasn't busy as their first physician at the Fort. He was able to make only a few short botanizing forays before diagnosing himself as tubercular. In 1835 he sailed with Nuttall (page 71) to Hawaii to take the sunshine and rest cure—unsuccessfully. He did manage to ship two notable specimens to England from Fort Vancouver. One, the trout that bears his name, is perhaps our most ardently sought creature. The other brings Gairdner no honor; it was the skull of Chief Concomly of the Chinook Confederation, which Gairdner stole from the grave at night, some five years after Concomly's death. A note accompanied it, offering to send the rest of the skeleton if so requested.

---

Cascade-Olympic Natural History

## Cutthroat Trout

*Salmo clarki* (clark-eye: after William Clark, p 234). Jaw lines streaked red or orange underneath; jaw longer than other trout, opening to well behind the eye; base of tongue has tiny teeth,* usually palpable; dorsal fin has 9–11 rays, pelvic rays usually 9, anal rays 9 (range 8–12); adults generally dark-speckled, but sometimes (esp on introduced fish from Rocky-Mtn stock) only near the tail; juveniles develop the throat streaks quite early, and have many small spots above and between the lateral parr marks but none or few (1–5) in a median line ahead of the dorsal fin. Salmonidae.

The reddish "cuts" on a cutthroat's throat serve to emphasize intraspecific gestures and displays. Though more reliable and distinctive of the species than the red "rainbows" on a rainbow trout, they may be obscure or lacking on cutthroats when very young, newly returned to freshwater, or dead. Coloring is notoriously variable in salmon and trout; fishermen recognize far more subspecies than scientists do. Like chameleons only much slower, trout alter their camouflage in response to visual stimuli; blinded trout eventually contrast sharply with their associates.

Cutthroats also vary in their migratory patterns. Some rivers with plenty of both spawning habitat and deepwater growing habitat have nonmigrating populations. More typically, the fry spend their first year in a small spawning stream and then swim down to a river or lake. Some return to the

**Cutthroat trout**

*Teeth on fish are used not for chewing but for gripping prey prior to swallowing it. They aren't confined to the jawbone area. Feeling for teeth inside the mouth of a caught trout can help identify it.

small streams seasonally—years before spawning—perhaps to escape turbidity during high water. At some age between three and ten (most often three in our region) those with access to the sea are likely to migrate there. Unlike steelhead, they rarely spend more than a year at sea nor go very far; many stay right in the estuary. Hence they don't grow as large as steelhead. However, freshwater cutthroat grow about as fast and as large as freshwater rainbow, and eat more fish and other large prey, using their bigger mouths and extra teeth. Cutthroat trout were probably as abundant here as rainbow trout originally, but are far rarer today since they are hard to raise in hatcheries.

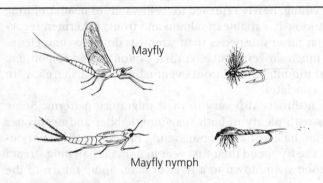

Mayfly

Mayfly nymph

## Fishing High-Country Lakes

Nearly inaccessible lakes can offer the most private fishing for the largest trophies. Some, however, may disappoint; the purest and highest lakes stunt or starve fish with their dearth of nutrients and short ice-free seasons. Some lakes may never have been stocked at all.

Due to gin-clear and usually peaceful water, the trout may be spooked by the plop of lures, bubble floats or big flies. Most insect prey here are quite small, midge pupae being perhaps commonest. Dry and wet flies imitative of midges (size 18 to 28) mayflies (size 10 to 16) and caddis (size 10) are recommended, on at least 8' of leader, with lightweight (3×-7×) tippets. Netting the hatches and imitating them specifically can help, but is very ambitious in some circumstances. For an all-purpose pattern,

# Dolly Varden and Brook Trout

*Salvelinus* spp. (sal-vel-**eye**-nus: the German term Latinized). Also **char**. Salmonidae.

**Dolly Varden,** *S. malma* (mal-ma: a Kamchatkan tribal term). Adults long and slender, olive-greenish back and sides regularly dotted with small pink to yellow or red spots; juveniles have unspotted dorsal fin, parr marks wider than the light spaces between; dorsal fin has 10 or 11 rays, anal fin usually 9.

**Brook trout,** *S. fontinalis* (fon-tin-**ay**-lis: of springs). Dark green back and sides with red spots surrounded by blue haloes, plus simple yellow spots; dorsal fin dark-spotted, with 8–10 rays, anal fin with 7–9 rays; juveniles have very broad greenish parr marks which come to be overlaid with lighter (± red and yellow) dots.

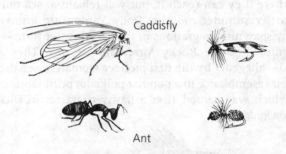

Caddisfly

Ant

Trueblood's Otter Shrimp (size 10 to 18) can be recommended.

If the shore is tree-lined, you may need to get out onto the lake on a logjam or rocks, or in hip-waders, in order to reach the action with your casts. If no jumping or swirling trout are visible, polarizing sunglasses (or a camera polarizer) can help spot fish.

Trout don't feed all day—only when the insects are most active. When the lake is first ice-free, feeding may go on from midmorning to midday, and by October they may feed all afternoon. If luck brings you a calm feeding period after an extended blow, look for frenzied feeding where floating insects are concentrated against the downwind shore. During early summer's spawning, look for trout near the inlet stream(s), if any, and in slabby shallows if there is no real inlet. A logjam at the outlet stream tends to be a good area at any time of year.

Inquire locally as to seasons and limits.

The Dolly Varden was perhaps our third most abundant native trout (after rainbow and cutthroat), whereas the brookie, native to Eastern North America, is our second commonest stocked trout (after rainbow). Though commonly called trout, fish of this genus are more properly distinguished as char, or charr, or charrr (in the American, English, and quasi-Scottish spellings). Mature char are distinguishable from true trout by having light spots against a darker background, rather than vice versa. They tend to be more naive than true trout about biting fishhooks, and less feisty after doing so, which makes them less popular with anglers out for a challenge, and more popular with ones simply out for fish. Flavor is excellent.

The brook is largely a surface feeder on insects, while the Dolly is more a bottom feeder. Either may run to sea if they live where they can reach it, but will remain at sea only for four to six summer months. Dolly Varden are known for lengthy spawning migrations in their freshwater form—over 100 miles in some Rocky Mountain streams. They were named—allegedly by the first pioneer woman to see them— for their resemblance to a popular polkadot print fabric of the day, which was named after a flashy character in Dickens' *Barnaby Rudge*.

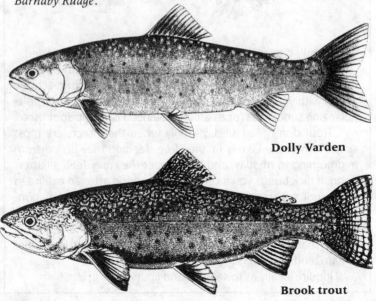

**Dolly Varden**

**Brook trout**

# Eulachon

*Thaleichthys pacificus* (thay-lee-ic-thiss: plentiful fish). Pronounced "you-la-kon." Also **candlefish.** 5–8″ fish, blue-gray above, silvery below; distinguished from salmon and trout by pelvic fin attached forward of dorsal fin and lacking an "axillary process, " a small flap or scale just above the base of each pelvic fin. Sporadically abundant on lower W-side rivers in brief Feb or March spawning runs. Osmeridae (Smelt family; formerly classed within Salmonidae).

Our common smelt, the eulachon or candlefish is so oily that it can be burned like a candle, once dried and threaded with wick. Its oil was the universal condiment in the cuisine of the Coastal tribes and of many inland tribes who went to the Coast to trade for it. (The name "eulachon" is from the Chinook

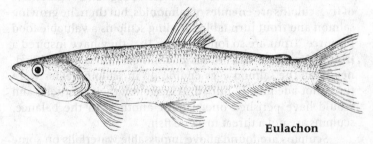

**Eulachon**

trading jargon.) It fetched a handsome cash price in British Columbia well into this century. The bitter froth of whipped soapberries suffused with eulachon oil was so popular it was nicknamed "Indian ice cream"; in later times, whopping doses of sugar were added.

Dipnetting for smelts is popular today among white people of the lower Columbia; during heavy smelt runs it requires so little skill or patience that bucketsful of smelt are often wasted by people whose enthusiasm for hauling them in exceeds their enthusiasm for cooking them or giving them away. Smelt runs are mysteriously sporadic and variable on each river; the Sandy had no smelt at all for over a decade, then had several good runs in recent years. Eulachon eggs have an outer membrane that bursts and sticks to the bottom on contact, leaving the egg in its inner membrane attached by a thread. The larvae drift to sea immediately on hatching, so the species is seen here only as spawning adults—about 95% three-year-olds, 5% making a second run as four-year-olds.

---

# Sculpins

*Cottus* spp. (cot-us: Greek term for some river fish). Also **muddler minnows, bullheads.** Scaleless, sometimes ± prickly, minnow-sized fishes with wide mouths, thick lips, depressed foreheads, very large pectoral fins, a ¾-length dorsal fin in 2 parts, and unforked tail fin. Widespread in streams. Cottidae (Sculpin family).

**Torrent sculpin,** *C. rhotheus* (rowth-ius: of noisy waters).

**Slimy sculpin,** *C. cognatus* (cog-nay-tus: related).

**Shorthead sculpin,** *C. confusus* (con-few-sus: clouded).

These funny-looking little fishes with their many funny names are the one group common in our mountain streams that's unrelated to salmon. As eaters of eggs and competitors of fry, sculpins are enemies of salmonids, but then the growing salmon and trout turn tables, finding sculpins a valuable food resource. Trout are so fond of them that they have inspired a trout fly pattern, the "muddler minnow." In all likelihood, nearly all the salmon eggs ending up in sculpin bellies were those not adequately buried in gravel by their mothers, and would have perished one way or another. In the balance, sculpins are not a threat to game fish.

Sculpins are found above impassable waterfalls on some of our rivers. They probably got there via waterways that crossed present-day drainage divides during unique geologic moments as the ice sheet retreated. Conspicuous sculpin traits are adaptations to life on the bottom: wide, depressed mouths for bottom-feeding; motley drab colors for camouflage against the bottom; eyes directed upward, the only direction there is to look; and huge pectoral fins to anchor them with little effort in strong current. The eggs, laid in spring, adhere to the underside of stones, and are guarded by the father.

**Shorthead sculpin**

*We are all in the gutter, but some of us are looking at the stars.* —O. Wilde

# 13

# Insects

Insects—six-legged animals with jointed external skeletons made of chitin—are far and away the most diverse and successful class of animals on earth. Over half a million species have been named, and at least as many remain to be discovered and described. Only a tiny sample of our insects can be discussed in this space—some of the ever charming butterflies, the invaluable pollinators—bumble bees and hover flies—and a rogues' gallery of our least loved insects.

A majority of insect life cycles entail metamorphosis from a wingless larva, which does most or all of the eating and growing, to a pupa or resting stage, and finally to an (often winged, often sexual) adult.

## Mosquitoes

Family **Culicidae,** Order Diptera (Flies)

A female mosquito has a most elaborate mouth. What we see as a mere proboscis is a tiny set of surgical tools—six "stylets" wrapped in the groove of a heavier "labium" flanked by two "palps." The operation begins with the two palps exploring your skin for a weakness or pore. There the labium sets down. Delicately and precisely, two pairs of stylets—one for piercing, one for slicing—set to work. They quickly locate and pierce a capillary, then bend and travel a short way within it. The remaining two stylets are tubes; one sucks blood out while the other pumps saliva in, stimulating bloodflow to the vicinity and perhaps inhibiting coagulation. Up to the point of blood stimulation, to which you may be allergic, you probably feel nothing. (Mosquito detection is the chief remaining "usefulness of hair on the legs," or so theorized poet Gary Snyder in his 1951 journal from a North Cascades fire lookout.)

Unlike bee stings, whose function is to inflict pain or injury, a mosquito bite raises its bump and itch—an allergic reaction to mosquito saliva—only incidentally. Research shows the sensitivity to be acquired; the first few times a given species of mosquito bites you, there's no pain or welt. As you develop sensitivity to that species, your response speeds up from a day or so at first, to one or two minutes. Eventually you may become desensitized to that species.

The females need at least one big blood meal (greater than their own weight) for nourishment to lay eggs; it takes the human-feeding species about two minutes to draw enough. Aside from that meal, they eat pollen and nectar; some plants probably depend on them as pollinators. Male mosquitoes eat nectar only, but contribute to our discomfort all the same—they hover around warm bodies in the reasonable expectation that that's where the girls are. If we wanted, we could get males to fly down our throats by singing the pitch hummed by female wings. Male antennae have evolved into fancy plumes that vibrate sympathetically with conspecific female wingbeats, and the males home in toward any steady source of this pitch. Each mosquito species has its own wing-beat frequencies, one for the males and a slightly lower one for the females; young adults speed up their wing beats as they reach sexual maturity.

Wing beats range up to 600 per second in mosquitoes, and peak at over 1,000 in their relatives, the midges. Anything over 50 beats per second is too fast to be triggered by individual nerve impulses. Instead, the thoracic muscles are in two groups, each stretched by the contraction of the other, and responding with a twitchlike contractive reflex. The thoracic shell is designed to snap back and forth between two stable shapes, like the lid of a shoe polish tin, one shape holding the wings up and the other down. The nervous system need supply only a slow, unsteady pulse of signals to keep this snapping vibration going. (Flies and many other insects fly similarly.)

Most mosquitoes are active only a short period each day. The time varies from species to species, but just before and after sunset, and just before dawn, are most popular.

The choice of victim, or "host," is also specialized. Certain varieties of the familiar domestic species *Culex pipiens* never bite people, but others rank among the peskiest. (**cue**-lex: the Roman term; **pip**-ee-enz: piping.) Most of our mountain mosquitoes are of genus *Aedes* (ay-**ee**-deez: repugnant). While other genera set their eggs afloat on bodies of still water, *Aedes* cannot; they lay eggs in the fall on spots of bare ground likely to be briefly submerged in the spring, when the larvae hatch and take off swimming. Alternatively, some boreal *Aedes* over-winter as already-mated females, and lay eggs on the melting snowpack in spring. In any case, the larvae, or "wigglers," feed by filtering algae and bacteria out of water.

Gruesome concentrations of mosquitoes typify the Far North, where musk-oxen and nesting ducks make ideal hosts. One stoical researcher in Manitoba counted 189 bites on a forearm exposed for one minute. From that he extrapolated 9,000 bites per minute for one entire naked person, who would thus lose one-fourth of his blood in an hour.

With their mosquito-borne diseases, the Tropics suffer even worse. Malaria has been detected in Oregon mosquitoes of genus *Anopheles* (an-**ah**-fel-eez: worthless), but isn't a major threat to humans in a climate this cool. Yellow fever entered our history indirectly. In 1803, it killed nine-tenths of a French army sent to conquer Haiti and the Mississippi Valley, leaving Napoleon in much more of a mood to sell "Louisiana" to Thomas Jefferson at a price Congress couldn't refuse. This purchase led directly to Lewis and Clark, and ultimately to the Oregon Territory ending up as part of the U.S. rather than Mexico, Canada or Russia. The two diseases together wielded enormous economic and military force, one indirect effect of which may have been to make mosquitoes the best-studied of all insect families.

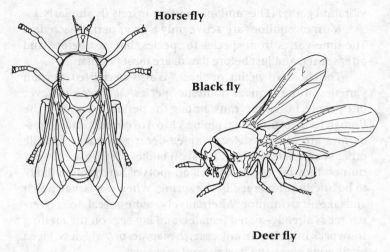

Horse fly

Black fly

Deer fly

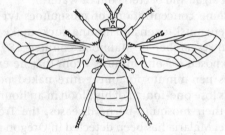

## Deer Flies

*Chrysops* spp. (**cry**-sops: gold-colored). Order Diptera (Flies).

About the size of house flies but much slower and softer, deer flies are maddeningly easy to kill—maddening because it's an irresistible exercise in utter futility. Where there's one deer fly, there are a thousand. On a hot July day in a North Cascades basin, the only respite may be nightfall, tent netting, your fastest stride, or rain—none of which are what you had in mind for this otherwise fine afternoon. Travel sometimes helps, since most deer flies stay within half a mile of the marsh or pond where they overwintered as larvae. Ranging from dull gray-brown to nearly black, the dozens of species of deer flies are distinguished partly by the patterns of pale brown blotches on their clear wings.

# Horse Flies

*Tabanus* spp. (ta-bay-nus: the Roman term). Order Diptera (Flies).

Horse flies are the biggest, fastest, and strongest of our biting flies, so we're lucky that they're sparse enough to be regarded as individuals—when you manage to swat one, you may actually be ahead of the game for awhile. They also obsess about the tops of our heads, often diverting them from our more vulnerable parts. Even so, they're hard to catch up with. With their black coloring and iridescent eyes, they resemble very large house flies.

Horse flies are commonest near large-mammal habitat. The nectar- and pollen-eating males are seen less often than the bloodsucking females. With larvae going through two winters before metamorphosing into adults, horse flies are longer-lived than smaller flies. Both horse and deer flies are also known as gad flies. Common usage would be "horsefly" or "gadfly," but entomologists prefer to keep "fly" separate in the names of true flies (order Diptera) to distinguish them from unrelated insects like dragonflies and butterflies.

# Black Flies

*Simulium* spp. (sim-you-lium: simulator). Order Diptera (Flies). Also buffalo gnats.

These vicious biters raise a welt way out of proportion to their size, sometimes drawing blood and reportedly causing bovine and human deaths when biting *en masse*. They render many of North America's boreal vacationlands uninhabitable for the month of June. Fortunately they tend to dissipate by midsummer. Here, they are less of a nuisance than the ubiquitous deer flies and mosquitoes.

"Black" flies are medium to dark gray, and stocky for their length (about ⅛"). They look humpbacked, and tilt steeply forward while biting. Only the females bite; once fed, they dive in and out of cold, fast streams attaching eggs singly to submerged stones. The larvae stay underwater, straining plankton from the stream, moving around and then reanchoring themselves with a suction disk. After emerging from pupation, the adults burst up through the water in a bubble.

---

# No-see-ums

*Culicoides* spp. (cue-lic-oh-eye-deez: gnat-like). Also **punkies, biting midges.** Order Diptera (Flies).

The name "no-see-ums" suffices to identify these pests. At up to ⅛" long, they are just big enough to see, but small enough to invade screened cabins. They are hard to make out in the waning light of dusk, when they do most of their biting. Surprise and defenselessness may irritate us more than the bites really hurt. No-see-ums are so localized around their breeding grounds that we can usually escape by walking 50'. Breeding habitats for various species include puddles, intertidal sands, and humus.

No-see-ums are in a whole family of bloodsuckers whose chief victims are mostly other insects ranging from their own size (midges) on up to dragonflies. In some species the females prey on the males.

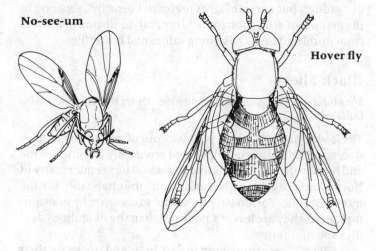

No-see-um

Hover fly

# Hover Flies

Family **Syrphidae.** Also **flower flies.** Order Diptera (Flies).

While hummingbirds, kestrels, and countless insects also achieve midair stasis, only certain hover flies adopt it as a normal stance. They are markedly more stationary than hovering bees or other insects, which inevitably bob and weave slightly.

Hover flies in the woods seemingly transfixed by an afternoon shaft of sunlight are males maintaining territory near a suitable spot, such as a rotting tree-hole, for a female to lay eggs. The female's last mating before laying eggs confers the best odds of prolific fatherhood. If you watch a stationary male, you may see him abruptly chase an approaching male or an unreceptive female. In the mornings, both males and females visit flowers in meadows, the females feasting on nectar, the males snacking lightly but mostly patrolling territory in hopes of a chance at a receptive female.

Only bees are more valuable than hover flies as pollinators for the plant world; the flies are especially crucial in our high mountain meadows, where bumble bees are the only common bees. The maggotlike larvae of many hover flies perform another enormous service to plants (and to agriculture) as the leading predators of aphids.

However dreadful the larva may appear to an aphid, hover flies are perfectly charming from our point of view, once we outsmart the adults' mimicry of black-and-yellow-banded bees and wasps. Hover flies even sound like bees, employing similar wingbeat frequencies. Mimicry is excellent protection against any predator that ever made the mistake of attacking a bee or wasp. The simplest way for us to tell hover flies from bees and wasps ought to be counting the wings; flies (Order Diptera, "two wings") have two, other flying insects have four. Unfortunately, the very narrow hindwings of bees and wasps (Order Hymenoptera, "membrane wings") are translucent and hard to see in a live insect. Long antennae, club-shaped ones especially, are a more useful stinger warning; most hover flies have antennae shorter than their heads, and a conspicuous, probing proboscis. Yellowjackets at rest fold their wings straight down their backs; hover flies and bees make a V shape. In my experience, hover flies and bumble bees are far more abundant here, especially in high meadows, than all other bees and wasps. If it looks like a yellowjacket (slender, not furry) but seems to like you and flowers, trust it.

# Bumble Bees

*Bombus* spp. (bom-bus: "buzz"). Large, rotund, furry, yellow-and-black bees; queens in most species are ¼–⅞" long, and fly in the spring; workers ¼–¾" long, appearing late spring to fall. Ubiquitous. Order Hymenoptera (Ants, wasps, bees).

Much "common knowledge" about bees is actually about honey bees, which aren't even native to the Americas. (The common European honey bee, *Apis mellifera*, has spread across America along with European culture, but is not abundant in our wilderness.) A honey bee worker stings only once,

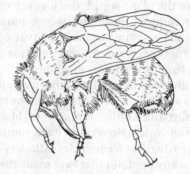

**Bumble bee**

losing her stinger and often her life with it, but other bees have repeat-use stingers. Honey bees have a highly evolved social order and elaborate communication rituals, but most wild bees are "solitary bees." (This term—an exaggeration, since many of them nest gregariously—means simply that they have no division of labor.)

Bumble bees, our commonest bees, are in the honey bee family, and do have a strong social order though their colonies are relatively small and short-lived. Each colony begins with a single queen who comes out from hibernating among dead leaves, having mated with several males the previous fall. Typically choosing a mouse or vole burrow for her nest, she secretes beeswax to make "pots." In some she lays eggs on a liner of pollen and nectar, sealing them over with more wax. Others she fills with honey (nectar concentrated within her body, chiefly by evaporation) which she sips for energy while working her muscles and pressing her abdomen onto her eggs—incubating them with body heat, like a warm-blooded bird.

The first brood is all workers, or small, nonmating females. Maturing in three or four weeks, they take over the
nectar- and pollen-gathering chores while the queen retires to
the nest to incubate eggs and feed larvae. Protein-rich pollen
nourishes the growing larvae, while pure carbohydrate nectar
and honey are all that's needed by the energetic adults. Bumble bee honey, by the way, is as delicious as honey bee honey
but can't be exploited by large omnivores like people and
bears because bumble bees don't store much for the future;
the queen keeps producing just as many broods of larvae as
the growing population of workers can feed. As the food supply allows, late broods will include increasing numbers of sex-
uals—male drones from unfertilized eggs, and new queens
produced simply by more generous and prolonged feeding of
female larvae. Only the queens eat well enough to survive the
onset of winter. The rest die in the fall, after the drones and
queens have mated. By feeding only a small minority of each
colony for hibernation, bumble bees conserve nectar and pollen resources which are scarce in cool climates.

Other key adaptations include the use of ready-made insulated nests; the relatively large, furry bodies; and the skill of
raising a near-constant body temperature for flight in a wide
range of weather conditions. (Arctic bumble bees have been
seen in flight in a snowstorm at 6° below freezing.) Generally,
bumble bees are the most valuable plant pollinators at high
latitudes and altitudes.

Some apparent bumble bee queens are false bumble bees,
genus *Psithyrus* (**sith**-ir-us: whisper). These nest parasites
murder queen bees and usurp their colonies. *Bombus* queens
may also usurp each other's colonies if they can, but *Psithyrus*
queens are genetically committed to it; they produce no workers of their own, and their legs carry no pollen. Worker bees on
guard duty may be able to get enough stingers through an
invader's armor to kill her, but once a false queen has established herself, the workers keep bringing home the bacon ob-
liviously. It is not unusual to find several dead queens in a
nest, or to find workers of mixed species in one colony.

# Ant Lion

*Myrmeleon immaculatus* (mer-me-lee-on: ant lion; im-ac-you-lay-tus: unspotted). Also **doodle bug.** Larvae found underneath ¼–⅜"-diam (rarely up to 1½"), perfect conical pits in fine dry sand, up to ½" long, sand-coated, pinkish gray with 6 black spots on stout body, 2 long heavy mandibles longer than the rest of the head; adults like frail, feeble damselflies, 1½" long, with 4 1"-long, narrow, little-used wings. Commoner E of Cas Cr. Order Neuroptera (Vein-winged insects).

Ant lion eggs are laid in the sand. As soon as they hatch, the larvae dig pits to trap ants and other crawling insects. Try dropping an ant into one if you go for that sort of entertainment, and you may catch some action—the ant lion's pit-digging motions. It flips its head and mandibles violently, tosses sand up the slopes of the pit, and rotates its body, keeping the prey trapped. As the larva grows, it moves on to bigger and better traps. Both common names refer to the ferocious predatory larvae. The adults, though large, are delicate, slow, innocuous, and rarely encountered.

# Cooley Spruce Gall Aphid

*Adelges cooleyi* (a-del-jeez: unseen; coo-lee-eye: after R. A. Cooley). Soft hemispherical bodies .04" long, covered at most stages with waxy, cottony white fluff; wings (if present) folded rooflike over body; on Douglas-fir needles (related species also on true firs or pines); more conspicuous are their conelike galls on branch tips of spruces, esp Engelmann. Widespread on and near spruces. Order Homoptera (Aphids etc.).

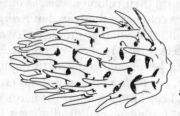

aphid gall on spruce twig

Many spruces seem to have an odd, spiky sort of cone in addition to their larger papery-scaled ones. Looking closer, we can see these aren't really cones because they are at branchlet tips, rather than several inches back, and because they are fused wholes, not a set of wiggleable scales. The "spikes" turn out to be simply spruce needles with a hard brownish or greenish skin drawn tight like shrinkwrap. This is a "gall," material secreted by the tree in response to chemical stimulation by an insect. Other examples of aphid galls include bright red marginal swellings on shrub leaves; the light, tan orbs on oak limbs are galls of tiny gall wasps.

Though each spruce gall cuts short the growth of one branchlet, galls in themselves are hardly ever a serious drain on their host plants. Living aphids sometimes are. They suck plant juices through minute piercing tubes nearly as long as their own bodies. They seem to suck in far more plant sugar than they can consume, since they pass copious sticky "honeydew" excretions. Some aphids are "herded" by ants or other insects that feed on aphid honeydew, but on our fir trees the honeydew is more likely to end up consumed by a dreadful-looking black smut fungus. Our forest trees are not much hurt, though, by either the feeding or the housing activities of this aphid.

Spruce gall aphids actually feed mostly on fir trees. Their life cycle includes no larvae, pupae, or males, as those terms are normally defined. Instead it is divided into five forms or castes, each egg-laying. One wingless form overwinters on spruce, then lays the eggs of the gall-making form, which emerges from the gall in late summer and flies to a Douglas-fir to lay eggs. These eggs produce the fir-overwintering form which in turn engenders two forms, one wingless and fir-bound, the other flying to spruce to beget either the spruce-overwintering form or a short-lived intermediary sexual generation. (The latter is unknown, or at least rare, in our region). The various female forms are perfectly able to perpetuate their clone "parthenogenetically," or without fertilization.

## Balsam Woolly Aphid

*Adelges piceae* (pie-**see**-ee: of spruces). Soft hemispherical bodies ½– 5" long, covered at most stages with waxy white "wool"; rarely found in winged stage. Sporadically epidemic on true firs. Order Homoptera (Aphids etc.).

The balsam woolly aphid coevolved in an unthreatening relationship with the fir species of Europe, but when accidentally introduced into North America, where it has no natural enemies, it proved deadly to several American firs. Subalpine firs have been devastated by it in many areas below 5,500', while grand and silver firs are moderately susceptible. The tree's leader may droop and break off early in an attack, then all the foliage may turn red-brown from the top downward and the trunk hemorrhage with resin.

This aphid resembles the innocuous Cooley spruce gall aphid (above) but feeds on true fir stems or twigs, which swell up in "gouty" knobs. Dispersing mainly on wind currents, this species rarely grows wings and never migrates to spruces or makes galls. Reproduction is asexual and males are unknown.

## Golden Buprestid

*Buprestis aurulenta* (bew-**pres**-tiss: Greek term for some beetle, meaning "cow-swelling"; or-you-**len**-ta: golden). ½–¾" beetle; metallic emerald green, coppery iridescent down center, edges, and underneath; back ridged lengthwise. Widespread but shy, adults may be found feeding on foliage, esp Douglas-fir. Order Coleoptera (Beetles).

This beetle beauty rivals our most glamorous butterflies. Foresters count it among our pests. Because it attacks trees in relatively small numbers it rarely does more than cosmetic damage, but that can add up to a lot of dollars annually. The larvae bore deep into seasoned heartwood, and may continue to do so as long as fifty years after the wood is milled and built into houses. Such records make the buprestid a contender for the title of longest-lived insect, though the larvae mature in less than a decade in their natural habitats. The eggs are laid in bark crevices, usually on fallen or injured trees.

Large fallen trees take several centuries to decompose, and would take longer still if the decomposers (fungi and bacteria) didn't have borers to make a rapid initial penetration of the protective bark. The forest ecosystem needs for the logs to decompose in order to make both space and nutrients available for new plants.

The ability of adult beetles to both fly and bore may be a key to their great success; they comprise easily the largest and most diverse order, not only among insects, but among all living things. Their name, Coleoptera, means "sheath wings" in Greek. The two forewings are modified into a hard sheath that encloses and protects the two hindwings when they are folded up, enabling beetles to bore into hard materials without jeopardizing their wings.

Rose chafers, genus *Dichelonyx*, (die-kel-oh-nix), are also metallic green to coppery purplish beetles, but are rarely as long as ½".

**Golden buprestid**          **Ten-lined June beetle**

 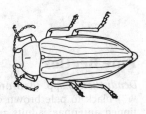

## Ten-lined June Beetle

*Polyphylla decemlineata* (poly-fil-a: many leaves; des-em-lin-ee-ay-ta: ten lined). ⅞–1¼" long, brown, with broad white stripes lengthwise on back; hairy underneath; males have large, thick, twisting antennae; adults found feeding on conifer foliage, or flying into lights at night. Order Coleoptera (Beetles).

The white, usually C-curved larvae (typical of the Scarab family) feed on plant roots, becoming occasional pests in conifer nurseries. The handsome adults huff and puff audibly through their breathing holes when disturbed.

## Ponderous Borer

*Ergates spiculatus* (er-ga-teez: worker; spic-you-lay-tus: with little spikes). Also **pine sawyer, spiny longhorn beetle**. Large beetles 1¾–2½" long; back minutely pebbled, dark reddish brown to black; thorax ("neck" area) may or may not bear many small spines; antennae long, jointed, curved outward. Order Coleoptera (Beetles).

Attracted to light, these clumsy nocturnal giants startle when they come crashing into camp. The equally ponderous larvae take several years to grow to a mature size of up to 3", chewing

1" to 2" diameter holes through pine or Douglas-fir heartwood. Loggers call the larvae "timber worms." One logger was inspired by their mandibles in inventing the modern saw chain design.

**Ponderous borer**

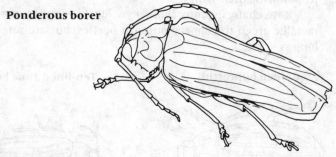

## Pine Beetles

*Dendroctonus* spp. (den-droc-ton-us: tree murder). Very small (⅛–¼") black to pale brown or red beetles with tiny, elbowed, club-tipped antennae; adults and larvae both live in inner bark layer, hence are little seen, but their excavation patterns in the bark and cortex are distinctive (see pp 459, 461). Ubiquitous. Order Coleoptera (Beetles).

**Mountain pine beetle,** *D. ponderosae* (ponder-oh-see: of ponderosa pine, though in fact this beetle is commoner on lodgepole while the following species is almost exclusive to ponderosa).

**Western pine beetle,** *D. brevicomis* (brev-ic-oh-mis: short hair).

**Spruce beetle,** *D. rufipennis* (roo-fip-en-iss: red wing).

**Douglas-fir beetle,** *D. pseudotsugae* (soo-doe-tsoo-ghee: of Douglas-fir).

Pine beetles are among the most devastating insect killers of western trees, especially lodgepole and ponderosa pines. The first signs of their attack are small round entrance holes exuding pitch and/or boring-dust. The pitch is a counterattack—the tree's attempt to incapacitate the beetles. The many exit holes, a generation later, look as if they could have been made by a blast of buckshot, and after the bark falls away you can see dramatic, branched engravings underneath.

The beetles, though less than ⅛" wide, chew much wider "egg galleries" through the tender inner bark layer, just barely cutting into the sapwood. The newly hatched larvae set off at right angles to the gallery—often in neat, closely spaced left/right alternation—growing as they proceed, and then pupating at the ends of their tunnels. From there they bore straight out through the bark upon emerging as adults. The tunnels are left packed with "frass" of excreted wood dust. In most species, the egg gallery runs from several inches to a yard, straight up and down the wood grain, and the tunnels of the larvae run straight to the sides or fan out slightly. However, western pine beetle galleries curl and crisscross all over the place like spaghetti, and the larvae leave little impression on the inner bark, preferring the outer bark. The related engraver beetles reveal themselves with further variations on the theme (page 460).

These beetles are all native here. In most years they benefit the forest by culling the damaged, diseased or slower-growing trees. Vigorously growing trees are unpalatable and unnutritious to beetles; they must overcome the tree's health, or it will overpower them. They do so by attacking in numbers and by bringing pathogenic fungi along for assistance. Single females scout, and then release scented pheromones to attract others to a vulnerable tree. Each female is followed into her entrance hole by a male, and they work as a pair.

Mountain and western pine beetles have occasional population outbreaks—typically after drought has stressed the region's forests for several years—when they kill trees regardless of health. They kill entire stands. At these time their economic equation shifts, making it advantageous to them to go for the biggest and healthiest trees. Foresters have found that the best preventive is to thin pine forests to keep them vigorous. Ponderosa pine stands used to be kept thin by naturally occurring fires; pine beetle outbreaks were presumably less frequent in nature, before people were suppressing fires.

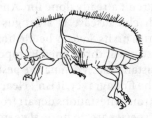

**Pine beetle**

# Engraver Beetles

*Scolytus* spp. (sco-lie-tus: truncated) ⅛" long, shiny blackish beetles; abdomen appearing "sawed off" a bit shorter than the end of the wing covers; known mainly by their egg galleries in the inner bark of dead and dying trees. Ubiquitous. Order Coleoptera (Beetles).

**Douglas-fir engraver,** *S. unispinosus* (you-ni-spy-no-sus: 1-spined).

**Fir engraver,** *S. ventralis* (ven-tray-lis: on the belly).

**Fir branch engraver,** *S. subscaber* (sub-scay-ber: somewhat rough).

The Douglas-fir engraver leaves its "signature" on a high proportion of fallen Doug-firs that lie in full or partial sunlight—a deep but rather short (4–6") egg gallery carved along the grain of the inner bark and sapwood, from which issue an array of larval tunnels that start out perpendicular but soon curve up or down, finally running with the grain. (If the tree falls into heavy shade, the related Douglas-fir pole beetle, *Pseudohylesinus nebulosus* may cut nearly identical marks in it.)

The fir engraver carves true firs, and its signature is 90° different; the main gallery runs straight across the grain, the larval tunnels with the grain. Still more distinctive is the mark of the fir branch engraver, which looks like a rounded E branded into the inner bark of dead lower (or fallen) branches of true firs. This species rarely attacks the main trunk, and can be thought of as abetting the tree's self-pruning efforts.

In the Northwest alone, dozens of species in the Scolytidae family of beetles take on specialized roles in recycling woody waste, a process crucial to the forest's vitality. Many of them have special organs ("mycangia") for holding and disseminating spores of fungi which, along with bacteria, are essential partners in this great symbiosis. A majority of them, like the Douglas-fir engraver, rarely attack trees or branches that aren't already done for. And then again, some are killers. Dutch elm disease is a fungus inadvertently introduced to America on its *Scolytus* beetle host, with disastrous results; the disease never threatened European elms, which are genetically resistant to it. With native pests like the western pine beetle and the fir engraver, it isn't clear to what extent the occasional disastrous infestations depart from the natural order. Even the killer species are, in normal years, cullers of infirm individuals,

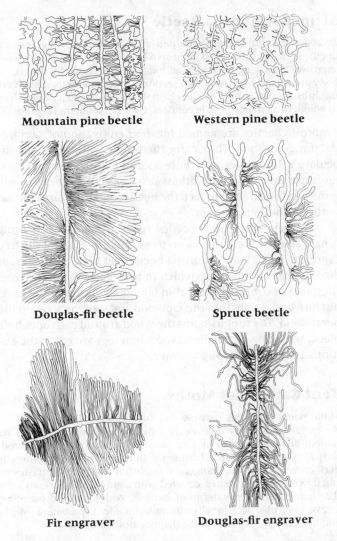

**Mountain pine beetle**

**Western pine beetle**

**Douglas-fir beetle**

**Spruce beetle**

**Fir engraver**

**Douglas-fir engraver**

benefitting the overall health of their "victim" population like wolves culling an elk herd.

If it weren't for its attackers, a tree would be virtually immortal—just as long as it could keep growing—since trees do not age in the sense that animals do. Without beetles and fungi, would the forest stop growing short of impenetrability? It's hard to imagine such a world.

---

# Striped Ambrosia Beetle

*Trypodendron lineatum* (try-po-den-dron: bore tree; lin-ee-ay-tum: lined). ⅛" long, shiny black to brown beetles with faint paler stripes lengthwise on wing covers; head hidden from above by thorax, so body appears to have only two sections; antennae shaped like very fat clubs; known mainly by their heavily black-stained pinholes deep in wood. In all major W-side conifers. Order Coleoptera (Beetles).

Ambrosia beetles are named for their cultivation of "ambrosia" fungi for food. They carry the fungus with them and inoculate it into holes they bore in downed or dying trees. If moisture and other conditions are perfect, the fungus will grow just fast enough to feed the beetles and their larval brood without smothering them.

Though other bark beetles (pages 458–61) also bring fungal allies to their attacks on trees, their food is still the tree cambium itself. Only ambrosia beetles get their nourishment directly the fungus, which in turn gets it from the tree; they are not confined to the thin tender layers under the bark, but burrow straight into the sapwood and sometimes even the heartwood. The fungus stains the wood around each bore hole black, wreaking economic havoc when logs are cut in the autumn and left out through winter.

# Tent Caterpillar Moths

*Malacosoma* spp. (ma-la-co-so-ma: soft body). Adults (moths) ws ¼–1½", variably brown, forewing (the one in view when wings are folded) divided in thirds by two parallel fine lines; larvae (caterpillars) 2" when fully grown, bristling with tufted long hairs, generally dark brown with blue, orange and reddish markings; egg masses on small twigs, esp at crotches, covered with a gray to dark brown foam that hardens to a ± waterproof coating. Widespread on broadleaf trees and shrubs; occasionally on conifers. Order Lepidoptera (Moths and butterflies), Lasiocampidae (Lappet moth family).

**Western tent caterpillar moth,** *M. californicum.* Egg masses plastered against twigs.

**Forest tent caterpillar moth,** *M. disstria* (dis-try-a: variously striped). Egg masses evenly encircling twigs.

Our region has an outbreak of tent caterpillars on red alder every 10 years or so. Many trees are defoliated and achieve little growth for a year or two, but few are killed. The "tent" is a big erratic web of silk that affords caterpillar groups some protection and insulation during resting periods between feeding sprees. Our abundant forest tent caterpillar is the only species in the genus that does *not* build tents, though the resting colony may spin (and lie on) a rudimentary silken web.

Tent caterpillars pupate within silken cocoons under curled leaves. The cocoons are coated with a skin-irritating dust. Adult tent caterpillar moths have no functional mouthparts; they survive purely on what they ate as caterpillars, living just a few days to mate and, if female, lay eggs. (It could be said of insects generally, to varying degrees, that adults are merely ephemeral bridges from one generation of larvae to the next.) In tent caterpillars, larvae do most of the traveling as well as all the eating and growing; the chief function of their bristles is to help the wind carry them. Tiny larvae are already formed within the egg cases by winter, ready to chew their way out when the leaf buds burst in spring.

Trees are far from defenseless against foliage grazers. They need only slow the caterpillars' growth by a few percent to increase dramatically the number taken by predators and parasites. They can do this by loading their leaves with tannin and other hard-to-assimilate chemicals. However, producing the tannin requires an investment of energy that the trees do well to avoid, so they may load only some of their leaves with tannin, forcing the caterpillars to expose themselves to predatory birds while searching for palatable leaves. Or they may wait until attacked, and only then step up tannin production. Remarkably, trees may increase their tannin in response to an insect attack on a *nearby* tree of the same species, apparently sensing distress pheromones cast upon the breeze by the attacked tree. This is the clearest evidence to date of communication among plants. In the words of one enthusiastic researcher, "plants, after all, are really just very slow animals."

## White-lined Sphinx Moth

*Hyles lineata* (**hi**-leez: of woods; lin-ee-**ay**-ta: lined). Ws 2¼–3½"; large heavy-bodied moths with rapid, buzzing wingbeats suggestive of hummingbirds; forewings dark olive brown with a broad pale stripe to the tip, and white-lined veins; hindwings pink and black, much smaller. Most often seen diurnally seeking nectar. Order Lepidoptera, Sphingidae (Hawk moth family). Color p 504.

Moths of this family are unusual in being nectar feeders, active both day and night. Their deep-pitched buzzing flight enables them to hover, and they seem on first acquaintance to resemble other nectar feeders—bumble bees or, in the case of this species, hummingbirds—rather than other moths. The bee-like species have hairy bodies and largely clear wings.

## Swallowtails

*Papilio* spp. (pa-pil-ee-o: butterfly) and
*Pterourus** spp. (tair-oo-rus: wing tail). Large butterflies with "tail" lobe trailing each hindwing (tails 4 or none on some other species), dramatically black-patterned, often with a few blue and/or orange spots near tail. Order Lepidoptera, Papilionidae (Swallowtail family).

**Anise swallowtail,** *Papilio zelicaon* (zel-ic-ay-on: mythic name). Ws 2½–3½", yellow with black markings ± following veins; body all black. Widespread near desert-parsley, cow-parsnip and related plants, the food of larvae and adults both. May-Sept. Color p 504.

**Western tiger swallowtail,** *Pterourus rutulus* (root-you-lus: red—). Ws 3½–5", yellow with black "tiger" stripes across veins back from leading edge of wing, and along body. Often abundant, usually near water; eats willow and cottonwood. May-Sept. Color p 504.

*Older texts include all of these species in genus *Papilio*.

---

## Moths vs. Butterflies

Moths and butterflies together comprise the order Lepidoptera ("scaly wings"). The difference between the two, though of less importance in taxonomy than in common usage, is fairly clear:

**Moths:**

are mostly nocturnal;

have slender-tipped, or else fernlike, antennae;

perch with wings spread to the sides, either flat or angled rooflike, forewings covering hindwings;

have bigger bodies, for their wing size;

may pupate in "cocoons" of silk.

---

**Pale tiger swallowtail,** *Pterourus eurymedon* (you-rim-e-don: broad middle). Ws 3½–4½", cream yellow (females) to virtually white (males) with very heavy black "tiger" stripes. Sporadic, near Snowbrush or other shrub food sources. May-July. Color p 504.

Anise and pale swallowtails are often seen "hilltopping" on Hurricane Ridge and elsewhere; certain butterfly species use tall landmarks (trees, buildings and even TV towers will serve, where hills are lacking) to help locate mates. The males arrive first, staking out sections of ridgeline territory and defending them against conspecific males while waiting for the females. Similarly, western tiger swallowtails use riverbanks or other lineations of trees as their "singles bars."

The larvae (caterpillars) flourish by late summer, and pupate over winter. Caterpillars of the anise swallowtail have striking black/yellow back bands against green, and are found on parsley-family plants such as garden fennel or anise. Caterpillars of both tigers have eyespots; these mock eyes apparently function defensively, as do the antlerlike scent glands (on all swallowtail caterpillars) which they stick out when disturbed.

# Parnassians

*Parnassius* spp. (par-nas-ius: of Mt Parnassus). Ws 2–3"; white butterflies with gray to black markings and small red, black-rimmed eyespots. Order Lepidoptera, Papilionidae (Swallowtail family).

*P. phoebus* (fee-bus: a Greek god). Fore- and hindwings usually with several red spots; antennae ringed black/white. Washington, above timberline, typically on or near stonecrop, the larval food. June-July. Color p 504.

---

**Butterflies:**

are diurnal;

have club-tipped antennae;

perch with left and right wings pressed together vertically, except when basking in sunlight for energy;

have smaller bodies;

pupate in a shell, or "chrysalis."

---

*P. clodius* (cloh-dius: a mythic name). Red eyespots on hindwings only, usually 4; antennae all-black. Widespread in mtn clearings. June-July. Color p 504.

Named after Mount Parnassus, a symbol of everything elevated, these are strictly montane butterflies. Female parnassians are usually seen encumbered with a waxy blob on the rear end, much larger on *clodius* than on *phoebus*. Facetiously called a natural chastity belt, this structure hardens from a liquid extruded by the male during mating, and blocks other males from mating with that female.

Seemingly grayish areas on Parnassian wings simply lack the minute scales that otherwise cover butterfly wings, leaving them translucent like the wings of bees or flies. Scales give butterfly and moth wings their opacity, color, softness and sometimes iridescence. Hundreds of them dust your fingertips if you attempt to grasp a butterfly.

# Veined White

*Artogeia napi** (ar-toe-jye-a: narrow—; nap-eye: of rape, a host plant). Ws 1½–1⅞", white with veins ± darkened, esp beneath; dark-stained near body; underside sometimes yellowish. Widespread in forest; feeds on mustard family, incl toothwort. April-Aug; first of two yearly broods flies early in spring. Order Lepidoptera, Pieridae (Sulphur family). Color p 504.

# Western White

*Pontia occidentalis** (pon-shia: a name of Aphrodite; ox-i-den-tay-lis: western). Ws 1¼–2", white with variable degrees of gray checking, esp on top of forewing, and pastel gray-green to yellow-brown vein linings, esp beneath hindwing. Abundant on mountaintops, ± ubiquitous exc in deep forest. July-Aug. Order Lepidoptera, Pieridae (Sulphur family). Color p 504.

Pairs of butterflies, especially whites, are often seen following each other upward in tall spirals. It is tempting to imagine

---

*Some authorities include one or both of *Artogeia* and *Pontia* in a broad genus *Pieris*.

these spiral flights as prenuptial stairways to heaven, but close study suggests a countererotic function—an already-mated female's last-ditch effort to escape an unwanted suitor by wearing him out in flight.

## Western Sulphur

*Colias occidentalis* (coh-lius: a name of Aphrodite). Ws 1¼–2″, yellow (rarely orange or albino) usually with a dark dot in center of forewing and a white, ± dark-ringed dot on hindwing; male has a heavy black border above and a faint pink fringe all around wings. Widespread, esp streambanks and roadsides; larvae feed on legumes, e.g., lupine, sweet-pea. June-Aug. Order Lepidoptera, Pieridae (Sulphur family). Color p 505.

Of several very similar sulphurs here, this one is commonest, and probably the only one in the Olympics.

## Sara's Orange-tip

*Anthocharis sara* (an-thoc-a-ris: flower grace: sah-ra: after Sara). Ws 1¼–1¾″, white to (females only) ± yellow with very distinctive deep orange forewing tips, and gray-green marbling under hindwing. Widespread in sunny habitats; larvae feed on mustard family plants. Among the earliest butterflies—late March (at low elevs) to August (high). Order Lepidoptera, Pieridae (Sulphur family). Color p 505.

## Alpines

*Erebia* spp. (er-ee-bia: underworld, i.e., darkness). Ws 1½–2″, dark gray-brown with marginal rows of orange-haloed black eyespots on all 4 wings, the haloes coalescing into an orange band on the forewings esp in Vidler's. Order Lepidoptera, Satyridae (Satyr family).

**Common alpine,** *E. epipsodea* (ep-ip-so-dia: upon itself?). Alpine to low E-side marshes and meadows, Cas only. June-July. Color p 505.

**Vidler's alpine,** *E. vidleri* (vid-ler-eye: after Vidler). Strictly alp/subalpine, in Wash and BC only. July-Aug. Color p 505.

Rarely attracted to flowers, alpines and arctics feed on grasses. They bask in preparation for flight, augmenting their metabolic energy with solar energy collected in their spread wings.

# Arctics

*Oeneis* spp. (ee-nee-iss: a mythic king). Orange-brown exc silvery gray-brown under hindwing (for camouflage, with wings folded, against bark or soil); 1–3 smallish black eyespots; forewing long and angular. Order Lepidoptera, Satyridae (Satyr family).

**Sierra Nevada arctic,** *O. nevadensis* (nev-a-den-sis: of the Sierra Nevada). Ws 2–2½"; in Cas, esp E-slope. May-Sept of even-numbered years only, with rare exceptions. Color p 505.

**Chryxus arctic,** *O. chryxus* (crix-us: gold, misspelled?). Ws 1¾–2"; sporadic at mid to high elevs in Wash and BC. June-Aug. Color p 505.

Arctics, like anise swallowtails (page 464) are "hilltoppers"— they emerge into adulthood with a powerful drive to travel uphill. The drive reverses (in females only) upon mating; eggs are laid at lower elevations so the next generation may have an uphill path available. Ridgelines thus acquire dense, evenly

---

## Eyespots

The abundance of "eyespots"—pale dots ringed with dark, and often again with pale—as decorative devices on butterfly wings attests to some selective advantage, presumably protection from predators. Birds are fooled into pecking at the fake eyes, thinking them vital targets when the opposite is the case. Butterflies have little trouble flying with somewhat tattered hindwings—where most eyespots and all "tails" (similar diversionary ornaments) are located.

Eyespots have another, more remarkable, protective use. All butterflies go through a few flightless hours when their wings are soft and useless. Wings first take form all wadded up inside the pupal shell, or "chrysalis"; as soon as the adult butterfly emerges, fluids pumped in through the veins extend the wings to their proper shape. Then they need several hours to dry into rigidity. Apparently, large eyespots on the upper wing surface save many butterflies by actually frightening birds away during those vulnerable hours. The butterfly responds to an exploratory first peck by abruptly opening its folded wings; between the two "eyes" suddenly revealed, the bird imagines a larger face than it had meant to take on.

---

distributed populations of hilltopping males, with scattered cameo appearances by eligible maidens. Along the leading edge of each forewing, males have a big dark stainlike patch consisting of "scent scales" which emit pheromones attractive to females.

## Ochre Ringlet

*Coenonympha ampelos\** (see-no-nim-fa: common nymph; am-pel-os: a youth loved by Dionysus). Ws 1¼–1¾", white-steaked rich ochre all over exc under hindwing, where ± gray-green; sometimes with 1 or more eyespots near margins underneath. Cascades, abundant around grasses, the larval food. May-Sept. Order Lepidoptera, Satyridae (Satyr family). Color p 505.

## Wood Nymphs

*Cercyonis* spp. (ser-**sigh**-o-nis: after a legendary thief?). Drab black-brown above, barklike brown beneath; 2 black eyespots (sometimes indistinct on males) on each forewing; spots usually yellow-rimmed beneath, and sometimes above on females; up to 6 tiny eyespots often on hindwings. Cascades; abundant in ponderosa pine zone; larvae eat grasses, adults various flowers. Order Lepidoptera, Satyridae (Satyr family).

**Least wood nymph,** *C. oetus* (ee-tus: doom, i.e., blackness). Also dark wood nymph. Ws 1½", almost as dark beneath as above; second forewing eyespot small and often inconspicuous. July-Aug.

**Large wood nymph,** *C. pegala* (peg-a-la: mythic name). Ws 1¾–2¾", much paler beneath. June-Aug. Color p 505.

## Lorquin's Admiral

*Basilarchia lorquini†* (bay-sil-**ark**-ia: royal chief; lor-**kwin**-eye: after Pierre J. M. Lorquin ). Ws 2¼–2¾", black above exc red-orange forewing tips and a heavy white band (broken by black veins) across both wings parallel to margin; dramatically striped beneath, mostly russet and white; flies with twitchlike beats alternating with gliding. Widespread, notably on streamsides, the larvae feeding on willow

\*Some authorities regard *ampelos* as a subspecies of *C. tullia*.
†Older texts place this species in genus *Limenitis*.

and cottonwood leaves. June-Sept. Order Lepidoptera, Nymphalidae (Brush-footed butterfly family). Color p 505.

The handsome admiral is no sipper of nectar, but a scavenger of dung and dead things. In the large and colorful family Nymphalidae, or Brush-footed butterflies, the front two legs are short, brushlike appendages useless for walking or perching.

## Painted Lady

*Vanessa cardui* (va-nes-a: derivation moot; car-dew-eye: of thistles). Ws 2–2¼″, salmon to orange above with short white bar and black mottling near forewing tip, row of 4–5 tiny eyespots near hindwing margin; larvae are spiny, lilac with yellow lines and black dots, and typically spin nets around thistle tops. Occasionally ubiquitous. May-Aug. Order Lepidoptera, Nymphalidae (Brush-footed butterfly family). Color p 505.

The painted lady normally resides in climates warmer than ours. Occasionally the population explodes, and painted ladies fly thousands of miles in all directions, becoming the most nearly worldwide of all insect species. They do not survive winter here, and very few make the trip up from Mexico in nonoutbreak years, so they have been rare here except (most recently) in 1958, 1966, and 1973. The next outbreak looks overdue. Preferred foods are thistles and other composites, but when overcrowded these larvae become omnivorous.

## Mourning Cloak

*Nymphalis antiopa* (nim-fay-lis: nymph—; an-tie-a-pa: a leader of Amazons). Ws 2–3¼″; ± iridescent deep red-brown above with full-length cream yellow border lined with bright blue spots; drab beneath, barklike gray-brown with dirty white border; margins ragged. Widely scattered on streamsides with cottonwood and willow. Any season. Order Lepidoptera, Nymphalidae (Brush-footed butterfly family). Color p 506.

The handsomely distinctive mourning cloak is known throughout the Northern Hemisphere. The adults hibernate—occasionally coming out to fly on sunny winter days—and sometimes aestivate through the dry late summer. They may breed during both spring and fall active seasons.

# California Tortoiseshell

*Nymphalis californica*. Ws 1¾–2¼"; mostly yellow-orange above with a continuous dark margin and several large dark blotches along leading edge; mottled gray-brown beneath for camouflage against bark—very similar to anglewings (below) but with more regular wing outline. Widespread, abundant in occasional years; larvae eat snowbrush. Any season. Order Lepidoptera, Nymphalidae (Brush-footed butterfly family). Color p 506.

In cold regions, tortoiseshells and anglewings may overwinter as either adults or pupae, employing glycerol as antifreeze for the bodily fluids. Getting an early start in spring, California tortoiseshells may breed two or even three generations a year here. Their populations are volatile, going through wild boom-and-bust cycles for unclear reasons. During boom years, smashed butterflies clog car radiators and coat the pavement on mountain passes, while the larvae defoliate snowbrush far and wide.

# Milbert's Tortoiseshell

*Aglais milberti* (a-glay-iss: Brightness, one of the Graces; mil-bert-eye: after Milbert, a friend of the namer). Ws 1½–2", dark brown above with 2 ± distinct orange blotches on each leading edge and a broad orange and yellow (to white) band just inside the narrow dark-brown margins; gray-brown beneath with ± light/dark patterns mirroring the upperside. Widespread, common in high meadows; larvae eat nettles, adults often visit daisies; any season. Order Lepidoptera, Nymphalidae (Brush-footed butterfly family). Color p 506.

Pale green egg clusters—or bristly small black web-spinning caterpillars—on nettles are likely those of Milbert's Tortoiseshell.

# Anglewings

*Polygonia* spp. (poly-go-nia: many angles). Ws 1¾–2¼", wing margins lobed fancifully, as if tattered; burnt orange above with dark margins and many scattered black blotches; silvery gray-brown beneath. Ubiquitous; larvae eat currants, willows, rhododendron, nettles, etc. Any season (exc over snowpack). Order Lepidoptera, Nymphalidae (Brush-footed butterfly family). Color p 506.

The odd-shaped anglewings are astonishingly well camouflaged against bark. One moment the creature flits about almost too fast to follow, and the next moment it disappears by abruptly alighting with wings folded. A tiny but distinct pale mark centered beneath the hindwing is variously boomerang, comma, C, or V-shaped on different species of anglewing.

## Greater Fritillaries

*Speyeria* spp. (spy-ee-ria: after Adolph Speyer). Ws 1½–3", orange with ± checkerlike black markings above; hindwing gray to golden tan beneath, with many round silvery-white to cream spots; forewing pale ochre to golden brown beneath, with blackish checks sometimes giving way to a few marginal white spots. Ubiquitous; larvae eat violets, adults suck nectar from composites, penstemons, dogbane, stonecrop, etc. May–Sept. Order Lepidoptera, Nymphalidae (Brush-footed butterfly family). Color p 506.

This gorgeous genus is behind many supposed monarch butterfly sightings in our mountains. Actual monarchs, *Danaus plexippus,* depend on milkweed as a larval host, and hence are rather rare here. *Speyeria* is considered a difficult genus even among specialists; the species, varieties, and hybrids are far from being sorted out yet. One species, *S. cybele,* is sexually dimorphic, with females black and pale yellow quite unlike the males, and unlike any other fritillaries. The name fritillary (in both lilies and butterflies) refers to a Roman dice-box with checkered markings.

## Checkerspots

*Occidryas** spp. (oc-sid-rius: Western nymph) and
*Charidryas** spp. (ca-rid-rius: grace nymph). Ws 1¼–2", generally 3-colored above—blackish, red-orange and pale yellow in a fine check pattern (sometimes just black and cream in *O. colon* but never just black and orange as in fritillaries, and the black rather than the orange tends to predominate); mostly brick-red with yellow to white checks and black veins beneath. Often abundant in sunny dry and rocky habitats; larvae feed on paintbrush, other figwort-family plants, and snowberry (*O. colon*), or on asters (*C. hoffmanni*). April–Aug. Order Lepidoptera, Nymphalidae (Brush-footed butterfly family). Color p 506.

*Checkerspots have lately gone through much taxonomic splitting; *Charidryas* was previously within genus *Chlosyne,* *Occidryas* previously within genus *Euphydryas.*

Identification of these notoriously variable species usually requires studying their genitalia under a microscope. One predominantly black above, with a trace of red-orange near the tips and leading edge, is likely a snowberry checkerspot, *O. colon,* if snowberry grows nearby.

## Field Crescentspot

*Phyciodes campestris* (fis-eye-o-deez: seaweed-red; cam-**pes**-tris: of the fields). Ws 1¼", largely black above with rows of red-orange and yellow-orange spots, the outermost spots crescent-shaped; patterns similar beneath but much paler, orange and yellow-brown with little or no dark brown. Common in sun at mid to high elevs. June-Sept. Order Lepidoptera, Nymphalidae (Brush-footed butterfly family). Color p 506.

## Purplish Copper

*Epidemia helloides\** (ep-id-ee-mia: ubiquitous; hel-oh-**eye**-deez: like species *helle*). Ws 1", males dark brown above with purplish iridescence sometimes catching the light, and an orange zigzag along hindwing margin; females yellow-orange above with brown speckles and borders; both yellow-orange-brown beneath with dark speckles and an orange zigzag on hindwing. Widespread. May-Oct. Order Lepidoptera, Lycaenidae (Copper family). Color p 507.

## Mariposa Copper

*Epidemia mariposa\** (mair-ip-oh-sa: butterfly in Spanish). Ws 1", males dark brown above, with purplish iridescence and orange zigzag (see above) often obscure or not evident; females lighter brown to orange, with dark speckles and borders; hindwings ashy gray beneath, forewings orange with speckles. Often abundant in mid- to high-elev meadows, bogs, and clearings; larvae feed on bistort and knotweed. July-Aug. Color p 507.

Iridescence on butterflies usually involves two types of microscopic scales, regularly interspersed on the wing surface, the different scales held at different angles—not unlike the way fabric designers combine contrasting warp and weft threads to make an iridescent satin weave. Iridescence is common in

---

\**Epidemia* was recently separated out from a large genus *Lycaena.*

tropical butterflies; here we find it among coppers, a large group of small, mostly orange-brown butterflies.

## Sooty Hairstreak

*Satyrium fuliginosum* (sa-tee-rium: satyr—; foo-li-jin-oh-sum: sooty). Ws 1", uniformly dark gray above, ashy gray beneath with faintly lighter and darker speckles. E slope of Cascades, where locally abundant in high meadows with lupine, the larval food plant. June-Aug. Order Lepidoptera, Lycaenidae (Copper family). Color p 507.

This species lacks the "hair" that hairstreaks are named for—a pair or quartet of hair-thin wing "tails"—but it sure has the "streak." Notoriously hard to keep track of when streaking around, it can sometimes be seen when flowers hold its attention. The sooty is similar (including a shared attraction to lupines) to females of the large group known as blues. While male blues are indeed usually blue, the unblue female blues tend to be a bit browner than the sooty hairstreak on top, a bit whiter and more strongly speckled underneath.

## Common Blues

*Lycaeides** spp. (lie-see-id-eez: like genus *Lycaena*) and
*Icaricia** spp. (ic-a-ree-shia: after Icarus) Ws 1", males blue above, females brown, darkening at margins but then white-fringed; grayish white beneath, dark-speckled esp on forewing. Order Lepidoptera, Lycaenidae (Copper family).

**Northern blue,** *L. argyrognomon*† (ar-jy-rog-no-mon: silver mark). May have rows of orange crescents parallel to margin. Abundant at mid to high elevs; larvae eat lupine and other legumes. July-Aug. Color p 507.

**Common western blue,** *I. icarioides* (ic-airy-o-eye-deez: like *Icaricia*—a nonsensical name now that it is placed in this genus). Widespread, always near lupine, the larval food. May-Aug.

Blues often congregate to sip water from mud. Caterpillars of many species, including the common western, are attended by ants who eat "honeydew" secreted by the caterpillars. In this mutualistic symbiosis, the "stock-herding" ants sometimes attack and repulse beetles that parasitize caterpillars.

*These genera were recently separated out from a large genus *Plebejus*.
†A recent revision recognizes this species in Europe but reidentifies the American populations as a different Eurasian species, *L. idas*.

# Arctic Blue

*Agriades franklinii\** (a-gry-a-deez: wild—?; frank-lin-ee-eye: after Sir John Franklin, below). Ws ¾", males silvery blue above, females rusty gray-brown with white-haloed dark spots; both sexes ± white-fringed; light gray-brown beneath with white-haloed spots. Strictly high-elev (6,500'+) here; larvae eat shooting stars. Late June-Aug. Order Lepidoptera, Lycaenidae (Copper family). Color p 507.

Scarce and rather drab, this doughty blue is nonetheless conspicuous from time to time as the only butterfly flying in gray weather on high windswept ridges. It ranges around the globe in Arctic lowlands.

# Dotted Blue and Square-spotted Blue

*Euphilotes†* spp. (you-fill-oh-teez: after genus *Philotes*). Ws ½–¾", males blue above, females brown, both with a row of ± distinct

\*In various texts, the arctic (or "high mountain") blue is species *glandon* or *aquilo* in genus *Plebejus* or *Agriades*.
†*Euphilotes* was recently separated out from a large genus *Philotes*.

---

**Sir John Richardson** and **Thomas Drummond** went on **Sir John Franklin**'s second expedition into Arctic Canada. Richardson had also been on the first, barely returning alive and certainly without his specimens, so it was game of him to try again. Twice was enough, though; he lived on as an eminent naturalist of the day and author of a fauna of boreal America (1829–36), rather than perishing on Franklin's ill-fated third expedition.

Young Drummond, meanwhile, soon separated from the party and spent the winter alone—two months without seeing a soul, six months without a person he shared a language with—in an improvised brush hut in Alberta, relying on game for food. The doughty Scot summed up his relations with grizzly bears thus: "The best way of getting rid of the bears, when attacked by them, was to rattle my vasculum or specimen-box, when they immediately decamp." He had known David Douglas (page 18) in Scotland; chance brought them together again in the Rockies in 1827, and they took the same ship back to England. They had a great time talking botany. Before his unexplained death in Cuba in 1835, Drummond spent two years on the first extensive botanical exploration of Texas.

---

orange spots near hindwing margin; marginal fringes of alternate black/white dashes; bluish white beneath, with dark speckles. E slope of Cascades; larvae feed on sulphur (and other) buckwheat. Order Lepidoptera, Lycaenidae (Copper family).

**Dotted blue,** *E. enoptes* (ee-nop-teez: mirror).

**Square-spotted blue,** *E. battoides* (bat-oh-eye-deez: like species *baton*). Color p 507.

The square-spotted breeds once a year and flies in May and June; the dotted has two annual generations, one flying in April and May and the second in July. Other than that, specialists have found puzzlingly little to separate these two species behaviorally, geographically, or physically, except for microscopic differences in their genitalia.

## Elfins

*Incisalia* spp. (in-sis-ay-lia: cut—). Order Lepidoptera, Lycaenidae (Copper family).

**Brown elfin,** *I. augustus* (aw-gus-tus: majestic, a perverse name for this species). Ws ¾–1″, plain brown above; no white fringe; reddish brown beneath with vaguely-defined dark inner portion of hindwings. Widespread; larvae feed on various shrub and tree leaves. April-June. Color p 507.

**Western pine elfin,** *I. eryphon* (er-if-un). Ws ¾–1″, light brown above with black/white-dashed marginal fringe usually visible; sharply zigzag-patterned beneath, light and dark reddish brown with black; older larvae pine-needle green, with paired whitish lengthwise stripes. Often abundant near pines, the larval food plant; adults visit willows, spiraea, buckwheats, etc. April-June. Color p 507.

**Early elfin,** *I. fotis mossii* (fo-tiss: a warming, i.e. spring; moss-ee-eye: after Moss). Ws ¾″, all brown (females ± reddish) with black/white-dashed marginal fringe ± visible; two-toned beneath, the paler gray-brown outer sections sharply, irregularly divided from the darker inner sections. Sporadic in ± low rocky, sunny habitats with stonecrop, the larval food plant. March-early May. Color p 507.

The early elfin is conspicuous only by its earliness; it is the first butterfly to emerge from a pupa in the spring, sometimes while snow still covers much ground. Any other butterflies flying then (e.g., anglewings or tortoiseshells) have almost surely hibernated as adults.

# 14

# Other Creatures

## Banana Slug

*Ariolimax columbianus* (airy-o-lie-max: compound of two Greek-derived slug genera; co-lum-be-ay-nus: of the Columbia River). Very large slugs (commonly 4–6", up to 10¼", when crawling); back netted with fine, dark, mostly lengthwise grooves, exc for the smooth mantle capping about ¼ the length, near the front; large breathing-hole near right-hand edge of mantle; color variable, typically olive, with blackish blotches varying from covering nearly the entire (hence blackish) slug, to merely a single spot on the mantle or no spots at all. W-side forest floor. Arionidae. Color p 503.

Most slugs evolved from snails, rather than vice versa, by losing part or all of their shells. (Banana slugs retain a small vestigial shell, buried in the mantle or "cap.") Shell atrophy occured repeatedly in snails of regions short on both the chief raw material (calcium) and the chief need (drought) for snail shells. The primary function of a land-snail shell is to provide a handy moisture-conserving shelter when it gets a bit warm and dry outside. Slugs are still found almost exclusively in moist, heavily shaded regions, often with calcium-poor volcanic soils. When they need to conserve water they retreat under vegetation or earth in lieu of a shell, and shrink to a fraction of their crawling length. Water loss through the skin is a problem on dry days, but water expenditure in slime for crawling is far greater, even on wet days. Slime varies from slippery to tacky, and is used for traction and for lubrication of the slug's smooth, tender sole. Slug activity is restricted to nighttime and dim, humid daytime conditions, because only then can enough humidity be absorbed to replenish the water consumed in slime.

Protection from predators is only a secondary function of snail shells. Many slugs make up for that loss by loading their slime with bitter or caustic chemicals. (Fried banana slugs are sometimes eaten, after a vinegar bath to remove the slime, by Northwesterners of French and German extraction. They aren't all that different from escargots.) Some are also brightly colored to advertise their repulsiveness; "banana slug" refers to the chrome yellow color of the southernmost races (if the word is not too strong) of this species and its close kin, *A. californicus*. Other Northwest slugs, genus *Prophysaon*, have a lizardlike ability to self-amputate their tails to decoy predators.

Slugs are hermaphrodites. In some types, each individual produces its sperm and its eggs asynchronously, making self-mating impossible. In others fertility is synchronous, and two slugs may fertilize each other in one 12-to-30-hour entwining. Banana slugs are notorious for chewing off their penises to conclude mating (both partners chew), probably because their unusually large organs are more difficult for them to withdraw than to regenerate later. Since touch is the only well-developed sense in snails and slugs, by touch they must identify potential mates. To make themselves recognizable, the species have evolved a bizarre assortment of palpable structures—sharp little jabbing needles, delicately branched sperm packets, and overdeveloped penises, all with dimensions peculiar to the species. Banana slug penises are large, but nothing like those of one rare race in the Alps—32½" tumescences dangling from 6" slugs.

A slug eats a wide variety of fungi and plants, rasping away at them with a tongue covered with several thousand minute teeth.

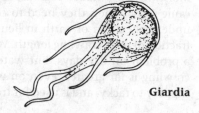

**Giardia**

## Snow Worms

*Mesenchytraeus** spp. (mez-en-kye-tree-us: after genus *Enchy-traeus*). Also **glacier worms**. Slender worms, average 1" long, body segmented or ringed by fine constricted bands. In late-lying snowfields or glacier surfaces, spring and summer. Enchytraeidae.

Snow worms are "segmented worms" (oligochaete annelid worms) related to earthworms and, more closely, to the enchytrae cultured and sold as aquarium food. They eat snow algae such as watermelon-snow (page 482) and may be eaten by birds like dippers and rosy finches. Some kinds of spiders, insects, and nematodes (more primitive worms) are also adapted for life in snow. All are active at temperatures close to 32°, and able to survive even colder temperatures either by dehydrating to prevent tissue freezing, or by restricting freezing to their intercellular spaces. Both strategies require synthesis of special proteins to lower the protoplasmic freezing point to well below 32°.

## Giardia

*Giardia duodenalis†* (jar-dia: after Alfred Giard; du-od-en-ay-lis: of the digestive tract).

One single-celled "animal" of intense interest to hikers is this intestinal parasite. Much remains to be learned about giardia, but two points seem clear: first, it can cause extremely unpleasant symptoms, colorfully described by the Oregon State Health Division as "sudden onset of explosive, watery, foulsmelling diarrhea with nausea and anorexia and marked abdominal distention with flatulence" perhaps accompanied by

---

*A 1971 taxonomic revision of the Oligochaete worms declared the widely used name *Mesenchytraeus* invalid, replacing it with *Analycus*, but this change has not been widely adopted.

†Taxonomy of *Giardia* is poorly understood; *Giardia* in humans is often referred to as *G. lamblia*, but that name is of dubious validity.

The five-kingdom system outlined on p 572 places *Giardia* in the (non-Animal) Kingdom Protoctista. In traditional two-kingdom systems it is a protozoan animal.

"chills and low-grade fever, vomiting, headache and malaise"; second, those who wish to incur no risk of picking it up in the mountains must treat all drinking and toothbrushing water.

Of the many *un*clear aspects of giardia, the burning issue for nonscientists is how to avoid it. There are ways, but all of them sacrifice the immediacy, flavor, and spiritual fulfillment of prostrating ourselves to the mountain stream goddess and drinking deeply. (If there is a safe place to do that still, it must be at the foot of a glacier or snowfield—almost as pure as the driven snow. Springwater isn't necessarily safe, since neither passage through earth nor freezing are any problem for giardia cysts, a communicable dormant phase of the creature.) Bringing water to a full boil, even for an instant, is 100% effective at any elevation in our range. Disinfectant tablets (especially tetraglycine hydroperiodide) are quite effective, but only if given enough time to work; the colder the water, the more time they need. The better filtering devices on the market are effective, albeit tedious and expensive. We can expect improved technology in this field over the next few years.

Symptoms usually appear, when they appear at all, one to three weeks after exposure. They typically last three or four days at a time, but may recur *ad nauseam*. If you think you've got them, see a doctor, who will probably send a stool sample to the lab. If it comes back negative, try again and again, since the cysts pass out of your system only at cyclic intervals. The three medications currently used in the U.S. for giardiasis each have unpleasant side effects and many contraindications, including pregnancy; each is also a carcinogen suspect. Atabrine is the most effective, Flagyl the close second and a bit less nasty. A cure often requires more than one weeklong course. Some people (pregnant ones especially) may prefer giving their own systems more time to bring the giardia into balance.

The deeper puzzle about giardia is why we weren't hearing about it 20 or 50 years ago. Many people assume that giardia arrived recently in the back country, brought by the hordes of backpackers and their dogs (both species often negligent of burying their scats) but this is impossible to prove, and several specialists doubt it. The journals of Northwest explorers like Meriwether Lewis and David Douglas are rife with bowel complaints sounding not unlike giardiasis. Giardia is distributed worldwide, in both underdeveloped and industrialized lands, and abounds within feces of dogs, cats, cows, horses, moose, coyotes, beaver, muskrats, water voles, etc. Some giardia strains (of undetermined taxonomic rank) appear highly host-specific, while others may be transmitted from one mammal species to another, including humans. (Beaver have been misleadingly singled out as culprits.) Human populations *not* reporting giardiasis turn up many giardia carriers in random samples; an estimated 10–20% of Americans are now hosting giardia asymptomatically—never experiencing symptoms. I'll bet I am, for one, along with most of you who drank from as many mountain streams as I did in years past.

Almost certainly, part of the explanation is tolerance of at least some strains of giardia. In one study in Colorado, odds of coming down with symptoms were inversely correlated with length of residency there. (Children are also more susceptible than adults; however, infants, who might otherwise be at greatest risk, seem to be protected by a giardicidal toxin in normal human milk—an argument against the use of infant formula in the Third World, where diarrheal dehydration is a leading cause of infant mortality.) The mix of Cascade backcountry sheepherders, loggers, hunters and other recreationists of 20 to 50 years ago presumably included a much higher percentage of long-term area residents than today's backpackers, and a higher proportion of "rugged types" unlikely to report or seek treatment for giardiasis. Uneven reporting and scarce records hinder appraisal of an alleged increase in the disease.

It is still possible that we rarely catch giardia from other mammals, after all, and mainly have human slobs to blame. If this proves true, we can at least have the satisfaction of feeling a little bit safer in very remote areas and earlier in the season, since the cysts are not thought to live more than a few months outside of hosts. In the meantime, don't count on it. And since it *is* known that humans can transmit the parasites to other humans, always defecate well away from any body of water, and cover up with a few inches of dirt.

# Watermelon Snow

*Chlamydomonas nivalis* (clam-id-o-**mo**-nus: mantled unit; ni-**vay**-lis: of snow).

Pinkness in late-lying snowfields consists of pigments in living snow algae. These algae are the producers in an entire food cycle that operates in snow. Snow worms, protozoans, spiders and insects are grazers on the algae, and are eaten in turn by predators like the rosy finch. Droppings from the predators, pollen blown from downslope, and bodies of producers and grazers provide food for decomposers (bacteria and fungi) that complete the cycle. The algae live on decomposition products, and minerals that fall in snow.

Most of the pinkness is in energy-storing oils within "resting spores" that sit out the winter wherever they happen to end up in the fall. In spring, under many feet of new snow, they respond to increasing light and moisture by releasing four "daughter cells" that swim up to the surface. The year's crop becomes visibly pink only where most concentrated. A footstep concentrates algae into visibility by compressing the top inch of snow to a thin film, or algae may gravitate into random depressions in the surface. Once concentrated, they melt their depressions deeper by absorbing more infrared radiation than the surrounding white. By late summer, this circular process (windblown dirt also contributing color) creates the texture we call "sun-cups." Notice how footprints across a snowfield are incorporated into the pattern.

Some say watermelon snow tastes like watermelon; others warn of diarrhea.

Over 100 species of snow algae have been named. *C. nivalis* is our most abundant by far, but the watermelon-snow you see may include other red species. Yellow, green, and purple snow are also known. Taxonomically, *Chlamydomonas* is a green alga. Its chlorophyll and celluloselike cell wall would seem to make it a sort of plant, but its swimming ability and well-developed eyespot are more animallike—a duality that argues for recognizing more kingdoms than just the plants and animals (see page 572).

Autumn galerina, p 268.  Destroying angels (*virosa* and *bisporigera*), p 267.

Fly amanita, p 265.  Honey mushroom, p 268.

Violet cortinarius, p 269.  Woolly chanterelle, p 269.

**Fungi:** poisonous (or not recommended)

Hedgehog mushroom, p 271.

Yellow chanterelle, p 270.

Admirable boletus, p 273.

King boletus, p 273.

Suillus, p 274.

Warted giant puffball, p 279.

Oyster mushroom, p 272.

Sulphur shelf fungus, p 278.

Angel-wings, p 272.

Purple-tipped coral, p 276.

Snow morel, p 281.

Bear's-head coral, p 277.

Cauliflower mushroom, p 277.

Map lichen, p 285.

Jewel lichen, p 288, and Rock tripe, p 288.

Worm lichen, p 292.

Corkir, p 286, and Puffed lichen, p 290.

*neglecta*
Imperfect lichens, p 287.

*candelaris*

Wolf lichen, p 290.

Globe lichen, p 291.

Iceland-moss, p 292.

Matchstick lichen, p 293.

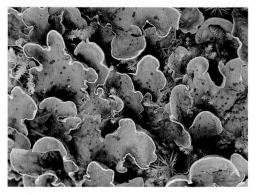

Green dog lichen, p 289.

Oregon lungwort, p 289.
(underside)

Reindeer lichen, p 293.

British soldiers, p 294.

Pixie goblets, p 294.

Thallose liverwort, p 259.

Horsehair lichen, p 291, Wolf lichen, p 290, and Old-man's-beard, p 291.

Shrew, p 297.

Boomer, p 307.

Pika, p 304.

Snowshoe hare, p 302.

Red tree vole, p 326.

Northern flying squirrel, p 316.

Heather vole nest, p 328.

Gopher core, p 317.

Olympic marmot, p 307.

Beaver, p 318.

Douglas squirrel, p 311.

Porcupine, p 329.

Golden-mantled squirrel, p 314.

Jumping mouse, p 328.

Western gray squirrel, p 314.

Yellow pine chipmunk,
p 313.

Coyote, p 331.

Bear scratchings, p 338.

           **Mammals**

Otter, p 347.

Striped skunk, p 348.

Bobcat scratchings, p 350. Wolverine, p 346.

Marten, p 344.

Fisher, p 345.

Short-tailed weasel, p 342.

Badger, p 348.

Cougar, p 352.

Mule deer buck ''in velvet,'' p 354.

Elk, p 358.

Mountain goat, p 362.

**Mammals**

Mallard, p 372.

Harlequin duck, p 372.

Common merganser, p 373.

Pied-billed grebe, p 371.

Prairie falcon, p 382.

Osprey, p 376.

Great blue heron, 371.

Goshawk, p 379.

Bald eagle, p 376.

American kestrel, p 384.

Red-tailed hawk, p 380. Golden eagle, p 381.

Vaux's swift, p 395. White-tailed ptarmigan, p 387.

Downy woodpecker, p 398. Blue grouse, p 386.

Northern flicker, p 398.

Common nighthawk, p 394.

Spotted sandpiper, p 389.

Rufous hummingbird, p 396.

American dipper, p 408.

Townsend's solitaire, p 410.

Western flycatcher, p 400.

Solitary vireo, p 412.

Tree swallow, p 400.

Yellow-rumped warbler, p 412.

Red-breasted nuthatch, p 407.

Cassin's finch, p 415.

Hermit thrush, p 411.

Song sparrow, p 414.

Pine siskin, p 416.

Dark-eyed junco, p 414.

Red crossbill, p 416.

Western red-backed salamander, p 425.

Mountain bluebird, p 410.

Oregon slender salamander, p 425.

Brown creeper, p 407.

Western toad(s), p 426, and
Cascades frog, p 427.

Rough-skinned newt, p 423.

Long-toed salamander, p 424.

Pacific giant salamander, p 424.

Larch Mountain salamander, p 425.

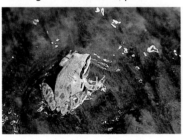

Pacific treefrog, p 427.

Olympic salamander, p 423.

Red-legged frog, p 426.

Banana slug, p 477.

Butterfly photos are male upperside unless otherwise stated.

Anise swallowtail, p 464.
Western tiger swallowtail, p 464.
Pale tiger swallowtail, p 464.

White-lined sphinx moth, p 463.
Parnassians, p 465: *phoebus*,
*clodius*.
Western white, p 466
Veined white, p 466.

Western sulphur, p 467.
Sara's orange-tip, p 467: male, female.
Alpines, p 467: Common, Vidler's.
Ochre ringlet, p 469.

Arctics, p 468: *nevadensis*, *chryxus*.
Large wood nymph, p 469 (underside).
Lorquin's admiral, p 469.
Painted lady, p 469.

Mourning cloak, p 469.
Tortoiseshells, p 471: California,
Milbert's,
Anglewings (*P. faunus*), p 471:
    male upperside,
male underside.

Greater fritillary, p 472 (*S. cybele*),
    male upperside,
male underside,
female underside.
Checkerspot, (*O. colon*), p 472.
Field crescentspot, p 473.

Purplish copper, p 473.
Mariposa copper,
    p 473: male,
female underside.
Sooty hairstreak, p 474.

Northern blue,
    p 474: male,
female upperside,
male underside.
Arctic blue, p 475.
Square-spotted blue,
    p 475.

Elfins, p 476: brown,
Western pine (male upper),
Western pine (female under).
Early (male underside).

Basalt "cinder."
Andesite with hornblende crystals.
Dacite, p 535.

Basalt, p 531.
Andesite, p 534.
Rhyolite, p 535.

Pumice, p 536.
Tuff.

Obsidian, p 537.
Tuff, p 537.

Granite, p 542.
Gabbro, p 544.

Granodiorite, p 543.
Periodotite, p 544.

Rocks: intrusive igneous

Mudstone, p 545.
Arkose, p 546.

Shale, p 545.
Graywacke, p 546.

Limestone, p 550.
Quartz-rich vein rock.
Conglomerate, p 547.

Chert, p 549.
Quartz crystal, p 549.
Breccia, p 548.

Slate, p 551.
Hornfels, p 553.

Phyllite, p 551
Greenstone, p 553.

Gneiss.
Gneiss, p 554.

Greenschist, p 553.
Schist with garnet, black mica, p 552.

**Rocks:** metamorphic

# 15

# Geology

## Cascade Volcanoes

Few Northwesterners in the first three quarters of the twentieth century thought of the Cascades as active volcanoes. Even the Encyclopedia Britannica said, rather self-contradictorily, that they were "considered extinct, but must have been active within historical time." Nineteenth-century settlers left many reports of eruptions, some factual and others fanciful, but most sounding curiously blase. That attitude passed, over the generations, into inexplicable obliviousness. Rivers were dammed close to Mt. Baker and Mt. St. Helens—known to have erupted 120 years ago—and papers were published asserting that Mt. Hood, Mt. Rainier, and Glacier Peak show no sign of activity for thousands of years past.

Then in 1975, Mt. Baker heated up, melting some of its crater ice into a lake and emitting steam and traces of ash. Geologists, now finding an attentive audience, saw no strong sign of imminent eruption, but pointed out the obvious—Mt. Baker is not extinct. Further, they found signs of many ash and/or dome eruptions, and many mudflows with and without eruptions, within the last 600 years on all five above-named cones. U.S. Geological Survey hazard evaluations singled out mudflows as a hazard and Mt. St. Helens as a likely site of hazardous eruptions in the near term.

It was a good call. St. Helens burst into activity in 1980,

---

and persisted in more or less activity for several years. The climactic eruption of May 18 did not involve a great amount of magma (by world standards or even Mt. St. Helens' own) but it was the most violent and destructive eruption yet beheld in an admitted State of the Union. (Alaska's 1912 eruption of Mt. Katmai was far bigger). It took an interesting form, though hardly unique. The eruptive mechanism had five main steps:

1. Over the two months prior to May 18, dacite magma ascended into the mountain. An old plug dome filling the central vent diverted the magma toward a point on the upper north slope between and above two more recent lateral plug domes, the Goat Rocks and the Dog's Head. Groundwater in the cone vaporized and caused small steam eruptions. The upper north slope gradually bulged outward over the intruding magma, weakening the rock and oversteepening the slope.

2. At 8:32:10 a.m. on May 18, a minor quake triggered the oversteepened bulge into an enormous landslide. This took the lid off the underlying gas pressure.

3. Seconds later, a mixture of volcanic gas, steam, some magma, and much preexisting rock exploded horizontally northward with tremendous force, scouring and leveling forests as far as 15 miles (the Blast Area). Though early reports assumed this blast to be a "glowing avalanche" (a rapid gravitational flow of incandescently hot new ash suspended in magmatic gases), in fact it was merely "warm," and contained little fresh magma. Toutle River mudflows were of this blasted material mixed with glacial melt. Glaciers on St. Helens' other three sides didn't melt, but disappeared under a heavy coat of ash.

4. An eruption proper, of purer dacite fragments and gas, ensued, lasting nine hours. Its mainly convective (vertical) force raised an ash column 15 miles high, which the wind carried eastward around the globe. At least one small glowing avalanche descended the south slope.

5. A dacite lava dome began to extrude from the vent, terminating the eruption.

A new pulse of gas-rich magma exploded the May 18 dome into ash and formed a new dome on May 25. This cycle was

## Errata and Update

**Drawing Credits, lines 12, 21:** *Should read* "Willis L. Jepson"

**page 15, line 2:** *Should read* "is on pages 572-79.)"

**page 200, line 21:** One variety of this species is native here.

**page 231, line 20:** *Should read* "*S. missurica.*"

**page 239, art:** The sword fern drawing is labeled "deer fern," and vice versa.

**page 243, art:** These four ferns are (clockwise from top left) bracken, oak fern, lady fern, and wood fern.

**page 265, line 3:** *Should read* "(See page 326.)"

**page 290, bottom line:** Color p 487 shows *H. imshaugii.* The branches on *H. enteromorpha* are ± flattened and drooping.

**page 292, line 18:** Color p 488 shows *C. nivalis,* one of several *Cetraria* species that are yellow, unlike *C. islandica* and *C. ericetorum.*

**pages 412-15, footnotes:** "O.A.U." *should read* "A.O.U."

**page 431, art:** The line from "axillary process" should point to the flap at the front corner of the pelvic fin.

**page 441, add footnote:** A 1978 taxonomic revision splitting the Dolly Varden into two species is gaining wide acceptance. According to this view, the native charr in Oregon and most of Washington are bull trout, *Salvelinus confluentus,* whereas Dollies are found here only from the Quinault River north; Dolly Varden are commonly anadromous, bull trout very rarely so.

**page 457, art:** The golden buprestid drawing is labeled "ten-lined June beetle," and vice versa.

**page 499, line 1:** The red-tailed hawk is immature.

**page 501, line 3:** The photo labeled "hermit thrush" is of a song sparrow.

**page 503, line 4:** The red-legged frog tadpole is metamorphosing.

**page 549, footnote:** *Should read* "(peridotites, p 544)"

**page 578:** "Lycopodiaceae" (actually a family name) *should read* "Lycopodiophyta" as a division name.

repeated on June 12, July 22, August 7, and October 16, 1980. The dome has since grown to partly replace what was blasted away on May 18. Just possibly, the mountain will regain its former elevation or higher during the current eruptive cycle. Or activity may lapse for centuries before the volcano grows again from a new vent, perhaps outside the present crater. Most of our stratovolcanoes, including pre–1980 St. Helens, show that their main vents shifted at least once.

## Mudflows

Mudflows may be the worst direct hazard we face from our volcanoes, because they are relatively common, hard to predict, and fast-moving compared to eruptions themselves. Clocked at over 16 miles per hour on the upper reaches of the Toutle River on May 18, a St. Helens mudflow was doubtless slower than many, yet you could not have outrun it on rough terrain.

A mudflow can be described as a cross between a flash flood and a landslide of volcanic ash mud, typically having a mortarlike consistency. The mud is so dense that boulders float on it rather than tumbling along. Our volcanoes are susceptible to mudflows because their steep slopes are made of loose (or merely frozen solid) ash covered with snow or ice. A classic cause of large mudflows is an eruption of very hot ash melting a glacier; the mud is then a mix of new ash, glacier ice and water, and underlying old ash and rocks.

Other mudflows originate on dormant volcanoes. Mudflows are usually considered distinct from ordinary (unheated) floods and avalanches, but it is hard to be certain whether fluctuating volcanic heat is a cause of a glacial "outburst flood" on a volcano. Unheated flash floods carrying copious loose ash act much like typical mudflows, though they rarely get quite as big. Mudflow-type deposits from recent outburst floods can be seen on Dusty Creek of Glacier Peak and on Kautz and South Tahoma creeks and the Nisqually River of Mt. Rainier.

Mudflows are slow to revegetate because of their great depth and chemical uniformity. Red alders are important pioneers on mudflows here; their nitrogen-fixing ability is crucial in the pure mineral mud.

# Plate Tectonics

Beneath all the erupting and intruding and folding and fault-ing that produces our mountains there is a Grand Design or mechanism. The upheavals (tectonics) that raise mountains are caused by collisions between segments (plates) of the Earth's crust, which grind along at speeds up to two or three inches a year. New plate material is generated out of fresh basalt lava at mid-ocean ridges, where two plates move apart; material plunges back into the depths at "subduction zones," where plates collide. Geologists recognize about ten big plates covering the Earth currently, the largest one underlying most of the Pacific.

Most freshly generated plate material is "oceanic crust," a sheet of basalt two to five miles thick. Under an ocean for millions of years, the sheet accumulates thick beds of marine sediments on top that gradually become sedimentary rocks. "Continental crust" is a thicker but less dense (i.e., lighter weight) crust of many kinds of rock, but with a predominant composition of granite rather than basalt. Due to its thickness and low density, continental crust rides high upon the semi-fluid subcrustal base; very little of it is covered with sea water, at least in this day and age. With a few exceptions—the Olym-pic Peninsula being one—nearly all oceanic crust does lie un-der the oceans.

Each of the continents is built around a "craton" area of very old granitic crust that has existed for a good part of the Earth's 4.7 billion years. When continental crust collides with oceanic crust, the former usually stays on top while the heavi-er oceanic basalt is forced underneath ("subducted"), to be resorbed into the mantle. Some marine sediments, also low in density, get scraped off on the continental edge and become a part of it.

Islands and "microcontinents" raft ashore with the oce-anic plate but refuse to subduct with it, and end up jammed onto the continent. In this fashion the continents have grown for the last 3 billion years. North America, for example, has only the Canadian Shield for its ancient craton, while the West Coast states and British Columbia seem to be a collage of small "terranes" accreted within the last 200 million years—less than 5% of the earth's history. Vancouver Island, the North

Cascades (Microcontinent), and the Okanogan Highlands, for example, are all different pieces of the collage. Fossil seashells in the North Cascades are so different from nearby fossils of the same Mesozoic age as to suggest this microcontinent came from the far side of the Pacific.

Subduction produces mountains in several ways. The scraped-off marine sediments can pile up and produce a Coast Range like California's. The Oregon and Washington Coast Ranges are similar in origin, but retain great slabs of unsubducted oceanic basalt as well as marine sediments. Where the basalt and sediments were rammed into a tight corner between two firm microcontinents (Vancouver Island and the North Cascades) they were folded and faulted more violently and raised much higher, creating the Olympics. Layer after sedimentary layer was "underthrust" beneath the next older layer. The original underlying basalt sheet ended up on top, bent into a huge arc—the Olympic Basaltic Horseshoe—reflecting the corner it was pushed into.

Inland from a coast range (or other subduction complex) there typically lies a line or "arc" of volcanoes. The oceanic plate descends at a 20–60° angle, putting it well underneath the edge of the continental plate, where heat and pressure melt it partially. (This is far shallower than the depths where the Earth as a whole is thought to be molten.) Only some of the minerals (rock-composing chemicals) melt; the mineral mixtures that melt are usually light enough to rise as fluid magma in "pulses." The chemical proportions in the magma determine what kind of rocks it will produce. Andesitic and basaltic magmas—typical of arc volcanism—reach the surface to erupt as stratovolcanoes, shield volcanoes, or cinder cones. (Page 533.) Remnants of the same magma are left behind down in the plumbing, where they slowly crystallize as dioritic, gabbroic, or other dark intrusions.

Sometimes chemical evolution of the rising magma raises the silica content enough to make a dacite or rhyolite. These lighter magmas (in both color and weight) are extremely viscous, like cold peanut butter, and reach the surface relatively rarely. When they do, the event may be explosive, as at Mt. St. Helens on May 18, 1980, producing copious ash and tuff; or it may be the gentler growth of a lava dome.

More often, pale magmas as they near the surface solidify as "batholiths" or smaller "plutons" of granitic or other pale intrusive rock. A mountain range may still result eventually, due to the thickening or deepening of continental crust along the line of magma formation; the thickened line, as an area of continental crust, floats higher. The overlying volcanoes (if any) and surrounding materials erode off the top, exposing the intrusions, which persist for millions of years as a granitic mountain range.

The volcanic Cascades fit the volcanic arc model quite well. Different phases of Cascade volcanism have been active over the last 40 million years, sometimes far more active and effusive than our present cones, which date from the last half-million years or so. Geologists suspect that shifts in the direction, speed, and nature of subduction caused the changes between the three phases of activity—the Western Cascades, Columbia Basalt Floods, and High Cascades.

The North Cascades are a different story. To be sure, they include two recent stratovolcanoes and several somewhat older granitic intrusions that are part and parcel of the Cascade volcanic arc, as described above. These materials were added to the North Cascades Microcontinent after it was superimposed, 40 to 50 million years ago, on the volcanic arc. Aside from them, however, the North Cascades are made of rocks laid down, contorted, metamorphically squeezed, and in places remelted, in the depths of and off the shores of an earlier volcanic chain probably located on an island or chain of islands in the Pacific. Subsequent uplift may result from residual buoyancy of those island mountains, or from the force of the microcontinent colliding with North America. When two pieces of continental crust collide (as opposed to one continental and one oceanic) both plates are too buoyant to subduct. Such rammings have produced spectacular mountains, including the Alps and the Himalayas.

For as long as there have been world globes to display the "fit" of the continents on opposite sides of the Atlantic, there have been theories of "continental drift." These were considered fanciful by most scientists until after 1915, when Alfred Wegener elaborated them with enough supporting evidence to spark several decades of intense debate. In the 1950s and 1960s, new technology for locating deep earthquakes, deep-sea contours, magnetic polarity of sea-floor basalts, etc., escalated the sophistication of the debate as well as turning it in Wegener's favor. The more inclusive term "plate tectonics" has supplanted "continental drift."

Geology has been an exciting branch of scientific inquiry for the last 30 years, and the prospect of the next 10 has me on tenterhooks. This chapter contains several theories that may face harsh reevaluation as geologic knowledge progresses. Questions of recent interest in the Northwest include:

Has subduction of the Juan de Fuca Plate slowed, or come to a virtual halt, or "locked up" in one segment only, or none of the above?

Is this then an area of infrequent (hence unrecorded) but very severe earthquakes?

Is there a shallow magma pool hundreds of miles long under the Cascades?

Which volcano will go off next?

What's beneath the east-west line of great volcanic centers from Central Oregon through Idaho's volcanic plain to the Yellowstone Caldera?

Where did our various terranes originate? What are their boundaries? How did they rotate and/or slip northward after they reached North America?

And the grander question: what force drives the clashing plates?

One thing is clear: even as new data raise thorny new questions, the basic picture of subducting plates is ever more firmly corroborated. It has become possible, with radio-telemetry, to measure the distance between Africa and South America accurately enough to confirm that they are moving two to three inches farther apart each year, roughly as Wegener predicted.

# Geologic Time
(years ago)

| | |
|---|---|
| 4.6 billion | Earth formed. |
| 3.8 billion | Oldest rocks yet found were formed. |
| | Oldest fossilized cells yet found lived. |
| 2-1.5 billion | Oldest rocks in our region (Yellow Aster Complex, N. Cascades) formed—far from North America. |
| 900 million | Multicelled life evolved. |
| 440 million | Life moved onto land. |
| 500-200 million | Volcanism on Pacific Islands produced rocks now found in North Cascades. Core rocks of North Cascades metamorphosed. |
| 225 million | Dinosaurs became predominant land animals. |
| 200 million | Western North America began to accrete terranes comprising Pacific Coast States and B.C. |
| 100 million | Flowering plants became predominant. |
| 90 million | Mt. Stuart and Black Peak Batholiths formed at depth. ("Older intrusions" on map, p 540.) |
| 65 million | Dinosaurs went extinct; mammals took over. |
| 60 million | Uplift of Rocky Mountains was at its most intense. |
| 55 million | Olympic basalts formed as undersea lava flows. |
| 50 million | North Cascades terrane joined North America. |
| 40-18 million | Main Western Cascades volcanic period; North Cascades' younger intrusions formed at depth under similar volcanoes (since eroded away). |
| 17-12 million | Columbia River Basalts poured out. |
| 8 million | Current North Cascades uplift began. |
| | High Cascade volcanism began. |
| 2 million | (Current) Ice Age began. (Estimate range: 3-1.6 million years ago.) |
| 1,800,000 | Ancestral hominids learned to shape stone tools. |
| 730,000 | Earliest eruptions of present High Cascade stratovolcanoes (limited to this date). |
| 500,000 | First *Homo sapiens* lived in Africa. |
| 100,000 | Latest great glacial advance in America began. |

4.6 billion — 3.8 billion — 2 billion — 900 million — 500 million — 200 million — NOW

# Recent Events
(date)

| | |
|---|---|
| 29,000 B.C. | People first migrated to North America, according to the "Early arrival" school of thought. (Estimates range to a little over 100,000 years ago.) |
| 16,000 | Last maximum of alpine glaciers in Cascades. |
| 12,000 | Maximum advance of Cordilleran Ice Sheet's Puget and Okanogan Lobes. |
| | Bretz Floods scoured Columbia Basin and Gorge. |
| 11,000 | The last Bretz Flood; glaciers rapidly retreating. |
| 10,000 | Glacier Peak erupted massively. |
| | People first migrated to, and quickly spread over, America, according to the "Late arrival" school. (Date of spear points found near Wenatchee. |
| | Woolly mammoth, giant beaver, saber-tooth tiger, and other American large mammals died out. |
| 8,000 | Salmon fishing cultures active in Columbia Gorge. |
| 4,700 | Mt. Mazama erupted massively, then collapsed to form a caldera eventually filled by Crater Lake. |
| 3,000 | Osceola Mudflow from Mt. Rainier buried Puyallup area. |
| 0 | Mt. Rainier built its present summit. |
| | Mt. St. Helens' eruptions built its present flanks. |
| A.D. 500 | Extensive Belknap lava flows at McKenzie Pass. |
| | Mt. Hood (Crater Rock) erupted, producing the large debris fan underlying Timberline Ski Area. |
| 1260 | Huge landslide temporarily dammed the Columbia, leaving Cascades the Mtns. were named for. |
| 1500; 1800 | Two eruptive periods on Mt. Hood. |
| 1600; 1820 | Minor glacial advances peaked. |
| 1830-80 | Minor activity on Mts. Baker, Rainier, St. Helens. |
| 1914-17 | Major eruptions of Mt. Lassen. |
| 1975 | Mt. Baker emitted steam—a false alarm. |
| 1980 | Mt. St. Helens erupted explosively. |
| Today | Glaciers retreat; mountain-building continues. |

29,000 B.C. — 16,000 — 12,000 — 10,000 — 8,000 — 4,700 — 3,000 — 0 — TODAY

This entire time span would be an invisible speck within the "now" tick on the timeline on the opposite page.

# Glaciers

Wherever an average year brings more new snow than can melt, snow accumulates and slowly compacts into ice. Eventually, the ice gets so thick and heavy it flows slowly downhill until it reaches an elevation warm enough to melt it as fast as it arrives. This flowing ice is a glacier, a mechanism that balances the snow's "mass budget." Mt. Olympus' Blue Glacier is America's most intensively studied glacier, and its students claim for it the distinction of having the fattest mass budget— most snow, and most melt—of all glaciers measured.

Ideally, the rate of flow is equal to both the excess snow accumulation in the upper part of the glacier and the excess melting in the lower part; this ideal glacier would neither advance nor retreat. Few glaciers are actually so stable. Instead, the elevation where the glacier terminates in a melting "snout" advances and retreats (drops and rises) in response to climatic trends. (Retreating glaciers don't turn around and flow back uphill, of course, but simply melt back faster than they flow down.) A global warming trend of the last 100 years caused widespread glacial retreat. In our region this intensified in the warm/dry 1930s and '40s, and again in the '80s. In the 1950s and '60s, many glaciers advanced.

The formula for advance or retreat is complex. Even neighboring glaciers vary greatly in the time lag from climatic trends to glacial ones. On Glacier Peak, for example, the Chocolate and Dusty glaciers advanced during the cool 1950s and '60s while the Whitechuck, a low-angle and relatively low-altitude glacier, inexorably wasted away, approaching stagnation. A glacier "stagnates" after shrinking so much that it no longer has enough mass and slope to keep flowing.

You can see the difference between advancing and retreating glacier snouts. Ideally, the former advance upon fields or forests like giant bulldozers, while the latter are surrounded by barren bouldery expanses from which they just retreated. Around here, unfortunately, both types of glaciers terminate in barrens, since even our advancing glaciers haven't yet made up the ground they lost 50 to 130 years ago, too recently for much revegetation. Still, a retreating snout typically tapers off, and is dirty and rocky with surface debris concentrated over recent decades; it may even be hard to tell where the glacier leaves off and the rock rubble or "till" ensues. An advancing snout, in contrast, may present a clifflike face where ice blocks come crashing down, or a chaotic expanse of towering ice chunks called "seracs."

Flanking the sides of a glacier or paralleling the arc of its snout, you often find low, rather smooth ridges of pure rock rubble, perhaps in parallel series. These are recessional moraines. They mark lines where the retreating glacier advanced or held a line for a few years before receding again. Many of these were created during the last 300 years' retreat. Older ones obscured by vegetation reveal to the practiced eye a map of the end of the latest Ice Age, 12 to 18 thousand years ago.

Small glaciers in pockets on the steep faces of North Cascade peaks seem nearly vertical; gnarly blue wrinkles show that they are glaciers. Often they terminate over vertical rock. Ice blocks that break off and avalanche down the rock band may recoalesce into a lower glacier, if the basin where they collect is high and/or shaded enough for them to stay frozen. Some basin glaciers, with or without hanging glaciers above them, are fed more by avalanches than by direct snowfall.

Regardless of whether it forms from fresh snow, avalanches, or huge blocks of older ice, a glacier's ice has a consistency utterly unlike a snowbank. Underneath a layer of last winter's snow, older snow is now recrystallized into coarse granules with hardly any air space. Eventually the granular

---

## Crevasses

As flow accelerates in a glacier, the semisolid mass stretches, leaving stretch marks in the form of deep cracks or crevasses. Other crevasses open where glaciers bend or compress. Crevasses are often bridged by recent snow which may or may not be solid enough to walk across. Crossing glaciers is reasonably safe for parties of three or more that follow these rules:

**First rehearse self-arrest** and belaying technique with ice axes;

**Rope up** properly;

**Include** at least one member experienced on glaciers.

---

texture will fade into massive blue ice. (A microscope will still reveal a texture more granular than, say, frozen lake ice.)

Our temperate-zone glaciers are "warm" glaciers at close to 32° throughout, in contrast with "cold" arctic glaciers. Warm glaciers, like ice skaters, glide on a film of pressure-melted water. They pour around bedrock outcrops by melting under pressure against the upstream side of the knob and refreezing against the downstream side (repaying the heat debt incurred by the first change of state). They erode rock ferociously, not because either the water layer or the ice itself is abrasive, but because sand, pebbles, and boulders are gripped and ground along over the substrate. It's the rock sediment

## Rock Flour

A milky-white color betrays streams that originate from the snouts of glaciers. At least three White Rivers and a Whitechuck River in the Cascades are all named for their "rock flour"—silt-sized sediment pulverized by the sole of a glacier and carried in the meltwater. Ice preserves the rock from alteration into darker mud. Streams get milkier as summer progresses; clear runoff from rain and snowmelt reaches a peak during the warm rains of spring, then dwindles while glacial melting increases in summer heat.

In lakes, rock flour gives a chalky green hue. This is conspicuous in tarns near glaciers, and also in Lake Diablo as contrasted with Ross Lake. Ross Lake drains a large basin east of the climatic Cascade Crest, where glaciers are relatively few and small. Lake Diablo, despite dilution by all of Ross Lake's clear outflow, is chalky green with Thunder Creek's rock flour from the huge North-face glaciers of Mts. Logan, Boston and Eldorado.

In the Olympics, most rock flour is in the Hoh and Queets Rivers, and Silt Creek. It is also visible in some Elwha tributaries. The snowfinger at the source of the Elwha was described as a glacier at the turn of the century, and the stream is still a bit milky.

Some writers find straight milky meltwater unpalatable, but I like it just fine. I figure it comes with a mineral supplement, and an exceptionally remote risk of giardia.

load that does the grinding in both glacier and river erosion; the difference between the two is a little like that between a power belt-sander and a sandstorm. The glacier's grit leaves parallel grooves or "striations" across bedrock convexities. Look for these at the lip of any high basin or "cirque" carved by an alpine glacier. In the area east of Ross Lake, there are even striations crossing high divides—proof that the Cordilleran Ice Sheet crossed here, since alpine glaciers necessarily originate below the high ridges.

## Ice Ages

If you love dramatic high-relief scenery in the style of the North Cascades or the Alps, count yourself lucky to have been born in an interglacial stage of an ice age; mountain erosion *without* ice usually results in tame, monotonous slope angles and ridge patterns.

Estimates of the number of ice ages range from just three up to maybe thirty in the 4,700,000,000 years of Earth history. The present Ice Age has been going on for two or three million years, encompassing the evolution of *Homo sapiens* and much of the rise of today's Northwest mountains. It has cycled many times between glacial and interglacial stages. (The latter, tending to last longer than the glacial stages, are not returns to pre-Ice Age conditions, which included neither large Polar ice caps nor alpine glaciers at temperate latitudes.) The ice caps and glaciers oscillate between advance and retreat, in accord with still subtler climatic cycles within the interglacial.

During a typical glacial stage, about half of North America was covered by two huge ice sheets, one on either side of the Canadian Rockies, similar to those blanketing Greenland and Antarctica today. The western one (the Cordilleran Ice Sheet) covered parts of Alaska, the Yukon, and British Columbia, and entered our range briefly at its most recent maximum, 14,000 years ago. At that point it shoved up against the Olympics, extended long tongues down the Puget Trough to just past Olympia, and ground across North Cascade ridges east of Ross Lake. The North Cascades must have looked the way southeast Alaska's mountains do today—great banded riverlike glaciers with branching tributaries filling the valleys, and marginal peaks protruding as "nunatak" islands.

Not the vast ice sheet, but the hundreds of alpine valley glaciers, gave our region its dramatic facelift. Wherever you see a U-shaped valley cross-section today, picture a thick glacier flowing for at least 1,000 years. The alpine glaciers marked their farthest advance about 18,000 years ago, when the climatic cycle must have reversed. It took another 4,000 years for melting to reverse the advance of the Cordilleran Ice Sheet, with its far greater mass. By 12,000 years ago all ice was in full rout, the glacial stage over.

One effect of the great ice sheets was to lower sea level as much as 300' by retaining ice that would otherwise be water in the oceans. This turned a wide area of shallow sea between Siberia and Alaska into habitable dry land, allowing many mammals, including humans, to migrate to the New World.

Theories of the cause of Ice Ages are controversial and

## The Bretz Floods

The basalt cliffs of the Columbia Gorge are strikingly clean of loose rock rubble. They got this way by scouring action from river floods 400' to 1,000' deep. That's right—the Columbia ran 1,000' deep at The Dalles where it backed up against the Cascade Range, and was still over 400' deep where it broke out from the confining gorge at Chanticleer Point near Corbett. This represents 200 times as much water as in any Columbia River floods in the historic record, and doubtless more than any other river flood interpreted from the geologic record. What's more, there were dozens of these deluges; a series of 15 of them at 35–55-year intervals shows clearly in a single pile of sediments.

The floods were a phenomenon of the rapid retreat of Ice Age glaciation. The last one was a mere 13,000 years ago; there could have been people living in the area, according to archaeologists of the "Early Arrival" persuasion. Source of the flood water was Lake Missoula, an ice-dammed lake in western Montana with about half the volume of modern Lake Michigan. The lake filled several long mountain valleys of the Clark Fork drainage when that river's route north to the Columbia was blocked by a south-flowing lobe of the Cordilleran Ice Sheet. Such lakes were not uncommon in the Ice Age; the Skagit Gorge, for example,

complex. The very occurrence of an Ice Age in geologic history, obvious though it seems today, was controversial 150 years ago. The intensity of solar radiation hitting land surfaces on earth varies with at least five known aspects of planetary motion, of which day/night and summer/winter are the most drastic and rapid. At very long but calculable intervals, the five cycles coincide to reduce solar heating to minimum effectiveness over long periods. This astronomical compound cycle is widely held to play a role in ice ages. It must be augmented by other, possibly random causes, dozens of which have been suggested: sunspots; volcanic dust; dust from meteor or comet impacts; carbon dioxide levels; ocean salinity levels; and continental drift as it affects the configuration of ocean currents, the sizes of continents and their locations relative to the poles, and the height and configuration of mountain ranges. The ice

was sometimes dammed by the Puget Lobe. But in Lake Missoula's case, the ice dam must have been slender enough to float when the lake rose high enough. With the dam floated or otherwise failed, the entire lake drained rapidly. Small-scale periodic floods due to ice-dam floating occur today in Iceland, source of the exotic term "jökulhlaups" sometimes applied to Lake Missoula's floods.

The floods explain several previously mystifying features of the Northwest landscape, including Eastern Washington's Dry Falls, Coulees, Potholes, and Channeled Scablands, and the Willamette Valley's erratics. A glacial erratic is a boulder transported by glaciers and dropped far (typically 10–100 miles) from its bedrock source. The Willamette erratics are puzzling because they are found where no glaciers have been, let alone the particular glaciers that plucked these boulders 500 miles away in Montana or British Columbia. They must have ridden the floods in icebergs that drifted out to the margins of a backwater lake filling the Willamette Valley, where they beached and melted. Other erratics indicate huge backwater lakes east of the Cascades.

The floods are named after geologist J Harlen Bretz, who first understood the evidence and collected enough of it to convince others. They have also been called Spokane Floods (Bretz' original term) and Missoula Floods.

itself may set up feedback loops to either sustain or reverse an ice age, once started.

At 12,000 years old, the current interglacial has neither lasted longer nor grown warmer than previous ones of this Ice Age. Nothing in the geologic record offers grounds to doubt that another glacial stage will come. However, scientists are looking hard at the possibility that the next reglaciation—or total deglaciation—may not resemble past fluctuations at all. It may instead be triggered by the chemical and energetic excesses of human civilization, which cause increasing carbon dioxide and decreasing ozone in the atmosphere.

---

## Geology: rocks

---

Classification of the earth's materials begins with the three ways that material can be transformed into new kinds of rock:

**Igneous rocks** solidify from a molten state.

**Sedimentary rocks** settle as fragmentary material (such as mud) and are then compacted and/or chemically cemented.

**Metamorphic rocks** recrystallize from other rocks under intense (but not liquefying) heat and pressure.

Igneous rocks are divided into two textural classes—fine-grained and coarse-grained—and then graded by chemical (or mineral) composition, mainly according to greater and lesser amounts of silica. High-silica rocks tend to be "light," low-silica rocks "dark." Each "darkness" grade on the compositional scale can occur with either a fine or a coarse texture. The two textures strongly correlate with two modes of origin:

**volcanic,** from magma that erupted upon the continental surface or the ocean floor; and

**intrusive,** from magma that solidified into rock somewhere beneath the surface.

Though not 100% accurate, it is traditional, and will serve our purposes, to assume that fine-grained igneous rocks originated volcanically, and coarse-grained ones intrusively.

# Basalt

Color p 508.

The most abundant rock in Oregon and Washington and, though less visibly, in the Earth's crust as a whole, basalt is the darkest of our major lava mixes. We think of it as normally black, but it ranges down to medium gray and is often altered to greenish or reddish. Its surface is drab and massive, usually without conspicuous crystals or other features except, frequently, bubbles, or "vesicles." These show that the lava came up from the depths full of dissolved gases. Just as uncapping a bottle of soda releases bubbles of carbon dioxide which, under pressure, had been dissolved invisibly in the liquid, so lavas often foam up as they near the surface. Abundant vesicles indicate that the rock congealed near the surface of a particular lava flow. In older basalts (e.g., those of the Olympics) vesicles may appear as solid polka-dots, the holes having filled with water-soluble minerals. Polygonal (6-, 5-, or 4-sided) basalt columns, often neatly vertical at one level and splayed-out at another, are another well-known characteristic. They result from shrinkage during cooling of large flows. Vesicles and/or columns also occur in andesite, though less commonly.

Basalt lava erupts in several styles. One of them, the basalt flood, is a volcano, broadly speaking, but not a mountain. The Columbia Gorge and the smaller canyons of the Klickitat, Clackamas and Sandy Rivers all cut through multilayered, often columnar formations of Columbia River Basalt—vast outpourings that covered much of eastern Oregon and Washington and flowed down the river valley to the ocean between 18 and 12 million years ago.

Originally flat and close to sea level, the flood basalts were subsequently arched to over 2800' in the Gorge, 5500' near Mt. Adams, and possibly over 7000' near Mt. Rainier. (Under Mt. Hood, however, faults dropped them to below sea level; the High Cascades stand in a trough rather than upon a geologic arch.) The folding and faulting were gradual enough that rivers like Columbia and Klickitat were able to keep pace in their downward erosion.

Two kinds of mountains usually made of basalt are shield volcanoes and cinder cones. Both are plentiful in the Cascades. Shield volcanoes, being made of basalt just slightly less fluid than the flood kind, are generally too low-relief to be dramatic peaks in themselves, but the entire High Cascades Plateau, on which the prominent volcanoes stand, is a montage of overlapping basaltic shields.

The prominent High Cascade stratovolcanoes, often generalized as andesitic, are in fact layerings of flows of various compositions ranging all the way from rhyolite to basalt. Their relief was beefed up by tuff-forming ashflows, an eruptive style that can happen with any type of lava, though basalt is relatively uncommon as a compostion of ash. Nevertheless, a few stratovolcanoes, like Middle Sister, are largely basalt.

Yet another type of effluence—oceanic basalt—is far and away the most extensive of all. Most of this erupts under the ocean and remains there forever. A bit of oceanic basalt crops up in the North Cascades (much of it more or less altered into greenstone, page 553), and a great deal of it in the Olympics, which seem to belong geologically to the Pacific. The entire east and north rim of the Olympics consists of a huge slab of basaltic sea-floor bent into a horseshoe shape and wedged up 90° from underneath, the flow-tops facing east and north. (Flow-tops here are revealed by the glassy-rinded, blob-shaped "pillow" structures that basalt characteristically forms when it erupts underwater.) A few of the uppermost tops display subaerial texture instead, showing they erupted on land, perhaps a mid-ocean ridge island like Iceland. Mt. Constance, The Brothers, Mt. Angeles, Mt. Washington, and all the other peaks conspicuous from Puget Sound are of the Olympics' basaltic horseshoe, and Mt. Deception, the second highest Olympic peak, is of a smaller basalt arc. These Crescent Basalts are among the hardest, most erosion-resistant rocks in the Olympics, and certainly the most popular with rock climbers.

### Stratovolcanoes

include all the prominent Cascade volcanoes. They build up over long periods (up to one million years) by alternating lava flows (andesite, dacite, and basalt) with copious tuff flows and ashfall. Mt. Rainier—207 cubic miles of it, rising more than 10,000', is a large one. Vents move around over the centuries; "satellite" vents erupt cinder cones, and late-stage vents often produce lava "plug" domes.

### Cinder cones

typically erupt from a single vent within a span of 100 years or far less, and rarely achieve 1,000' of relief. They erupt as fountains of small, foamy basalt or andesite "cinders," and often extrude tonguelike basalt flows from their bases. Examples abound in Central Oregon.

### Lava domes

superficially resemble cinder cones, but at heart they are massive extrusions of pasty, viscous rhyolite or dacite. The surface cracks into blocks, some of which tumble down the steep sides to form a rubbly cone. The vent plugs up within a few years, rarely producing 1,000' of relief.

### Shield volcanoes,

made by multiple flows of basalt, produce only gentle relief. A few build for millions of years; "the world's biggest mountain" Mauna Loa, is 10,000 cubic miles rising 29,000' above the sea floor. Cascade shield volcanoes like Larch Mtn. and the Simcoes are 3,000'–5,000' high.

### Flood basalts

are not mountains; they are so fluid when molten that they lie almost flat. They produce no relief, but have the greatest volumes and flow rates of any terrestrial volcanoes. The Columbia River Basalt, a ten-million-year sequence of flows contains more than 60,000 cubic miles of lava.

---

# Andesite

Color p 508.

Andesite lavas make rough gray, greenish, or sometimes reddish brown rocks. Often they are speckled with crystals up to
¾" across, scattered throughout a fine matrix otherwise lacking crystals of visible size. These were already crystallized in
the molten magma when it erupted. With close examination
and a little practice, you can easily tell them from the shards of
noncrystalline whole rock jumbled up in tuff.

The different lava rocks, however, can be hard to tell
apart. Color descriptions are unavoidably vague; andesite is
"medium-dark," between "dark" basalt and "light" dacite and
rhyolite, but actual tints overlap, and green and reddish hues
are unpredictable. The technical definitions are ranges on a
graph of "bulk chemistry"; positive identifications may require specimens to be pulverized and analyzed by a lab.
"Light" color correlates with high silica ($SiO2$) content—
around 50% in basalt, 60% in andesite, 70+% in rhyolite.
(Though the mineral quartz is $SiO2$, most magmatic silica crystallizes in other minerals, especially feldspars; only the $SiO2$
excess over about 55% crystallizes as quartz.)

The chemistry of Cascade volcanism has shifted many
times. Andesite predominated in making the Western Cascades, between 40 and 8 million years ago. Since then (i.e., in
the High Cascades) basalt has contributed the most lava, but
predominantly andesitic sequences produced most of the outstanding peaks; in other words, most rocks you see on Mt.
Rainier, Mt. Hood, and the other major cones, are andesite.
The rock is named after the Andes Mountains, which include
the world's highest volcanoes and are the model of everything
a subduction zone mountain range is supposed to be. The Cascades were closer to that model during Western Cascade activity than they are today.

# Dacite

("day-site"). Color p 508.

Dacite lavas are intermediate in silica content, viscosity and color—more or less medium-gray, sometimes pinkish, brownish or buff. They include light and dark scattered crystals often, flow-streaking occasionally, and bubbles or holes rarely. They are common in Rumania, known to the ancient Romans as Dacia, hence the word dacite. Here, they built most of Glacier Peak plus the plug domes on several peaks. Disappointment Peak on Glacier Peak and Crater Rock on Mt. Hood are dacite domes whose eruptions melted enough iced-up ash and pumice to create enormous debris fans downslope. Mt. Hood's youngest debris fan, only 200–300 years old, underlies the Palmer Glacier, Timberline Lodge, and the entire smooth south slope as seen from Portland. A dacite lava dome is building within St. Helens' crater.

# Rhyolite

Color p 508.

Rhyolite lavas are pale, typically pinkish tan rocks, usually easy to tell from other lavas but sometimes confused with tuff, which is often made of rhyolite magma that exploded rather than flowing as lava. Rhyolite is named for the Greek word for flow; streaky or swirling flow patterns may help to distinguish it, especially from tuff or basalt. But as a lava it is much less fluid than others. Lava viscosity increases dramatically with silica content, and flowing rhyolite may be up to 1000 times more viscous than flowing basalt. At this point "flowing" seems like the wrong word—it squeezes out of the ground more like dried-up peanut butter. It so resists flowing that it may plug up a given vent forever; it marks the final phase of activity for some volcanoes that previously erupted other kinds of lava. This "constipating" viscosity of rhyolite helps to explain the abundance of granite—rhyolite magma that never managed to erupt. (See page 520.)

Alternatively, gas-rich rhyolite magma may explode, showering the area with pumice, ash and larger fragments; rhyolite tuff and pumice are superabundant in some volcanic provinces, including Nevada. Then again, sometimes rhyolite erupts in the glassy form that produces obsidian (page 537). The southernmost major Cascade volcano, Mt. Lassen, is about as big a dome as rhyolite can produce.

## Pumice

Color p 509.

More often than not, magmas reach the surface bearing a gaseous component that expands, sometimes explosively, as it is released from subterranean pressure, blasting globs of glassy froth sky-high—a little like a well-shaken bottle of warm Guinness. In mid-air, the froth may harden into pumice, a volcanic glass so full of gas bubbles that it's light enough to travel on the wind, even in sizable chunks. Most of our recent volcanoes and much terrain downwind of them are mantled with pumice fragments. In particular, ridgelines for a good 20 miles east of Glacier Peak (including the misspelled Pomas Pass) are covered with pumice from a huge explosive eruption about 12,000 years ago. Like St. Helens' more modest ash plume of May 18, 1980, Glacier Peak's extends in only one direction; it was a single, intense event so brief that the wind made no shifts. Dust-sized ash (much of it pumice) from that eruption is recognized in soil profiles all across Montana—a layer somewhat below the one from the great Crater Lake eruption of 6,700 years ago.

Pumice has unusual uses, such as grinding away at toes or kitchen grills, and mixing into concrete as soundproofing. Breaking or grinding a piece of pumice may release a sulfurous smell—the original volcanic gas (largely sulfur dioxide and carbon dioxide) still trapped in the little bubbles. These are much tinier than the vesicles in basalt and andesite, and make pumice lighter than vesicular lava.

## Obsidian

Color p 509.

Obsidian is commonly called "volcanic glass," and that's exactly what it is. It is distinguished by breaking like glass, in a pattern of concentric arcs. It is ideal for chipping into sharp arrowheads and other tools, so it was a major trade commodity for all Western tribes. The main source near here was Central Oregon's Newberry Crater—actually a caldera, like Crater Lake. There are smaller obsidian flows among the Three Sisters, but there isn't much of it in the rest of our range.

Technically, glass is noncrystalline. Obsidian is lava that cooled without organizing its ions into crystalline minerals. (Many lavas lack visible crystals, but are mixtures of microscopic crystals with a lesser glassy component.) Even though most obsidian is chemically equivalent to the paler lavas, rhyolite and dacite, it is usually black or dark red-brown.

## Tuff

Color p 509.

Much of the volcanic rock produced by our volcanoes was never lava flowing out of the ground, but instead erupted violently into the air as pyroclastic ("fire-broken") fragments. These range in size from "bombs" bigger than basketballs to dustlike "ash." (Named for its frequent resemblance to fine wood ashes, volcanic ash is no more burned than any other volcanic material.) The fragments may quickly wash away or otherwise fail to cohere, but some consolidate into rocks called "tuff."

Tuff has two distinct ways of consolidating from ash. Ash*fall* is material that shot up into the air and settled back to earth at modest temperatures; decades must pass before sufficient groundwater full of dissolved silica percolates through to cement it into an "ashfall tuff." Ashfall is typically found in graded beds, each bed coming from a single ash eruption and having coarse fragments at its base (they fall to earth fastest) and finer fragments toward its top. Ash*flow*, on the other hand, comes barreling down the slopes suspended in hot whirling volcanic gas; much of the gas exsolves out of the fragments even as they fly. These events move like ava-lanches, and are sometimes called "glowing avalanches," or *nuees ardentes*. The ashflow is laid down fast, without grading and often with forcible compaction, and is quickly cemented by its own heat into a "welded tuff." Ashfall tuff is often crumbly, and vulnerable to erosion, but welded tuff is hard; ashfall tuff's closest resemblance may be to sandstone (tuff is sometimes classed as a sedimentary rock), but welded tuff resembles lava. The easiest way to tell tuff from lava is if there are inclusions; in lavas these are crystals, often of a contrasting color. In tuff they are usually coarse, irregular fragments about the same color and composition as the rest of the tuff.

Tuff makes up a substantial volume of old Western Cascades rock, as well as newer High Cascades rock. The highest Cascade peaks are stratovolcanoes—structures of lava flows interlayered with pyroclastic deposits both loose and tuffaceous. Lava is the part with greater strength, but lava flows alone cannot build volcanoes that combine the height, steepness, and graceful conical form of ours; tuff's contribution is needed. Particles dumped in a heap settle at their own proper "angle of repose"; watch sand settle in a large hourglass for an elegant demonstration of this. The angle varies with particle roughness, angularity, and moisture—in a word, cling. It ranges from 23° for most scree, to 30° for cinder cones, and even higher for welded tuffs, which cheat. Basalt and andesite flows cannot create such steep angles. Pyroclastics on a stratovolcano fill in gaps, gouges and irregularities left by viscous lava flows and subsequent erosion, until the mountain approximates a 30° cone steepening a bit toward the top. Mt. St. Helens, prior to 1980, was more perfect than our other cones because it had erupted enough since the Ice Age to repair all the disfigurements of glaciation.

# Volcanic Cascade Geology

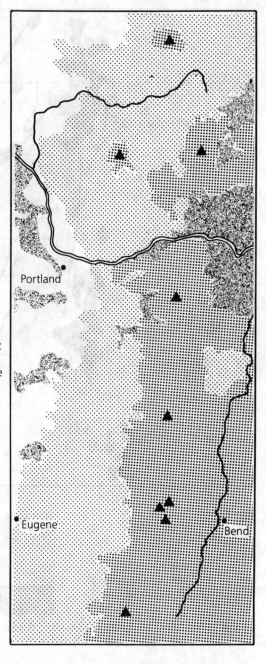

Columbia River Basalt
(where exposed; also
underlies much of the
intervening area)

Western Cascade
Volcanic rocks

High Cascade
Volcanic rocks

Mt. Rainier
Mt. St. Helens
Mt. Adams
Mt. Hood
Mt. Jefferson
The Three Sisters
Diamond Peak

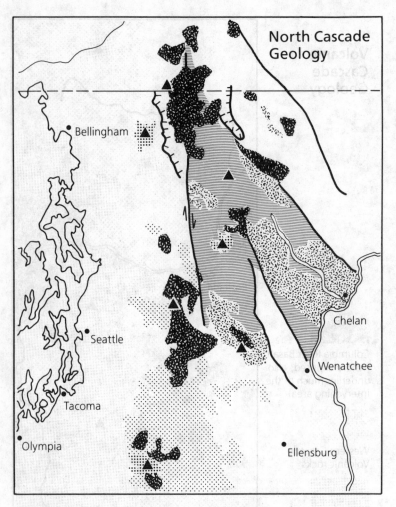

# North Cascade Geology

Bellingham

Chelan

Seattle

Wenatchee

Tacoma

Olympia

Ellensburg

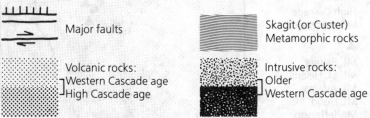

<u>𝗅𝗅𝗅𝗅𝗅</u>
⇄
Major faults

Skagit (or Custer)
Metamorphic rocks

Volcanic rocks:
⎱ Western Cascade age
⎰ High Cascade age

Intrusive rocks:
⎱ Older
⎰ Western Cascade age

▲ Slesse Mtn., Mt. Baker, Goode Mtn., Glacier Peak, Mt. Index, Mt.
Stuart, and Mt. Rainier (North to South)

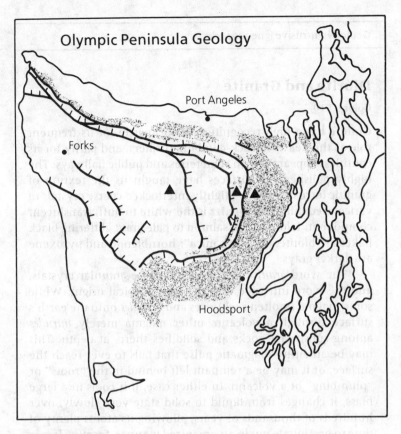

## Olympic Peninsula Geology

Port Angeles

Forks

Hoodsport

 Major faults

Basalt: the two large belts comprise the Olympics Basaltic Horseshoe; small broken segments are the inner basaltic rings.

▲ Mt. Olympus, Mt. Deception, and Mt. Constance (left to right), the three highest Olympic peaks.

## Diorite and Granite

Color p 510.

Granite has a high recognition factor, partly for its frequent role as the heavy in rock-climbing thrillers, and partly for its countless appearances in cemeteries and public hallways. The highly polished cut surfaces have taught us the texture of granitic intrusive rocks—tightly interlocked coarse crystals of varicolored minerals. Quartz is the white to buff, translucent component; feldspars are salmon to pale gray; glittering black flakes are biotite, or "black mica"; hornblende and pyroxene are darker grays.

The word *gran*itic derives from these *gran*ular crystals, and the word intrusive describes their typical origin. While some magma (molten rock) rises and *extrudes* onto the earth's surface through a volcano, other magma merely *intrudes* among subsurface rocks and solidifies there, at depth. This may be an entire magmatic pulse that fails to ever reach the surface, or it may be a remnant left behind in the "roots" or "plumbing" of a volcano. In either case, if it cools in a large mass, it changes from liquid to solid state very slowly, over hundreds of thousands of years, allowing its atoms plenty of time to precipitate out in an organized manner, forming larger and larger mineral crystals. Large (coarse) crystal size characterizes the granitic class of igneous rocks.

Since magma under pressure exploits weaknesses in the rock it intrudes, there are many narrow, more or less flat intrusions, or "dikes," that were originally forced into cracks. Dikes can be an inch wide, or many yards; narrower ones, if intruded into relatively cool surroundings, may cool nearly as fast as lava, producing a microcrystalline rock like andesite. Otherwise, dikes are typically granitic or coarse.

Of the several shapes intrusions come in, the ones that supply the most volume of granitic rocks are large, essentially bottomless ones called "plutons," of which the very biggest (those over 38 square miles at the surface) are called "batholiths" ("deep rocks" in Greek). Plutons aren't necessarily salient topographic features; they're simply areas on and under the earth surface, where the bedrock is intrusive.

Now, there's a reason these paragraphs have spoken often of granitic rocks, and only once of granite. Rock climbers and tombstone shoppers may speak of any sort of coarse-grained igneous rock as granite, but geologists do not. Hedging by substituting "granitic" is loose but not incorrect. Geologists use "granitoid" to include true granite and all other granitic rocks containing at least a 1:5 quartz/feldspar ratio. Our commonest granitoids are granodiorite and tonalite; we have a fair amount of true granite as well, particularly in the Golden Horn Batholith, an area on the North Cascades Hightway including the headwaters of Granite Creek, the Early Winters Spires and Liberty Bell—beloved of rock climbers and photographers.* Granite is typically pinkish or yellowish white with darker speckles. Also fairly abundant in the North Cascades is diorite, a darker, typically salt-and-pepper speckled gray rock otherwise resembling granite. By definition, diorite has less than 20% quartz, but in many batholiths it grades into other granitoids. Intrusive rocks are rare in our range outside of the North Cascades.

"Light," "medium," and "dark" igneous rocks are sorted out on the same chart of quartz and feldspars regardless of whether their texture places them among the intrusives or the volcanics; intrusive types are each chemically equivalent to volcanic types. Granite is chemically equivalent to rhyolite, and the Golden Horn Batholith, when it was still magma, probably fed a rhyolitic volcano. Granodiorite and tonalite are equivalent to dacite, and diorite to andesite, and sure enough, those types of lava predominated in the volcanic periods (120–20 million years ago) when most of our batholiths were emplaced.

*This nomenclature follows the International Union of Geological Sciences' 1973 definitions, which arbitrated between American and European geologists' usage. Widespread assertions that our range has little granite outside of the Golden Horn are based on the earlier, American definitions. The abundant N. Cascade rocks called "quartz monzonite" and "adamellite" in papers by Misch, Tabor, Crowder, etc., are included in I.U.G.S. granite, and those called "quartz diorite" and "leuco-trondhjemite" are mostly I.U.G.S. tonalite.

## Gabbro

Color p 510.

Gabbro is the intrusive counterpart of basalt; it contains the same mix of minerals, but in somewhat coarser crystals. Though basalt is the most abundant volcanic rock, gabbro is rather uncommon, cropping up in the North Cascades as medium-sized dikes, never as large batholiths. In places (lining Fourth-of-July Basin, for example) extensive bands of dark gray gabbro alternate dramatically with pale granitic bands. Gabbroic bands are usually well over 4" wide, considerably wider than the bands typical of schist or gneiss. The minerals within the gabbro show little contrast in color, making the crystals relatively inconspicuous.

## Peridotites

("Per-id-o-tights"). Color p 510.

Unusual intrusive rocks loosely termed peridotites are characteristically green, often flecked with black crystals, with a greasy-looking luster. They weather into red soils difficult for most plants to grow in. An abrupt transition to reddish soil with sparse, dwarfed plant cover may be a clue to these "serpentine" soils underlain by peridotites. They are found in a long arc around Mt. Stuart that includes Tumwater Canyon and much of Ingalls Creek; all over the Twin Sisters near Mt. Baker; and in smaller outcrops scattered through the North Cascades.

In general, peridotites are restricted to environs of deep faulting within subduction-zone mountain ranges. They are thought to be pieces brought up from the upper mantle, the subcrust layer of the earth. Peridotite is defined by a high proportion (40%+) of the yellow-green mineral olivine, which predominates in the mantle. Olivine-rich material is too heavy to reach the surface without exceptionally deep and rapid rock movement. Peridotites containing over 90% olivine are "dunite," the least adulterated mantle specimens. Geologists fascinated by dunite mostly have to content themselves with little bits and pieces. The Twin Sisters, probably the largest outcrop of dunite in the Western Hemisphere, attest to the violent crustal upheaval that produced the North Cascades.

Peridotites have no fine-grained (volcanic) equivalent. Though they occasionally shoot out of volcanoes as inclusions in lava, their melting temperature is so much hotter than lava that their crystals are assumed to originate long before (and far below) the melting of the magma.

---

---

## Shale

Color p 511.

Abundant throughout the Olympics, shale seems like just your basic sedimentary rock—clayey river mud carried out to sea, and piled up in ocean basins to such depths that through thousands or millions of years under the weight of subsequent layers it was compacted into more or less solid rock. Most kinds of sediments have to be cemented together with water-soluble minerals to become sedimentary rocks, but clay particles are so minute and flaky that they can become shale through compaction alone, or with minor amounts of cement. Shale is still close enough to simple clay that you can breathe on it and smell clay, or break a piece in your hands, or scrape it with your knife and spit on the scrapings to mix up a batch of fresh clay mud.

Shale is generally gray, and has a strong tendency to break into flattish leaves along its original bedding planes. A similar rock that doesn't break in flat leaves (due to different mineral content) is "mudstone." Where we find shales and mudstones, we may also find less-processed and/or more-processed beds nearby—dense clays that lay on the ocean floor too briefly and shallowly to get compacted into rock, and others that came out metamorphosed into slate or phyllite.

---

# Sandstones

**Arkose** and
**Graywacke** ("gray-wacky"). Color p 511.

Sandstone is a broad term for sedimentary rock made up of compacted sand-size (.06–2 mm) particles cemented together with water-soluble minerals like silica or calcite. The sand grains may be largely quartz, as in the classic blond sandstone. Quartz sandstone is uncommon here, though there is some near Mt. Olympus. It's hard to imagine how beds or beaches of nearly pure quartz sand are ever collected, since the waters that bring them are eroding all the diverse rocks of a watershed. The answer is a bit technical; quartz is both hard and chemically stable, tending to break down to sand-size particles and stop there. Feldspars and other abundant minerals are subject to mechanical fragmentation, as quartz is, and additionally to chemical weathering with water; they end up largely as silt and clay. Eventually, most sand-size particles in a body of water are quartz, as the finer silt washes away; this takes a long time, preferably in pounding surf, and it helps to have a quartz-rich source.

The common sandstones of our mountains had neither the quartz-rich source nor the time. Geologists give them other names, reserving "sandstone" unmodified for quartz sandstone. Feldspar-rich arkose, common in the North Cascades, looks coarse, motley, and sometimes pinkish, resembling granite and diorite, from which it derives. It forms in lakes, riverbeds and shallow seas near active mountain ranges where granitic rocks are rapidly uplifting and eroding; transport and deposition is too swift to allow chemical weathering to remove the abundant feldspars. North Cascades arkoses formed in river basins nestled among the edges of ancestral Cascade ranges. Westside outcrops are largely among foothills, but also include a few such peaks as Gothic and Mt. Hagan, which have resisted erosion admirably. Eastside exposures are in the Wenatchee-Chwah River trough and a swath along the Methow Valley. Sands in the arkose around Leavenworth derive from erosion, 50–60 million years ago, of the same Mt. Stuart Batholith that still stands high above them.

Coast ranges (including the Olympics) typically contain a

lot of "dirty" sandstone called graywacke. A belt of graywacke and some paler sandstone (arkose and quartzose both) runs through the heart of the Olympics from the upper Soleduck to the North Fork Quinault, including parts of Mt. Olympus. Graywacke resembles shale in its dark color and clayey odor when damp, and is held together mainly by compacted clay rather than by soluble minerals. With its mixed-size, angular fragments, often of andesite, it may equally resemble volcanic tuff; but tuff is as rare in the Olympics as graywacke in the Oregon Cascades.

Graywackes are thought to be deposited in deep ocean trenches above subduction zones, offshore from chains of volcanoes, when avalanchelike bursts of mud-and-sand slurry course down the slopes of these trenches at intervals of several hundred years, dropping many inches or feet of sediment at a time. These "turbidity currents" are the only well-accepted mechanism for depositing unsorted sediments (i.e., a range of particle sizes in each bed) on the ocean floor. Though unsorted, "turbidite" beds are more or less "graded"; each single bed, or layer deposited by one turbidity avalanche, starts at its bottom with coarser sediments, which settled quickly, and grades upward to fine sediments, which may have taken weeks to settle. Of course, the original top and bottom are most likely turned some other way by the time you see them, due to radical faulting and folding in the Olympics. However that may be, the finest sediments abruptly adjoin the coarsest sediments of the subsequent bed.

# Conglomerate

Color p 512.

Conglomerates are defined as sedimentary rocks composed at least 50% of erosion-rounded stones over 2 mm in diameter; most of them are pebble beaches and river bars turned to stone. They have to include both smaller rock particles to fill in the spaces, and soluble minerals as glue. Well-cemented conglomerates will break straight through pebbles and interstices alike—a fine sight. Weaker ones break through the cement (or "matrix") only, leaving the pebbles sticking out just like from any old gravel bank. There are good conglomerates around

Devil's Dome and Holman Peak, east of Ross Lake. Ludden Peak in the Olympics is a spectacular knob of hardened "metaconglomerate"—compressed so hard during metamorphism that the pebbles were all flattened in one direction.

## Breccia

("bretch-ia"). Color p 512.

Breccia consists of unrounded rock pieces bound up in a finegrained matrix other than volcanic tuff. The distinction from conglomerate, in which the inclusions are erosion-rounded, may seem trifling, but on further thought it is puzzling, since only eroded rocks are normally deposited as sediments.

The various brecciating processes all involve breakage, from which the Italian word *breccia* derives. Some are sedimentary, some volcanic, and some more or less metamorphic:

**Sedimentary breccia** is often a coarser version of graywacke containing shaley shards ripped from the surface of deep-sea sediments by a turbidity current and then deposited among the turbidite sediments. Alternatively, ordinary landslides in shaley, slaty, or limestone terrain can end up compacted into breccia.

**Volcanic breccia** (the commonest type here) can be simply the margins of basalt flows, where chunks of crust that formed as a skin on the slowly flowing lava are still visible after being shattered and reincorporated by the continuing flow. In this case, matrix and inclusions are of the exact same rock.

**Tectonic breccia** is a mix of coarse and fine rock fragments broken by shearing forces within a fault, and then compacted under moderate pressure; pairs of adjacent large fragments can often be matched up visually.

There is a lot of breccia out there of one type or another, but much of it looks like an obscure mess and fails to draw attention. It takes rock saws and polishers to reveal the splendid cross-sections of breccia that appear in rock books and as architectural facings.

# Quartz and Chert

Color p 512.

The most familiar kind of "rock crystal" is quartz—transparent (though sometimes tinted) with six generally unequal sides and a six-faceted point. It's the only stone in this chapter that forms single crystals; it's the only one that's a mineral rather than a rock. In geologists' terms, a rock is a mixture of minerals whose proportions to each other vary only within limits that define that kind of rock; the respective mineral crystals, whether visible or microscopic, are discrete. A mineral, on the other hand, is a single chemical compound or continuum of compounds with one characteristic crystal shape. Quartz is silicon dioxide ($SiO2$, usually simply called silica) adulterated with too few impurities to break up its proper crystal form. It is hard, too hard to scratch with a knife, lightweight and light-colored, and very abundant as a rock ingredient; it is in nearly every light- to medium-colored rock you see, except limestone. Its abundance at the earth's surface is thought to result from its light weight—it has risen among heavier materials while powerful forces stirred the earth around for some four billion years.*

Large, free-sided quartz crystals are not abundant, but they do catch the eye. You can find them in several parts of the Olympics, including crevices in greenstone on Crystal (!) Peak and Chimney Peak. They form in cavities in rocks (often volcanic) that spent a long time filled with moving groundwater rich in dissolved silica. The silica precipitated out as "rock crystals" on the cavity lining, just like sugar crystallizing on a string

---

*Conversely, the superabundant heavy metals nickel and iron accumulate toward Earth's core. The mantle, in between, is made up largely of the mineral olivine, consisting of silicate anions ($SiO4$) weighed down by iron and magnesium atoms. Olivine-rich rocks (peridotites, p 000) occur at the surface rarely, in extreme contrast to their abundance down below. At Earth's continental surface, quartz is rivaled in abundance only by the feldspars, a group of silicate/aluminum minerals, making silicon the second most abundant element at the surface, after oxygen. Specific gravities of quartz (2.65) and feldspars (2.5–2.8) compare with averages of 2.85 for Earth's crust as a whole, 4.5 for the mantle, and 10 for the core.

---

as "rock candy" (named after quartz.) Quartz veins in intrusive and metamorphic rock bodies reflected as a gleam in a prospector's eye, since precious metals occur in or near them; the same hot subterranean streams carry and deposit both silica and minerals.

Virtually pure quartz may also occur as the opaque rock chert, a cryptocrystalline or microcrystalline (in either case, not visibly crystalline) form. Chert resembles porcelain while crystalline quartz is transparent like glass; ground rock silica is the main ingredient in both glass and porcelain. Porcelain can be used to finish a fine edge on a knife, and a hard variety of chert called Arkansas novaculite makes famous whetstones. Like glass and obsidian, chert chips with an even, shallowly concave fracture; its dark form, flint, rivaled obsidian as an arrowhead and blade material among stone-tool cultures.

Cherts form in very different ways. The commonest is similar to limestone—crumbled marine shell material settling to the ocean floor or precipitating there from solution in sea water. Calcite shells are most familiar to us, but there is also a large group of one-celled organisms that make a sort of shell out of silica, and these are the main source of chert.

## Limestone

Color p 512.

Many marine sediments, including those that become sandstone or shale, are ground-up rocks from the continents. Others are deposits of ground-up and/or dissolved seashells. Hard-shelled marine organisms (don't picture shellfish, since the majority are microscopic, and many are plants) make shells out of various compounds, the most abundant being calcite ($CaCO_3$). Sedimentary rock characterized by over 50% calcite content is limestone. (You can test for limestone with a drop of hydrochloric acid or vinegar, which will effervesce in reaction with the alkaline calcite.) Some limestones are virtually pure crystalline calcite precipitated upon the ocean floor from completely dissolved shells of dead organisms. Others, such as chalk, form from marine algae that secrete calcite without making a substantial shell. Typically, precipitated calcite cements the pulverized shell material.

Limestones underlie vast regions of the midwestern U.S. and many other parts of the globe. Here, too, they may be the commonest shelly rocks, but that's not to say they're truly common. Chowder Ridge's 150 million-year-old limestone bouillabaisse, cropping out from the young flanks of Mt. Baker, is one example. Another is limestone that formed in small interstices between undersea basalt flows that are now high in the eastern Olympics; minerals from the hot lava altered much of it to deep red. Above the Napeequa River there are belts of marble, which is metamorphosed limestone.

---

---

## Slate and Phyllite

Color p 513.

Slate is recognizably metamorphosed shale; even the names look alike, and the rocks share dark gray colors (varying through reds and greens in slates) and platy textures that tend to break into flat pieces. Surprisingly, the flat cleavage of slate is not at all the same thing as the flat bedding plane of shale. Bedding planes started out horizontal, being simply the gradual layering of mud or other sediment as it settled out from the water. Slaty cleavage planes, in contrast, are perpendicular to the direction of pressure that metamorphosed the rock. There may have been two or more such directions at different times, producing slate that fractures into slender "pencils." Some slates metamorphosed from mudstone, and never had distinct bedding planes; in others the bedding has faded. Commonly, though, the parent shale's bedding planes are visible as streaks or bands in slate, either happening to align with the new cleavage planes or crossing them, but no longer causing a plane of weakness.

Slate gets a slight satiny luster from microscopic crystals of

---

mica and other minerals. Metamorphism has aligned these all parallel, creating the primary cleavage plane and making it relatively shiny. Higher-pressure metamorphism of the same rock produces larger crystals (just barely visible without a lens), a much stronger, glittery but perhaps wavy gloss, and a somewhat weaker tendency to cleave flat. Such rock is called "phyllite" by geologists, though it is often quarried as roofing "slate." Slate and phyllite are abundant, often together, in the Olympics, and sporadic in the North Cascades. Slate is shatter-prone and weak in landforms, but phyllite is erosion-resistant enough to endure as such outstanding peaks as Whitehorse and Three Fingers.

## Schist

Color p 514.

Conspicuous fine-layered texture in a rock may be of either sedimentary or metamorphic origin (see Slate, above). Metamorphic layering alters with increasing heat and pressure; the parallel layers may get tightly crimped, crumpled or curved, while the rock's strength in most cases increases. At the same time, the flaky crystals (mainly mica) first grow larger as "slaty cleavage" converts to "schistose foliation," and then are replaced altogether by coarse crystals that may show "gneissose banding." Though "schist" derives from the Greek for "split" (as in schizophrenia), schists aren't nearly as easy to split as slates. Schistose texture, which defines schists, is tricky to describe; worse, it is often dulled and obscured by weathering or rust. Unweathered specimens glimmer in the sun, showing the parallel orientation of their mica flakes.

Schists vary widely in color. A typical variety just east of Glacier Peak is garnet-biotite schist. The garnet appears as scattered, reddish, smeared crystals ¼" or larger, with schistose foliation making almond shapes around them. Black mica gives the rock an overall blackish hue. Garnet did not exist in these rocks until they reached the exact degree of heat and pressure that produces garnet schist. Metamorphism not only recrystallizes preexisting minerals, it makes new minerals by recombining atoms into new molecules. The mineral content of metamorphic rocks is determined partly by the overall proportions of elements in the parent material, and partly by the "metamorphic grade" of heat and pressure they underwent.

Schists are widespread in the North Cascades. Olympic rocks are younger and have seen only lower-grade metamorphism, producing much slate and a little low-grade schist.

## Greenstone and Greenschist

Color pp 513, 514.

Greenstone is an informal term for slightly metamorphosed basalt or andesite. It looks like basalt, but with a light to dark greenish cast. It is much harder than basalt, as attested by the sheer relief of Hozomeen and Jack Mountains, part of a belt of old metamorphosed sea-floor basalt in the North Cascades. Basalts in the Olympics are also metamorphosed in places (Mt. Ferry to Mt. Christie, for example), but few exposed basalts of the Oregon Cascades have ever been buried deep enough for metamorphism.

Mt. Shuksan is composed of greenschist, an equally durable metabasalt distinguished from greenstone by its schistose texture—lustrous crystals smeared out into layers. This metamorphism may result from sideways-shearing stress superimposed on the compressive stress that makes greenstone.

## Hornfels

Color p 513.

If you ever notice that you have just left an area of granitic rocks, you may be in the vicinity of hornfels. It is a "contact metamorphic" rock, meaning that it was baked by close contact with hot intrusive magma, but not substantially altered by pressure. It was also altered by the addition of volatile chemicals suffusing out from the magma. Where hornfels occurs in shale, it may grade into spotted slate and then shale as you walk a few yards farther from the granitic intrusion. Hornfels is a nondescript, fine-grained, often dark gray to black crystalline rock resembling anything from basalt to sandstone; the easiest way to tell it from basalt is by its location near granitics.

# Gneiss

("nice"). Color p 514.

North Cascades climbers know the Skagit Gneiss intimately. Redoubt, Fury and Terror, Snowfield and Eldorado, Forbidden and Formidable, Boston, Buckner, Logan, Goode, Johannesberg, Magic, Mixup, Maude, Fernow, Fortress and Bonanza—peaks forming a backbone of the range are made of this Mesozoic metamorphic complex. Plainly, gneiss stands up to erosion every bit as well as the intruding granitics and better than nearby schists.

Gneiss is intermediate between schist and granitic rocks in appearance, in grade of metamorphic intensity, and often in location. Much gneiss started out as shale or arkose and was metamorphosed by heat and pressure past the phyllite and schist stages, nearly to the point of melting or granitization. Other gneiss was actually granitic before metamorphism; gneisses of either type of parent rock look much alike in the field. Coarse gneiss may overlap fine-grained granitics in crystal size, but the grains are at least slightly flattened in gneiss, never in granite. The plane of flattening may also show color banding, often contorted. Very distinctly segregated mineral bands may occur, as in schist, and many pieces of gneiss are broken along planes of glittery mica flakes. For the rock to be gneiss rather than schist, flaky minerals must be outweighed by fatter grains overall. (Gneiss is a loose field term primarily denoting a texture and a metamorphic grade; geologists often prefer more mineralogically specific names.)

Skagit Gneiss is part of the Skagit Metamorphic Suite,* a broken and convoluted group including many of the oldest surface rocks in our range. The eastern North Cascades—the Okanogan, Chelan Crest, Chelan and Entiat Ranges—are also largely made up of old gneissic and intrusive rocks. The Skagit

---

*The "Skagit Gneiss" is not a variety of rock but a "unit," or patch on a geologic map indicating apparent shared origins and history for the component rocks—many of which happen to be gneiss. Some geologists argue that the Skagit Gneiss and Skagit Metamorphic Suite properly belong within either the Custer Gneiss or the Swakane Gneiss; the "Skagit" terms may fall from use.

Suite has been in at least three successive mountain ranges, each including volcanism as well as folding and faulting. Metamorphism took place deep in their churning, grinding roots. The "microcontinent" theory of the North Cascades suggests that the earliest range was a large tropical island, perhaps resembling New Guinea. About 50 million years ago the microcontinent, by then eroded down to gentle terrain, rafted onto North America. The "new kid on the block" arrived in time for a great mountain-building episode, the Western Cascades volcanism. Beginning about 8 million years ago, the present North Cascades uplift caused all the northern volcanoes of that period to erode off the top, exposing both their "plumbing"— our young batholiths (Snoqualmie, Cloudy Pass, Golden Horn, Chilliwack and others)—and the Skagit Metamorphics. Judging by the amount of pressure required to achieve the Skagit's metamorphic grade, the net uplift of these rocks has far exceeded ten vertical miles, several miles in the current uplift alone. Erosion must have worked fast to keep the mountains from being any higher than they are.

## Migmatite

Heat and pressure at depth during mountain-building can go beyond metamorphism and exceed the melting point of rocks, producing a liquid magma that solidifies into an igneous rock such as granite.* Where metamorphism proceeds just to that point before cooling off again, it produces spectacular rocks straddling the metamorphic/igneous boundary. They often show beautiful swirling bands, and are prized facing stone for buildings. Called migmatite, from the Greek for "mix," they contain metamorphic rocks in a granitic matrix. The metamorphic inclusions are gneiss or schist—the last metamorphic stops on the road to igneous melt. The migmatite may have

---

*Some geologists believe that "granitization," or extreme metamorphism just short of melting, can also produce granitic rocks virtually identical to ones that did melt. (They are distinguishable only by where they occur in relation to other rocks.)

---

previously been a single rock body, only parts of which melted; or the granitic matrix may have been injected along cracks from a nearby magma body.

Migmatites are not uncommon within the gneiss core of the North Cascades. Some appear in roadcuts in the Skagit Gorge and at Chelan Falls. Migmatite differs from metamorphic breccia in the coarse crystalline (granitic) texture of the matrix portion.

# 16

# Climate

Weather may be critical to the success of your trip to the mountains, and it is at least as crucial in determining the landforms and life forms you will see.

One weather fact is well known—it rains a lot here. More significantly:

It rains a lot in winter, almost as much in fall and spring, and relatively little in June through September.

It rains a lot West of the Olympic and Cascade Crests, but much less on the East slopes, approaching desert conditions at the foot of the Cascades.

It rains a lot and snows even more at higher elevations.

The resulting snowpack is slow to melt due to its sheer mass rather than to really cold weather; it persists longer in summer, and produces more glaciers and permanent snowfields, than snow in other mountains with equally mild or somewhat colder climates, such as the U.S. Rockies.

# When It Falls

Our summer "drought" is without equal among the world's rainy climates. Most climates with strong wet seasons and dry seasons are tropical, and lack summers and winters. Most of the Temperate Zone, on the other hand, gets precipitation all year round, or with a slight emphasis in summer. Our climate is a midpoint between California's Mediterranean type (dry summers and subtropical temperatures) and Southeast Alaska's West-Coast Marine type (copious rain, slightly reduced in the summer, and cool temperatures year-round). Our closest analogs are certain narrow locales in Southern Chile, Western Scotland, the Northern Honshu coast on the Sea of Japan, and Norway's Fjord Country. Each of those wet West-Coast Marine climates produces 1½ times as much rain and snow in its wettest winter month as in its driest summer month. That's nothing; here, most weather stations record between 6 and 20 times as much precipitation in December as in July.

Causes for our unique weather lie in the border tension between the cold polar air mass and the warmer subtropical air mass. The front between those masses runs around the world between roughly 40° and 60° North latitude; rather than running due east it wanders in giant lobes directly under the polar front jet stream, a high belt of high-velocity west wind circling the globe.

Though the jet stream's lobes or loops vary somewhat unpredictably, causing year-to-year weather eccentricities, nearly every autumn brings a southward migration of the jet and expansion of the polar air mass. On average, the polar jet crosses Anchorage in September, Ketchikan in October, Bella Coola in November, Astoria (and all of our range) in December, and Eureka in January. Those are the respective wettest months when each locale is directly in line for rainy "storms" —low-pressure centers spun off from the edge of winter's great Aleutian Low.

The air in the storms, originally quite cold, travels for days across relatively warm (45°) ocean, to reach our shores modestly warmed and immodestly moistened. On a satellite photo the storms look like whorls of clouds spiraling counterclockwise. On a weather map they look like steep, eastward-drifting waves written in heavy "front" lines. The leading edge of the wave crosses us as a warm front, and the trailing edge, 8 to 48 hours later, as a cold front. Lulls in between are typically filled with partly cloudy, showery weather comprised of unstable (busily convecting) moist warm air.

Rarely, the jet stream develops a sharp northward swing offshore, dragging warm, sunny California weather up into our area for days or weeks at a time. Alternatively, a strong southward swing of the jet, farther inland, may drag Arctic cold pressure over the midcontinent; some of this air pours gravitationally over the Cascades from the east (especially down the Columbia Gorge) sliding in under the warmer marine air to produce lowland snow and/or freezing rain. If the southward swing is a little closer to us, Arctic air and high pressure can take over the Northwest completely, for a sunny cold period.

As summer approaches, the jet stream snakes its way back north, taking the frontal systems with it. Its progress is more uneven than in fall, so there is no northward progression of wet months. The Aleutian Low weakens while a subtropical North Pacific High strengthens and stabilizes, typically with a large, benign bulge in our direction. The Polar Front continues to bring rain to the coast, but now mostly in British Columbia and Alaska. All along the West Coast, including our range, summer drought gets less and less consistent northward.

Our dominant summer air consists of NW breezes spinning clockwise out from the center of the High. Since these travel across ocean colder (55°) than themselves, they are unable to evaporate much moisture from it. They are chilled by contact with the water, often creating vast ocean fogbanks, but on hitting the warmer land surface they warm, and the fog soon dissipates. Sometimes during high pressure, moist marine air is strong enough to blanket Olympic Westside valleys and much of the Puget-Willamette Trough with low clouds and even drizzle, while peaks and slopes above 3,000' remain crystal clear.

All this means that it's certifiably possible to enjoy summer recreation anywhere in our range, despite rumors to the contrary. It does not mean that you can safely go unprepared for bad weather. Several recent Julys have brought blizzards to North Cascade locales as low as 4,000'.

# Where It Falls

In contrast to our precipitation's unique summer-winter split, its imbalance between west and east slopes is familiar in mountain ranges the world over; it is typical of mountains as rainmaking devices and as barriers between moist marine air and drier continental air. When a prevailing air flow crosses mountains, the air must inevitably rise. In rising, it cools; air temperature drops with air pressure, which drops with increasing altitude because higher air has less atmosphere above it to weight down on it. All other factors being equal, clear air cools 5.5° with each 1,000' of altitude. Other factors are rarely equal, so the actual temperature lapse rate is usually less.

The air's capacity to carry water vapor also rises and falls with temperature. When air containing water vapor cools, it may sooner or later reach a point where it can no longer carry all of its moisture, which will condense into clouds or fog and then may fall as precipitation. Where the air descends after crossing mountains, the effect is reversed. If there was just enough moisture to make clouds and drizzle on the windward slope, the clouds may abruptly vanish as they cross the crest; enough for heavy weather on the windward slope may turn to scattered showers on the downwind slope.

Sometimes the condensation and then reevaporation of moist air flowing over a mountain becomes a graphic image in the sky—a "lenticular" ("lens-shaped") cloud. These pure white slivers or crescents hang motionless over high peaks. They may stack up two or three deep, and they may stand downwind instead of—or in addition to—directly over the peak. A lenticular cloud is that portion of a uniform layer of air which, as it flows over the mountain, is cooled enough to condense some of its water vapor into cloud droplets. After descending abruptly, the layer may bounce upward once or twice, in waves, to form downwind lenticular clouds.

Since the prevailing wind here is from the WSW in winter and the SW in spring and fall, and the Olympics are a more or less round mountain range, their downwind side, known as the "Rain Shadow," is on the NE. The Elwha and North Fork Skokomish Valleys are rain-shadowed, receiving probably less than a third as much precipitation as the Hoh and Quinault; and the Dungeness and Dosewallips are even drier. A crooked chain of ridges, running from Sourdough Mtn. to Prospect Ridge via the Bailey and Burke Ranges, constitutes the Olympic Crest as far as climate is concerned, though the term is not in common usage.

The Cascades comprise a long north/south divide, so their wet side is their west side even though prevailing winds are southwesterly. The barrier is broken by our two biggest rivers, the Columbia and Skagit, creating two unique biogeographic locales astride the climatic fence.

The Columbia Gorge cuts through the Cascades nearly at sea level, funneling wind and weather, as well as fish, plant species, and human traffic, between the Eastside and Westside lowlands. Some Westside species, like Oregon white oak, reach the Eastside in and near the Gorge. Many plant varieties are found only in the Gorge, or found there disjunctly (far from the closest similar populations.) On the other hand, the river itself is a barrier to North-South spread of some small animals, like red-backed voles; many butterfly species are also bounded by it.

In fall and winter, relatively high pressure east of the Cascades pours down the Gorge in the form of a protracted east wind strengthening toward the west; the Northwest's record windspeed was clocked at Crown Point, the funnel's mouth. When a freezing east wind slides down the Gorge under the edge of a rain-bearing warm front, it produces freezing rain, colloquially called a "silver thaw." In summer, when high pressure blankets the Northwest, prevailing westerlies are focussed by the Gorge into a steady west wind, strongest toward the east end, where it has made Hood River into a world-renowned windsurfing center. Evidence of both Gorge Winds takes the form of "flagged" trees; those at the east end sweep eastward, those at the west, westward.

The Skagit rises within the Cascades rather than east of them, so the Cascade Crest is customarily drawn (along with county lines and the Pacific Crest Trail) along the eastern edge of the Skagit drainage. This book does not follow that custom, because the upper Skagit Valley (above Ross Dam) is in rainfall and biotic regimes that fit the Eastside type better. To put that another way, the sharpest divide in terms of rain shadow runs along the Picket Range and down the center of North Cascades

---

National Park; that divide is the Cascade Crest—in terms of climate, plant and animal life, and the purposes of this book. The dramatic contrast between this true crest's glacier-mantled peaks and the drier Eastside peaks (from viewpoints like Crater Mtn.) exactly resembles that between Crest and Eastside peaks elsewhere in the North Cascades.

The Puget-Willamette Trough is moderately rain-shadowed by the Olympic and Coast Ranges. Though urbanites of Seattle, Portland, and Vancouver share a sometimes hysterical belief that they see a lot of rain, their actual recorded precipitation of 33–42" per year is precisely in a league with Peoria, Providence, Chattanooga, Calistoga, and the worldwide average. The number of gray and rainy days per year is indeed high in Northwest cities, but the bulk of the rain—60 to 250"—falls on the Cascade and Coast Range flanks. One indication that the Trough isn't very wet is the ponderosa pines and the climax grand firs in the Willamette Valley, resembling the dry east slope of the Cascades.

The rain shadow effect can be put to use in planning outings. If the day for your trip arrives gray and showery, consider shifting to an Eastside destination. You might not find clear skies, but you will almost certainly get less wet than on the Westside. One common exception is a summer high-pressure system with fair mornings and a line of afternoon or evening thunderheads developing along the eastern front of the Cascades; this is decent weather on either slope, producing only light rainfall. Just move away from salient trees or peaks if lightning begins to strike nearby.

If you want to reach high country in June or July, but snow still covers most of it, head for the northeast corners of the North Cascades (the Sawtooth or Pasayten Wildernesses) and Olympics. Less snow accumulates there, and it melts off weeks earlier each summer. In Oregon you have to go to the Wallowas or Strawberries to make use of that principle, since no high Cascade ridges extend far east of the Crest.

## Forecasting Rain

Instruction in do-it-yourself weather forecasting seems called for at this point. But it could be guaranteed to dissatisfy—often. All the same, there is some value in the old saws about types of clouds to watch out for:

"Mare's-tails." These are cirrus clouds—the very high, thin, wispy family—arrayed in parallel, all or most of them up-turned at one end like sled runners. Especially if blowing northward, they may presage a weather front by 12 to 24 hours. Broad sheets of cirrus whiteness, if northbound and/or thickening and lowering, may have the same meaning. However, scattered shreds of cirrus resembling pulled-out cotton puffs are typical of good weather.

"Cloudcaps." Small clouds sitting all day around the heads of outstanding peaks are common, and not necessarily ominous. If they thicken steadily for hours, they may be the most visible part of an increasing cloudiness presaging rain. Portentous or not, they are already bad weather ahead if the peak they cap is your goal. Reconsider: the view will be erased, the wind strong and cold. Also watch lenticular cloud streamers above or downwind of high peaks; their arrival or increase displays an increase in wind speed and/or moisture of the air. But the moist air flow may be modest, especially in summer, and may come and go without producing heavy weather.

Cumulus clouds—the white, rounded puffy kind—increasing in the afternoon are most often a stable fair weather pattern. Along a mountain crest they may by evening build and darken into cumulonimbus, perhaps generating thunderstorms. Even if this pattern repeats for days, it may be followed by either better or worse weather. Morning cloudiness filling Westside valleys is not necessarily a bad sign either, even if it brings palpable moisture or if it rises rapidly and temporarily engulfs the ridges.

It takes years of familiarity with the weather patterns of a particular mountain area to develop a really good eye for weather signs. Whether you think you have such an eye or not, it is vital to carry into the high country enough insulation, shelter and food to keep you alive, and enough navigation aids

---

and skills to get you out again, should the weather happen to turn bad.

In the short run, the most useful instruction is to catch a complete weather forecast. In between network TV weathermen, there is continuous, frequently updated weather information on cable TV weather stations, or on weather radio. The latter can't be picked up on ordinary AM/FM radios; it's on the Public Service Band at 162.4, 162.475 or 162.55 MHz. (Small weather radios can be taken backpacking, but mountain topography blocks these wavelengths, so the signal can only be picked up on peaks high enough to receive a nearly straight beam.) If you lack cable TV or weather radio, look for a telephone listing (under "Airports" or "US Dept. of Commerce") for 24-hour recorded weather advisories for pilots. Recordings for the general public are often less useful—generalized, short-range, focussed on the urban area. Try phoning the National Park or Forest Ranger Station closest to your destination; they usually have a local weather forecast, and they can always look out the window.

## Forest Microclimate

All of the above climatic subjects—the ones weathermen talk about—are macroclimatic. Of equal concern to hikers and other creatures is the microclimate, or narrow climate near the ground. A microclimate may be much warmer or cooler than its surroundings for several reasons:

**Hot air rises** and cold air sinks. Where protected from wind, small masses of cold air may descend along stream drainages, or settle in valleys and slight depressions. Some caves collect cold air drainage intensely, even retaining year-round ice at moderate elevations.

**Ground and lakes heat up** in the sun, even on cloudy days, and heat the air next to them. Dark surfaces heat much more than pale ones, south-facing slopes much more than north-facing ones. East slopes heat best in the morning, west slopes in the afternoon. High peaks are subject to especially intense solar radiation, including heat, by way of reradiation from surrounding clouds, snow or ice. In one extreme example, dark, dry humus soil on a high south-facing slope was found to reach 175° while the surrounding air was only 86°. Summer afternoons at 175° are simply the facts of life—or death—there for a seedling or a crawling invertebrate.

**Vegetation insulates;** the tree canopy, the shrub, herb, and moss layers, and the snowpack are all blankets, keeping everything under them warmer in cold weather, and vice versa. The combination of ground heat retention and snowpack insulation create a winterlong 30–32° environment utilized by many rodents who would otherwise either hibernate or live elsewhere. Deer and elk take "thermal cover" in deep old forests during cold spells of winter. In summer, on the other hand, the forest stays chillier than clearings.

**Vegetation and rough topography impede wind.** This effect allows cold air collected by sinking, or air heated by warm ground, to stay put longer than they otherwise would.

A forest canopy may also make the community under it either drier or moister. You may notice that rainfall descending from a forest canopy ("throughfall," consisting of drips from foliage and branches) starts and ends much later than individual showers in nearby clearings, and it falls as much bigger drops. Measured during a rain shower, it amounts to less; much of the rain is absorbed by the canopy and the epiphytic plants on the trunk, reevaporating eventually without ever reaching the ground. Light drizzle around here often fails to wet the forest floor at all.

An opposite effect also works here; when fog sweeps the forest canopy, moisture condenses on foliage, and some drips to the ground as throughfall while a rain gauge in the open would recieve no precipitation. Since low vegetation can catch only a fraction of the the fog that a tall forest can, clearcutting a high Cascade watershed is estimated to immediately reduce its total precipitation by about half. Rain, fog throughfall, wind and sun all hit different parts of a tree differently, so that each tree offers several microclimates for small plants that would grow on it (page 258).

Effects of a tree canopy on evaporation are also mixed, and even harder to measure. Trees shade the forest floor from the drying sun, but they also suck huge volumes of water up through their roots and transpire it into the air, leaving soils parched. Gravelly, underdeveloped soil and dependence in summer on residual snowmelt both exacerbate understory drought in our mountains. For small plants, having to compete with overstory trees may be a worse handicap in terms of water than of light; many of our sparsest understory communities are found under somewhat open canopies. Non-green plants (page 180), with their fungal lifelines borrowing water back from the tree roots, are at a particular advantage there.

Snowpack depth is reduced under forest cover. Though most snow that settles in the canopy does reach the forest floor, some of it melts first, and most of it, falling in big clumps, is much compacted on impact. Winter melting is greater in the forest. The dark canopy absorbs solar radiation and reradiates some heat downward into the insulated forest microclimate, while the bright white of snow-covered clearings reflects nearly all of the solar radiation that hits it, and doesn't heat up as much.

On the other hand, when warmer air masses arrive in spring and summer, the insulating canopy impedes their warmth from reaching the forest floor. Now melting proceeds faster in the open than in the forest, and the snowpack disappears from clearings several days sooner, on average, than from adjacent forest. Within meadows, individual trees dramatically hasten snowmelt because, while their insulative value is negligible without companions, their heat-absorbing effect is maximized. All the above microclimatic effects on temperature and moisture determine which plants can live on a microsite.

## Mountain Microclimate

Mountain and valley forms have broader effects. (Climate measured on a scale of a few kilometers rather than meters is sometimes distinguished as "mesoclimate.") Just as the sun shines hotter at noon than in morning and evening, at the equator than here in the mid latitudes, and in summer than in winter, it shines hotter on south-facing slopes than on other aspects. South slopes have hotter, drier plant community types than north slopes. Bottoms of steep-sided valleys have relatively cool, moist microclimates not only due to cold air drainage but also because the sun shines there few hours a day; the effect is strongest in east/west-running valleys, weakest in south-draining valleys. Exposed peaks and ridges are at the other extreme in receiving copious sunlight, but they are unable to hold the resulting heat. High thin air tends to be chilly.

The warmest levels in mountains comprise a mid-elevation "thermal belt" subject to neither cold air drainage nor thin air heat loss. If you're concerned about sleeping warm, you may gain as much as 15° by leaving a stream bottom and camping on a slightly higher.

Mountains' chief effect on broad prevailing winds is to deflect them from deep valleys and to strengthen them on salient peaks, major divides, and through long gorges aligned with the wind. The strengthening happens because the air flow, like a river in a narrowing gorge (or like air over a wing), is constricted; it has to speed up just to maintain a constant volume.

The air contained in valleys expands in the daytime heat and contracts again at night. This is palpable as "valley winds" and "slope winds," often stronger than the broader or prevailing windflow. The valley wind is a main trunk flow parallel to the main creek or river of the valley, while the slope wind is a thin sheet of air moving up or down the valley walls. Up in the day and down at night is the basic rule for both winds, but the valley wind, being larger, lags behind the slope wind. In early morning, for example, the upslope wind begins while the night's downvalley wind continues in the valley's center. Occasionally the flow gets almost violent in pulselike cycles just after sunset, when downslope and downvalley winds join forces. Valley and slope winds are strongest in clear summer weather.

# Appendixes

## Pronouncing Scientific Latin

Pronunciations of genus and species names are suggested in this book simply to keep the names from seeming too intimidating to use at all. If you want to pronounce them some other way, feel free. Biologists are far from unanimous in their pronunciation of scientific Latin, and few are stuffy about it.

The general consensus is, when in America, pronounce vowels as Americans do, but do as the Romans did in regard to consonants and to syllable stressing. The consonant *x* is phonetically a "z," final *es* is "eez," *ch* is "k," and *j* is "y." The one consonant sound not clarified by Americanized spelling is *th*, which is always soft as in "thin," never hard as in "then." No airtight phonetic system was devised for this book. The attempt was simply to break each name into units that, as English, would be hard to mispronounce.

Syllable stressing causes much difficulty and variation in American scientists' pronunciations. This book respects the basic Latin rule that the second-to-last syllable is stressed if its vowel is long, is a diphthong (vowel pair), or is followed by two consonants before the next vowel; otherwise the third-to-last syllable, if there is one, gets the stress. When the length of the vowel was not known, we consulted *Webster's Third New International Dictionary* and *Gray's Manual of Botany*.

Exceptions are made for a few names that have entered the English language; *Anemone* and *Penstemon*, for example, are in common speech stressed on their third-to-last syllables, rather than on their second-to-last as they would have been in ancient Rome. Other exceptions include the many non–Latin proper names with Latin endings tacked on. Consonants and diphthongs in these are pronounced as in English, like the *j* and *ey* in *jeffreyi*, but their stress still falls according to Latin rules. Even non–American proper names are Americanized, however unfair that may be, if their native pronunciations are difficult for Americans. For example, *Castilleja,* named after a Spanish botanist, is "cas-ti-**lay**-a" here, while in Mexico it is pronounced "cah-stee-**yeh**-ha."

According to Latin rules, there is a syllable for every vowel except in the diphthongs *ae, oe, au, eu, ui, ei,* and occasionally *oi,* as in *Dendroica* ("den-**droy**-ca"). Thus, the common ending *-ii* is properly "ee-eye," *-eae* is "eh-ee," and *-oides* is "oh-**eye**-deez." Many biologists aren't so proper, and simplify those endings to "eye," "ee," and "**oy**-deez," respectively. On the other hand, some biologists are exceedingly proper and end many species names with "**ah**-ta," a pronunciation taught in Latin classes, rather than the more popular "**ay**-ta." This book takes a middle course.

Scientists differ in stressing syllables in Latinized proper names. Contradictions are found even within excellent books offering pronunciations. For example, *The Audubon Society Encyclopedia of North American Birds* (Terres 1980) says "nut-**ALL**-ih-eye" on page 635, "**NUT**-all-ih-eye" on page 1023. *Wildflowers of British Columbia* (Clark 1973) says "lýallii" on page 14, "lyállii" on pages 143 and 274. The book in your hands may have better luck, as to consistency. The bottom line, though, is that no pronunciation is "wrong."

(As for those pesky *-ii* endings, controversy embroils even their spelling. There is a movement afoot in some of the life sciences to replace every *-ii* with a single *-i* for the reason that the liberal strewing about of the double *-ii* by nineteenth-century namers was not well grounded in Classical Latin usage. For example, most recent writing on the rainbow trout refers to it as *Salmo gairdneri,* a correction of the original *gairdnerii.* However, the idea of "correcting" scientific nomenclature flies in the face of one of its best-established principles—that the most practical route to unanimity is to always honor the first validly published name, warts and all. I suspect the ichthyologists will return to the fold where the botanists have stood fast—dotting every old double *-ii* whether they pronounce it or not.)

In this book, a genus or species entry with its pronunciation and translation omitted is either the same name as the preceding entry, or clearly similar to an English word of essentially the same meaning, e.g., *americanus.* A family name with its English equivalent omitted is either the same as the preceding *and* visibly similar to its English version, e.g., Orchidaceae; or so obscure that the only way to come up with an English equivalent would be to fabricate one out of the Latin name.

No pronunciations are supplied for names of families and orders. Only a few different endings appear in those names, so a few rules should suffice. Family names end in *-ae,* and the third-to-last syllable is stressed; Pinaceae is "pie-**nay**-seh-ee" and Canidae is "**can**-id-ee." Bird orders end in *-formes,* "**for**-meez." Insect orders end in *-ptera* with the *p* pronounced and stressed, as in "**Dip**-ter-a."

# Chronology of Early Naturalists

**1741**     Georg Steller, sailing from Kamchatka under Captain Vitus Bering, explored islands now in Alaska. Page 402.

**1778**     Captain Cook visited Vancouver Island; little is known of anything William Anderson, the naturalist on board, accomplished there.

**1787**     Archibald Menzies visited Vancouver Island briefly.

**1791**     A grand Spanish expedition under Alejandro Malaspina reached British Columbia; Bohemian botanist Thaddeus Haenke was aboard, but his important collections were from earlier in the voyage.

**1792**     Menzies returned to the Northwest under Captain Vancouver. Page 94.

**1803–06**     Georg Heinrich von Langsdorf's attempts to preserve plants from Alaska and California were mostly thwarted by his Russian captain, Nicolai Petrovich von Rezanoff; rough seas on the Columbia Bar turned them back from landing in Washington.

**1805–06**     Lewis and Clark canoed the lower Columbia. Page 234.

**1815–18**     Chamisso and Eschscholtz studied Alaska and California under Russian captain Kotzebue. Page 219.

**1825–27**     David Douglas made his first (and more successful) exploration of the Northwest for the Royal Horticultural Society. Page 18.

**1825**     John Scouler, arriving with Douglas, explored on his own. Page 67.

**1825–27**     Drummond and Richardson collected in western Canada on Franklin's second expedition. Page 475.

**1826–29**     Karl H. Mertens sailed around the world; his stop at Sitka yielded many new plant species. Page 210.

**1830–33**     David Douglas returned to the Northwest. Page 18.

**1833**     Meredith Gairdner assumed the post of physician at Fort Vancouver. Page 438.

**1833**     William F. Tolmie began a long career in the Northwest with the Hudson's Bay Co. Page 198.

**1834**     Nuttall and Townsend reached the Northwest with Nathaniel Wyeth. Pages 71 and 313.

**1844**     Karl A. Geyer crossed the Rockies. Page 151.

**1851–53**     John Jeffrey hunted plants for the "Oregon Association" of Scotland. Page 208.

**1855**     John S. Newberry studied the east flank of the Oregon Cascades. Page 223.

**1858–59**     David Lyall and George Gibbs studied the 49th Parallel while surveying for the Boundary Commission. Page 51.

**1874–77**     Major Charles Bendire was stationed in Oregon. Page 297.

# Five-Kingdom Taxonomy

This chart's purpose is to show the relative positions in the living world of organisms discussed in *Cascade-Olympic Natural History;* these, and the higher categories they belong to, are in boldface type. A few undiscussed organisms are shown for comparison, in light type.

Broad relationships among organisms are of perennial interest, even though they can never be nailed down once and for all. The chart follows Margulis and Schwartz (1982) at the kingdom, phylum, and (where possible) class levels.

The weakest kingdom in this particular system is the Protoctista, which can be defined only by exclusion of the other four kingdoms.

| Kingdom | Phylum | Class |
|---|---|---|
| **Monera** (single cells without nuclei) | **Cyanobacteria** <br> 15 other phyla | |
| **Protoctista** | **Zoomastigina** <br> **Chlorophyta** <br> 25 other phyla | |
| **Fungi** | **Zygomycota** <br> **Ascomycota** <br> **Basidiomycota** <br> Deuteromycota <br> **Mycophycophyta** | |

Some earlier five-kingdom systems had instead a Kingdom Protista, consisting strictly of one-celled organisms with nuclei—an easier concept to grasp, perhaps. Protoctista combines such diverse things as amoebas, one-celled algae, and large algae such as kelp. These creatures obstinately resist organization under any simple concept.

In the Fungi, both Deuteromycota and Mycophycophyta, or lichens, are growth-form categories rather than proper lineages. The Latin binomial we use for a lichen is technically the name of its fungal component alone, and if this fungus were growing independently it would belong to one of the other phyla—usually Ascomycota.

| Examples |
| --- |
| Blue-green "algae" in lungwort and dog lichen |
| Giardia, trichomonas<br>Green algae in most lichens, watermelon-snow |
| Most underground-fruiting fungi, bread mold<br>Truffles, snow morel, snow mold, yeasts<br>All gilled mushrooms, boletes, chanterelles<br>Penicillium, monilia yeasts on people<br>All lichens |

The "vertebrates" (animals with backbones) are the five most advanced Chordate classes—Osteichthyes through Mammalia—and represent a single lineage.

| Kingdom | Phylum | Class |
|---------|--------|-------|
| Animalia | 28 primitive phyla | |
| | Annelida | |
| | Mollusca | Pelecypoda<br>**Gastropoda**<br>Cephalopoda<br>4 other classes |
| | Arthropoda | Arachnida<br>Crustacea<br>Diplopoda<br>Chilopoda |
| | | **Insecta** |
| | Chordata<br>(vertebrates<br>and their<br>closest<br>relatives) | 4 primitive classes<br>Cyclostomata<br>Chondrichthyes<br>**Osteichthyes** |
| | | Amphibia |
| | | Reptilia |

The six boneless classes of Chordata together with the 31 other phyla are loosely termed "invertebrates," a term that can be defined only by exclusion.

| Order | Examples |
|---|---|
| | |
| | **Snow worms,** earthworms |
| | Clams, scallops, oysters<br>**Slugs,** snails<br>Octopi, squids, nautilus |
| | Spiders, mites, ticks<br>Crabs, shrimp, pillbugs<br>Millipedes<br>Centipedes |
| Neuroptera<br>Homoptera<br>Coleoptera<br>Lepidoptera<br>Diptera<br>Hymenoptera<br>many other orders | **Ant lion,** lacewings<br>**Aphids,** scale insects, cicadas<br>**Beetles**<br>**Moths, butterflies,** skippers<br>**Flies, mosquitos**<br>**Bees,** wasps, ants |
| | Lampreys (jawless invertebrate fishes)<br>Sharks, rays (cartilaginous fishes)<br>**Trout, most other fishes (vertebrate)** |
| Caudata<br>Salientia | **Salamanders, newts**<br>**Frogs, toads** |
| Crocodylia<br>Testudines<br>**Squamata** | Crocodiles, alligators<br>Turtles, tortoises<br>**Lizards, snakes** |

| Kingdom | Phylum | Class |
|---|---|---|
| Animalia (continued) | Chordata (continued) | Aves (birds) |
| | | Mammalia |

| Order | Examples |
|---|---|
| Gaviiformes | Loons |
| Podicepediformes | Grebes |
| Ciconiiformes | Herons, egrets, storks, flamingos |
| Anseriformes | Ducks, geese |
| Falconiformes | Hawks, eagles, New World vultures |
| Galliformes | Grouse, ptarmigan, quail, chickens |
| Charadriiformes | Sandpipers, gulls |
| Columbiformes | Pigeons, doves |
| Strigiformes | Owls |
| Caprimulgiformes | Nighthawks, whip-poor-wills |
| Apodiformes | Swifts, hummingbirds |
| Coraciiformes | Kingfishers |
| Piciformes | Woodpeckers |
| Passeriformes | Jays, swallows, finches, many others |
| many other orders | |
| Marsupialia | Opossum, kangaroos |
| Insectivora | Shrews, moles |
| Chiroptera | Bats |
| Primates | Monkeys, lemurs, orangutan, person |
| Lagomorpha | Rabbits, hares, pikas |
| Rodentia | Mice, voles, squirrels, beaver |
| Cetacea | Whales, porpoises |
| Carnivora | Dog, bear, raccoon, weasel, cat, civet, and hyena families (Fissipedia) and seals, walrus, sea lions (Pinnipedia) |
| Perissodactyla | Horses, rhinoceros |
| Artiodactyla | Deer, Cow/sheep, and 7 other families |

Botanists use the term "division" instead of "phylum." The first four divisions are spore plants and the last five divisions are seed plants. The bryophytes are primitive, lacking vessels, whereas the other eight divisions are all vascular. The ferns, clubmosses, and horsetails thus represent an intersection of these two ways of splitting the plants, and were long known as vascular cryptogams; but that

| Kingdom | Division | Class |
|---------|----------|-------|
| Plantae | Bryophyta | Hepaticae |
|         |          | Anthocerotae |
|         |          | Musci (mosses) |
|         | Lycopodiaceae | |
|         | Sphenophyta | |
|         | Filicinophyta | |
|         | Coniferophyta | |
|         | Angiospermophyta (flowering plants) | Magnoliopsida Liliopsida |
|         | 3 other divisions | |

category is no longer considered a natural taxon, i.e., one with a common ancestor. "Flowering plants" is synonymous with Angiosperms, and the four non-flowering divisions of seed plants are Gymnosperms ("naked seeds"); some authorities consider the Gymnosperms a division and separate the conifers from the ginkgos, etc., at a lower taxonomic level. There are also many views on the proper level for separating the yews from the pines and other conifers.

| Order | Examples |
|---|---|
| Jungermanniales<br>Marchantiales | Leafy liverworts<br>Thallose liverworts |
| | Hornworts |
| Andreaeidae<br>Sphagnidae<br>Polytrichidae<br>Bryidae<br>2 other subclasses | Granite moss<br>Peat moss<br>Haircap moss<br>Most other mosses |
| | Clubmosses, spikemosses |
| | Horsetails |
| | Ferns |
| Pinales<br><br>Taxales | Pine, cypress, sequoia, and<br>monkey-puzzle families<br>Yew and 2 other families |
| | Dicots<br>Monocots |
| | |

# Abbreviations and Symbols

| | |
|---|---|
| " | inches |
| ' | feet |
| ° | degrees Fahrenheit |
| + | or more |
| ± | more or less |
| × | by (as in length by width) |
| C | central |
| **Cas Cr** | Cascade Crest |
| **diam** | diameter (diameter at breast height in the case of trees, defined as 4'6" above the ground) |
| **E** | east(ern) |
| **elev(s)** | elevation(s) above sea level, in feet |
| **E-side** | the area east of the Cascade Crest |
| **esp** | especially |
| **exc** | except |
| **incl** | including |
| **mm** | millimeters |
| **mtn(s)** | mountain(s) |
| **N** | north(ern) |
| **Oly(s)** | Olympic(s) |
| **Ore** | Oregon |
| **p(p)** | page(s) |
| **PNW** | the Northwest: Ore, Wash, W British Columbia, and SE Alaska |
| **S** | south(ern) |
| **spp.** | species plural: any and all of the species of a genus, not distinguished at the species level |
| **W** | west(ern) |
| **Wash** | Washington |
| **ws** | wingspread: the measurement across outspread wings |
| **W-side** | the area west of the Cascade Crest |

# Glossary

**Abundant:** present in great numbers—even more numerous, at least within some habitats, than what is implied by "common."

**Alevin:** a hatchling fish, especially a salmon or trout in the stage when it is still attached to and nourished by an egg yolk sac.

**Alpine:** of or in the elevational zone above where tree species grow numerously in upright tree form. Some texts define "alpine" as above the growth of tree species in any form, even **krummholz**; but in our mountains krummholz is found at the same elevations as most of the vegetational characteristics thought of as alpine elsewhere.

**Alternate:** arranged with only one leaf at any given distance along a stem. Contrast with **opposite** or **whorled**.

**Anadromous:** participating in a life cycle of birth and breeding in streams or lakes separated by an extended period of growth at sea.

**Angle of repose:** the steepest slope angle that a given loose sediment is able to hold.

**Anomalous:** differing from a norm, such as characteristic color or form.

**Ascending:** held in positions generally well above perpendicular to the stem or axis, but not nearly parallel or aligned with it; intermediate between **spreading** and **erect**.

**Ash:** fine particles of volcanic rock (usually glassy in structure) formed when **magma** is blasted out of a volcano in a fine spray.

**Aspect:** the compass direction a slope faces.

**Axil:** the crotch between a stem and a leaf.

**Awn:** a stiff, hairlike extension of the tip of a bract in the floret of a grass.

**Basal:** (said of leaves) attached to a plant at its root crown, not at any higher point on the stem.

**Batholith:** a large formation of intrusive igneous rock with no known bottom; i.e., a large, essentially monolithic intrusion exposed at the surface. In technical usage, a batholith must include at least 100 square km of exposed or subsoil bedrock.

**Biomass:** total living matter, usually expressed as a measure of dry weight (or sometimes volume) per unit area.

**Bisexual:** with functioning stamens and pistils in the same flower.

**Bloom:** a pale, powdery coating on a surface.

**Boreal:** of the Northern Hemisphere—wide belt dominated by conifer forests; transitional between Arctic and Temperate climate.

**Bract:** a modified leaf or leaflike appendage, often subtending a flower or inflorescence, and generally smaller and/or more specialized than a leaf.

**Broadleaf:** common term for all trees and shrubs other than conifers; very few have leaves narrow enough to be confused with conifer needles.

**Buttress:** a wide flaring-out at the base of a tree.

**Call:** any vocal communication common to birds of a given species. Compare song.

**Calyx:** a flower's sepals spoken of collectively, or else a ring of sepal-like lobes that would be sepals if they were separate all the way to their bases; i.e., the outermost whorl or circle of parts of most flowers. Plural: "calyces." Compare corolla.

**Canopy:** in forest structure, the uppermost more or less continuous mass of tree branches and foliage. All those trees of such stature that their tops form part of the canopy may be collectively called the "canopy layer."

**Cap:** the broad top portion of a typical mushroom, supported by a stem and supporting spore-bearing organs (gills, tubes, spines, etc.) on its lower surface only.

**Capsule:** a seed pod, technically a nonfleshy fruit that splits to release seeds; in mosses, the spore-containing organ.

**Catkin:** the form of **inflorescence** of certain trees and shrubs, consisting of a dense **spike** of minute, dry, petal-less flowers.

**Cirque:** a head of a mountain valley having a characteristic amphitheater shape due to glacial erosion.

**Class:** a **taxonomic** group broader than an **order** or **family** and narrower than a (plant) division or (animal) phylum. Examples: mammals, fishes, birds, spiders, insects, dicots.

**Clastic:** formed out of fragmented bits of rock (a description of most **sedimentary rocks** except those formed by chemical precipitation).

**Climax:** a more or less hypothetical condition of stability in which all **successional** changes resulting from plant community growth have taken place, so that future changes can only follow destructive **disturbances.**

**Cloaca:** an anal orifice (in birds, reptiles, fishes, etc.) through which the urinary, reproductive, and gastrointestinal tracts all discharge.

**Clone:** a group of genetically identical progeny produced by vegetative or other asexual reproduction from a single progenitor. (In nonscientific use, this term sometimes refers to the seeming individual, rather than the group, as a clone.)

**Clone:** to reproduce asexually.

**Composite:** a member of the plant family Asteraceae (formerly Compositae), characterized by composite flower heads which resemble single flowers but are actually well-organized inflorescences of tiny flowers. Examples: daisies, thistles, dandelions.

**Compound leaf:** a structure of three or more leaflets on stalklike ribs, resembling as many leaves on a branchlet except that it terminates in a leaflet rather than a flower, a bud, a growing shoot, etc., and that it grows from a node on a stem but its leaflets do not each grow from nodes of their own.

**Congener:** a member of a species in the same genus.

**Conifer:** a tree or shrub within a large group generally characterized by needlelike or scalelike leaves and by seeds borne naked between the woody scales of a "cone," or (in the yew family) borne naked but cupped within a berrylike "aril." (Some writers do not include the yew family as conifers. See page 578 for one of the many **taxonomic** placements that have been assigned to the conifers.)

**Conspecific:** (an individual) of the same species.

**Convect:** to move upward and downward due to temperature contrasting with that of other fluid masses in the same system—hot air rises, cold air sinks.

---

**Cordillera:** the entire mountain system comprising the western half of North America, from the Olympics to the Front Range of the Rockies, including most of Alaska, the Yukon, British Columbia, and Mexico. In Spanish, *cordillera* simply means "mountain range."

**Corolla:** the whorl or circle of a typical flower's parts lying second from the outside; i.e., the **petals** spoken of collectively, or else a ring of petallike lobes that would be called petals if they were separate all the way to their bases. Compare **calyx**.

**Cosymbiont:** a "partner" of a given organism in a **symbiotic** relationship.

**Cotyledon:** a seed leaf—the specialized leaf first produced by a plant after germinating from a seed, significant particularly in that the most consistent distinction between the **dicot** and **monocot** classes is whether two simultaneous seed leaves are produced, or just one.

**Crevasse:** a large crack in a glacier, expressing flow stresses.

**Crown:** the leafy (top) part of a tree.

**Crust:** the surface layer of the earth, averaging about 6 miles thick under the oceans and 20 miles thick in the continents, defined primarily by lower-density (lighter) rock than what's underneath.

**Crystal:** the characteristic structure of many chemical compounds (including all **minerals**) in which the atoms position themselves in a particular geometry that repeats indefinitely, so that it is usually visible in broken surfaces of the material.

**Deciduous:** (a tree or shrub) shedding its leaves annually; i.e., not evergreen. ("Deciduous" can also refer to any shedding part, for example, petals that fall off before the pistil and stamens mature, or sepals that fall off as the fruit matures.)

**Dicot:** any member of one of the two huge classes of flowering plants; includes all of our **broadleaf** trees and shrubs and virtually all of our terrestrial **herbs** except lilies, orchids, and grasslike plants (the **monocots**). Short for "dicotyledon."

**Dike:** a body of more or less homogeneous rock that is very much longer and deeper than it is thick: usually an **igneous** rock body from fluid **magma** intruded into a **fault** or crack in a contrasting rock body.

**Disk flower:** one of the tiny, often dry and drab-colored flowers making up a dense circle either in the center of a **composite** flower head (the "eye" of a daisy) or comprising an entire composite flower head such as a pearly-everlasting. Only in the family Asteraceae.

**Disturbance:** an external force which causes abrupt changes in a plant community. Clearing by humans, extraordinary floods, and fires are usually considered disturbances, while "normal" weather effects are not.

**Disjunct:** separated (said of a local population of a species separated from the main population by an area in which that species is not found).

**Dominant:** a plant species contributing easily the greatest percentage of cover within a given layer of a plant community. When no layer is specified, the uppermost layer is assumed, e.g., the **canopy** layer in a forest or the **herb layer** in a meadow.

**Duff:** matted, partly decayed litter on the forest floor.

**Erect:** aligned nearly parallel with the stem, if describing a leaf, a flower-stalk, etc.; or more or less vertical, if describing a main stem.

**Evergreen:** bearing relatively heavy leaves which normally keep their form, color, and function on the plant through at least two years' growing seasons. Opposite of **deciduous**, with **persistent** describing an in-between leaf type.

**Family:** a taxonomic group broader than a genus or species, and narrower than an **order** or **class**. Latin names for families end in *-aceae* if they are plants and *-idae* if they are animals.

**Fascicle:** a bundle, especially the characteristic cluster of 1–5 pine needles sheathed at their base in tiny dry, membranous **bracts**.

**Fault:** a fracture in rock or earth (or the location of that fracture) where relative movement of earth on the two sides has taken place; it may be of any size whatsoever.

**Floret:** a small, inconspicuous flower within a compact **inflorescence** such as the **head** of a **composite** or the **spikelet** of a grass.

**Fruit:** among the seed plants, an **ovary** wall matured into a seed-bearing structure; also, more broadly, any seed-bearing or **spore**-bearing structure.

**Fry:** young fish (both singular and plural). Among salmon and trout, the fry stage can be understood to follow the **alevin** stage and precede the **parr** stage; some species migrate to sea as fry.

**Fused:** unseparated; fused **petals** (or **sepals**) form a single **corolla** (or **calyx**) ring or tube, usually with lobes at the front edge that can be counted for identification purposes.

**Gills:** (1) in aquatic animals, the respiratory organs where oxygen suspended in the water passes into the animal; (2) on some mushrooms, **spore**-bearing organs consisting of paper-thin or sometimes wrinklelike radii on the underside of the cap.

**Granitic rocks:** coarse-grained **igneous rocks**; see **intrusive rocks**. (Recent geologic texts generally prefer a more techical term such as "phaneritic" to describe coarse textures, reserving "granitic" for a narrower group of high-silica igneous rocks.)

**Head:** a tight, compact **inflorescence**.

**Herbaceous:** not **woody**; normally dying or withering (at least in all its aboveground parts) at the end of the growing season. (Some low, soft-stemmed plants classed as herbs may persist through milder winters and into summer.)

**Herb layer:** all the **herbaceous** plants and low shrubs of comparable size, as a group comprising a structural element of a plant community. Typical forest communities here have a **canopy** layer, one or two shrub/small tree layers, an herb layer, and a moss layer.

**Here:** in our range; i.e., in the Olympic Mountains and the portion of the Cascade Range from Diamond Peak north. See page 5.

**High-grade:** (said of metamorphic rocks) altered by relatively high degrees of heat and pressure.

**Hybrid:** an offspring of parents of different species.

**Hyphae:** the countless, minute filaments that carry on all the normal nutritive and growth functions of a fungus. (Singular: "hypha.")

**Igneous rocks:** rocks that reached their present **mineral** composition and texture while cooling from a liquid into a solid state; may be either volcanic or, if they solidified underground without ever erupting, **intrusive**.

**Inflorescence:** a cluster of flowers from one stem, or the pattern of the cluster.

**Intergrade:** to vary along a continuum between one well-defined type (such as a species or a rock type) and another.

**Introduced:** not thought to have lived in a given area (the Northwest, in this book) before the arrival of white people; opposed to **native**.

**Intrusive rocks:** igneous rocks that did not surface as fluid lava in a volcanic eruption, but solidified underground. Best-known example is granite; most intrusives are coarser-grained than most volcanic rocks and may, in nontechnical usage, be lumped under the loose term granitic.

**Inuit:** Eskimo(s).

**Involucral bracts:** small leaves circling a stem immediately beneath an inflorescence (most often a flower head in the composite family).

**Irregular flower:** one made up of petals (or of sepals) that are conspicuously not all alike in size and/or shape; usually they make a bilateral symmetry.

**Juvenile plumage:** feathers (and their color pattern) of birds old enough to fly, but not yet entering their first breeding season; the two sexes are more or less alike at this stage.

**Key:** a document enabling readers to identify specimens by proceeding through a branching set of alternative descriptions.

**Krummholz:** ground-hugging shrubby growth, near alpine timberline, of conifer species that grow as upright trees under more moderate conditions. Translatable from German as "crooked wood"; sometimes called "elfinwood."

**Landlocked:** (said of fish) completing their life cycle in fresh water, where some barrier (usually a dam, but sometimes natural) prevents the anadromous (seagoing) life cycle the species is capable of.

**Larva:** an insect, amphibian, or other animal when in a youthful form that differs strongly from the adult form. Caterpillars are the larvae of moths and butterflies, maggots are the larvae of flies, etc. Also "grub."

**Lava:** rock that is flowing, or once flowed, in a more or less liquid state across the earth in a volcanic eruption; see magma.

**Leader:** the topmost central shoot of a plant.

**Leafstalk:** a narrowed stalk portion of a leaf, distinguished from the leaf "blade." (Synonym of "petiole.")

**Linear:** (leaves) very narrow, the two sides nearly parallel.

**Low-grade:** (said of metamorphic rocks) altered by relatively mild degrees of heat and pressure.

**Lumpers:** scientists inclined to reduce the numbers of species, genera, etc., by recognizing any given distinction at a lower taxonomic level. Opposed to **splitters**.

**Magma:** subsurface rock in a melted, liquid state. If and when it reaches the surface, its name will change depending on whether it flows out in a stream or paste of **lava** or explodes into the air as **pyroclastic** fragments.

**Mantle:** the major portion of the earth's interior, which lies below the **crust** and above the core; it is known mainly by indirect methods like seismic studies that indicate its high density.

**Margin:** outer edge, as of a leaf.

**Matrix:** in some rocks, the fine-grained material in which much larger grains, crystals, or contrasting pieces of rock are embedded.

**-merous:** a suffix for numerals, derived from "numerous." A 5-merous plant species is one whose **petals**, **sepals**, and **stamens** each number 5, 10, 15, or 20.

**Meta-:** a prefix sometimes used in naming **metamorphic rocks** in terms of the parent rock they are thought to have metamorphosed from. Examples: metaconglomerate, metaigneous.

**Metamorphic rocks:** rocks whose **mineral** composition and texture was established when they were "cooked" at great heat and/or pressure within the earth.

**Metamorphism:** the heat-and-pressure process that makes **metamorphic rocks**.

**Metamorphosis:** the process by which **larval** animals transform into adults.

**Milt:** fish semen.

**Mineral:** any natural ingredient of the earth having a well-defined chemical formula and a characteristic crystalline structure. Rocks and soils are largely mixtures of minerals. (The formula may express a range of compositions rather than a single uniform one.)

**Monocot:** any member of the huge **class** of flowering plants (the other subclass being the **dicots**) generally characterized by parallel-veined leaves and flower parts in threes or sixes. No monocots here are **woody**, and most are either lilies, orchids, aquatic or grasslike plants. Short for "monocotyledon."

**Moraine:** a usually elongate heap or hill of mixed rock debris lying where it was deposited by a glacier when the glacier retreated from that point.

**Mycology:** the branch of biology dealing with fungi.

**Mycophagy:** the eating of fungi; the collecting of wild fungi for food.

**Mycorrhiza:** a tiny organ formed jointly by plant roots and fungi for passing nutritive substances between plant and fungus. Plural: "mycorrhizae" and "mycorrhizas" are both used.

**Native:** (species) thought to have been living in a given area (the Northwest, for our purposes) before and independently of travel there by people. Generally this refers to travel by white people, since knowledge of pre–Columbian introductions to the New World is highly speculative. Opposed to **introduced**.

**Nature:** the universe as it was preceding or is outside of human civilization.

**Nectar:** a sweet nutritious liquid secreted in some kinds of flowers to attract pollinating animals.

**Niche:** the role of a species within an ecological community, in terms especially of how it exploits the habitat and other community members to meet its life requirements there.

**Northwest, the:** Oregon, Washington, Western British Columbia, and Southeast Alaska.

**Nurse log:** a rotting tree serving (and in many situations virtually required) as a seedbed for tree reproduction.

**Offset:** a short propagative shoot from the base of a plant, or a small bulb from the side of a plant's bulb.

**Old-growth:** mature, essentially natural forest, having recovered with little or no human interference for at least 175 years since the last clearing **disturbance**; opposed to either "second-growth" (following logging) or natural submature forest.

**Opposite:** arranged in pairs (of leaves) along a stem. The stem terminates in a bud, flower(s), or growing shoot. If it appears to terminate in a leaf or **tendril**, these are not opposite leaves, but leaflets of **compound leaves** which may be either opposite or **alternate**.

**Order:** a taxonomic group broader than a **family**, genus or species, and narrower than a **class**. Among plant taxonomists today, the orders are largely disregarded or superseded by other grouping concepts; but orders are very familiar and important groups in both birds and mammals. Examples: Rodents, carnivores, owls, woodpeckers, perching birds.

**Ovary:** the egg-producing organ; in flowers, this is generally the enlarged basal portion of the pistil, at the base of the flower.

**Overstory:** in plant community structure, whatever layer is highest; in a forest this would be the tree canopy.

**Our:** of or in the range explicitly covered by this book, namely, the Cascade Range from Diamond Peak North and the Olympic Mountains. See page 5.

**Outwash:** fragmental rock debris deposited by a glacier and subsequently carried and more or less sorted by a stream. Distinct from till, which is still lying, unsorted, where the glacier melted away from it.

**Palmately lobed:** shaped (like a maple leaf) with three or more main veins branching from one point at the leaf base, and the leaf outline indented between these veins.

**Panicle:** an inflorescence symmetrical around a main stem or axis with at least some of the side branches again branched to bear several flowers. Compare raceme, spike.

**Parasite:** an organism that draws sustenance for at least part of its life cycle out of an organism of another kind, more or less detrimentally to the latter (the "host") but without ingesting the host or any whole part of it.

**Parr marks:** vertical blotches that appear on the sides of anadromous salmon and trout of "parr" age, which comes when actively feeding in fresh water, before migrating to sea.

**Pendent:** attached by a downward stem from a larger stalk.

**Persistent:** (leaves) tending to stay on the stem through fall and winter even though functionally dead, and lacking the heavy weight and gloss characteristic of true evergreen leaves.

**Petal:** a modified leaf, typically non-green and showy, in the inner of two or more concentric whorls of floral leaves. If there is only one whorl, no matter how showy, its members are considered sepals.

**Pheromone:** a chemical produced by an animal (or plant; see p 463) that stimulates some behavior in others of the same species.

**Photosynthesis:** the synthesis of carbohydrates out of simpler molecules as a means of converting or storing sunlight energy—the basic function of chlorophyllous or green parts of plants and algae.

**Pinna:** a segment of a fern or a compound leaf, branching directly off of the main fern stalk or leaf axis; the pinna may be again branched before reaching the terminal pinnule or leaflet. Plural: "pinnae."

**Pinnately compound:** (said of a leaf) composed of an odd number (5 or more) of leaflets attached, except for the last one, in opposite pairs to a central leaf axis. If they are attached to mini-axes paired in turn along a central axis, the leaf is pinnately twice-compound.

**Pioneer:** a species growing on freshly disturbed (e.g., burned, clear-cut, deglaciated or volcanically deposited) terrain.

**Pistil:** the female organ of a flower, including the **ovary** and **ovules** and any **styles** or **stigmas** that catch **pollen**.

**Plate tectonics:** the dynamics of the Earth's crust in terms of the relative motion of more or less continent-sized plates of crust "floating" on an effectively fluid layer underneath. Also, the major unifying theory (known in its simpler days as "Continental Drift") that has dominated geophysical thought since the 1960s, using those dynamics to explain much of Earth history.

**Pollen:** a fine dust of male reproductive cells (pollen "grains") borne on the **stamen** tips of a flower, each capable, if carried by wind, animal "pollinators" or other agents to the **pistil** of a **conspecific** flower, of fertilizing a female reproductive cell.

**Pore:** a minute hole; especially, one allowing passage of substances between an organism and its environment.

**Propagate:** to reproduce, either sexually or not.

**Propagule:** (in lichens and mosses) any multicelled structure more or less specialized to break off and grow into an organism independent of (but genetically identical to) the one it grew on; a means of asexual reproduction or cloning.

**Prostrate:** growing more or less flat upon the **substrate**.

**Pupa:** an insect in a generally quiescent life stage transitional from a **larva** to an adult. Plural: "pupae." Verb: "pupate."

**Pyroclastic:** composed of rock fragments formed by midair solidification of **lava** during an explosive volcanic eruption. "Volcanic **ash**" is fine pyroclastic material; the rock "tuff" consists of consolidated pyroclastic material.

**Raceme:** a slender **inflorescence** with each flower borne on an unbranched **petiole** from the central axis. (Compare **panicle**, **spike**.)

**Radial symmetry:** an arrangement of several more or less identical elongate members (e.g., **petals**) from a central point.

**Rank:** a vertical row (e.g., one parallel to the stem).

**Raptor:** any bird of the Hawk and Owl orders. Not exactly synonymous with "bird of prey"; see page 393.

**Ray flower:** in a composite flower head such as a daisy or dandelion, a member which is petallike and strap-shaped for most of its length but tubular at its base, enclosing a pistil or, in dandelionlike composites, both pistil and stamens. (Technically, ray flowers never have stamens—the dandelionlike flowers are "ligulate"—but the two look much alike and are both considered "rays" in this book.)

**Regular:** with all of its sepals and with all of its petals (if any) essentially alike in size, shape, and spatial relationship to the others.

**Relief:** the vertical component of distance between high and low points.

**Resin blisters:** horizontally elongate blisters conspicuous on the bark of younger true fir trees, initially full of liquid pitch.

**Rhizine:** a tough, threadlike appendage on certain lichens, serving as a holdfast to the substrate, not a vessel.

**Rhizome:** a rootstalk, or horizontal stem just beneath the soil surface connecting several aboveground stems; sometimes thickened for storage of starches.

**Rhizosphere:** the layer of soil permeated and affected by roots and hyphae, considered as a stratum of the "biosphere."

**Rut:** an annual period of sexual activity in certain mammals.

**Saprophyte:** an organism (generally a fungus or bacterium) that gets its carbohydrate nutrition by decomposing dead organisms without first ingesting them.

**Scats:** feces. (Wildlife biologists' slang derived by back-formation from "scatology.")

**Schistosity:** a metamorphic rock texture in which the minerals are arranged in layers of visible-sized crystal grains.

**Scree:** loose rock debris lying at or near its angle of repose upon or at the foot of a steeper rock face from which it broke off. (Regarded by some as synonymous with talus, but gravel-sized debris is more often called "scree," boulder-sized more often "talus," and debris in gullies high on a cliff usually "scree" as opposed to "talus" at the bottom.)

**Sedimentary rocks:** rocks formed by slow compaction and/or cementation of particles previously deposited by wind or water currents or by chemical precipitation in water.

**Sepal:** a modified leaf within the outermost whorl (the calyx) of a flower's parts. Typically, sepals are green and leaflike and enclose a concentric whorl of petals, but in many cases they are quite showy and petallike. (Sepallike leaves enclosing a composite flower head are involucral bracts.)

**Smolt:** a young salmon or sea trout when first migrating to sea, typically losing its **parr marks** and becoming silvery. (Derived from the same Old English word as "smelt.")

**Softwood:** any **conifer**, in industry jargon; most, but not all, conifers have softer wood than most hardwoods.

**Song:** a relatively long and variegated form of bird **call**, generally distinctive of a species, practiced mostly by males in the Perching Bird order (Passeriformes).

**Songbirds:** a popular term for the bird order Passeriformes, comprising over half of all bird species, or for a large suborder.

**Sorus:** a small clump of **spore** -bearing organs visible as a raised dot line, crescent, etc. on a fern or other leaf. Plural: "sori."

**Spawn:** to breed (said of fish and other aquatic animals).

**Spike:** an **inflorescence** of flowers attached directly (i.e., without conspicuous stems of their own) to a central stalk. Compare **raceme**, **panicle**.

**Spikelet:** in the **inflorescence** of a grass, a compact group of from one to several **florets** and associated scalelike **bracts**, on a single axis branching off of the central stem.

**Splitters:** scientists inclined to increase the numbers of species, genera, etc., by recognizing any given distinction at a higher taxonomic level. Opposed to **lumpers**.

**Sporadic:** irregularly distributed over a range, perhaps common in some locales but not reliably so for any zone or major community type.

**Spore:** a single cell specialized for being released and later growing and differentiating into a new multicelled individual, among the lower plants and fungi. It is thought of as corresponding functionally to the seed in higher plants, since it travels; but in ferns and mosses it is comparable morphologically and genetically to the **pollen** grain, since it is not the sexually produced stage in the life cycle.

**Spreading:** (leaves, branches, etc.) growing in a horizontal to slightly raised position. Compare **erect**, **ascending**, **pendent**.

**Steppe:** unforested arid regions or plant communities dominated by sparse bunchgrasses and/or sparse shrubs. Instead of the common-parlance "sagebrush desert," ecologists say "sagebrush steppe," reserving "desert" for sites so barren that no plant community can be described. True deserts cover large areas of the world, but in the U.S. they persist only on such forbidding **substrates** as active dunes, or dry lakebeds too saline or alkaline for plant growth.

**Stamen:** a male organ of a flower; typically, several of them surround a central **pistil,** and each consist of a **pollen**-covered tip on a delicate stalk.

**Stigma:** the **pollen**-receptive tip of the **pistil** of a flower.

**Stolon:** a stem that trails along the ground, producing several upright stems and thus enabling the plant to spread vegetatively.

**Stomata:** minute **pores** in leaf surfaces for the **transpiration** of gases. Singular: "stoma." Sometimes anglicized to "stomates."

**Style:** the stalklike part of a flower's **pistil,** supporting the **stigma** and conveying male sexual cells from it to the **ovary.**

**Subalpine:** of or in the elevational zone lying below the **alpine** zone where tree species are absent, dwarfed, or **prostrate,** and above the continuous forest. See page 26.

**Subduction:** the hypothesized process in which the edge of a "plate" of the Earth's **crust** bends and sinks underneath the edge of an adjacent plate, eventually to be reconsumed in the underlying mantle.

**Subduction zone:** an elongate region including the margins of both a subducting plate and an overriding plate. Most geologists consider our region the subduction zone of the Juan de Fuca Plate subducting under the North American Plate, and our mountains effects of the subduction process.

**Subshrub:** a perennial plant with a persistent, somewhat **woody** base, thus intermediate between clearly an "herb" and clearly a "shrub."

**Substrate:** the base on which an organism lives. Examples: soil, rock, bark, skin.

**Subtend:** to be immediately below and next to.

**Succession:** gradual change in the species composition of an area, resulting from shifting relationships amongst community members and developing soil, in the absence of abrupt **disturbances.**

**Symbiosis:** intimate association of unlike organisms for the benefit of at least one of them. In scientific use the term includes not only "mutualism," where both or all partners benefit, but also **parasitism** and all other permutations of benefit, harm, or indifference to the respective participants. In common use many people mistakenly think of symbiosis as mutualistic by definition.

**Talus:** rock debris lying at the foot of a rock face from which it broke off. Often understood to mean cobble- to boulder-sized fragments, in contrast to finer **scree**.

**Taxa:** species, genera, and any other units of classification. Singular: "taxon."

**Taxonomy:** the scientific naming and classifying of organisms, particularly within the Linnaean system (species, genus, etc.) which uses Latin and Latinized names and endeavors to reflect lines of evolutionary descent.

**Tectonic:** related to large-scale geologic deformation such as folding, faulting, volcanism, and regional **metamorphism**.

**Temperate:** of the climatically moderate parts of the world. Opposed to Tropical on the one hand and Arctic on the other.

**Tendril:** a slender organ that supports a climbing plant by coiling or twining around something.

**Tepal:** a **petal** or **sepal** on a flower (such as some lilies) whose petals and sepals are identical.

**Terminal:** located at an end, such as a growing tip of a plant or the lower end of a glacier.

**Terrane:** in current **plate tectonics** jargon, any geologically mapped area whose rock formations originated without any known relation to those of adjacent terranes, suggesting that at some time it may have moved independently of the latter.

**Till:** Rock debris of mixed sizes, once transported by a glacier and still lying where the glacier deposited it in receding. Compare **outwash.**

**Tolerance:** the relative ability of various plant species to thrive under a given stressful or limiting condition. In this book, as in most technical writing on Northwest forests, **understory** tolerance (of which shade tolerance is the major component) is to be understood where no other condition is specified.

**Transpiration:** emission of water vapor into the air from plant parts.

**Turf:** an upper layer of soil permeated by a dense, cohesive mat of roots, especially of grasses and/or sedges.


**Ubiquitous:** found everywhere (within **our** range or a given part of it). In notes on animal ranges, this means that the species may inhabit any part of **our** mountains, with the possible exception of the **alpine** zone, at least at some time of year.

**Umbel:** an inflorescence in which several flower pedicels diverge from a single point atop the main stem. Example: carrot tops or "Queen Anne's Lace."

**Understory:** in plant community structure, any community layer except the highest one, or all such layers collectively.

**Veil:** a membrane extending from the edge of the cap to the stem, in certain mushrooms when immature, soon rupturing and often persisting as a ring around the stem; also (the "universal veil") in most *Amanita* mushrooms when they are very young "buttons," an additional, outer membrane extending from the edge of the cap all the way around the underground base of the mushroom, soon rupturing and often visibly persisting as a more or less cup-shaped enlargement of the mushroom's base.

**Vesicle:** a small hole in a volcanic rock, which originated as a gas bubble in lava; vesicular porosity characterizes all pumice and many basalt and andesite specimens.

**Whorl:** an arrangement of three or more leaves (or other parts) around the same point along a stem or axis.

**Widespread:** More or less common (but not necessarily abundant or dominant) over a large portion of our range. Used in this book to imply more consistent occurrence than sporadic, but less so than ubiquitous.

**Woody:** reinforced with fibrous tissue so as to remain rigid and functional from one year to the next. Woody plants are trees or shrubs. Opposed to herbaceous.

**Zooplankton:** nonplant aquatic organisms that drift with the currents, lacking effective powers of either locomotion or attachment. A loose term including anything from one-celled organisms up to jellyfish and larvae of some insects. Pronounced "zoh-oh-**plank**-ton." Singular: "zooplankter."

# Selected References

## Plants (Chapters 2–6)

Arno, Stephen F. 1977. *Northwest Trees*. Seattle: The Mountaineers.

Campbell, Alsie G., and Franklin, J. F. 1979. *Riparian vegetation in Oregon's Western Cascade Mountains*. Bulletin No. 14, Coniferous Forest Biome. Seattle: U of Wash Press.

Clark, Lewis J. 1973. *Wild Flowers of British Columbia*. Sidney: Gray's Publishing. This went out of print and was replaced after the author's death with an expanded and revised version, listed below. Many of the same photographs, with cursory identifying descriptions, were also published as a six-volume paperback set, divided according to habitat. The photos are among the best on Northwest flowers.

———. 1976. *Wildflowers of the Pacific Northwest from Alaska to Northern California*. Ed. by John G. S. Trelawny. Seattle: U of Wash Pr.

Conard, Henry S., and Redfearn, Paul L. Jr. 1979. *How to Know the Mosses and Liverworts*. 2nd Edition. Dubuque: Wm. C. Brown. The book for serious moss-identifiers who can't find Schofield (1969), and certainly the book for liverwort identifiers.

Gilkey, Helen M., and Dennis, La Rea J. 1975. *Handbook of Northwestern Plants*. Corvallis: Ore St U Bkstrs. Offers a less intimidating alternative to Hitchcock and Cronquist; it uses fewer abbreviations, and it weighs and costs considerably less since its range extends eastward only to the Cascade Crest, as opposed to the Montana Rockies.

Franklin, J. F., and Dyrness, C. T. 1973. *Natural Vegetation of Oregon and Washington*. USDA FS Gen. Tech. Report PNW-8. Portland: PNW Forest and Range Experiment Sta. The authoritative text on plant community types and zones in Ore and Wash; not edited for popular reading, but well worth consulting.

Harris, Larry D. 1984. *The Fragmented Forest.* Chicago: U of Chicago Pr. An overview of new understandings of forest ecology.

Harthill, Marion P., and O'Connor, Irene. 1975. *Common Mosses of the Pacific Coast.* Healdsburg: Naturegraph.

Haskin, Leslie L. 1949. *Wild Flowers of the Pacific Coast.* Portland: Binford & Mort. (Reprint, 1977. New York: Dover.) A charming old-fashioned flower book; it retells countless reported Indian uses, some of which are doubted by scholarly ethnobotanists.

Hayes, Doris W., and Garrison, George A. 1960. *Key to Important Woody Plants of Eastern Oregon and Washington.* USDA Agriculture Handbook No. 148.

Hitchcock, C. Leo, and Cronquist, A. 1976. *Flora of the Pacific Northwest.* 3rd Printing, with corrections. Seattle: U of Wash Press. **Our authority** on scientific names of all seed plants and ferns. The bible for serious plant identifiers in the Northwest, it takes some getting used to. A condensation of the five-volume edition listed below, it contains all of the species, with reduced and cropped drawings of each, but with the descriptions and nomenclatural data much abridged.

Hitchcock, C. Leo; Cronquist, A.; Ownbey, M.; and Thompson, J. W. 1955–69. *Vascular Plants of the Pacific Northwest.* In 5 parts. Seattle: U of Wash Pr. At full size, the exceptional beauty of Jeanne R. Janish's botanical illustrations is striking.

Horn, Elizabeth L. 1972. *Wildflowers 1: the Cascades.* Beaverton: Touchstone.

Jolley, Russ. 1988. *Wildflowers of the Columbia Gorge.* Portland: Ore. Hist. Soc. Over 700 species photographed.

Kirk, Ruth. 1966. *The Olympic Rain Forest.* Seattle: U of Wash Pr.

Kruckeberg, Arthur R. 1982. *Gardening with Native Plants of the Pacific Northwest: an illustrated guide.* Seattle: U of Wash Pr.

Larrison, Earl J.; Patrick, G. W.; Baker, W. H.; and Yaich, J. H. 1974. *Washington Wildflowers.* Seattle: Seattle Audubon Society.

Lawton, Elva. 1971. *Moss Flora of the Pacific Northwest.* Nichinan, Japan: Hattori Botanical Laboratory.

Lellinger, David B. 1985. *A Field Manual of the Ferns and Fern-Allies of the United States and Canada.* Washington, D.C.: Smithsonian.

Little, Elbert L., Jr. 1980. *The Audubon Society Guide to North American Trees: Western Region.* New York: Knopf.

Niehaus, Theodore F., and Ripper, Charles L. 1976. *A Field Guide to the Pacific States Wildflowers.* Boston: Houghton.

Radford, Albert. 1986. *Fundamentals of Plant Systematics.* New York: Harper.

Randall, Warren R., and Keniston, Robert F. 1981. *Manual of Oregon Trees and Shrubs.* Revised by D. N. Bever and E. C. Jensen. Corvallis: Ore St U Bkstrs.

Roberts, Anna. 1983. *A Field Guide to the Sedges of the Cariboo Forest Region, BC.* Victoria: BC Ministry of Forests.

Ross, Robert A.; Chambers, Henrietta L.; and Stevenson, Shirley A. 1988. *Wildflowers of the Western Cascades*. Portland: Timber Pr. Exceptionally sharp photography.

Schofield, W. B. 1969. *Some Common Mosses of British Columbia*. Victoria: BC Provincial Museum. Any volume from this long, ongoing series of BC Provincial Museum publications is a good buy; this one in particular is the only thorough, inexpensive, guide to PNW mosses. Unfortunately, these volumes fall in and out of print, over the years, in a leisurely manner. An expanded, revised edition of *Mosses* is in preparation.

Sudworth, George B. 1908. *Forest Trees of the Pacific Slope*. USDA Forest Service. (Reprint, 1967. New York: Dover.)

Szczawinski, Adam F. 1969. *The Orchids of British Columbia*. Victoria: BC Provincial Museum.

———. 1970. *The Heather Family of British Columbia*. Victoria: BC Provincial Museum.

Taylor, Ronald J., and Douglas, George W. 1975. *Mountain Wild Flowers of the Pacific Northwest*. Portland: Binford & Mort.

Taylor, T. M. C. 1973a. *The Ferns and Fern-allies of British Columbia*.

———. 1973b. *The Lily Family of British Columbia*.

———. 1973c. *The Rose Family of British Columbia*.

———. 1974a. *The Pea Family of British Columbia*.

———. 1974b. *The Figwort Family of British Columbia*. (All) Victoria: BC Provincial Museum.

Underhill, J. E. 1974. *Wild Berries of the Pacific Northwest*. Saanichton, B.C.: Hancock House.

## Plant Articles

American Forestry Association. 1986 (and subsequent addenda). National register of big trees. *American Forests* 92(4): 21–52.

Arno, Stephen F., and Habeck, J. R. 1972. Ecology of alpine larch in the PNW. *Ecol. Monographs* 42: 417–50.

Campbell, Alsie G., and Franklin, J. F. 1979. *Riparian vegetation in Oregon's Western Cascade Mountains*. Bulletin No. 14, Coniferous Forest Biome. Seattle: U of Wash Press.

Christy, John A.; Lyford, John H.; and Wagner, David H. 1982. Checklist of Oregon Mosses. *The Bryologist* 85: 22–36. **Our authority** on scientific names of mosses.

Douglas, George W. 1972. Subalpine plant communities of the western North Cascades, Wash. *Arctic & Alpine Research* 4: 147–66.

Douglas, George. W., and Bliss, L. C. 1977. Alpine and high subalpine communities of the North Cascades Range, Wash. and BC. *Ecol. Monographs* 47: 113–50.

Dyrness, C. T.; Franklin, J. F.; and Moir, W. H. 1974. *A preliminary classification of forest communities in the central portion of the Western Cascades in Oregon*. Bulletin No. 4, Coniferous Forest Biome. Seattle: U of Wash Press.

Franklin, J. F.; Cromack, K.; Denison, W. C.; et al. 1981. *Ecological characteristics of old-growth Douglas-fir forests.* USDA FS Gen. Tech. Report PNW-118. Portland: PNW Forest and Range Exp. Sta.

Fonda, R. W. 1974. Forest succession in relation to river terrace development in Olympic National Park, Wash. *Ecology* 55: 927–42.

Fonda, R. W., and Bliss, L. C. 1969. Forest vegetation of the montane and subalpine zones, Olympic Mtns., Wash. *Ecol. Monographs* 39: 271–301.

Furman, T. E., and Trappe, J. M. 1971. Phylogeny and ecology of mycotrophic achlorophyllous angiosperms. *Q Rev of Biology* 46: 219–25.

Hawk, Glenn M., and Zobel, D. B. 1974. Forest succession on alluvial landforms of the McKenzie R. Valley, Ore. *NW Science* 48: 245–65.

Hickman, James C. 1976. Non-forest vegetation of the Central Western Cascade Mtns. of Oregon. *NW Science* 50: 145-55.

Kuramoto, R. T., and Bliss, L. C. 1970. Ecology of subalpine meadows in the Olympic Mtns., Wash. *Ecol. Monographs* 40: 317–47.

McKee, Arthur, LaRoi, G.; and Franklin, J. F. 1980. Structure, composition, and reproductive behavior of terrace forests, S Fork Hoh River, ONP. In *Proceedings of the 2nd Conference on Scientific Research in National Parks, San Francisco, 1979.*

Minore, Don. 1972. *The wild huckleberries of Ore. and Wash.–a dwindling resource.* USDA FS Research Paper PNW-143. Portland: PNW Forest and Range Exp. Sta.

Nadkarni, Nalini M. 1985. Roots that go out on a limb. *Natural History* 94 (5): 42–49.

Swedberg, Kenneth C. 1973. A transition coniferous forest in the Cascade Mtns. of N. Oregon. *Am. Midland Naturalist* 89: 1–25.

Thornburgh, Dale Alden. 1969. Dynamics of the true fir-hemlock forests of the west slope of the Wash. Cascade Range. Ph.D. diss., U of Wash.

Waring, R. H., and Franklin, J. F. 1979. Evergreen coniferous forests of the PNW. *Science* 204: 1380–86.

# Fungi (Chapter 7)

Ahmadjian, Vernon. 1967. *The Lichen Symbiosis.* Waltham: Blaisdell Publishing. Fairly technical; Ahmadjian was the man who first achieved the lichenologists' Holy Grail—inducing free-living fungi and algae to combine and form a lichen *in vitro.*

Ahmadjian, Vernon, and Paracer, Surindar. 1986. *Symbiosis: an Introduction to Biological Associations.* Hanover: U Pr of New England.

Arora, David. 1986. *Mushrooms Demystified: a Comprehensive Guide to the Fleshy Fungi.* 2nd Edition. Berkeley: Ten Speed Press. **Our authority** on scientific names of nonlichenized fungi, and an excellent, very thick, mushroom guide.

Bandoni, R. J., and Szczawinski, A. F. 1964. *Guide to Common Mushrooms of British Columbia.* Victoria: BC Provincial Museum.

Bland, John H. 1971. *Forests of Lilliput: the Realm of Mosses and Lichens.* Englewood Cliffs: Prentice.

Bolton, Eileen M. 1972. *Lichens for Vegetable Dying.* McMinnville: Robin and Russ Handweavers.

Denison, W. C., and Carpenter, S. M. 1973. *A guide to air quality monitoring with lichens.* Corvallis: Lichen Technology, Inc.

Hale, Mason E., Jr. 1983. *The Biology of Lichens.* 3rd Edition. Baltimore: E. Arnold.

————. 1979. *How to Know the Lichens.* 2nd Edition. Dubuque: Wm. C. Brown. **Our authority** on scientific names of lichens, except for subsequent revision of *Alectoria*.

Hawksworth, D. L., and Hill, D. J. 1984. *The Lichen-forming Fungi.* London: Academic Pr.

Howard, Grace E. 1950. *Lichens of the State of Washington.* Seattle: U of Wash Press. Out of print.

Lincoff, Gary H. 1981. *The Audubon Society Field Guide to North American Mushrooms.* New York: Knopf.

Marteka, Vincent. 1980. *Mushrooms Wild and Edible: a Seasonal Guide to the Most Easily Recognized Mushrooms.* New York: Norton.

McKenny, Margaret, and Stuntz, Daniel E. 1987. *The New Savory Wild Mushroom.* Rev. by Joseph F. Ammirati. Seattle: U of Wash Press. The best PNW-only mushroom guide.

Miller, Orson K., Jr. 1972. *Mushrooms of North America.* New York: Dutton. An outstanding mushroom guide.

Richardson, D. H. S. 1974. *The Vanishing Lichens: Their History, Biology, and Importance.* New York: Macmillan, Hafner Press.

Savonius, Moira. 1973. *All Color Book of Mushrooms and Fungi.* London: Octopus Bks.

Smith, Alexander H. 1963. *The Mushroom Hunter's Field Guide.* Ann Arbor: U of Mich Pr.

————. 1975. *A Field Guide to Western Mushrooms.* Ann Arbor: U of Mich Pr.

Smith, A. H., and Smith, H. V. 1973. *How to Know the Non-gilled Fleshy Fungi.* Dubuque: Wm. C. Brown.

Smith, A. H., Smith, H. V., and Weber, Nancy. S. 1979. *How to Know the Gilled Mushrooms.* Dubuque: Wm. C. Brown.

Smith, A. H., and Weber, N. S. 1985. *A Field Guide to Southern Mushrooms.* Ann Arbor: U of Mich Pr.

**Fungi Articles**

Ahmadjian, Vernon. 1982. The nature of lichens. *Natural History* 91(3): 31–37.

Denison, William C. 1973. Life in tall trees. *Scientific American* 228(6): 74–80.

Douglas, George W. 1974. Lichens of the North Cascades Range, Wash. *The Bryologist* 77: 582–92.

Fogel, Robert, and Trappe, J. M. 1978. Fungus consumption (my-cophagy) by small mammals. *NW Science* 52: 1–31.

Hacskaylo, Edward. 1972. Mycorrhiza: the ultimate in reciprocal parasitism? *BioScience* 22: 577–83.

Hoffman, G. R., and Kazmierski, R. G. 1969. An ecological study of epiphytic bryophytes and lichens on *Pseudotsuga menziesii* on the Olympic Peninsula, Wash.; I. A description of the vegetation. *The Bryologist* 72: 1–19.

Maser, Chris; Trappe, J. M.; and Nussbaum, R. A. 1978. Fungal-small mammal interrelationships with emphasis on Oregon coniferous forest. *Ecology* 59: 799–809.

Pike, Lawrence H.; Denison, W. C.; Tracy, D. M.; Sherwood, M. A.; and Rhoades, F. M. 1975. Floristic survey of epiphytic lichens and bryophytes growing on old-growth conifers in western Oregon. *The Bryologist* 78: 389–402.

Ryan, Bruce D. 1985. Lichens of Chowder Ridge, Mt. Baker, Wash. *NW Science* 59: 279–93.

Trappe, J. M., and Fogel, R. D. 1977. *Ecosystematic Functions of Mycorrhizae.* Range Science Dept. Series No. 26, pp 205–14. Fort Collins: Colo. State U.

# Mammals (Chapter 8)

Burt, William H., and Grossenheider, R. P. 1976. *A Field Guide to the Mammals.* 3rd Edition. Boston: Houghton.

Chadwick, Douglas H. 1983. *A Beast the Color of Winter: the Mountain Goat Observed.* San Francisco: Sierra Club.

Cowan, Ian McT., and Guiguet, C. J. 1965. *The Mammals of British Columbia.* Victoria: BC Provincial Museum.

Honacki, J. H.; Kinman, K. E.; and Koeppl, J. W., eds. 1982. *Mammal Species of the World.* Lawrence: Allen Pr and Assn of Systematics Collections. **Our authority** on the scientific names of mammals.

Ingles, Lloyd G. 1965. *Mammals of the Pacific States.* Stanford: Stanford U Pr.

Kritzman, Ellen B. 1977. *Little Mammals of the Pacific Northwest.* Seattle: Pacific Search.

Larrison, Earl J. 1976. *Mammals of the Northwest.* Seattle: Seattle Audubon Society.

Maser, Chris; Mate, Bruce R.; Franklin, J. F.; and Dyrness, C. T. 1981. *Natural History of Oregon Coast Mammals.* USDA FS Gen. Tech. Report PNW-133. Portland: PNW Forest and Range Exp Sta. Among regional mammal books, this one stands out for thorough literature search combined with wide personal experience. However, the Forest Service has not thus far chosen to make it widely available; they should.

Murie, Olaus J. 1975. *A Field Guide to Animal Tracks.* Boston: Houghton. A treasure, written by a master.

Savage, Arthur, and Savage, Candace. 1981. *Wild Mammals of North-west America*. Baltimore: Johns Hopkins. This is a fine coffee-table book written for nonscientists. Large, showy mammals strongly emphasized; focus on British Columbia.

Vaughan, Terry A. 1978. *Mammalogy*. Philadelphia: Saunders.

Whitaker, John O., Jr. 1980. *The Audubon Society Field Guide to North American Mammals*. New York: Knopf.

## Mammal Articles

American Society of Mammalogists. *Mammalian Species*. An ongoing series.

Barash, David P. 1974. The social behavior of the hoary marmot. *Animal Behavior* 22: 256–61.

Clutton-Brock, T. H. 1982. The functions of antlers. *Behavior* 79: 108–23.

de Vos, A.; Brokx, P.; and Geist, V. 1967. A review of social behavior of the North American cervids during the reproductive period. *Am Midland Naturalist* 77: 390–417.

Hooven, Edward F., and Black, Hugo C. 1976. Effects of some clear-cutting practices on small-mammal populations in western Ore. *NW Science* 50: 189–208.

Jenkins, Kurt H., and Starkey, Edward E. 1984. Habitat use by Roosevelt elk in unmanaged forests of the Hoh Valley, Wash. *J of Wildlife Mgmt.* 48: 642–46.

Muller-Schwartz, D. 1971. Pheromones in the black-tailed deer. *Animal Behavior* 19: 141–52.

Rogers, Lynn. 1981. A bear in its lair. *Natural History* 90(10): 64–70.

Roze, Uldis. 1985. How to select, climb, and eat a tree. *Natural History* 94(5): 63–68.

# Birds (Chapter 9)

American Ornithologists' Union. 1983 (and supplements). *Check-list of North American Birds.* 6th Edition. Lawrence: Allen Press. **Our authority** on both scientific and common names of birds.

Beebe, Frank L. 1974. *Field Studies of the Falconiformes of British Columbia.* Victoria: BC Provincial Museum.

Farner, Donald S. 1952. *The Birds of Crater Lake National Park.* Lawrence: U of Kansas Pr.

Farrand, John, Jr., ed. 1983. *The Audubon Society Master Guide to Birding.* In 3 volumes. New York: Knopf. For now, it's the ultimate birder's guide—if you're ready for a three-volume set.

————. 1988. *Western Birds: an Audubon Handbook.* New York: Mc-Graw-Hill.

Gabrielson, Ira N., and Jewett, S. G. 1940. *Birds of Oregon.* (Reprinted, 1970, as *Birds of the PNW.* New York: Dover.) Oregon needs an update on this atlas of bird occurrence.

Guiguet, C. J. 1955–78. *The Birds of British Columbia.* 10 parts. Victoria: BC Provincial Museum.

Larrison, Earl J. 1981. *Birds of the Pacific Northwest.* Moscow: U of Idaho Pr.

National Geographic Society. 1983. *Field Guide to the Birds of North America.* Washington, D.C.: Natl Geog. Bulkier than Robbins et al., this guide is preferred by some for its detailed drawings, which often distinguish geographic races of a species.

Nehls, Harry B. 1981. *Familiar Birds of the Northwest.* Portland: Audubon Society of Portland.

Peterson, Roger Tory. 1961. *A Field Guide to Western Birds.* Boston: Houghton. An update is said to be in the works.

Ramsey, Fred L. 1978. *Birding Oregon.* Corvallis: Audubon Society of Corvallis.

Robbins, Chandler S.; Bruun, B.; Zim, H. S.; and Singer, A. 1983. *Birds of North America.* New York: Golden. Still the popular choice for a clear, easy-to-use national bird guide.

Terres, John K. 1980. *The Audubon Society Encyclopedia of North American Birds.* New York: Knopf.

Udvardy, Miklos D. F. 1977. *The Audubon Society Field Guide to North American Birds: Western Region.* New York: Knopf.

## Bird Articles

Bjorklund, Jonathon, and Drummond, David. 1986. *Bird checklist.* Sedro Woolley: North Cascades Natl Park Services Complex.

Gordon, Steve. 1984. Lane County breeding bird atlas. *Oregon Birds* 10: 126–33.

Manuwal, D. H.; Huff, M. H.; Bauer, M. R.; Chappell, C. B.; and Hegstad, K. 1982. Summer birds of the upper subalpine zone of Mt. Adams, Mt. Rainier, and Mt. St. Helens, Wash. *NW Science* 61: 82–92

Medin, Dean. 1985. Densities and nesting heights of breeding birds in an Idaho Douglas-fir forest. *NW Science* 59: 45–52.

Rickard, W. H., and Fitzner, R. E. 1978. Avifaunal survey of the Trojan nuclear power station. *NW Science* 52: 61–6.

Stalmaster, Mark V., and Gessaman, James A. 1984. Ecological energetics and foraging behavior of overwintering bald eagles. *Ecol. Monographs* 54: 407–28.

# Reptiles and Amphibians (Chapters 11–12)

Behler, John L., and King, F. Wayne. 1979. *The Audubon Society Field Guide to North American Reptiles and Amphibians*. New York: Knopf.

Carl, G. Clifford. 1968. *The Reptiles of British Columbia*. Victoria: BC Provincial Museum.

———. 1973. *The Amphibians of British Columbia*. Victoria: BC Provincial Museum.

Nussbaum, R. A.; Brodie, E. D., Jr.; and Storm, R. M. 1983. *Amphibians and Reptiles of the Pacific Northwest*. Moscow: U of Idaho Pr.

St. John, Alan D. 1980. *Knowing Oregon Reptiles*. Salem: Salem Audubon Society.

Stebbins, Robert C. 1985. *A Field Guide to Western Reptiles and Amphibians*. 2nd edition. Boston: Houghton. **Our authority** on scientific names of reptiles and amphibians.

# Fishes (Chapter 12)

Bailey, R. M., and Bond, C. E. 1963. Four new species of freshwater sculpins, genus *Cottus*, from Western North America. Occ. Papers Mus. Zool., U of Mich. 634: 1-27.

Brown, Bruce. 1982. *Mountain in the Clouds: A Search for the Wild Salmon*. New York: Simon.

Carl, G. Clifford; Clemens, W. A.; and Lindsey, C. C. 1959. *The Freshwater Fishes of British Columbia*. Victoria: BC Provincial Museum.

Lusch, E. A. 1985. *Comprehensive Guide to Western Gamefish*. Portland: F Amato Pubns.

Simpson, James C., and Wallace, R. W. 1982. *Fishes of Idaho*. Moscow: U of Idaho Pr.

Wydoski, Richard S., and Whitney, Richard R. 1979. *Inland Fishes of Washington*. Seattle: U of Wash Press. **Our authority** on scientific names of fishes.

# Insects (Chapter 14)

Arnett, Ross H. 1985. *American Insects*. New York: Van Nos Reinhold. **Our authority** on scientific names of insects.

Borror, Donald J.; DeLong, D. M.; and Triplehorn, C. A. 1976. *An Introduction to the Study of Insects*. 4th Edition. New York: Holt.

Christensen, James R. 1981. *A Field Guide to the Butterflies of the Pacific Northwest*. Moscow: U of Idaho Pr.

Cole, Frank R. 1969. *The Flies of Western North America*. Berkeley: U of Calif Pr.

Dornfeld, Ernest J. 1980. *The Butterflies of Oregon*. Portland: Timber.

Essig, E. O. 1958. *Insects and Mites of Western North America*. 2nd Edition. New York: Macmillan.

Ferris, C. D., and Brown, F. M. 1981. *Butterflies of the Rocky Mountain States*. Norman: U of Okla Pr.

Furniss, R. L., and Carolin, V. M. 1977. *Western Forest Insects*. USDA FS Misc. Publ. No. 1339. Focus on insects destructive to trees.

Gillett, J. D. 1971. *Mosquitoes*. London: Weidenfeld and Nicolson.

Heinrich, Bernd. 1979. *Bumblebee Economics*. Cambridge: Harvard U Pr. Shows how fascinating field research can be.

Hughes, Dave, and Hafele, Rick. 1981, *The Complete Book of Western Hatches*. Portland: F Amato Pubns. Entomology applied to dry-fly fishing.

Milne, Lorus, and Milne, Margery. 1980. *The Audubon Society Field Guide to North American Insects and Spiders*. New York: Knopf.

Neill, W. A., and Hepburn, D. J. 1976. *Butterflies Afield in the Pacific Northwest*. Seattle: Pacific Search.

Pyle, Robert M. 1974. *Watching Washington Butterflies*. Seattle: Seattle Audubon Society.

———. 1981. *The Audubon Society Field Guide to North American Butterflies*. New York: Knopf.

Scott, James A. 1986. *The Butterflies of North America*. Stanford: Stanford U Pr.

Swan, L. A., and Papp, C. S. 1972. *The Common Insects of North America*. New York: Harper.

Thornhill, Randy, and Alcock, John. 1983. *The Evolution of Insect Mating Systems*. Cambridge: Harvard U Pr.

Tilden, James W., and Smith, Arthur Clayton. 1986. *A Field Guide to Western Butterflies*. Boston: Houghton.

## Insect Articles

Deyrup, Mark. 1981. Deadwood decomposers. *Natural History* 90(3): 84–91.

Pettinger, L. F., and Johnson, D. W. 1972. A field guide to important forest insects and diseases of Ore. and Wash. Portland: USDA FS Div of Timber Mgmt.

Raffa, K. F., and Berryman, A. A. 1983. The role of host plant resistance in the colonization behavior and ecology of bark beetles (Coleoptera: Scolytidae). *Ecol. Monographs* 53: 27–49.

Shaw, David C., and Taylor, R. J. 1986. Pollination ecology of an alpine fell-field community in the North Cascades. *NW Science* 60: 21–31.

Voegtlin, David J. 1982. *Invertebrates of the H. J. Andrews Experimental Forest.* Forest Res. Lab. Special Publ. No. 4. Corvallis: OSU School of Forestry.

# Other Creatures (Chapter 15)

Garric, Richard K. 1965. The cryoflora of the PNW. *Am J of Botany* 52: 1–8.

Meyer, Ernest A. 1985. The epidemiology of giardiasis. *Parasitology Today* 1(4): 101–5.

Pilsbry, Henry A. 1948. *Land Mollusca of North America (North of Mexico).* Philadelphia: Academy of Natural Sciences of Phila.

Solem, Alan. 1974. *The Shell Makers: Introducing Mollusks.* New York: Wiley.

# Geology (Chapter 15)

Allen, John Eliot. 1979, *The Magnificent Gateway: a Layman's Guide to the Geology of the Columbia River Gorge.* Portland: Timber. Excellent popular-level presentation of the Gorge.

Allen, John Eliot, Burns, M., and Sargent, S. C. 1987. *Cataclysms on the Columbia.* Portland: Timber.

Alt, David D., and Hyndman, Donald W. 1978. *Roadside Geology of Oregon.* Missoula: Mountain.

————. 1984. *Roadside Geology of Washington* Missoula: Mountain. Alt and Hyndman are unequalled in their ability to make geology easy and fun to read about. Some will find the result glib; in each book the latest speculations (no sources cited) on plate tectonics are presented as fact—and they change from one book to the next.

Baldwin, Ewart M. 1981. *Geology of Oregon.* 3rd Edition. Dubuque: Kendall-Hunt.

Dietrich, Richard V. 1980. *Stones: their Collection, Identification and Uses.* San Francisco: W. H. Freeman.

Dietrich, R. V., and Skinner, Brian J. 1979. *Rocks and Rock Minerals.* New York: Wiley.

Harris, Stephen L. 1980. *Fire and Ice: the Cascade Volcanoes.* Revised edition. Seattle: Mountaineers and Pacific Search.

McKee, Bates. 1972. *Cascadia: the Geologic Evolution of the Pacific Northwest.* New York: McGraw. The only good first-year-level text on the entire region; unfortunately, mountain-building processes are described in pre–Plate Tectonics terms.

Mottana, A.; Crespi, R.; and Liborio, G. 1978. *Simon and Schuster's Guide to Rocks and Minerals.* New York: Simon.

Tabor, Rowland, W. 1975. *Guide to the Geology of Olympic National Park.* Seattle: U of Wash Pr. Excellent popular-level presentation of the Olympics.

Williams, Howel, and McBirney, Alexander R. 1979. *Volcanology.* San Francisco: Freeman Cooper.

## Geology Articles

Beget, James E. 1982. Recent volcanic activity at Glacier Peak. *Science* 215: 1389–90.

Blackwell, David D.; Bowen, R. G.; Hull, D. A.; Riccio, J.; and Steele, J. L. 1982. Heat flow, arc volcanism, and subduction in northern Oregon. *J of Geophysical Research* 87B: 8735–54.

Brown, E. H. 1987. Structural Geology and accretionary history of the Northwest Cascades system, Wash and BC. *Geol Soc of Am Bulletin* 99: 201–14.

Cater, Fred W. 1982. Intrusive rocks of the Holden and Lucerne quadrangles, Wash. USGS Prof. Paper No. 1220.

Drake, Ellen T. 1982. Tectonic evolution of the Oregon continental margin. *Oregon Geology* 44(2): 15–21.

Hammond, Paul E. 1979. A tectonic model for evolution of the Cascade Range. In *Cenozoic Paleogeography of the Western U.S.*, edited by J. M. Armentrout *et al.*, pp. 219–36. Los Angeles: Soc. of Economic Paleontologists and Mineralogists.

Jones, David L.; Cox, Allan; Coney, Peter; and Beck, Myrl. 1982. The growth of western North America. *Scientific American* 247(5): 70–84. An overview of the terrane interpretation of the region, written by principals in the debate and addressed to nongeologists.

Lipman, P. W., and Mullineaux, D. R., eds. The 1980 eruptions of Mt. St. Helens, Wash. USGS Prof. Paper No. 1250.

McBirney, Alexander R. 1978. Volcanic evolution of the Cascade Range. *Ann. Rev. of Earth and Planetary Sci.* 6: 437–456.

Misch, Peter H. 1966. Tectonic evolution of the Northern Cascades of Wash. State. Canadian Inst. of Mining and Metallurgy Special Vol. 8: 101–48.

Priest, George R., and Vogt, Beverly F., eds. 1982. *Geology and geothermal resources of the Mount Hood area, Oregon.* Special Paper 14. Portland: Ore. State Dept. of Geology and Min. Resources.

Tabor, R. W., and Crowder, D. F. 1969. On batholiths and volcanoes: intrusion and eruption of Late Cenozoic magmas in the Glacier Peak Area, N. Cas., Wash. USGS Prof. Paper No. 604.

Taylor, Edward M. 1981. Central High Cascade roadside geology. In *Guides to Some Volcanic Terranes in Wash., Ida., Ore., and N. Cal.*, ed. by David Johnston and Julie Donnelly-Nolan. USGS Circular 838.

Waitt, Richard B. 1977. Evolution of glaciated topography of upper Skagit River drainage basin, Wash. *Arctic and Alpine Research* 9: 183–92.

———. 1980. About forty last-glacial Lake Missoula jökulhlaups through southern Wash. *J of Geology* 88: 653–79.

# Climate (Chapter 16)

Barry, Roger G., and Chorley, Richard J. 1985. *Atmosphere, Weather and Climate*. 5th Edition. Mew York. Methuen.

Calder, Nigel. 1974. *The Weather Machine*. New York: Viking.

Critchfield, Howard J. 1960. *General Climatology*. Englewood Cliffs: Prentice Hall.

Geiger, R. 1965. *The Climate Near the Ground*. Rev. edition. Cambridge: Harvard UP.

Lydolph, Paul E. 1985. *The Climate of the Earth*. Totowa: Rowman and Allanheld.

Reifsnyder, William E. 1980. *Weathering the Wilderness: the Sierra Club Guide to Practical Meteorology*. San Francisco: Sierra Club.

# Cross-disciplinary and Miscellaneous

Arno, Stephen F. 1984. *Timberline: Mountain and Arctic Forest Frontiers*. Seattle: Mountaineers.

Beckey, Fred. 1977, 1981, 1987. *Cascade Alpine Guide: Climbing and High Routes*. 3 volumes. Seattle: Mountaineers. The standard climbers' route reference; also contains a wealth of research on history and geology; Wash and BC Cascades only.

Douglas, David. 1980. *Douglas of the Forests: the North American Journals of David Douglas*. Ed. by John Davies. Seattle: U of Wash Pr.

Eells, Myron. 1985. *The Indians of Puget Sound*. Seattle: U of Wash Pr.

Franklin, J. F.; Hall, F. C.; Dyrness, C. T.; and Maser, C. 1972. *Federal Research Natural Areas in Ore. and Wash.: a Guidebook for Scientists and Educators*. Portland: USDA FS PNW Forest and Range Exp Sta.

Guenther, Erna. 1973. *Ethnobotany of Western Washington*. Seattle: U of Wash Pr.

Judson, Katherine Berry. 1910. *Myths and Legends of the Pacific Northwest*. Chicago: A. C. McClurg.

Kelly, David, and Braasch, Gary. 1988. *Secrets of the Old Growth Forest*. Layton, Utah: Peregrine Smith.

Kimerling, A. Jon, and Jackson, Philip L. 1985. *Atlas of the Pacific Northwest*. 7th Edition. Corvallis: Ore St U Pr.

Kozloff, Eugene N. 1976. *Plants and Animals of the Pacific Northwest*. U of Wash Pr. A good guide to Puget-Willamette lowland species, giving rare attention to molluscs, arthropods, mosses and liverworts; but no birds, few insects or large mammals.

Lewis, Meriwether, and Clark, William. *The Journals of Lewis and Clark*. Ed. by Bernard DeVoto, 1953. Boston: Houghton.

Maser, Chris, and Trappe, James M. 1984. *The Seen and Unseen World of the Fallen Tree*. USDA FS Gen. Tech. Report PNW-164. Portland: PNW Forest and Range Exp Sta.

Margulis, Lynn, and Schwartz, Karlene V. 1982. *Five Kingdoms*. San Francisco: W. H. Freeman. **Our authority** on taxonomy at the kingdom and phylum levels; R. H. Whittaker's five-kingdom system as revised by Margulis, a prominent microbiologist.

McKelvey, Susan Delano. 1955. *Botanical Exploration of the Trans-Mississippi West, 1790--1850*. Jamaica Plain: Arnold Arboretum of Harvard U.

Ramsey, Jarold. 1977. *Coyote Was Going There: Indian Literature of the Oregon Country*. Seattle: U of Wash Pr. Enjoyable compendium of authentic Indian tales.

Snyder, Gary. 1969. *Earth House Hold*. New York: New Directions. Contains interesting journal material from summers (1951 and '52) as a fire lookout in the North Cascades. Snyder's early poetry (*Myths and Texts; Riprap; Mountains and Rivers Without End; The Back Country*) is also rich with wilderness experiences in the Northwest.

Spring, Ira, and Fish, Byron. 1981. *Lookouts: Firewatchers of the Cascades and Olympics*. Seattle: Mountaineers. Entertaining lore of earlier days of the Forest Service.

Storer, Tracy I., and Usinger, R. L. 1963. *Sierra Nevada Natural History*. Berkeley: U of Cal Pr. As there is no popular guide to invertebrates (other than butterflies) of the PNW, this book is the best substitute. It was the inspiration for *Cascade-Olympic Natural History*.

Turner, Nancy J. 1975. *Food Plants of British Columbia Indians: Coastal Peoples*. Victoria: BC Provincial Museum.

———. 1978. *Food Plants of British Columbia Indians: Interior Peoples*. Victoria: BC Provincial Museum.

———. 1979. *Plants in British Columbia Indian Technology*. Victoria: BC Provincial Museum. These three volumes are the most compact, thorough, authoritative, inexpensive reference on Northwest ethnobotany that we could ask for.

Welch, Lew. 1973. *Ring of Bone: Collected Poems 1950–1971*. Bolinas, California: Grey Fox.

Zwinger, Ann H., and Willard, Beatrice E. 1972. *Land Above the Trees: a Guide to American Alpine Tundra*. New York: Harper.

# Index

Species-level modifiers are omitted from index entries unless they affect the page numbers given. For example, Pacific ninebark: both that species of ninebark and the other one are on page 80, so the index entries are "Ninebark," "*Physocarpus*," and "Sevenbark" (an alternate common name). Of these entries, only "Ninebark" gives the color photo page number, 117, since the Latin and alternate names do not appear on the color page. A footnote is designated by "n."

Abbreviations, 580
*Abies*
    *amabilis*, 28
    *concolor*, 34
    *grandis*, 33
    *lasiocarpa*, 25
    *magnifica*, 33
    *procera*, 31
*Accipiter*, 379
*Acer*
    *circinatum*, 62
    *glabrum*, 61
    *macrophyllum*, 59
*Achillea*, 186
*Achlys*, 193, **132**
*Aconitum*, 204
*Actaea*, 234
*Actitis*, 389
Adder's-tongue, 166
*Adelges*
    *cooleyi*, 454
    *piceae*, 455
*Adenocaulon*, 189
*Adiantum*, 241
Admiral, 469, **505**
*Aedes*, 447
*Aegolius*, 390

Agaric, fly, 265
*Agaricus*, 263
*Aglais*, 471
*Agoseris*, 191, **131**
*Agriades*, 475
*Agropyron*, 156
*Agrostis*, 156
*Aira*, 155 n
Alaska-cotton, 151
*Alces*, 360 n
Alder
    red, 62, 286, 462, 517
    Sitka, 65
    slide, 65
*Alectoria*, 291
Algae
    aquatic, 430, 573
    bluegreen, 282, 289
    lichen, 282-83
    snow, 482
Allelopathy, 102
All-heal, 200
*Allium*, 168-69
*Allotropa*, 181
*Alnus*
    *rubra*, 62
    *sitchensis*, 65

Alpine (butterfly), 467, **505**
Alpine plant
    adaptations, 25, 28-29, 50-51, 149, 187, 205, 220, 227
Altruistic behavior, 305
Amanita
    fly, 265, **483**
    panther, 266
*Amanita*
    *muscaria*, 265
    *bisporigera*, 267
    *pantherina*, 266
    *phalloides*, 267
    *verna*, 267
    *virosa*, 267
*Ambystoma*, 424
*Amelanchier*, 78
Amphibians, 422-28, 417
Anacardiaceae, 88
*Analycus*, 479 n
*Anaphalis*, 187
*Anas*, 372
Andesite, 534, **508**, 519, 543, 547

*Andreaea*, 253
*Anemone*
  *deltoidea*, 222, **139**
  *drummondii*, 236
  *lyallii*, 222
  *occidentalis*, 236
  *oregana*, 222
Angel-wings (mush-
  room), 272, **484**
Angle of repose, 538
Anglewing (butterfly),
  471, **506**
Anguidae, 417
Annelid worms, 479
*Anopheles*, 447
Anseriformes, 372-73
Ant lion, 454
Antelopebrush, 81
*Antennaria*, 187
*Anthocharis*, 467
Anthocyanin, 227, 253
Antlers, 356
*Aphelocoma*, 403
Aphid
  balsam woolly, 455,
    27
  spruce gall, 454, 37
Apiaceae, 224-26
*Aplodontia*, 306
*Apocynum*, 207
Apodiformes, 395-96
Apple, Oregon crab-, 76
*Aquilegia*, 219-20
*Aquilo*, 381
Araceae, 161
Araliaceae, 90
*Arborimus*, 326
Arborvitae, 52
*Arbutus*, 98
Arctic (butterfly), 468,
  **505**
*Arctostaphylos*
  *columbiana*, 102
  *nevadensis*, 108
  *patula*, 102
  *uva-ursi*, 108
*Ardea*, 371
*Arenaria*, 215
*Ariolimax*, 477
Aristolochiaceae, 232
Arkose, 546, **511**
*Armillaria*, 268 n
*Armillariella*, 268

*Arnica*, 184, **130**
Arrowhead, 160
Arrowwood, 79
*Artemisia*
  *douglasiana*, 188
  *dracunculus*, 188
  *ludoviciana*, 188
  *tridentata*, 90
  *trifurcata*, 188
Artiodactyla, 354-64
*Artogeia*, 466
*Aruncus*, 219
*Arvicola*, 324 n
Arvicolidae, 323, 322-28
*Asarum*, 232
*Ascaphus*, 428
Asexual reproduction,
  187, 199, 455-56; *see
  also* Propagation
Ash
  mountain-, 77-78, **116**
  Oregon, 74
Aspect, 564
  effect on habitat, 355,
    565
  on trees, 258
Aspen, 69
Asphodel, false-, 168
*Aster*, 184, **130**
Asteraceae, 90, 182-91
*Astragalus*, 201
*Athyrium*, 242
Avalanche track
  community, 65, 69
Avens, purple, 218, **141**
Azalea, Cascades, 93

Bacteria, 252, 265, 327,
  456, 482, 575
  cyano-, 282, 289-90
Badger, 348, **494**
Baker, Mt., 24 n, 94,
  515, 523, 551
*Balsamorhiza*, 184
Balsamroot, 184, **130**
Baneberry, 234, **146**
Barberry family, 105,
  200, 233
Bark
  as food, 22, 330, 337
  slashed (animal sign)
    337-38, 350, 361

Basalt, 531-33, **508**, 518,
  522, 534, 542, 553
  floods, 533, 522
*Basilarchia*, 469
Basketry, 61, 62, 67, 75
Bat, 300
Batholiths, 522, 542,
  546, 555, 541
*Batrachoseps*, 425
Bear
  black, 337, **cover**
  grizzly, 340 n,
  skunk-, 346
Beargrass, 171, **124**
Beaver fever, 479-81,
  526
Beaver, 318, **491**
  mountain-, 306
Bedstraw, 230
Bee, 452
Beech family, 70, 99
Beetles, 456-62
Beetle
  ambrosia, 462
  bark, 458-62
  buprestid, 456
  chafer, 457
  Douglas-fir, 458, 460
  engraver, 460
  june, 457
  longhorn, 457
  pine, 458
  ponderous borer, 457
  sawyer, 457
  spruce, 458
Bellflower, 210, **138**
Bender, Karl Emil, 297
Bendire, Charles, 297
Berberidaceae, 105, 200,
  233
*Berberis*, 105
Betulaceae, 62-66
Big-ears, 176
Bilberry, 95
Biomass, 11, 21, 64
Birch family, 62-66
Birdbeak, 196, **134**
Birds 365-416
Birthwort family, 232
Bistort, 223, **142**
Bitterbrush, 81, **118**
Bitterroot, 234, **146**
Black body effect, 28,
  112

Blackberry, 83-84
Blackcap, 83
*Blechnum,* 238
Bleedingheart, 203, **133**
Blue (butterfly)
    arctic, 475, **507**
    dotted, 476
    northern, 474, **507**
    western, 474
    square-spotted, 476,
        **507**
Blue-eyed Mary, 197,
    **133**
Blue-eyed-grass, 173
Bluebells, 210, **138**
Bluebelly, 418
Blueberry, 95-96, **119**
Bluebird, 410, **502**
Bobcat, 349
Bogs, 13, 152, 252
Boletes, 273-76
*Boletus,* 273, **484**, 179,
    275, 276
*Bombus,* 452
*Bonasa,* 388
Boomer, 306, **490**
Borer, 457
Bovidae, 359 n, 362-64
*Bovista,* 279 n
Boxwood, 104, **120**
Bracken, 241
Bramble, 217, **139**
Brassicaceae, 229
Breccia, 548, **512**, 556
Bretz floods, 528, 523
Bretz, J Harlen, 529
British soldiers, 294,
    **489**
Brush-footed butterfly
    family, 469-73
*Bryoria,* 291
*Bubo,* 390
Buckbean, 210, **139**
Buckthorn, 73
Buckthorn family, 73,
    103, 113
Buckwheat, 232-33, **145**
Buckwheat family, 199,
    223, 232
*Bufo,* 426
Bugbane, false-, 193,
    **132**
Bullhead, 444

Bunchberry, 231, **143**
Buprestid, 456
*Buprestis,* 456
Buttercup, 219, **138**
Buttercup family, 193,
    203-4, 222, 230,
    234-46
Butterflies, 464-76
Buttresses (on trees), 34,
    53
Buzzard, 374

*Calamagrostis,* 156
*Calbovista,* 279
Calla-lily family, 161
*Callospermophilus,* 314 n
*Calocedrus,* 56
*Calochortus,* 168
*Caloplaca,* 288 n
*Caltha,* 235
*Calvatia,* 279
*Calypso,* 177, **128**
*Calyptridium,* 202 n
Camas, 170, **126**
    death, 171, **126**
*Camassia,* 170
Camouflage, 295, 303,
    366, 387, 392, 407,
    439, 471
Camp-robber, 401, 403
*Campanula,* 210
Campion, moss-, 215,
    **139**
*Canachites,* 386 n
Candlefish, 443
Candyflower, 221
Candystick, 181, **129**
Canidae, 331-36
*Canis*
    *latrans,* 331
    *lupus,* 332
*Cantharellus,* 270
Caprifoliaceae, 91-92
Caprimulgiformes, 394
*Cardamine,* 229
*Carduelis,* 416
*Carex,* 148
    *aquatilis,* 151
    *geyeri,* 151
    *nigricans,* 149
    *phaeocephala,* 150
    *spectabilis,* 149
Caribou, 285, 293, 357

Carnivora, 331-53
*Carpodacus,* 415
Caryophyllaceae, 214-
    15
Cascade Crest, 8, 561
Cascara, 73
Cashew family, 88
*Cassiope,* 111
*Castanopsis,* 99
*Castilleja,* 195
*Castor,* 318
Cat family, 349-53
Cat tracks, 352
Cat's-ears, 168, **127**
Catchfly, 215, **139**
Caterpillar, tent, 462, 64
*Cathartes,* 374
*Catharus,* 410
Cattails, 159
Cattle family, 359 n,
    362-64
*Ceanothus*
    *prostratus,* 113
    *sanguineus,* 104
    *velutinus,* 103
Ceanothus, redstem, 104
Cedar
    Alaska-, 54, **115**
    eastern red-, 55, 57,
        332
    hinoki, 55
    incense-, 56, **115**
    Port Orford-, 54-55
    true, 55
    western red-, 52, **115**
    white-, 55
    yellow-, 54
*Cedrus,* 55
Celastraceae, 104
Cepe, 273
*Cephalanthera,* 179 n
*Cerastium,* 214
*Cercyonis,* 469
*Certhia,* 407
Cervidae, 359 n, 354-62
*Cervus,* 358
*Ceryle,* 397
*Cetraria,* 292
*Chaetura,* 395
Chafer, rose, 457
*Chamaecyparis,* 54-55
*Chamerion,* 228 n
Chamisso, Adelbert, 219

Chanterelle
  white, 270
  woolly, 269, **483**
  yellow, 270, **484**
Chaparral, 102
Char(r), 441
Charadriiformes, 389
*Charidryas*, 472
Checkerspot, 472, **506**
Chehalis, 306
*Cheilanthes*, 240
Cherry, 75, **117**
Chert, 549-50, **512**
Chestnut blight, 100
Chickadee, 406
Chickaree, 311
Chicken-of-the-woods,
  278
Chickweed, 214, **140**
*Chimaphila*, 108, 179 n
Chinook jargon, 170,
  443
Chinquapin, 99
Chipmunk, 313, **492**
Chiroptera, 300
*Chlamydomonas*, 482
*Chlosyne*, 472 n
Chocolate-tips, 226, **142**
Chokecherry, 75
*Chordeiles*, 394
Christmas trees, 32
*Chrysops*, 449
Cicely, sweet, 225
Ciconiiformes, 371
*Cicuta*, 224
*Cinclus*, 408
Cinder cones, 533, 9,
  532
Cinquefoil, 218
  shrubby, 86, **117**
*Circus*, 378
Cirques, 527
*Cirsium*, 186
*Citellus*, 314 n
*Cladina*, 294
*Cladonia*, 294, 293 n
Clark, William, *see*
  Lewis and Clark
*Clarkia*, 228
Classification of living
  organisms, 572-79
*Claytonia*, 220
Clearcuts, 82, 103, 326,
  410, 565

*Clethrionomys*, 325
Climate
  macro-, 557-64, 11
  micro-, 564-67
  meso-, 566-67
Climax forest, 12-15, 20,
  24, 34, 44
*Clintonia*, 164
Cloning, *see* Propa-
  gation; Asexual re-
  production
Clouds, 560, 562, 563
Cloven-hooved
  mammals, 354-64
Clubmoss, 244
*Coenonympha*, 469
Coevolution, 43, 46,
  157, 174-75, 312,
  403, 406, 456, 460
*Colaptes*, 398
Coleoptera, 456-62
*Colias*, 467
*Collinsia*, 197
*Collomia*, 207, **137**
Coloring, protective,
  295, 303, 366, 387,
  392, 439, 468, 471
Coltsfoot, 189, **131**
*Coluber*, 420
Colubridae, 419-20
*Columba*, 389
Columbia River Gorge,
  231, 287, 523, 561
Columbine, 219-20, **141**
Compositae, 182 n
Composite flowers, 182-
  91
Condor, 374 n
Conglomerate, 547-48,
  **512**
Conifers, 11-58
*Conium*, 224
*Conocephalum*, 259
Continental drift, 520,
  529
*Contopus*, 400
Cony, 304
Coolwort, 212
Copper (butterfly), 473,
  **507**
Copper family, 473-76
Copperhead (squirrel),
  314

Coprophagy, 304
*Coptis*, 235
Coraciiformes, 397
Coral (fungus)
  bear's-head, 277, **486**
  purple-tipped, 276,
  **485**
*Corallorhiza*, 178
Coralroot, 178, **128**
Corkir, 286, **487**
*Cornus*
  *canadensis*, 231
  *nuttallii*, 71
  *stolonifera*, 72
Corpse-plant, 180
*Cortinarius*, 269, **483**
*Corvus*, 404
*Corydalis*, 203, **133**
*Corylus*, 66
Cottongrass, 151, **123**
Cottonwood, 68
*Cottus*, 444
Cotyledons, 56, 159
Cougar, 352, **495**
Cow-parsnip, 224
Coyote, 331, **492**
Coypu, 325
Crabapple, 76
Crassulaceae, 213-14
Craton, 518
Crazyweed, 201, **135**
Creambush, 79
Creeper, brown, 407,
  **502**
Crescentspot, 473, **506**
Crevasses, 525
Cricetidae, 323, 320-21
Crossbill, 416, **502**
*Crotalus*, 421
Crow, 404
Crowberry, 113, **122**
Cruciferae, 229 n
Cryptogams, 237
*Cryptogramma*, 240
Cudbear, 286
*Culex*, 447
Culicidae, 445
*Culicoides*, 450
*Cupressaceae*, 52-58
*Currant*, 86-87, **119**
Cushion plants, 205-9
Custer, Henry, 51
Cyanobacteria, 282,
  289-90, 575

Cyanocitta, 402
Cyperaceae, 148-51
Cypress family, 52-58
Cypresses, 55
Cypripedium, 177, **128**
Cypseloides, 395

Dacite, 535, **508**, 516, 519, 543
Daisy, 183
Dandelion, mtn.-, 191
Death cap, 267
Deer
    blacktail, 354
    mule, 354, **495**
    red, 360
    whitetail, 354 n, 358
Deer family, 354-62
Deerfoot, 193
Delphinium, 203
Dendragapus, 386
Dendroctonus, 458
Dendroica, 412
Dentaria, 229 n
Dentinum, 271
Deschampsia, 155
Destroying angel, 267, **483**
Devil's club, 90, **119**
Dewberry, 83
Dicamptodon, 424
Dicentra, 203
Dichelonyx, 457
Dicots, 159, 179-236
Dicranum, 254
Digitalis, 198
Dikes, 542, 544
Diorite, 542, **510**, 519
Dipper, 408, **500**
Diptera, 445-51
Dirt heaps (animal sign), 299, 306, 317
Dirty socks, 232
Disporum, 163
Disturbances, 13
Divisions, plant, 572-79
Dodecatheon, 208
Dog family, 331-36
Dogbane, 207, **138**
Dogwood
    creek, 72
    ground-, 231
    Pacific, 71
    red-osier, 72

Dolly Varden, 441
Doodle bug, 454
Douglas, David, 18-19, 2, 17, 31, 48, 67, 87, 101, 169, 170, 172, 173, 210, 474, 480
Douglas-fir, 16, **115**, 12-15
Douglasia, 208, **137**
Drummer, 388
Drummond, Thomas, 475, 169
Dryocopus, 398
Dryopteris, 243
Duck, 372-73, **497**
Dunite, 543
Dyes, 63, 91, 286

Eagle
    bald, 376, **498**
    golden, 381, **499**
Ears
    harrier, 378
    mole, 299
    owl, 392
    salamander, 424
Earthquakes, 521
Eburophyton, 179
Echolocation, 300
Edible plants (major), 75, 96, 102, 160, 161, 166, 170, 186, 193, 224, 226, 235
Elderberry, 73-74, **116**
Elephant's head, 196, **134**
Elfin, 476, **507**
Elgaria, 417 n
Elk, 358, **496**
Elymus, 156
Empetrum, 113
Empidonax, 400
Ephemerals, spring, 220-21, 173
Epidemia, 473
Epilobium, 228
Epiphytes, 60, 174, 247, 258
Equisetum, 245-46
Erebia, 467
Erethizon, 329
Ergates, 457
Ericaceae, 93-102, 107-12, 179-81

Erigeron, 183
Eriogonum, 232-33
Eriophorum, 151
Eriophyllum, 184
Ermine, 342
Erratics, 6, 529
Erysimum, 229
Erythrocoma, 218 n
Erythronium, 165
Eschscholtz, J. von, 219
Eulachon, 443
Euphilotes, 475
Euphydryas, 472 n
Eurhynchium, 255 n
Eutamias, 313
Evening primrose family, 228
Everlasting, 187, **130**
Evolution, 157, 110, 174-75, 417, 522
Eyespots, 468

Fabaceae, 201-2
Fagaceae, 70, 99
Fairy rings, 261
Fairy slipper, 177
Fairy-bells, 163, **125**
Fairy-lanterns, 163
Falco
    mexicanus, 382
    peregrinus, 382
    sparverius, 384
Falconiformes, 374-385
Falcon, 382-85
    peregrine, 382
    prairie, 382, **497**
False-bugbane, 193, **132**
False-morel, snowbank, 281, **486**
Farewell-to-spring, 228, **144**
Fat metabolism, 310, 361, 431
Feathers, 365-66
Feldspar, 546, 549 n
Felidae, 349-53
Felis
    concolor, 352
    lynx, 351 n
    rufus, 349 n
Fern
    bracken, 241
    brake, 241
    deer, 238

Fern (continued)
   lace, 240
   lady, 242
   licorice, 239
   lip, 240
   maidenhair, 241
   oak, 243
   parsley, 240
   shield, 243
   sword, 238
   wood, 243
Ferns, 237-43
*Festuca*, 154
Fiddleheads, 241
Figwort family, 195-98,
   231
Filbert, 66
Finch, 415, **501**
Fir
   Douglas-, 16, **115**, 12-
      15
   grand, 33
   hem-, 32
   lovely, 28
   lowland white, 33
   Pacific silver, 28, 12-
      15, 24
   Shasta red, 33
   subalpine, 25, 456
   white, 34
Fire, role in succession,
   13, 20, 26-27, 34,
   43-45, 56, 67, 70,
   95, 97, 102, 103, 157
Fireweed, 228
Fisher, 345, **494**
Fishes, 429-44
Fishing, 440-41, 429-30,
   435-44
   by Indians, 22, 87
   by predators, 347,
      371, 376, 377, 398
Flagged vegetation, 26,
   561
Flat leaf arrangement,
   22, 164-65, 214
Fleabane, 183, **130**
Fleeceflower, 223
Flicker, 398, **500**
Flowering plants
   herbs, 147-236
   shrubs65-67, 70-114
   trees, 59-77

Fly
   black, 449
   deer, 448
   flower, 450
   gad, 449
   horse, 44
   hover, 450
Flycatcher, 400, **500**
Foamflower, 212
Fool-hen, 386
Forest line, 26
Fox, 336
Foxglove, 198, **135**
*Fragaria*, 217
Franklin, Sir John, 475
*Fraxinus*, 74
Fringecup, 212, **140**
*Fritillaria*, 166
Fritillary (butterfly),
   472, **506**
Fritillary (lily), 166
Frog
   Cascades, 427, **502**
   red-legged, 426, **503**
   tailed, 428
   tree-, 427, **503**
Fumariaceae, 203
Fumitory family, 203
Fungi, 261-94
   hallucinogenic, 266,
      268
   Kingdom, 262, 574-75
   pathogenic, 47, 48, 87,
      100, 263, 459, 460,
      462
   poisonous, 261-68,
      275, 281
   underground-fruiting,
      264, 306, 314, 316,
      325, 326-28
Fungus
   bear's-head coral, 277,
      **486**
   purple-tipped coral,
      276, **485**
   shelf, 278, **485**
Fur trade, 319, 324, 325,
   336, 343, 344, 345,
   347, 401

Gabbro, 543, **510**, 519
Gairdner, Meredith,
   438, 169

*Galerina*, 267-68, **483**
*Galium*, 230
Galliformes, 386-88
Galls, 37, 70, 454-55
*Gaultheria*
   *humifusa* 107
   *ovatifolia*, 107
   *procumbens*, 107
   *shallon*, 101
*Gavia*, 370
Gemmae, 248
Gentian, 236, **146**
*Gentiana*, 236
Geologic time, 522
Geology, 515-56
   North Cascades, 7-8,
      519-20, 532, 540,
      543-46, 551-56
   Olympics, 5-6, 519,
      532, 541, 545-53
   volcanic Cascades, 8-
      10, 515-17, 519-21,
      532-39
*Gerrhonotus*, 417
*Geum*, 218
Geyer, Karl Andreas,
   151
Ghost-plant, 180
*Giardia*, 479, 526
Gibbs, George, 51
*Gilia*, 206
Gills
   fish, 431
   mushroom, 264, 271,
      272
   salamander 424
Ginger, wild-, 232, **144**
Ginseng, 90
Glacier Peak, 515, 517,
   523, 524, 535
Glaciers, 524-27, 559
   erosion by, 6-10, 526
   plant succession after,
      63, 284, 285
   travel on, 525
Glass, volcanic, 537, 536
*Glaucidium*, 390
*Glaucomys*, 316
Gleaning, 409
Gnat, buffalo, 449
Gneiss, 554, **514**
Goat Rocks, 10
Goat, mountain, 362

Goatsbeard, 219, **142**
Goldenrod, 185, **130**
Goldthread, 235
*Gomphus*, 269, **483**
*Goodyera*, 177
Gooseberry, 86-87
Gopher, 317
Gramineae, 154 n
Granite, 542, **510**, 520,
  555 n, 556
Grape, Oregon-, 105,
  **120**
Grass
  bear-, 171, **124**
  bent, 157
  blue-eyed-, 173
  cotton-, 151, **123**
  fescue, 154
  hair-, 155, **123**
  needle-, 155
  oat-, 154
  -of-Parnassus, 211,
    **140**
  panic, 156
  pine-, 156
  purple-eyed-, 173,
    **127**
  redtop, 157
  squirreltail, 155
  wheat-, 156
  -widows, 174
  wildrye, 156
Grass family, 148, 154-
  58
Grasslike plants, 148-59
Graywacke, 546, **511**
Grebe, 371, **497**
Greenschist, 553, **514**
Greenstone, 553, **513**,
  549
*Grimmia*, 253
Grossulariaceae, 86
Ground-pine, 244
Groundsel, 185
Grouse
  blue, 386, **499**
  Franklin's, 386
  ruffed, 388
  snow, 387
  sooty, 386
  spruce, 388
Grouseberry, 96
*Gulo*, 346

*Gymnocarpium*, 243
*Gymnogyps*, 374 n
*Gyromitra*, 281

*Habenaria*, 174
Hairstreak, 474, **507**
*Haliaeetus*, 376
Hardhack, 85, **118**
Hare, 302, **490**, 304 n
Harebell, 210
Harrier, 378
Hawk
  Cooper's, 379
  duck, 382
  gos-, 379, **498**
  marsh, 378
  night-, 394, **500**
  red-tailed, 380, **499**
  sharp-shinned, 379
  sparrow, 384
Hawks and eagles, 374-
  85
Hawkweed, 191, **131**
Hazel, 66
Heath family, 93-102,
  107-12, 179-81
Heather, 111-12, **122**
Hedge-nettle, 200, **132**
Hellebore, false-, 167
Hemlock
  black, 23
  mountain, 23
  lowland, 20
  poison, 224
  water-, 224
  western, 20, 12-15
Hemp, Indian-, 207
Herbs, flowering, 147-
  236
*Hericium*, 277
Heron, 371, **498**
*Herpotrichia*, 24, 291
Hibernation, 308-11,
  314, 328, 338-39,
  470, 476
*Hieracium*, 191
*Hirundo*, 401
*Histrionicus*, 372
Holly-grape, 105
*Holodiscus*, 79
Homoptera, 454-55
Honeysuckle
  bush, 91
  orange, 92, **118**

Hood, Mt., 94, 523, 535
Hooker, Sir Joseph, 169
Hooker, Sir Wm. J., 169,
  19
Hooter, 386
Hornfels, 553, **513**
Horns, 356, 364
Horsetail, 245-46
Hovering, 384, 396, 397,
  450
Huckleberry
  Alaska, 97
  Cascades blueberry,
    96, **119**
  fool's-, 94, **119**
  grouseberry, 96
  oval-leaved, 97
  red, 97
  thinleaf, 96
Hummingbird, 396, **500**
Hunting, 314, 347, 355,
  358, 363, 372, 389
Hybrids
  coy-dog, 335
  fir, 33, 34
  hemlock, 24
  manzanita, 103
Hydrangeaceae, 216
*Hydrophyllum*, 209
*Hyla*, 427
*Hyles*, 463
*Hylocomium*, 255
Hymenoptera, 452
*Hypericum*, 216
Hyphae, 178, 261-65
*Hypnum*, 257
*Hypogymnia*, 290
*Hypopitys*, 179

*Icaricia*, 474
Ice Age, 527-30, 6-10,
  27, 33, 429, 444,
  523, 525
Iceland-moss, 292, **488**
Igneous rocks, 530-545,
  **508-10**
Iguanidae, 418
Implantation, delayed,
  345-46
*Incisalia*, 476
Indian-hemp, 207
Indian Paintbrush, 195,
  **136**

Indian-pipe, 180, **129**
Indian-plum, 77, **116**
Inkberry, 91
Innocence, 197
Insectivora, 297-99
Insects, 445-476
Inside-out flower, 233, **145**
Intrusions, 542, 553
Intrusive rocks, 542-44, **510**
Inuit ("Eskimos"), 285, 293, 308
*Ipomopsis*, 206 n
Iridaceae, 173
Iridescence, 473
*Iridoprocne*, 400 n
*Iris*, 173, **127**
Ironwood, 79
*Isothecium*, 257
Ivy, poison-, 88
*Ixoreus*, 411

Jacob's-ladder, 207
Jay
  blue, 403
  Canada, 401
  gray, 401
  scrub, 402
  Steller's, 402
Jeffrey, John, 208
Jet stream, 558-59
Juncaceae, 148, 152-53
*Junco*, 414, **501**
*Juncus*, 152
Jungermanniales, 260
Juniper
  common, 58
  Rocky-Mtn., 57
  western, 57
*Juniperus*
  *communis*, 58
  *occidentalis*, 57
  *scopulorum*, 57
  *virginianus*, 55-57

*Kalmia*, 112
Kestrel, 384, **498**
King's crown, 214
Kingdoms, taxonomic, 572-79
Kingfisher, 397
Kinglet, 409

Kinnickinnick, 108, **121**
Kittentails, 231, **143**
Knotweed, 223
Krummholz, 25, 26, 45, 46, 50, 58

Labiatae, 200 n
Laceflower, 212
Ladies-tresses, 176
Lady's-slipper, 177, **128**
*Laetiporus*, 278
Lagomorpha, 302-5
*Lagopus*, 387
Lamiaceae, 200
Larch
  "Oregon," 32
  subalpine, 50
  western, 49, **115**
  woolly, 50
*Larix*
  *lyallii*, 50
  *occidentalis*, 49
Larkspur, 203, **134**
Lasiocampidae, 462
*Lathyrus*, 201
Latin, scientific, 569
Laurel
  alpine, 112, **121**
  mountain, 112
  sticky-, 103
Lava domes, 533, 516, 519, 535
Lava, 531-35, 548
Layering, 25, 102
LBMs, 268
Legume family, 201-2
Lepidoptera, 462-76
*Lepraria*, 287
*Lepus*, 302
*Letharia*, 290
*Leucolepis*, 251
*Leucosticte*, 415
Lewis and Clark, 234, 66, 81, 160, 319, 413, 447, 480
*Lewisia*
  *columbiana*, 233, **145**
  *rediviva*, 234, **146**
  *tweedyi*, 233, **145**
*Libocedrus*, 56 n
Lichen
  beard, 291, **489**
  clot, 285

dog, 288-89, **488**
globe, 291, **488**
horsehair, 291, **489**, 316
imperfect, 287, **487**
jewel, 288, **487**
map, 285, **487**
matchstick, 293, **488**
powdery paint, 287
puffed, 290, **487**
reindeer, 293, **489**
wolf, 290, **489**
worm, 292, **487**
Lichens, 282-94, 247
  air quality and, 285
  crustlike, 283, 285-88, 63
  dyes from, 286
  leaflike, 283, 288-90
  plant succession and, 284
  shrublike, 283, 290-94
Life form classes, 106
*Ligusticum*, 225
Liliaceae, 162-72
*Lilium*
  *columbianum*, 166
  *washingtonianum*, 167
Lily
  avalanche, 165, **124**
  bead, 165, **125**
  Cascades, 167, **126**
  chocolate, 166, **126**
  corn, 167, **127**
  fawn, 165
  glacier, 165, **127**
  mariposa, 168
  may, 162, **125**
  -of-the-valley, 162
  pond-, 235, **145**
  rice-root, 166, **126**
  tiger, 166, **126**
  trout, 166
Lily family, 162-72
*Limenitis*, 469 n
Limestone, 550, **512**
*Linnaea*, 114
Linnaeus, Carolus, 114
Linne, Carl von, 114
*Listera*, 176
Liverwort
  leafy, 260
  thallose, 261, **489**

Lizard
  alligator, 417
  fence, 418
*Lobaria,* 289
Locoweed, 201
*Lomatium,* 226
Longevity, exceptional,
  51, 53, 55, 57, 261
*Lonicera*
  *ciliosa,* 92
  *conjugialis,* 91
  *involucrata,* 91
Loon, 370
Lousewort, 196, **134**
Lovage, 225
*Loxia,* 416
*Luetkea,* 216
*Luina,* 190
Lumber trade, 16, 18,
  43, 47, 64
Lungwort (flower), 210
Lungwort (lichen), 289,
  **488**
Lupine, 202, **133, 135**
*Lupinus,* 202
Lütke, Fedor, 210
*Lutra,* 347
*Luzula,* 153
Lyall, David, 51
*Lycaeides,* 474
*Lycaena,* 473 n
Lycaenidae, 473-76
*Lycoperdon,* 279 n
*Lycopodium,* 244
Lynx, 351
*Lynx*
  *canadensis,* 351
  *rufus,* 349
*Lysichitum,* 161

Mackenzie, A., 234
Madder family, 230
Madroño, 98
*Mahonia,* 105 n
*Maianthemum,* 162, 164 n
*Malacosoma,* 462
Mallard, 372, **497**
*Malus,* 76 n
Mammals, 295-364
Manzanita
  green, 102, **120**
  hairy, 102
  pinemat, 108

Map, 4, 539-41
Maple
  bigleaf, 59
  Douglas, 61
  vine, 62
  Rocky-Mtn., 61
*Marchantia,* 259
Marigold, marsh-, 235,
  **146**
Marmot, 307, **491**
*Marmota,* 307
Marshes, 53, 151, 159,
  252, 319
Marshmarigold, 235,
  **146**
Marten, 344, **494**
*Martes*
  *americana,* 344
  *pennanti,* 345
Mating and courtship
  bald eagle, 377
  bird, 366
  butterfly, 466, 467
  cougar, 353
  coyote, 334
  deer, 357
  Douglas' squirrel, 311
  elk, 360-62
  gopher, 317
  grouse, 387
  hare, 303
  harrier, 378
  hover fly, 451
  hummingbird, 397
  marten, 344
  mosquito, 446
  mountain goat, 364
  newt, 423
  nighthawk, 394
  red fox, 336
  salamander, 425
  salmon, 434
  slug, 478
Mazama, Mt., 536, 523
Meadowrue, 230, **144**
*Megaceryle,* 397 n
*Melospiza,* 414
*Menyanthes,* 210
Menzies, Archibald, 94,
  169
*Menziesia,* 94
*Mephitis,* 348
Merganser, 373, **497**

*Mergus,* 373
Mertens, Franz K., 210
Mertens, Karl H., 210
*Mertensia,* 210
*Mesenchytraeus,* 479
Metamorphic rocks,
  551-56, **513-14**
Mica, 552, 554
Microcontinents, 8, 518-
  20, 522, 555
*Microseris,* 191
Microtines, 323, 322-28
*Microtus*
  *longicaudus,* 322
  *oregoni,* 322
  *richardsoni,* 324
  *townsendii,* 323
Middens, 312
Midge, biting, 450
Midsummer-men, 214
Migmatite, 555
Migration
  seasonal 309, 370,
    377, 387, 407, 414,
    431, 435, 440
  to North America,
    522, 528
Mimicry
  visual, 420
  vocal, 401, 402
*Mimulus,* 195
Miner's-lettuce, 221,
  **142**
Mink, 343
Minnow, muddler, 444
Mint Family, 200
Mires, 252
Mission bells, 166
*Mitella,* 211
Mitrewort, 211
*Mnium,* 251 n
Mobbing, 402
Mold, 264
  snow, 24, 291
Mole, 299
Mollusks, 477
Molting, 303, 364, 366,
  372
Monarch butterfly, 472
*Moneses,* 110 n
Monkeyflower, 195,
  **134-35**
Monkshood, 204, **133**

Monocots, 159, 147-79
*Monotropa,* 180, 179 n
Monotropaceae, 179 n
*Montia,* 221
Moose, 360 n
Moraines, 525
Morel, snow, 281, **486**
Mosquito, 445
Moss
  badge, 251
  beaked, 255
  bearded, 250
  big shaggy, 256
  broom, 254
  club-, 244
  curly-leaf, 257
  fern, 256
  frayed-cap, 254
  granite, 253
  *Grimmia,* 253
  haircap, 249
  heron's-bill, 254
  Iceland-, 292, **488**
  icicle, 257
  peat, 251-53
  rope, 255
  Spanish, 258
  spike-, 244
  tree, 251
  yellow shaggy, 257
Mosses, 247-58
Moss-campion, 215, **139**
Moth
  hawk, 464
  lappet, 462
  sphinx, 463, **504**
  tent caterpillar, 462
Mountain goat, 362, **496**
Mountain-ash, 77-78, 116
Mountain-beaver, 306
Mountain-dandelion, 191
Mountain-heather, 111, 122
Mountain-lover, 104
Mourning cloak, 470, **506**
Mouse
  deer, 320
  field, 322
  house, 320, 322
  jumping, 328, **492**

Mouse families, 323, 320-28
Mudflows, 517, 45, 516, 523
Mudstone, 545, **511**
Muridae, 323
Mushroom
  cauliflower, 277, **486**
  chicken, 278
  hedgehog, 271, **484**
  honey, 268, **483**
  oyster, 272, **484**
  tooth, 271
Mushroom-eating with caution, 270-71, 274-75, 277, 278, 280
Muskrat, 324
Mustard family, 229
*Mustela*
  *erminea,* 342
  *frenata,* 342
  *vison,* 343
Mustelidae, 342-49
*Myadestes,* 410
Myco-cuisines, 280
*Mycoblastus,* 285
Mycorrhizae, 262-65, 178-81, 36, 110, 269, 327
*Myocastor,* 325
Myomorphs, 322
*Myrmeleon,* 454

Naturalists in PNW, 571
Necks, volcanic, 9
Nectar feeders, 446, 449, 451, 464-76
Nematodes, 273
*Neotoma,* 321
Nests
  black bear, 338
  bumble bee, 452
  chickadee, 406
  Douglas' squirrel, 312
  eagle, 377, 381
  harlequin duck, 373
  kingfisher, 397
  loon, 370
  packrat, 321
  raven, 405
  swallow, 401
  vireo, 412

vole, 327, 328, **490**
woodpecker, 399
wren, 408
Nettle, 193, **132**
  hedge-, 200, **132**
Neuroptera, 454
*Neurotrichus,* 299
Newberry, John S., 223
Newt, 423, **503**
Night vision, 351, 392
Nighthawk, 394, **500**
Nightjars, 394
Ninebark, 80, **117**
Nitrogen fixing, 63, 103, 202, 289, 327, 517
No-see-um, 450
Non-green plants, 178-81, 110, 566
*Nothocalais,* 191 n
*Nucifraga,* 403
*Nuphar,* 235
Nurse logs, 34-37,
Nutcracker, Clark's, 403
Nuthatch, 407, **501**
Nutria, 325
Nuttall, Thomas, 71
Nymph, wood (butterfly), 469, **505**
Nymphalidae, 469-73
*Nymphalis*
  *antiopa,* 470
  *californica,* 471

Oak
  Garry, 70
  oregon white, 70, 285
  poison-, 88
Obsidian, 537, **509**
*Occidryas,* 472
Ocean-spray, 79, **117**
*Ochotona,* 304
*Ochrolechia,* 286
*Odocoileus,* 354
*Oemleria,* 77
*Oeneis,* 468
Old-growth, 17-18, 393
Old-man's-beard, 291, 355, **489**
Old-man's-whiskers, 218
Olympic Crest, 561
Olympus, Mt., 24n, 288n

Onagraceae, 228
*Oncorhyncus*, 432-36, 435
*Ondatra*, 324
Onion, 168-69, **126**
*Oplopanax*, 90
*Oporornis*, 413
Orange-tip, 467, **505**
Orchid
  bog, 174, **128**
  calypso, 177, **128**
  deer's-head, 177
  phantom, 179, **128**
  rein, 174
Orchidaceae, 174-79
*Oreamnos*, 362
Oregon-boxwood, 104, **120**
Oregon-grape, 105, **120**
Oregon-tea, 104
*Osmaronia*, 77 n
Osoberry, 77
Osprey, 376, **498**
Otter, 347, **493**
Outburst floods, 517
Outwash, 6
Ouzel, 408
*Ovis*, 364 n
Owl, 390
*Oxalis*, 214
*Oxyria*, 199
*Oxytropis*, 201

*Pachistima*, 104
Packrat, 321
Paintbrush, Indian, 195, **136**
Painted lady, 470, **505**
*Pandion*, 376
*Panicum*, 156
*Papilio*, 464
Papilionidae, 464-66
Parasites, 111, 180-81, 195, 262, 479
Parks or parkland (meadows), 29, 319
*Parnassia*, 211
Parnassian, 465-66, **504**
*Parnassius*, 465-66
Parrotbeak, 196
Parsley, desert-, 226, **142**
Parsley family, 224-26
Parsnip, cow-, 224

Partridgefoot, 216, **140**
*Parus*, 406
Pasqueflower, 236
Passeriformes, 400-16
*Pcilocybe*, 268
Pea, sweet-, 201, **135**
Peat, 13, 251
*Pedicularis*, 196
*Peltigera*, 288-89
*Penstemon*, 197, **134**
Perching birds (Order), 400-16, 405
Perching, 398
Peridotite, 543, **510**
*Perisoreus*, 401
*Peromyscus*, 320
Pesticides, 376, 383
*Petasites*, 189
*Petrochelidon*, 401 n
*Phacelia*, 209, **137**
*Phenacomys*, 328, 326 n
Pheromones, 298, 319, 321, 325, 332-33, 343-47, 350, 356-57, 361, 364, 463
*Philotes*, 475 n
*Phlox*, 205, **137**
Phlox family, 205-7
Photosynthesis, 12, 50, 110, 221
*Phyciodes*, 473
Phyla, animal, 572-79
Phyllite, 551, **513**, 545
*Phyllodoce*, 111
Physiographic provinces, 3-4
*Physocarpus*, 80
*Picea*
  *engelmannii*, 37
  *sitchensis*, 35
Piciformes, 398
*Picoides*, 398
Pieridae, 466-67
*Pieris*, 466 n
Pigeon, 389
Pika, 304, **490**
*Pilophoron*, 293
Pinaceae, 16-51
Pine
  bristlecone, 51
  eastern white, 47
  lodgepole, 44, 458-59, **115**

"Oregon," 32
  ponderosa, 42, **115**, 458-59
  shore, 45
  sugar, 48
  western white, 47, **115**
  western yellow, 42
  whitebark, 46, 28, 403
Pine family, 16-51
Pinedrops, 180, **129**
Pinesap, 179, **129**
Pink, cushion or moss, 215
Pink family, 214-15
Pinnae, 237
*Pinus*
  *albicaulis*, 46
  *contorta*, 44
  *lambertiana*, 48
  *monticola*, 47
  *ponderosa*, 42
Pipsissewa, 108, **122**
*Piranga*, 413
*Pituophis*, 420
Pixie goblets, 294, **489**
*Plagiomnium*, 251
Plantain, rattlesnake-, 177, **127**
Plate tectonics, 518-21
*Plebejus*, 474 n, 475 n
*Plethodon*, 425
*Pleurocybellus*, 272 n
*Pleurotus*, 272
Plum, Indian-, 77, **116**
*Pneumonanthe*, 236 n
Poaceae, 148, 154-58
Podicipediformes, 371
*Podilymbus*, 371
*Pogonatum*, 250
Poison-ivy, 88
Poison-oak, 88, **118**
Poisonous
  berries, 40, 234
  lichens, 290
  mushrooms, 261-68, 275, 281
  newts, 423
  plants, 90, 167, 171, 193, 198, 201, 203, 204, 224, 234
  snakes, 421
  toads 426

Polemoniaceae, 205-7
*Polemonium,* 207
Pollination, 71, 112,
    161, 174-75, 183,
    204, 206, 207, 220,
    451, 453
Polygonaceae, 199, 223,
    232
*Polygonia,* 471
*Polygonum,* 223
*Polyphylla,* 457
Polypodiaceae, 237-43
*Polypodium,* 239
*Polyporus,* 278 n
*Polystichum,* 238
*Polytrichadelphus,* 249 n
*Polytrichum,* 249
Pond-lily, 235, **145**
*Pontia,* 466
Population outbreaks,
    64, 322, 403, 459,
    462, 470, 471
*Populus*
    *tremuloides,* 69
    *trichocarpa,* 68
Porcupine, 329, **491,** 345
*Porella,* 260
Portulacaceae, 220-21,
    233-34
*Potentilla*
    *diversifolia,* 218
    *flabellifolia,* 218
    *fruticosa,* 86
Pouches, cheek, 314,
    315, 318
Prairie-smoke, 218
Precipitation, 558-65, 6,
    11, 24 n
Predator control, 331,
    353, 354, 376, 385
Primrose family, 208,
    233
Primulaceae, 208, 233
*Procyon,* 341
Pronunciation, 569
Propagation, vegetative
    alpine sedge, 149
    aspen, 69
    conifer, 25, 27-28
    lichen, 282
    manzanita, 102
    moss, 248
    willow, 67

*Prophysaon,* 478
Prospectors, 549
Protoctista, 479-82, 575
Protozoa, 479 n, **482**
*Prunella,* 200
*Prunus,* 75
*Pseudohylesinus,* 462
*Pseudotsuga,* 16
Ptarmigan, 388, **499**
*Pteridium,* 241
*Pterourus,* 464-65
Puffball, 279, **484**
*Pulsatilla,* 236 n
Pumice, 536, **509,** 45,
    172
Punky, 450
Purple-eyed-grass, 173,
    **127**
Pursh, Friedrich, 81
*Purshia,* 81
Purslane family, 220-21,
    233-34
Pussypaws, 202, **135**
Pussytoes, 187, **130**
Pyroclastic rocks, 537,
    **509**
*Pyrola,* 109, **121,** 179 n
Pyrolaceae, 179 n
*Pyrus,* 76

Quartz, 549, **512,** 534,
    546
Queen's-cup, 164
*Quercus,* 70

Rabbit Order, 302-5
Rabbits, 303, 304 n
Raccoon, 341
Racer, 420
Rain forest, Olympic, 6,
    35, 36, 60
Rain shadow, 560, 562
Rainfall, *see*
    Precipitation
Rainier, Mt., 24 n, 94,
    198, 517, 523
*Ramaria,* 276
*Rana*
    *aurora,* 426
    *cascadae,* 427
Range of this book, 3-5
Ranunculaceae, 193,
    203-4, 222, 230,
    234-46

*Ranunculus,* 219
Raspberry, 217
Rat
    black, 322
    musk-, 324
    Norway, 322
    pack-, 321
    water, 324
    wood-, 321
Rattlesnake, 421
Rattlesnake-plantain,
    177, **127**
Raven, 404
*Regulus,* 409
Reindeer, *see* Caribou
Reptiles 417-21
*Rhacomitrium,* 254
Rhamnaceae, 73, 103,
    113
*Rhamnus,* 73
*Rhizocarpon,* 285
*Rhodiola,* 214 n
Rhododendron
    Pacific, 100, **120**
    white, 93, **119**
*Rhododendron*
    *albiflorum,* 93
    *macrophyllum,* 100
*Rhus,* 88
*Rhyacotriton,* 423
Rhyolite, 535, **508,** 519,
    534, 543
*Rhytidiadelphus,* 255
*Rhytidiopsis,* 255
*Ribes,* 86-87
Richardson, Sir John,
    475
Ringlet, 469, **505**
Robin, 411
Rock flour, 526
Rock tripe, 288, **487**
Rockchuck, 307
Rodents, 306-31
Romanzoff, Count, 219
Root rot, shoestring, 268
*Rosa,* 85
Rosaceae, 75-86, 211,
    216-19
Rose family, 75-86, 211,
    216-19
Rose, wild-, 85, **116**
Roseroot, 214, **141**

Rot
  dry-, 56
  heart-, 21, 278
  root-, 64, 268
Rubiaceae, 230
Rubus
  discolor, 84
  laciniatus, 84
  lasiococcus, 219
  leucocermis, 83
  nivalis, 217
  pedatus, 217
  parviflorus, 82
  spectabilis, 82
  ursinus, 83
Rue, meadow-, 230, **144**
Rumex, 214
Runways (animal sign),
  323, 324, 347
Rush, 152
  bul-, 153
  scouring-, 246
  wood-, 153
Rush family, 148, 152
Rust, white pine blister,
  47, 48, 87

Sagebrush, 90
Sagittaria, 160
Saint Helens, Mount,
  515-17, 523, 94, 535
  effects of eruption,
  167, 205, 246, 320
Saint-John's-wort, 216,
  **141**
Salal, 101, **120**
Salamander
  giant, 424, **503**
  Larch Mountain, 425,
  **503**
  long-toed, 424, **503**
  Olympic, 423, **503**
  red-backed, 425, **502**
  slender, 425, **502**
Salicaceae, 67-69, 192
Salix
  cascadensis, 192
  nivalis, 192
  scouleriana, 67
Salmo
  clarki, 439
  gairdneri, 437
  salar, 432
  trutta, 437

Salmon, 432-36, 435
Salmonberry, 82, **117**
Salvelinus, 441
Sambucus, 73-74, **116**
Sandpiper, 389, **500**
Sandstone, 546, **511**
Sandwort, 215
Saprophytes, 180, 262,
  283
Satinflower, 173
Satyridae, 467-69
Satyrium, 474
Saussurea, 189
Saw-wort, 189, **131**
Sawyer, pine, 457
Saxifraga, 213
Saxifragaceae, 198, 211-
  213
Saxifrage, 213, **140**
Scapania, 260
Scapanus, 299
Sceloporus, 418
Scent glands; see
  Pheromones
Schist, 552, **514**
Sciurus, 314
Scleroderma, 279
Sclerophylls, 98
Scolytus, 462
Scouler, John, 67
Scouring-rush, 246
Scrophulariaceae, 195-
  98, 231
Scrub line, 26
Sculpin, 444
Sedge
  black alpine, 149
  dunhead, 150
  elk, 151
  showy, 149, **123**
  water, 151
Sedge family, 148-51
Sedimentary rocks, 545-
  51, **511-12**
Sedum
  divergens, 213
  roseum, 214
Selaginella, 244
Selasphorus, 396
Self-heal, 200, **133**
Senecio, 185
Serpentine, 45, 543
Serviceberry, 78, **116**

Silverback, 190, **131**
Silvercrown, 190, **131**
Simulium, 449
Siskin, 416, **501**
Sisyrinchium, 173
Sitanion, 155
Sitta, 407
Skunk, 348, **493**
Skunk-bear, 346
Skunk-cabbage, 161,
  **124**
Sky-pilot, 207, **137**
Skyrocket, 206, **138**
Slate, 551, **513**, 545, 553
Slippery jack, 274
Slug, 477, **503**
Smelowskia, 229, **144**
Smelt, 443
Smilacina, 164
Snails, 477
Snake
  garter, 419
  gopher, 420
  racer, 420
  rattle-, 421
Snow, pink, 482
Snowbed communities,
  149, 216
Snowberry, 92
Snowbrush, 103, **120**
Snow creep, 26, 167
Snowfall, see
  Precipitation
Snowpack, 15, 24, 25,
  36, 50, 58, 559, 566
Snow queen, 231
Solidago, 185
Solitaire, 410, **500**
Solomon's-seal, false-,
  164, **125**
Sonar, animal, 300
Sorbus, 77-78
Sorex, 297
Sorrel
  mountain-, 199, **136**
  wood-, 214, **141**
Sparassis, 277
Sparrow, 414, **501**
Speedwell, 231
Spermophilus, 314
Speyeria, 472
Sphaerophorus, 291
Sphagnum, 251, 13

Sevenbark, 80
Sewellel, 306
Shale, 545, **511**, 553
Sheep, bighorn, 364 n
Shield volcanoes, 533, 10
Shinleaf, 109-10
Shoestring root rot, 268
Shooting star, 208, **138**
Shorebirds, 389
Shrew, 297, **490**
Shrew-mole, 299
Shrubs, 25-26, 58, 65-67, 70-114, 197
*Sialia*, 410
*Sibbaldia*, 211
Sickletop, 196, **134**
*Silene*, 215
Sphingidae, 464
Spikemoss, 244
*Spilogale*, 348
Spiraea
  birchleaf, 84
  hardhack, 85, **118**
  rock-, 79
  steeplebush, 85
  subalpine, 84, **118**
*Spiraea*, 84
*Spiranthes*, 176
Spore plants, 237-60; *see also* Fungi
Spores, 237, 245, 248, 270, 282-83, 306
*Spraguea*, 202
Springbeauty, 220, **142**, 228
Spruce
  Engelmann, 37
  Sitka, 35
  tideland, 35
Squawcarpet, 113, **121**
Squirrel
  Douglas', 311, **491**
  golden-mantled ground, 314, **492**
  gray, 314, **492**
  flying, 316, **490**
  pine, 311
Squirrel family, 307-316
*Stachys*, 200
Staff-tree family, 104
Starflower, 233, **144**
Steelhead, 437

Steeplebush, 85
Steinpilz, 273
Steller, Georg, 402, 432
*Stellula*, 396
Stickcandy, 181, **129**
*Stipa*, 155
Stoat, 342
*Stokesiella*, 255
Stonecrop, 213, **141**
Strawberry, 217, **139**
*Streptopus*, 163
Strigiformes, 390
*Strix*, 390
*Struthiopteris*, 238 n
Subduction, 518; *see* Plate tectonics
Subshrubs, 106-114 (and others cross-referenced on 107)
Succession, plant, 12-15, 20, 21, 68, 94
*Suillus*, 274-76, **484**
Sulphur (butterfly), 467, **505**
Sulphur family, 466-67
Sulphur-flower, 232
Sun-cups, 482
Sunflower, woolly-, 184, **130**
Swallow, 400-1, **501**
Swallowtail, 464, **504**
Sweet cicely, 225
Sweetpea, 201, **135**
Swift, 395, **499**
*Sylvilagus*, 303
Symbiosis, mutualistic
  ant/caterpillar, 474
  beetle/fungus, 459, 460, 462
  bird/plant, 43, 46, 403
  lichen, 282-83
  mycorrhizal, 262
  rodent/fungus/ bacteria/plant, 326
*Symphoricarpos*, 92
*Symplocarpus*, 161
*Synthyris*, 231
Syrphidae, 450

*Tabanus*, 449
*Tachycineta*, 400
Tamarack, 49
*Tamiasciurus*, 311

Tanager, 413
*Taraxacum, 191*
*Taricha, 423*
Tarragon, 188
*Taxidea, 348*
Taxonomy, 572-79; *also* each chapter introduction
*Taxus, 40*
Tectonics, 518-21
Teeter-tail, 389
*Tellima, 212*
Terranes, 518-22
*Thaleichthys, 443*
*Thalictrum, 230*
*Thamnolia, 292*
*Thamnophis, 419*
Thermal cover, 355, 565
Thermals, 374-75
Thimbleberry, 82, **117**
Thistle, 186, **131**
*Thomomys, 317*
Throughfall, 565
Thrush
  hermit 411, **501**
  Swainson's, 410
  varied, 411
*Thuja, 52, 55*
*Tiarella, 212, **140***
Timberline
  subalpine, 26-29, 6
  lower, 26, 44
Toad, 426, **502**
Toadstools, 266
Tobacco-brush, 103
*Tofieldia, 168*
Tolerance
  chemical, 45, 285
  drought, 44, 566
  heat, 565
  understory (shade), 12-15, 21, 22, 24, 28, 32, 45, 70, 73
Tolmie, William F., 198, 169
*Tolmiea, 198*
Toothwort, 229, **143**
Torpor, 308-11, 314, 342, 395, 397
Tortoiseshell, 471, **506**
Towhead baby, 236, **138, 145**
Trail-plant, 189